SOURCES OF EUROPEAN ECONOMIC AND BUSINESS INFORMATION

SIXTH EDITION

SOURCES OF EUROPEAN ECONOMIC AND BUSINESS INFORMATION

6th Edition

Edited by
The British Library
Business Information Research Service

Gower

Published by
Gower Publishing Limited
Gower House
Croft Road
Aldershot
Hampshire GU11 3HR
England

Gower
Old Post Road
Brookfield
Vermont 05036
USA

The British Library Business Information Service has asserted their right under the Copyright, Design and Patent Act 1988 to be identified as the author of this work.

British Library Cataloguing in Publication Data

Sources of European Economic and Business
Information - 6Rev.ed
 330.94

ISBN 0-566-07487-7

Library of Congress Cataloging-in-Publication Data

Sources of European economic business information / [edited by]
    British Library Business Information Service. – 6th ed.
        p.   cm.
    Includes indexes.
    ISBN 0-566-07487-7
    1. Europe–Economic conditions–Bibliography. 2. Europe–Economic
conditions–Information services–Directories.   I. British Library.
Business Information Service.
Z7185.E8S87 1994
[HC240]
016.33094'055–dc20   94–11836
                                                               CIP

Database design and typesetting in Great Britain by Newsome, Southampton. Data input by Cavalier Business Associates, Southampton
Printed in Great Britain by the University Press, Cambridge.

CONTENTS

Introduction... vii

International sources
Sources covering Europe and the wider world...................... 1

Europe
Sources covering Europe - East and West.......................... 23

European Union
Sources covering the European Union as a group 41

Country sections
Individual sections on European countries

Albania 53
Austria 57
Baltic States................................. 65
Belgium....................................... 69
Bulgaria 77
C.I.S... 81
Cyprus 85
Czech and Slovak Republics.................... 89
Denmark....................................... 93
Finland 99
France 107
Germany....................................... 117
Gibraltar 131
Greece 135
Hungary 141
Iceland 147
Ireland....................................... 151
Italy... 159
Liechtenstein 165
Luxembourg 169
Malta... 173
Netherlands 177
Norway 183
Poland 189
Portugal 195
Romania....................................... 201
Spain 205
Sweden.. 211
Switzerland................................... 221
Turkey.. 227
United Kingdom 233
Yugoslavia 255

Publisher details
Publisher with Name / Address and other contact details.......... 259

Title Index
Sources listed by title.. 291

Subject Index
Sources listed by Subject and Country / Region. 313

Country Index
Sources listed by Country / Region and Subject. 327

INTRODUCTION

The sixth edition of the Sources of European Economic and Business Information has been revised at a time of change in Europe. Recent events are radically altering the political map of Europe. The institutions which would normally publish economic data are in many cases themselves in the process of transition and development, particularly in what was formally known as Eastern Europe. This sixth edition is essentially a snapshot of the position as it was in 1993.

For the purposes of this directory the former USSR, which is undergoing radical change, has been divided into the Commonwealth of Independent States (C.I.S.) and the Baltic States. The position in Czechoslovakia altered during the course of research and this has been reflected in the profile for it. Brief Country Profiles and quick reference maps have been added to the directory to remind readers of the geographical position of particular countries.

The aim of the directory is to provide a listing of the most useful sources of economic and statistical data. "Economic" and "data" have been broadly defined to encompass a range of sources, from economic indicators to data from market research publishers. Economic theory and "How to.." sources have largely been excluded. The publishers represented range from national statistical institutes and intergovernmental bodies to banks and financial institutions.

The emphasis has been on publications which are relatively well established and accessible. Questionnaires, in several languages, were sent to publishers appearing in the Fifth edition and in addition, enquiries were made with other bodies identified by the researchers. Where no response was received or where details could not be verified, the sources were excluded. Directories, trade/buyers' guides and monographs have been mostly excluded as the emphasis has been on sources of data.

The main body of the directory lists the sources, arranged by country and then alphabetically by title. The divisions by country refer to the country with which the source deals, not the country of publication - a source dealing with France published in the UK will therefore appear in the section covering France. The subject indexing used is the SCIMP/SCANP Thesaurus 8th edition 1991.

Sample Entry

01111 FOOD CONSUMPTION STATISTICS
Organisation for Economic Cooperation and Development (OECD)

International - English/French
annual 1990 540pp 30.00
ISBN 9264035133

Statistics on food consumption with latest figures available for 1979-1988.

Title:	Food Consumption Statistics
Publishing body:	Organisation for Economic Cooperation and Development (OECD)
Country/region covered:	International
Language(s) of publication:	English / French
Frequency:	annual
Date of first issue:	1990
Approx. number of pages:	540 pp
Cost:	£30.00
ISSN / ISBN:	ISBN 9264035133
Notes on subject coverage:	Statistics on food consumption with latest figures available for 1979-1988.

Key to currency symbols

Country	Currency	Symbol
Albania	Lek	Lek
Austria	Austrian Shilling	AS
Belgium	Belgian Franc	BF
Bulgaria	Leva	L
C.I.S.	Rouble	Rub
Cyprus	Cypriot Pound	C
Czechoslovakia	Koruna	K
Denmark	Danish Kroner	DKr
European Union	ECU	ECU
Finland	Markka	Fmk
France	Franc	FrF
Germany	Deutschmark	DM
Gibraltar	Gibraltar Pound	Gib £
Greece	Drachma	Dr
Hungary	Forint	HUF
Iceland	Kroner	IKr
Ireland	Punt	IR
Italy	Lira	L
Liechtenstein	Swiss Franc	SF
Luxembourg	Luxembourg Franc	LF
Malta	Lira	Lm
Netherlands	Florin	DF
Norway	Kroner	NKr
Poland	Zloty	Zl
Portugal	Escudo	Esc
Romania	Leu	Leu
Spain	Peseta	PTA
Sweden	Kroner	Skr
Switzerland	Swiss Franc	SF
Turkey	Lira	TL
United Kingdom	Pound	£
United States	Dollar	US$
Yugoslavia	Dinar	YD

The British Library Business Information Research Service

The British Library Business Information Research Service is a leading commercial supplier of business information in the United Kingdom and beyond. Using the service gives you access to one of the primary collections of business information. This is supported by a wide range of online and CD-ROM databases and expert research staff. You can contact the Business Information Research Service on 071 323 7457, Fax - 071 323 7453. For free quick enquiries call 071 323 7464.

The researchers involved in this publication were Daniel Barrett, Michael Leydon, Philip Ruston, Nigel Spencer and Elizabeth Traynor. The maps were designed by Ralph Canning.

Michael J. Leydon
April 1994

Part One
SOURCE PUBLICATIONS

INTERNATIONAL

01001 ABECOR COUNTRY REPORTS

Barclays Bank Economics Dept.

International - English annual/free

Covers the political and economic situation, economic outlook, balance of trade and payments for 100 plus countries.

01002 AGRIBUSINESS WORLDWIDE

Sosland Publishing Company

International - English 7 issues per year/US$42.00

International agribusiness; livestock, products and news.

01003 AGRICULTURAL POLICIES, MARKETS AND TRADE: MONITORING AND OUTLOOK 1992

Organisation for Economic Co-operation and Development (OECD)

International - English/French/ 488pp £28.00 ISBN No. 926413655X

Analyses the current economic outlook for agriculture and the implications for agricultural reform policy. Numerous graphs and charts are used throughout the text.

01004 ANNUAIRE DE L'INDUSTRIE AUTOMOBILE MONDIALE - OICA

Society of Motor Manufacturers and Traders Ltd.

International - English annual/ £22.00

Published by the Organisation Internationale Des Constructeurs d'Automobiles it includes the previous year's production, import and export figures for Europe, the Americas, Australasia, South Africa and Japan. Trade associations and regulatory bodies are also listed.

01005 ANNUAL BULLETIN OF STATISTICS: INTERNATIONAL TEA COMMITTEE

International Tea Committee Ltd.

International - English annual/ £160.00

Tea production, exports, imports for consumption, re-exports and stocks, auctions, quantities and country data.

01006 ANNUAL BULLETIN OF TRADE IN CHEMICAL PRODUCTS

United Nations Economic Commission for Europe - UN/ECE

International - English annual/ US$45.00 ISSN No. 0251-0081

Covers total value of production, gross fixed capital formation in the chemical industry, annual average increases in main aggregates, index numbers of production, employment, and production and trade of selected chemicals by SITC group, origin and destination.

01007 ANNUAL OIL MARKET REPORT

Organisation for Economic Co-operation and Development (OECD)

International - English annual/92pp £25.00

Major developments in world oil markets, with current statistical and historical data.

01008 ANNUAL REPORT OF THE INTERNATIONAL COCOA ORGANISATION

International Cocoa Organisation

International - English/French/ Spanish/Russian annual/1973/4 35-40pp free

Report on the work of ICCO and the world cocoa economy.

01009 ANNUAL REVIEW OF ENGINEERING INDUSTRIES AND AUTOMATION

United Nations Economic Commission for Europe - UN/ECE

International - English annual/ 275pp US$50.00

Covers Europe and the USA reporting on developments in telecommunications, electronics, machine-tools, and robotics in view of their economic performance. In two volumes.

01010 ANNUAL SUMMARY OF MERCHANT SHIPS COMPLETED IN THE WORLD

Lloyd's Register of Shipping

International - English annual/15pp free

Statistical summary of world completions and launchings for all ships of 100 gross tonnage and above.

01011 ARL RESEARCH REPORTS: CONSUMER DURABLES

Answers Research Ltd

International - English annual/ £200.00 - £12,000.00 per report

Market research reports which look at the international consumer durables market.

01012 ARL RESEARCH REPORTS: CONSUMER GOODS

Answers Research Ltd

International - English annual/ £200.00 - £12,000.00 per report

Market research reports which look at the international consumer goods market.

01013 ARL RESEARCH REPORTS: ELECTRONICS INDUSTRIES

Answers Research Ltd

International - English annual/ £200.00 - £12,000.00 per report

Market research reports which look at the electronics industry.

01014 ARL RESEARCH REPORTS: HEALTHCARE INDUSTRIES

Answers Research Ltd

International - English annual/ £200.00 - £12,000.00 per report

Market research reports which look at the international healthcare industry.

01015 ARL RESEARCH REPORTS: VETERINARY PRODUCTS

Answers Research Ltd

International - English annual/ £200.00 - £12,000.00 per report

Market research reports which look at the international veterinary products industry.

01016 BALANCE OF PAYMENTS STATISTICS YEARBOOK

International Monetary Fund

International - English annual/ US$45.00 ISSN No. 0250-7463

A two-part publication which gives balance of payments statistics for most of the world.

01017 BANK PROFITABILITY STATISTICAL SUPPLEMENT: FINANCIAL STATEMENT OF BANKS 1981-1990

Organisation for Economic Co-operation and Development (OECD)

International - English/French/ 214pp £21.00 ISBN No. 9264035311

01018 BASIC INTERNATIONAL CHEMICAL INDUSTRY STATISTICS

Chemical Industries Association Ltd.

International - English annual/ £70.00

A compilation of tables and graphs of basic economic series relating to the total chemical industry for EC and EFTA countries, Japan and the US. Also available on Lotus compatible floppy disc.

01019 BASIC SCIENCE AND TECHNOLOGY INDICATORS

Organisation for Economic Co-operation and Development (OECD)

International - English annual/ 372pp £23.00

01020 BOOK OF VITAL WORLD STATISTICS

Economist Books

International - English irregular/ 250pp £16.00

Gives basic economic and business statistics including cost of living indices.

01021 BULLETIN OF STATISTICS ON WORLD TRADE IN ENGINEERING PRODUCTS

United Nations Economic Commission for Europe - UN/ECE

International - English annual/ 286pp US$45.00 ISSN No. 0084-8174

Covers 98% of total world exports of engineering products, broken down by region, destination, exporter, and commodity group.

01022 BUSINESS INTERNATIONAL

Business International Ltd

International - English 50 issues per year/£355.00 P.A. ISSN No. 0007-6872

Current issues, analysis and case studies of global economic issues.

01023 BUSINESS WEEK INTERNATIONAL

Business Week International

International - English weekly/1929 £1.50 per issue

International business magazine with regular features/articles on economic data.

01024 CAPACITY CHANGES IN LEAD AND ZINC IN THE 1980'S

International Lead and Zinc Study Group

International - English/1990 26pp £50.00

Reviews the capacity changes in lead and zinc mines and metallugical works during the 1980's.

01025 CHARITY TRENDS

Charities Aid Foundation

International - English annual/

£12.95

Covers the voluntary sector in the UK with some international data.

01026 CHEMICAL INDUSTRY MAIN MARKETS 1989-2000

Chemical Industries Association Ltd

International - English 1990/£40.00 ISBN No. 090062342X

Looks at the present structure and past growth of the main chemical markets by area and product sector group, and on the basis of GDP growth rate assumptions, the future structure of the industry. Main markets considered are Western Europe, North America, Japan, Far East, Central & Southern Americas and Eastern Europe.

01027 CHEMSOURCE '93

Frost & Sullivan Ltd

International - English annual/ £295.00

Market shares, trends and forecasts for the international chemical industry.

01028 CIVIL AVIATION STATISTICS OF THE WORLD

International Civil Aviation Authority

International - English/French/ Spanish/Russian annual/US$18.00

01029 COAL INFORMATION

Organisation for Economic Co-operation and Development (OECD)

International - English annual/ 500pp £65.00

Comprehensive reference book on current world coal market trends and long term prospects. Covers OECD countries and selected other countries. Contains country specific and cross-country statistical analysis.

01030 COCOA CONSUMPTION

International Cocoa Organisation

International - English/French/ Spanish annual/price varies with report

A series of reports on cocoa consumption in individual countries. Reports date from 1990/1 and cover Spain, Czechoslovakia and Japan.

01031 COCOA NEWSLETTER

International Cocoa Organisation

International - English/French/ Spanish annual/1990 16pp free

General ICCO news and economic news of the world cocoa economy.

01032 COMMODITIY TRADE STATISTICS

United Nations Publications

International - English 28 issues per year/US$225.00 P.A. ISSN No. 0010-3233

Analyses more than 150 groups of commodities exported or imported by the world's leading trading nations.

01033 COMMODITY TRADE STATISTICS

United Nations Publications

International - English annual/ US$225.00 annual subscription

Annual trade statistics issued in instalments of about 250 pages and categorised according to 615 sub-groups of the UN SITC codes.

01034 COMMONWEALTH NOTES: FACTS IN BRIEF

Commonwealth Secretariat

International - English irregular/free

A series of short leaflets which give background notes on the political and economic situation in the Commonwealth.

01035 COMMONWEALTH SECRETARIAT: EXPERT GROUP REPORTS

Commonwealth Secretariat

International - English irregular/ £5.00

A series of studies on economic topics with Commonwealth and international relevance; typical titles include "Programme for sustainable forestry" and "Towards a new Bretton Woods".

01036 COMPENDIUM OF SOCIAL STATISTICS AND INDICATORS: SOCIAL STATISTICS AND INDICATORS - SERIES K

United Nations Publications

International - English/French irregular/685pp US$75.00

A compendium describing important social and related economic conditions and changes for 178 countries. Data is published 1-2 years in arrears.

01037 COMPETITION POLICY IN OECD COUNTRIES

Organisation for Economic Co-operation and Development (OECD)

International - English/French annual/332pp £22.00

Summarizes the main developments in competition policy and enforcement in OECD countries and developments resulting from EC competition policy. Latest edition published for 1988-89.

01038 CONSTRUCTION TODAY

Thomas Telford House

International - English monthly/ 1985 70pp £59.00 ISSN No. 0141-5999

Articles and news about projects, finance and business in the international construction industry with some data included.

01039 CONSUMER POLICY IN OECD COUNTRIES

Organisation for Economic Co-operation and Development (OECD)

International - English/French annual/248pp £18.00

Latest edition is for 1987-88.

01040 CONSUMPTION STATISTICS FOR MILK AND MILK PRODUCTS

International Dairy Secretariat

International - English annual/24pp BF1,000

Statistical coverage of dairy products such as butter and cheese.

01041 COUNTERTRADE OUTLOOK

Countertrade Outlook

International - English fortnightly/ 1983 various pages US$548.00

A fortnightly intelligence report on barter, counterpurchase, offset, switch, buyback and clearing.

01042 COUNTRY DATA FORECASTS

Bank of America

International - English twice a year/ US$495.00 P.A.

Strategic historic and forecast data for 80 plus countries with two updates per year.

01043 COUNTRY FORECASTS

Business International Ltd

International - English quarterly/ £360.00 per country

Formerly known as "Global forecasting service". Provides a range of medium forecasts covering main political, economic and business trends in 55 countries.

01044 COUNTRY GUIDES

Euromoney Books

International - English

A series of economic guides. Price varies with each guide.

01045 COUNTRY NOTES

The Market Research Society

International - English irregular/ 1993 £125.00 per set or £40.00 per region

Regional profiles covering Western Europe and Eastern Europe plus the Far East and North America. Gives key demographic characteristics, and features of the market research process in the region.

01046 COUNTRY OUTLOOKS

Bank of America

International - English five times per year/US$495.00 P.A.

Portrays in detail, the business, financial and economic environment expected over the coming two years.

01047 COUNTRY PROFILES

Business International Ltd

International - English annual/ £60.00 per country

Contains data from both official and non-official sources and gives 6 year statistical runs. Annual equivalent of "County Reports" by the same publisher.

01048 COUNTRY PROFILES

Department of Trade and Industry Export Publications

International - English irregular/ £10.00 per country

Basic economic profile of a country mainly intended for exporters or those doing business in the country.

01049 COUNTRY REPORTS

Business International Ltd

International - English quarterly/ £150.00 per country

Covers main political and economic events over the past quarter, with statistical appendices for 165 countries. An annual country profile is included.

01050 COUNTRY REPORTS

Dun & Bradstreet

International - English irregular/30-40pp £250-300

Gives a detailed description of the business and economic situation and a medium-term forecast. Payment issues and credit terms are also included. Mainly intended for those doing business with the country concerned.

01051 COUNTRY RISK MONITOR

Bank of America

International - English twice a year/ US$495.00 P.A.

Current and future risk evaluations for 80 countries with two updates per year.

01052 COUNTRY RISK SERVICE

Business International Ltd

International - English quarterly/ £13,776 P.A.

Monitors the financial and political conditions in 82 countries and gives short-term forecasts and risk assessments. There is a minimum subscription to seven countries.

01053 COUNTRY STATEMENTS ITMF

International Textile Manufacturers Federation

International - English annual/1977 SF100.00

A review of the current state of the textile industry in each member country. Included are data relating to the general economic situation, textile manufacturing capacities, activity levels and trade in textiles. Both short staple sector and pan-textile data are shown in tabular form with a commentary for each country.

01054 CUSTOMS AREAS OF THE WORLD - STATISTICAL PAPERS, SERIES M

United Nations Publications

International - English annual/27pp US$5.00

Statistics on the entry of goods for home use, warehousing of goods, free zones and inward processing concepts that determines the trade system.

01055 DAFSA INTERNATIONAL REPORTS

DAFSA

International - English loose-leaf with regular updates

Gives basic economic and financial information on a variety of international companies, divided by sector.

01056 DEMOGRAPHIC YEARBOOK

United Nations Publications

International - English/French annual/1020pp US$125.00

Worldwide demographic statistics and trends. Latest edition available covers 1990. Data is published 2-3 years in arrears.

01057 DENMARK REVIEW

Royal Danish Ministry of Foreign Affairs

International - English quarterly/ September 1986 60pp free ISSN No. 0418-6745

General review of Danish economy and society.

01058 DEVELOPMENT COOPERATION

Organisation for Economic Co-operation and Development (OECD)

International - English annual/ 254pp £18.00

Statistics on movement of resources to developing countries.

01059 DIGESTS OF STATISTICS: INTERNATIONAL CIVIL AVIATION AUTHORITY

International Civil Aviation Authority

International - English/French/ Spanish/Russian annual/1947 price varies with digest

A series of eight statistical digests which cover civil aviation in detail; an example being "Airport traffic".

01060 DIRECTION OF TRADE STATISTICS - QUARTERLY

International Monetary Fund

International - English quarterly/ US$86.00 P.A. ISSN No. 0252-306X

Provides data on the country and area distribution of countries' exports and imports as reported by them or their trading partners. Annual subscription includes a Yearbook.

01061 DIRECTION OF TRADE STATISTICS - YEARBOOK

International Monetary Fund

International - English annual/ US$25.00 ISSN No. 0252-3019

Provides data on the country and area distribution of countries exports and imports as reported by them or their trading partners. A quarterly journal is also available.

01062 DIRECTORY OF ORGANIZATIONS PROVIDING COUNTERTRADE SERVICES 1992-93

Countertrade Outlook

International - English annual/free with a subscription to the journal "Countertrade Outlook"

A directory of organizations providing facilities for barter, counterpurchase, offset, switch, buyback and clearing.

01063 DOING BUSINESS IN...

Price Waterhouse

International - English/free

Gives a description of the business and economic situation in a particular country. All developed countries are covered.

01064 DOING BUSINESS IN...

Ernst & Young

International - English 1-2 years/ free

An individual report on each country giving broad details of the business and economic climate. Payment issues, credit terms and business practice are also included. Mainly intended for those doing business with the country concerned.

01065 DREWRY SHIPPING REPORTS

Drewry Shipping Consultants

International - English/£300-£400 per year frequency varies

Market research type reports which look at the tanker and shipping markets worldwide.

01066 ECE ENERGY SERIES

United Nations Economic Commission for Europe - UN/ECE

International - English/price on application

Monographic series which deals with the energy industry in Europe and the USA. Energy efficiency, pollution, Eastern Europe energy industry structure, and design of energy facilities have been covered.

01067 ECONOMIC ACCOUNTS FOR AGRICULTURE

Organisation for Economic Co-operation and Development (OECD)

International - English/French annual/240pp £23.00

Latest data available is for 1976-89.

01068 ECONOMIC BULLETIN FOR EUROPE

United Nations Economic Commission for Europe - UN/ECE

International - English annual/ 125pp US$60.00

Examines world trade and the economies in Europe and North America based on the latest data. Also covers east-west trade with special chapters focusing on specialised topics in east-west trade. Volumes 33-40 were published quarterly, from volume 41 1991 it will be published annually.

01069 ECONOMIC DEVELOPMENT INSTITUTE SERIES

World Bank Publications

International - English irregular/ price varies with report

Informal documents prepared under the auspices of the Economic Development Institute (EDI). A department of the World Bank that offers courses, seminars and workshops relevant to development issues. EDI publications comprise the EDI development policy case series, (which consists of analyitical case studies, and teaching cases), EDI development studies, EDI policy seminar reports, EDI seminar papers, and EDI technical materials.

01070 ECONOMIC FORECASTS

Elsevier Science Publishers Ltd

International - English monthly/24pp £375.00 P.A. ISSN No. 0169-1767

Expert evaluations of forecasts and analysis of economic policies.

01071 ECONOMIC GROWTH AND LEAD AND ZINC CONSUMPTION

International Lead and Zinc Study Group

International - English/1990 90pp £90.00

Investigates the relationship between the consumption of lead and zinc, industrial production or gross national product, metal and energy prices in 16 major consuming countries over 1960 to 1988.

01072 ECONOMIST INTELLIGENCE UNIT: RESEARCH REPORTS: CHEMICALS AND PHARMACEUTICALS

Business International Ltd

International - English/price varies with report

Market research reports which look at the international chemical and pharmaceutical industries.

01073 ECONOMIST INTELLIGENCE UNIT: RESEARCH REPORTS: COMMODITIES

Business International Ltd

International - English/price varies with report

Market research reports which look at international commodity markets.

01074 ECONOMIST INTELLIGENCE UNIT: RESEARCH REPORTS: PAPER AND PACKAGING

Business International Ltd

International - English/price varies with report

Market research reports which look at the international paper and packaging industries.

01075 ECONOMIST INTELLIGENCE UNIT: RESEARCH REPORTS: TRAVEL AND TOURISM

Business International Ltd

International - English/price varies with report

Market research reports which look at the international tourism industry.

01076 EDUCATION IN OECD COUNTRIES: A COMPENDIUM OF STATISTICAL INFORMATION

Organisation for Economic Co-operation and Development (OECD)

International - English annual/ £17.00

Comparative statistics on education systems in all OECD countries, latest edition available is for 1987-88.

01077 EFTA
European Free Trade Association

International - English annual/free

A pocket leaflet with statistics on EFTA.

01078 EFTA ANNUAL REPORT
European Free Trade Association

International - English/French/German annual/1967 free ISSN No. 0531-4127

Looks at the work of EFTA for the preceding year.

01079 EFTA TRADE
European Free Trade Association

International - English/French/German annual/1112pp free ISSN No. 0531-4119

Specialist economic publication containing details of member countries trade.

01080 EFTA - WHAT IT IS AND WHAT IT DOES
European Free Trade Association

International - English/French/German irregular/free

Information pack type material which gives general information on the structure and functions of EFTA.

01081 ELECTRICITY SUPPLY IN OECD COUNTRIES
Organisation for Economic Co-operation and Development (OECD)

International - English annual/480pp £54.00

01082 ELSEVIER ADVANCED TECHNOLOGY REPORTS
Elsevier Science Publishers Ltd

International - English irregular/price varies with report

A series of market research reports covering different products in the electronics and telecommunications industry.

01083 EMERGING MARKET ECONOMIES REPORT
IMD

International - English annual/350pp SF700.00

Covers 20 countries and ranks them by 200 criteria.

01084 EMERGING STOCK MARKETS FACT BOOK
World Bank Publications

International - English annual/190pp US$60.00

International Finance Corporation. Provides ten years of basic data on emerging stock markets.

01085 ENERGY BALANCES AND ELECTRICITY PROFILES
United Nations Economic Commission for Europe - UN/ECE

International - English annual/484pp US$60.00

Statistics on energy data for developing countries and with overall energy balances and electricity profiles for over 200 countries.

01086 ENERGY BALANCES FOR EUROPE AND NORTH AMERICA 1970-2000
United Nations Economic Commission for Europe - UN/ECE

International - English/1988 US$45.00

Part of the ECE Energy monographic series which deals with the energy industry in Europe and the USA.

01087 ENERGY BALANCES OF OECD COUNTRIES
Organisation for Economic Co-operation and Development (OECD)

International - English/French annual/214pp £21.00

Contains a compilation of data on supply and consumption of coal, other solid fuels, oil, gas, electricity and heat. Latest edition available is for 1988-89.

01088 ENERGY BALANCES OF OECD COUNTRIES - HISTORICAL SERIES
Organisation for Economic Co-operation and Development (OECD)

International - English/French irregular

Contains a compilation of data on supply and consumption of coal, other solid fuels, oil, gas, electricity and heat. Two separate publications cover 1980-1989 and 1960-1979.

01089 ENERGY, ELECTRICITY AND NUCLEAR POWER FOR THE PERIOD UP TO 2010
International Atomic Energy Agency

International - English annual/1980 50pp AS 80.00 ISSN No. 1011-2641

Estimates of energy, electricity, and nuclear power.

01090 ENERGY POLICIES OF IEA COUNTRIES: 1990 REVIEW
Organisation for Economic Co-operation and Development (OECD)

International - English 1991/420pp £62.00 ISBN No. 9264135626

IEA review of member countries energy policies and the international energy situation. Latest edition available is for 1990.

01091 ENERGY PRICES AND TAXES
Organisation for Economic Co-operation and Development (OECD)

International - English/French quarterly/£90.00 P.A. ISSN No. 0256-2332

Contains a major compilation of energy prices at all market levels; import, industry, and consumer prices. Statistics cover main petroleum products.

01092 ENERGY STATISTICS OF OECD COUNTRIES
Organisation for Economic Co-operation and Development (OECD)

International - English/French annual/264pp £28.00

Contains a compilation of energy supply and consumption data for coal, oil, gas and electricity. Summary tables for all time periods are shown for the category "other solid fuels", such as wood and waste. Data is published 1-2 years in arrears.

01093 ENERGY STATISTICS OF OECD COUNTRIES - HISTORICAL SERIES
Organisation for Economic Co-operation and Development (OECD)

International - English/French irregular/£45.00

Separate volumes cover 1960-69 and 1980-89.

01094 ENERGY STATISTICS YEARBOOK
United Nations Publications

International - English/French annual/440pp US$75.00

A compilation of international energy statistics. It provides a global framework for comparable data on long-term trends and recent developments in the supply of all commercial forms of energy. Data on production, import, export, bunker, stock change, and consumption for 200 countries.

01096 ENVIRONMENT IN EUROPE AND NORTH AMERICA: ANNOTATED STATISTICS - STATISTICAL STANDARDS AND STUDIES

United Nations Economic Commission for Europe - UN/ECE

International - English annual/ US$54.00

Annotated statistics on some of the major environmental indicators and issues facing Europe and North America and statistics concerning agriculture in particular.

01097 ENVIRONMENT PAPERS

World Bank Publications

International - English irregular/ price varies with report

Informal documents on the World Bank's environmental work, intended to make the Bank's latest results available to the development community as quickly as possible.

01098 ERC INTERNATIONAL MARKET REPORTS: FOOD AND DRINKS PRODUCTS

ERC Statistics International

International - English annual/ £4,500-£6,000 per report

Market research reports on the food and drinks industry with international coverage.

01099 EUROPA WORLD YEARBOOK

Europa Publications

International - English annual/ 3,139pp £265.00

Basic source of detailed information on every country with brief surveys. Lists international organizations.

01100 EVALUATION RESULTS FOR 1990

World Bank Publications

International - English/1992 159pp US$9.95

A World Bank operations evaluations study. Covers 359 operations that were evaluated in 1990.

01101 FAO PRODUCTION YEARBOOK

United Nations Publications

International - English/French/ Spanish annual/283pp £23.00

Data on agricultural production, presented country by country.

01102 FAO TRADE YEARBOOK

United Nations Publications

International - English annual/ 379pp £23.00

Imports and exports of agricultural products of many of the countries of the world.

01103 FERRO-ALLOY DIRECTORY AND DATABOOK

Metal Bulletin Books Ltd.

International - English annual/1992 £58.00

Statistics of ferro-alloy production and world trade.

01104 FINANCE AND DEVELOPMENT

International Monetary Fund

International - English/French/ German/Spanish quarterly/free ISSN No. 0015-1947

Published jointly by the IMF and the World Bank it contains articles on current international monetary trends and economic development.

01105 FINANCIAL MARKET TRENDS

Organisation for Economic Co-operation and Development (OECD)

International - English 3 issues per year/£24.00 P.A. ISSN No. 0378-651X

Presents commentary and analysis of the major international financial markets including bonds, syndicated credits, and borrowing facilities, as well as for national markets.

01106 FINANCIAL MARKET TRENDS AND OECD FINANCIAL STATISTICS

Organisation for Economic Co-operation and Development (OECD)

International - English monthly/ £115.00 P.A. ISSN No. 0378-651X

Combined subscription which gives both statistical surveys.

01107 FINANCIAL STATISTICS OF OECD

Wirtschaftsstudio Des Österreich - Ischen Gesellschafts - und Wirtschaftsmuseums

International - English/French annual/price varies

Publication in 3 parts, covering flow-of-funds accounts, balance sheets and statements of income in 20 countries by institutional sectors and financial investments.

01108 FINANCIAL TIMES SURVEYS: COUNTRY SURVEYS

Financial Times - Surveys Department

International - English annual/ included with FT newspaper

Newspaper features which are published in the FT, countries being updated about every year.

01109 FINANCIAL TIMES SURVEYS: INDUSTRY SURVEYS

Financial Times - Surveys Department

International - English annual/ included with FT newspaper

Newspaper features which are published in the FT, industry surveys being updated every year.

01110 FINANCING AND EXTERNAL DEBT OF DEVELOPING COUNTRIES

Organisation for Economic Co-operation and Development (OECD)

International - English annual/ £19.50

Information on financial resources flows to developing countries. Data is published about 1 year in arrears.

01111 FOOD CONSUMPTION STATISTICS

Organisation for Economic Co-operation and Development (OECD)

International - English/French annual/540pp £60.00

Statistics on food consumption with latest figures available for 1979-1988.

01112 FOREIGN INVESTMENT ADVISORY SERVICE PAPERS

World Bank Publications

International - English irregular/ price varies with report

Informal documents prepared under the auspices of Foreign Investment Advisory Service (FIAS) to help countries attract beneficial foreign private capital, technology and managerial expertise. The FIAS is a joint venture of the Bank's International Finance Corporation and the Multilateral Investment Guarantee Agency.

01113 FOREIGN TRADE BY COMMODITIES

Wirtschaftsstudio Des Österreich - Ischen Gesellschafts - und Wirtschaftsmuseums

International - English/French annual/DM 74.00 ISSN No. 0474-540X

Matrix tables showing trade between OECD countries and partner countries for commodity groups at the 1 and 2 digit SITC level. Separate volumes are issued for exports and imports.

01114 FOREIGN TRADE BY COMMODITIES - SERIES C

Organisation for Economic Co-operation and Development (OECD)

International - English 5 volumes per year/£90.00 P.A. ISSN No.

0474-540X

Summary information on the value of trade flows of OECD members with all partners by commodity to the 2 digit SITC level.

01115 THE FOREST RESOURCES OF DEVELOPED, TEMPERATE-ZONE REGIONS

United Nations Economic Commission for Europe - UN/ECE

International - English/1992 price on application

Covers forests and forest products with data for forecasting.

01116 FOREST RESOURCES OF THE ECE REGION (EUROPE, THE FORMER USSR, NORTH AMERICA)

United Nations Economic Commission for Europe - UN/ECE

International - English/1992 220pp US$21.00 ISBN No. 9211163447

Published in 3 volumes, expanded to include tree biomass and environmental issues.

01117 FOREST RESOURCES OF THE TEMPERATE ZONES

United Nations Economic Commission for Europe - UN/ECE

International - English/1992 33pp US$13.00

Looks at the developed temperate zones and their stocks of forest resources and forest products.

01118 FOREST RESOURCES OF THE TEMPERATE ZONES: MAIN FINDINGS OF THE UN-ECE/FAO 1990 FOREST RESOURCE ASSESSMENT

United Nations Economic Commission for Europe - UN/ECE

International - English/1991 32pp US$13.00 ISBN No. 9211165482

Covers the temperate zones of Europe, including Cyprus, Israel and Turkey, the former USSR, North America, Australasia and Japan.

01119 THE FREE TRADE AGREEMENTS OF THE EFTA COUNTRIES AND THE EUROPEAN COMMUNITIES

European Free Trade Association

International - English/French/German annual/1967 free

01120 FT MANAGEMENT REPORTS: AUTOMOTIVE

FT Management Reports

International - English irregular/£150.00 - £250.00 per report

Management/market research type reports which cover the international and national automotive industry.

01121 FT MANAGEMENT REPORTS: BANKING & FINANCE

FT Management Reports

International - English irregular/£160.00 - £250.00 per report

Management/market research type reports which cover the international banking industry and related topics.

01122 FT MANAGEMENT REPORTS: ENERGY

FT Management Reports

International - English irregular/£150.00 - £250.00 per report

Management/market research type reports which cover the international energy industry; "North Sea operations" and "Natural gas in the EC" are typical titles.

01123 FT MANAGEMENT REPORTS: INSURANCE

FT Management Reports

International - English irregular/£150.00 - £250.00 per report

Management/market research type reports which cover the international insurance industry; "pension provision in the EC" is a typical title.

01124 FT MANAGEMENT REPORTS: MARKETING

FT Management Reports

International - English irregular/£150.00 - £250.00 per report

Management/market research type reports which cover the international marketing industry; "Direct marketing in Europe" is a typical title.

01125 FT MANAGEMENT REPORTS: PACKAGING

FT Management Reports

International - English irregular/£150.00 - £250.00 per report

Management/market research type reports which cover the international packaging industry; "World waste paper" is a typical title.

01126 FT MANAGEMENT REPORTS: PHARMACEUTICALS

FT Management Reports

International - English irregular/£150.00 - £250.00 per report

Management/market research type reports which cover the international pharmaceuticals industry.

01127 FT MANAGEMENT REPORTS: PROPERTY

FT Management Reports

International - English irregular/£150.00 - £250.00 per report

Management/market research type reports which cover the international property market.

01128 FT MANAGEMENT REPORTS: TELECOMMUNICATIONS

FT Management Reports

International - English irregular/£150.00 - £250.00 per report

Management/Market research type reports which cover the international telecommunications market.

01129 GLOBAL CURRENCY OUTLOOK

Barclays Bank Plc

International - English monthly/£150.00 P.A. or £30 per report

One year forecasts for over 20 currencies together with the outlook for interest rates on a monthly basis.

01130 GLOBAL ECONOMIC FORECASTING

Henley Centre for Forecasting

International - English quarterly with monthly updates/£900.00 P.A.

Covers the main economic activity indicators.

01131 GLOBAL ECONOMIC PROSPECTS AND THE DEVELOPING COUNTRIES

World Bank Publications

International - English annual/80pp US$10.95 ISSN No. 1014-8906

Offers a concise survey of long-term prospects for the global economy.

01132 GLOBAL HEALTH SITUATION AND PROJECTIONS: ESTIMATES

World Health Organisation

International - English annual/94pp US$11.70

Sets out the latest WHO global statistical estimates for morbidity and mortality from individual diseases and for trends demographic, socio-economic, environmental and other indicators of health status.

01133 GLOBAL OUTLOOK 2000: AN ECONOMIC, SOCIAL AND ENVIRONMENTAL PERSPECTIVE

United Nations Economic Commission for Europe - UN/ECE

International - English irregular/1990 340pp US$19.95

Provides an overview to the year 2000 and describes the likely course of the world economy in the 1990's with particular reference to the environment.

01134 GOLD AND SILVER SURVEY

World Reports Ltd

International - English 10 times per year/£215.00 P.A. ISSN No. 0196-2546

Covers the gold and silver market in the major economies of the world.

01135 GOVERNMENT FINANCE STATISTICS YEARBOOK

International Monetary Fund

International - English annual/ US$48.00 ISSN No. 0250-7374

Provides information on the various units of government, government accounts, enterprises and financial institutions that governments own and control and the national sources of data on government operations. World tables arranged by topic are also included.

01136 GRAIN MARKET REPORT

International Wheat Council

International - English monthly/ £100.00 P.A.

Includes an annual supplement on national policies.

01137 HANDBOOK OF INDUSTRIAL STATISTICS

United Nations Publications

International - English annual/ 482pp US$170.00

Covers world industry and major industrial activity and structure. Data is published 2-3 years in arrears.

01138 HANDBOOK OF INTERNATIONAL TRADE AND DEVELOPMENT STATISTICS

United Nations Publications

International - English/French annual/600pp US$80.00

Collection of statistical data relevant to the analysis of world trade and development. Data is published 1-2 years in arrears.

01139 HANDBOOK OF THE NATIONS

Gale Research International

International - English annual/ £86.00

Statistics on 250 plus countries and some political profiles.

01140 HINTS TO EXPORTERS

Department of Trade and Industry Export Publications

International - English irregular/ £5.00 per country

Basic economic and commercial information on a country mainly intended for exporters or those doing business in the country.

01141 IFC AND THE ENVIRONMENT

World Bank Publications

International - English annual/ US$6.95

Looks at the International Finance Corporation activities in the environmental area and the economic implications.

01142 IMF PAPERS ON POLICY ANALYSIS AND ASSESSMENT

International Monetary Fund

International - English occasional/ price varies

IMF staff studies prepared in non-technical language and aimed at operational staff.

01143 IMF STAFF PAPERS

International Monetary Fund

International - English/French/ Spanish quarterly/US$46.00 P.A. ISSN No. 0020-8027

Quarterly economic journal with a theoretical bias.

01144 IMF SURVEY

International Monetary Fund

International - English/French/ Spanish 23 times per year/free ISSN No. 0047-083X

A topical report of the IMF's activities including all press releases, communiques, major speeches and SDR valuations. Subscription includes occasional special supplements and an annual supplement.

01145 IMF WORKING PAPERS

International Monetary Fund

International - English irregular/ US$200.00

120 papers per year covering a wide range of economic and financial topics.

01146 INDICATORS OF INDUSTRIAL ACTIVITY

Organisation for Economic Co-operation and Development (OECD)

International - English quarterly/ £30.00 P.A.

Presents up-to-date production, deliveries, orders, prices and employment indicators on an annual, quarterly, and monthly basis for industrial sectors.

01147 INDUSTRIAL MINERALS

Metal Bulletin Books Ltd.

International - English monthly/ £132.00 P.A. ISSN No. 0019-8544

Includes surveys of trends and developments in the industry and statistics.

01148 INDUSTRIAL MINERALS DIRECTORY 1991

Metal Bulletin Books Ltd.

International - English irregular/ £94.00

Covers non-metallic mineral resources on a country by country basis, including a buyers guide and suppliers listings.

01149 INDUSTRIAL POLICY IN OECD COUNTRIES: ANNUAL REVIEW

Organisation for Economic Co-operation and Development (OECD)

International - English annual/ £20.00

Examines recent government initiatives and reviews trends in industrial performance. Also contains information on Central and Eastern European industrial policies.

01150 INDUSTRIAL STATISTICS YEARBOOK

United Nations Publications

International - English annual/ 660pp US$95.00

Compilation on world industry and major industrial activity and structure for 101 countries, classified by branches of activity. Data is published 2-3 years in arrears.

01151 INDUSTRIAL STRUCTURE STATISTICS

Wirtschaftsstudio Des Österreich - Ischen Gesellschafts - und Wirtschaftsmuseums

International - English annual/1982 DM 38.00

In 2 parts. Part 1 contains the data derived directly from the industrial statistics surveys and the foreign trade data. Part 2 presents the estimates provided by the national accounts disaggregated by industry. All data is classified to ISIC.

01152 INDUSTRIAL STRUCTURE STATISTICS

Organisation for Economic Co-operation and Development (OECD)

International - English annual/

168pp £15.00

Provides data covering a number of variables including production, exports and value added for 2-4 digit ISIC manufacturing classes and some one digit non-manufacturing classes. Data is published about 3 years in arrears.

01153 INFORMATION, COMPUTER AND COMMUNICATIONS POLICY - ICCP SERIES

Organisation for Economic Co-operation and Development (OECD)

International - English irregular/ various prices

A series which includes economic and business aspects of information technology. Typical titles include "Telecommunications equipment: Changing markets and trade structures 1991".

01154 INSURANCE AND OTHER FINANCIAL SERVICES: STRUCTURAL TRENDS

Organisation for Economic Co-operation and Development (OECD)

International - English/French annual/1992 156pp £17.00

Contains an overview of the salient regulatory and economic aspects of the relationship between insurance and other financial services in the OECD area.

01155 INTEREST RATE SERVICE

World Reports Ltd

International - English 10 times per year/£450.00 P.A. ISSN No. 0308-9002

Covers interest rates and related financial developments in the main economies of the world.

01156 INTERNATIONAL BUSINESS OPPORTUNITIES SERVICE (IBOS)

World Bank Publications

International - English irregular/ price varies with level of subscription

A subscription which gives a variety of publications, including 500 "Contracts awards" per year, 800 "General and specific procurement notices", "Monthly operational summary", and 250 "Technical data sheets" per year. Designed to give notice of contracts open to tender or being awarded by the World Bank.

01157 INTERNATIONAL CAPITAL MARKETS: DEVELOPMENTS AND PROSPECTS

International Monetary Fund

International - English annual/ US$20.00

Comprehensive review of the latest developments in international capital markets.

01158 INTERNATIONAL COTTON INDUSTRY STATISTICS (ICIS)

International Textile Manufacturers Federation

International - English annual/1958 SF100.00

Contains data on productive capacity, machinery utilization, and raw materials consumption in the short-staple sector of the textile industries.

01159 INTERNATIONAL CURRENCY REVIEW

World Reports Ltd

International - English quarterly/ £225.00 P.A. ISSN No. 0020-6490

Covers currencies, interest rates and other factors in the area of financial economics.

01160 INTERNATIONAL DEVELOPMENT POLICIES

Commonwealth Secretariat

International - English quarterly/ £40.00 P.A. ISSN No. 0964-699X

Review of the activities of international organizations.

01161 INTERNATIONAL ECONOMIC OUTLOOK

Blackwell Publishers

International - English twice a year/ £150.00 P.A. ISSN No. 0960-8869

Published by the London Business School this journal contains detailed forecasts for the major world economies, commodity prices, world exchange and interest rates.

01162 INTERNATIONAL ECONOMIC REVIEW

International Economic Review

International - English quarterly/ 250pp US$150.00 institutional subscription, US $45.00 individual subscription ISSN No. 0020-6598

International economics with an emphasis on quantitative economics.

01163 INTERNATIONAL FINANCIAL CORPORATION TECHNICAL OR DISCUSSION PAPERS

World Bank Publications

International - English irregular/ price varies with paper

Informal documents prepared by the International Finance Corporation (IFC) that focus on issues relating to the development of the private sector in developing countries.

01164 INTERNATIONAL FINANCIAL STATISTICS

International Monetary Fund

International - English/French/ Spanish monthly/US$188.00 P.A. ISSN No. 0250-7463

Covers exchange rates, international banking and liquidity, money and banking, interest rates, prices, production, international transactions, government accounts and national accounts.

01165 INTERNATIONAL FINANCIAL STATISTICS - SUPPLEMENT

International Monetary Fund

International - English/US$12.00 ISSN No. 0252-3027

Supplement series which provides expanded coverage on economic topics of special interest; trade statistics or balance of payments for example. Latest editions are for the late 1980's.

01166 INTERNATIONAL FINANCIAL STATISTICS - YEARBOOK

International Monetary Fund

International - English/French/ Spanish annual/US$50.00 ISSN No. 0250-7463

Contains international and national financial statistics for over 35 years for countries appearing in the monthly edition of the same work.

01167 INTERNATIONAL GIVING AND VOLUNTEERING SURVEY

Charities Aid Foundation

International - English with abstracts in French and Spanish/ 1992 £16.95 ISBN No. 0904757595

Surveys individual charitable donations and activity.

01168 INTERNATIONAL HERALD TRIBUNE

International Herald Tribune

International - English daily/1887 £00.75 per issue

International business and company news.

01169 INTERNATIONAL HOUSING FINANCE SOURCEBOOK

Building Societies Association/Council of Mortgage Lenders

International - English/1991 £30.00

Gives an overview of housing finance systems in the Middle East, the Americas, Asia and Australia, and Europe.

01170 INTERNATIONAL MANAGEMENT

Reed Business Publishing Ltd

International - English monthly/1946 US$65.00 P.A. ISSN No. 0020-7888

International business and management from a European perspective. Features include economic data.

01171 INTERNATIONAL MARKETING DATA AND STATISTICS 199-

Euromonitor Plc

International - English annual/ 700pp £150.00

Marketing source book.

01172 INTERNATIONAL MOTOR BUSINESS

Business International Ltd

International - English quarterly/ £405.00 P.A. ISSN No. 0267-8225

Each issue contains an average of five reports on the international motor business.

01173 INTERNATIONAL PRODUCTION COST COMPARISON, SPINNING/ WEAVING (IPCC)

International Textile Manufacturers Federation

International - English annual/ SF100.00

Examines the cost structure of yarn and fabric production in Brazil, Germany, India, Japan, Korea and the USA. Latest edition available is for 1991.

01174 INTERNATIONAL RISK AND PAYMENT REVIEW

Dun & Bradstreet

International - English monthly/ £390.00 P.A.

Risk analysis type profiles of 100 plus countries.

01175 INTERNATIONAL SEA-BORNE TRADE STATISTICS YEARBOOK

United Nations Publications

International - English annual/ 406pp US$45.00

Gives aggregates for worldwide sea-bourne trade. Data published 4-5 years in arrears.

01176 INTERNATIONAL STEEL STATISTICS

International Steel Statistics Bureau

International - English annual/ £1,250.00

22 annual country surveys showing production, materials consumed, apparent consumption, and detailed imports/exports by quality and market.

01177 INTERNATIONAL STEEL STATISTICS: SUMMARY TABLES

United Kingdom Iron and Steel Statistics Bureau

International - English annual/ £100.00 ISSN No. 0952-6803

Annual summary tables of major steel producing countries with iron, crude steel, imports and exports and finished steel products covered.

01178 INTERNATIONAL TEXTILE MACHINERY SHIPMENT STATISTICS (ITMSS)

International Textile Manufacturers Federation

International - English annual/ SF150.00

Compiled in cooperation with the world's leading manufacturers of spinning and weaving machinery, it shows shipments, by country of destination, of ring spindles, o-e rotors, shuttle and shuttle-less looms. Provides a picture of country spinning and weaving capacity and investment trends.

01179 INTERNATIONAL TIN STATISTICS

United Nations Publications

International - English quarterly/ US$90.00 P.A.

Covers the activities of the International Tin Council and data supplied by member countries. Production, consumption and trade are covered with data on tin prices and uses.

01180 INTERNATIONAL TOURISM REPORTS

Business International Ltd

International - English 4 issues per year/£280.00 P.A. ISSN No. 0269-3747

Contains reports on four countries in each issue, looking at their tourism markets.

01181 INTERNATIONAL TRADE STATISTICS YEARBOOK

United Nations Publications

International - English/French annual/239pp US$125.00

A two volume set covering 152 countries external trade, performance, and overall trends by current value, volume, and price. Data published 2-3 years in arrears.

01182 IP STATISTICAL SERVICE

Institute of Petroleum

International - English/£25.00 P.A.

Included are: oil data sheets with updates quarterly, UK petroleum industry statistics, world oil statistics annual booklet, petroleum statistics card annual, and oil data sheet monthly updates.

01183 IRON AND MANGANESE ORE DATABOOK

Metal Bulletin Books Ltd.

International - English irregular/ £81.00

International iron ore, mines, traders, price movements, production statistics and analysis.

01184 THE IRON AND STEEL INDUSTRY

Organisation for Economic Co-operation and Development (OECD)

International - English/French annual/50pp £10.00

Statistical publication containing tables on steel production, consumption and trade data, as well as other indicators of activity such as employment.

01185 LABOUR FORCE STATISTICS

Organisation for Economic Co-operation and Development (OECD)

International - English annual/ 500pp £48.00

1969-1989 data published in 1991.

01186 LABOUR FORCE STATISTICS

Wirtschaftsstudio Des Österreich - Ischen Gesellschafts - und Wirtschaftsmuseums

International - English/French annual/1968 DM 86.00

Historical time series of the evolution, population and labour force of the OECD. Part 1 contains general tables, part 2 figures by country and part 3 participation rates and unemployment rates by age and sex.

01187 LEAD AND ZINC STATISTICS: 1960-1988

International Lead and Zinc Study Group

International - English/1990 250pp £150.00

Long-term historical coverage of world production and consumption of lead and zinc, combining detailed annual tables with monthly series.

01188 LIVING STANDARDS MEASUREMENT STUDY (LSMS) WORKING PAPERS

World Bank Publications

International - English irregular/ price varies with paper

Informal documents prepared by the LSMS, which was begun by the World Bank to develop improved methods for collecting and analysing household and community data on living standards. Includes critical surveys covering different aspects of the LSMS data program and reports on improved methodologies for using LSMS data.

01189 LLOYD'S REGISTER OF SHIPPING ANNUAL REPORT

Lloyd's Register of Shipping

International - English annual/ c50pp free

Brief survey of developments in the shipping sector.

01190 LLOYD'S REGISTER OF SHIPPING STATISTICAL TABLES

Lloyd's Register of Shipping

International - English annual/ c80pp free

Details of the world's merchant fleets, ships built and lost. Fleets are analysed by type, size, age and tonnage.

01191 LONDON CURRENCY REPORT

World Reports Ltd

International - English 10 times per year/£450.00 P.A. ISSN No. 0307-0360

Covers the London currency market.

01192 MAIN ECONOMIC INDICATORS

Organisation for Economic Co-operation and Development (OECD)

International - English monthly/ £106.00 P.A. ISSN No. 0474-5523

Gives a detailed picture of the most recent changes in OECD member country economies. Statistics are given for GNP, industrial production, deliveries, stocks and orders, construction, wholesale and retail sales, employment, labour costs, prices, money supply and financial markets, trade balance and current balance.

01193 MAIN ECONOMIC INDICATORS - HISTORICAL STATISTICS

Organisation for Economic Co-operation and Development (OECD)

International - English irregular/ various prices

Series is published about two years in arrears; in 1990, statistics for 1969-88 were published.

01194 MAIN SCIENCE AND TECHNOLOGY INDICATORS

Organisation for Economic Co-operation and Development (OECD)

International - English/French twice yearly/£20.00 P.A. ISSN No. 1011-792X

Contains data on the scientific and technological performance of OECD member countries.

01195 MARITIME TRANSPORT

Organisation for Economic Co-operation and Development (OECD)

International - English annual/ 172pp £15.00

Coverage of international developments in maritime transport with particular reference to national shipping policies and longer term trends in international shipping and trade.

01196 MEAT BALANCES OF OECD COUNTRIES

OECD - Organisation for Economic Co-operation and Development

International - English/French annual/146pp £19.00

Contains international comparisons for production, trade and consumption for each category of meat. Data covers longer periods - 1984-90 for example.

01197 MERCHANT SHIPBUILDING RETURNS

Lloyd's Register of Shipping

International - English quarterly/ c20pp free ISSN No. 0261-1848

Statistical summaries of world shipbuilding for all self-propelled ships of 100 gross tonnage and above, which are under construction or on order.

01198 METAL BULLETIN

Metal Bulletin Books Ltd.

International - English bi-weekly/ £312.00 P.A. ISSN No. 0026-0533

Price quotations, market reports, trade and industry in the metals sector.

01199 METAL BULLETIN MONTHLY

Metal Bulletin Books Ltd.

International - English monthly/ £139.00 P.A. ISSN No. 0373-4064

Included with a subscription to "Metal Bulletin" - market reports, trade and industry developments and analysis of trends in the metal sector.

01200 METAL BULLETIN PRICES AND DATA

Metal Bulletin Books Ltd.

International - English annual/ £312.00

Price quotations, market reports, trade and industry in the metals sector.

01201 MIGRATION: THE DEMOGRAPHIC ASPECTS

Organisation for Economic Co-operation and Development (OECD)

International - English/1991 92pp £15.00 ISBN No. 9264134395

Part of the series "Demographic change and social policy" 1991.

01202 MONTHLY BULLETIN LEAD AND ZINC STATISTICS

International Lead and Zinc Study Group

International - English monthly/ £125.00 P.A. ISSN No. 0023-9577

Tables and graphs on world supply and demand; detailed figures on mine production, metal production, metal consumption, principal exports and imports of concentrates and refined metal, secondary recovery, prices, plus detailed country profiles.

01203 MONTHLY BULLETIN OF STATISTICS

United Nations Publications

International - English monthly/ 118pp US$225.00 academics, US $450.00 commercial institutions

Statistics on 60 subjects from over 200 countries. Quarterly data for significant world and regional aggregates are included regularly as well as special features on selected topics.

01204 MONTHLY COMMODITY PRICE BULLETIN

United Nations Publications

International - English monthly/ US$125.00 P.A. US $ 15.00 single issue

A series of monthly, quarterly, and annual average prices of the movements of all food; tropical beverages; vegetable oilseeds and oils; food; agricultural raw materials; minerals; ores and metals; cocoa and coffee; rubber and lead.

01205 MONTHLY EXCHANGE AND INTEREST RATE OUTLOOK

Barclays Bank Plc

International - English monthly/ £150.00 P.A. or £30.00 per report

Covers the outlook for the world's major currencies - US Dollar, Japanese Yen, D. Mark, and Sterling. Forecasts given for the year ahead.

01206 MONTHLY STATISTICAL SUMMARY: INTERNATIONAL TEA COMMITTEE

International Tea Committee Ltd.

International - English monthly/ 12pp £130.00 P.A. ISSN No. 0309-0477

Tea production, exports, imports for consumption, UK imports, re-exports and stocks, weekly auctions, quantities and prices.

01207 MONTHLY STATISTICS OF FOREIGN TRADE - SERIES A

Organisation for Economic Co-operation and Development (OECD)

International - English monthly/ £78.00 P.A. ISSN No. 0474-5388

Presents an overall picture of foreign trade of OECD countries, including analysis by flows with countries and country groupings by origin and destination, and seasonally adjusted foreign trade indicators. Summary monthly tables showing trade by main commodity categories are included.

01208 MOTORSTAT EXPRESS

Society of Motor Manufacturers and Traders Ltd.

International - English fortnightly/ £75.00 P.A.

Divided into two tables, Table I gives total production, new registrations, and exports for cars and commercial vehicles, showing the current month and year to date. Countries covered are Belgium, France, West Germany, Italy, Japan, Netherlands, Spain, Sweden, UK and USA. Table II gives new registrations in Austria, Denmark, Irish Republic, Finland, Luxembourg, Norway, Portugal and Switzerland.

01209 MULTILATERAL OFFICIAL DEBT RESCHEDULING; RECENT EXPERIENCE

International Monetary Fund

International - English biennial/ US$15.00

Reviews trends in official debt rescheduling and recent experience with debt re-negotiations.

01210 NARCOTIC DRUGS: ESTIMATED WORLD REQUIREMENTS - STATISTICS

United Nations Publications

International - English/French/ Spanish annual/188pp US$36.00

Data relates to the worldwide availability and use of narcotic drugs. Forecasts of production and utilization of drugs and analyses their trade patterns. Also updated monthly in English.

01211 NATIONAL ACCOUNTS OF OECD COUNTRIES

OECD - Organisation for Economic Co-operation and Development

International - English/French annual/£61.00

Published in two volumes - Main aggregates, and Detailed tables. Data is published 2-3 years in arrears.

01212 NATIONAL ACCOUNTS STATISTICS: ANALYSIS OF MAIN AGGREGATES

United Nations Publications

International - English annual/ 278pp US$125.00

Annual summary of the main national accounts aggregates presented in analytical tables. Gross domestic product or net material product by type of expenditure, kind of economic activity and cost components, and other income aggregates based on current and constant prices are included. Data is published 4-5 years in arrears.

01213 NATIONAL ACCOUNTS STATISTICS: MAIN AGGREGATES AND DETAILED TABLES

United Nations Publications

International - English annual/ 302pp US$110.00

A two volume set on detailed national account estimates of 165 countries categorized by type of economy; market economy or centrally planned.

01214 NATIONAL ACCOUNTS STATISTICS: STUDY OF INPUT-OUTPUT TABLES 1970-80

United Nations Publications

International - English 1991/321pp US$36.00

01215 NATIONAL POLICIES AND AGRICULTURAL TRADE

Organisation for Economic Co-operation and Development (OECD)

International - English irregular/150-300pp various prices

Country profiles of different national agricultural markets, updated every 2-3 years.

01216 NATO ECONOMICS COLLOQUIUM

North Atlantic Treaty Organisation

International - English/French annual/c300pp free

Economic development in Central and Eastern Europe.

01217 NATO FACTS AND FIGURES

North Atlantic Treaty Organisation

International - English/French/ Italian/German/Dutch/Spanish irregular/free

Looks at the history, structure and operation of NATO and includes basic statistics.

01218 THE NORDIC SECURITIES MARKET

Copenhagen Stock Exchange

Denmark/Sweden/Finland/Norway - English quarterly/April 1990 6pp free

Broad coverage of news/data about the Scandinavian securities market.

01219 NUCLEAR POWER REACTORS IN THE WORLD

International Atomic Energy Agency

International - English annual/66pp AS 100.00 ISSN No. 1011-2642

Most recent reactor data and reactor performance data.

01220 OECD ECONOMIC OUTLOOK

Organisation for Economic Co-operation and Development (OECD)

International - English twice yearly/ 230pp £13.50 P.A. ISSN No. 0474-5574

Latest economic forecasts for OECD member countries and selected other regions.

01221 OECD ECONOMIC OUTLOOK - HISTORICAL STATISTICS 1960-1990

Organisation for Economic Co-operation and Development (OECD)

International - English 1991/168pp £17.00 ISBN No. 9264035419

Looks at main economic indicators and chart changes over the historical period.

01222 OECD ECONOMIC STUDIES

Organisation for Economic Co-operation and Development (OECD)

International - English twice yearly/ £25.00 P.A. ISSN No. 0255-0822

Features articles on applied macroeconomic and statistical analysis.

01223 OECD ECONOMIC SURVEYS AND CCEET ECONOMIC SURVEYS

Organisation for Economic Co-operation and Development (OECD)

International - English annual/ £122.00 ISSN No. 0376-6438

Combines a subscription to the Economic Surveys with surveys covering Poland, Hungary and the Czech-Slovak Federal Republic.

01224 OECD EMPLOYMENT OUTLOOK
Organisation for Economic Co-operation and Development (OECD)

International - English/French annual/£26.00

Describes short-term employment prospects for the industrialized world, presents analysis of labour market topics, and sets out options for policy.

01225 OECD ENVIRONMENTAL DATA: COMPENDIUM 199-
Organisation for Economic Co-operation and Development (OECD)

International - English/French annual/320pp £33.00

01226 OECD FINANCIAL STATISTICS
Organisation for Economic Co-operation and Development (OECD)

International - English fortnightly/ £162.00 P.A.

Published in three parts. Part one covers domestic markets in twelve issues. Part two with international markets in twelve issues. Part three provides financial accounts of OECD member countries in twenty issues. Two annual publications are also included which provide non-financial enterprises financial statements and a methodological supplement.

01227 OECD NUCLEAR ENERGY DATA
Organisation for Economic Co-operation and Development (OECD)

International - English/French annual/44pp £7.00

OECD Nuclear Energy Agency's annual compilation of basic statistics on electricity generation and nuclear power in OECD countries.

01228 OECD OBSERVER
Organisation for Economic Co-operation and Development (OECD)

International - English bi-monthly/ £12.00 P.A. ISSN No. 0029-7054

Provides information and news of the latest OECD work and forecasts.

01229 OFFSHORE CENTRES REPORT
World Reports Ltd

International - English 10 times per year/£100.00 P.A.

Covers financial markets in offshore centres.

01230 OIL AND GAS INFORMATION 1988-1990
Organisation for Economic Co-operation and Development (OECD)

International - English/French/ 580pp £65.00 ISBN No. 9264035079

Current development in oil and gas supply and demand, including country specific statistics for OECD member countries on production, trade, demand, prices and reserves. Data on world production, trade and consumption of major oil product groups and natural gas is available in summary tables.

01231 OPERATING EXPERIENCE WITH NUCLEAR POWER STATIONS IN MEMBER STATES
International Atomic Energy Agency

International - English annual/AS 2,200

Gives annual performance data and outage information with a historical summary of production and outages during the lifetime of individual plants.

01232 OPERATION AND EFFECTS OF THE GENERALISED SYSTEM OF PREFERENCES: NINTH AND TENTH REVIEWS
United Nations Publications

International - English irregular/ 88pp US$35.00

Evaluates the effects of preferential tariff treatment granted to imports from developing countries by developed industrial countries.

01233 OPERATIONS EVALUATION PAPERS
World Bank Publications

International - English irregular/ price varies with paper

Systematic and comprehensive reviews of the World Bank's development experience.

01234 POLICY AND ADVISORY SERVICE PAPERS
World Bank Publications

International - English irregular/ price varies with paper

Informal documents that provide advice and technical expertise to developing countries on attracting foreign investment.

01235 POPULATION AND THE WORLD BANK: IMPLICATIONS FROM EIGHT CASE STUDIES
World Bank Publications

International - English irregular/ 1992 171pp US$10.95 ISBN No. 0821320815

Published as a World Bank operations evaluation study. Provides a broad view of the demographic, social and economic changes that have occurred since the bank began lending for population projects in 1968.

01236 POPULATION AND VITAL STATISTICS REPORT
United Nations Publications

International - English quarterly/ US$30.00 P.A. US $10.00 single issue

Latest census data, plus worldwide demographic statistics on birth and mortality.

01237 POPULATION BULLETIN OF THE UNITED NATIONS
United Nations Publications

International - English biannual/ US$13.00

Bulletin on population studies carried out by the United Nations, its specialized agencies and other organizations, aimed at those engaged in social and economic research.

01238 PRICE PROSPECTS FOR MAJOR PRIMARY COMMODITIES
World Bank Publications

International - English biennial with quarterly supplements/US$275.95

Published in two volumes. Volume one covers summary, energy, metals and minerals. Volume two covers agricultural products, fertilizers, and tropical timber. Periods covered include 1990-2005.

01239 PRICES AND PRICE INDICES IN FOREIGN COUNTRIES
W. Kohlhammer GmbH

International - English/German/ quarterly 1977 DM 4.50 P.A.

01240 PRICES OF AGRICULTURAL PRODUCTS AND SELECTED INPUTS IN EUROPE AND NORTH AMERICA
United Nations Economic Commission for Europe - UN/ECE

International - English annual/ 136pp US$38.00

Annual ECE/FAO price review which gives product, producer, and selected input prices.

01241 PRINCIPAL USES OF LEAD AND ZINC: 1960-1990
International Lead and Zinc Study Group

International - English/1992 112pp £95.00

Short-term review of trends in the main uses of lead and zinc during 1985-1990 in countries responsible for about 90% of Western World consumption, together with a long term historical review of trends since 1960 in main European countries, Japan, the USA, and the factors influencing them.

01242 PROCEEDINGS OF THE WORLD BANK ANNUAL CONFERENCE ON DEVELOPMENT ECONOMICS

World Bank Publications

International - English annual/ c360pp US$10.95 ISSN No. 1014-7268

The 1991 edition, published in 1992 contained papers addressing the transition in socialist economies, military spending, urbanization, and the role of governance in development.

01243 THE PULP AND PAPER INDUSTRY

Organisation for Economic Co-operation and Development (OECD)

International - English/French annual/100pp £18.00

Production and consumption of pulp and paper products with foreign trade included.

01244 QUARTERLY BULLETIN OF COCOA STATISTICS

International Cocoa Organisation

International - English/French/ Spanish/Russian quarterly/50-60pp £40.00 P.A. ISSN No. 0308-4469

Statistical data on the world cocoa economy.

01245 QUARTERLY LABOUR FORCE STATISTICS

Organisation for Economic Co-operation and Development (OECD)

International - English quarterly/ £22.00 P.A. ISSN No. 0255-3627

Provides current statistics on the short term developments of the major components of the labour force in the USA and 12 other OECD countries.

01246 QUARTERLY NATIONAL ACCOUNTS

Organisation for Economic Co-operation and Development (OECD)

International - English quarterly/ £36.00 P.A. ISSN No. 0304-3738

Provides the latest national income statistics for 11 OECD member countries and the USA.

01247 QUARTERLY OIL STATISTICS AND ENERGY BALANCES

Organisation for Economic Co-operation and Development (OECD)

International - English quaterly/ £90.00 P.A. ISSN No. 1013-9362

Statistics on oil and gas supply and demand in the OECD area.

01248 RECORD OF SHIPMENTS OF WHEAT AND WHEAT FLOUR

International Wheat Council

International - English annual/ £20.00

Commercial and non-commercial shipments of wheat and flour by origin and destination, including estimates for non-reporting countries. Includes separate tables for durum.

01249 RECYCLING LEAD AND ZINC: THE CHALLENGE OF THE 1990'S

International Lead and Zinc Study Group

International - English/1991 459pp £95.00

Includes assessments of expected trends in world supply for lead and zinc during the 1990's.

01250 REPORT OF THE CROP OF THE YEAR (WHEAT)

International Wheat Council

International - English annual/ £20.00

Incorporates the former "Annual report" and "Review of the world situation in wheat".

01251 RETAIL MONITOR INTERNATIONAL

Euromonitor Plc

International - English monthly/ £365.00 P.A. ISSN No. 0952-9594

Retail market research, includes data on sales figures, prices and consumer spending.

01252 REVENUE STATISTICS OF OECD MEMBER COUNTRIES

Organisation for Economic Co-operation and Development (OECD)

International - English/French annual/£26.00

Detailed and internationally comparable tax data in a common format for all 24 OECD countries covering the period from 1965 onwards. Latest data available is for 1990.

01253 REVIEW OF FISHERIES IN OECD MEMBER COUNTRIES

Organisation for Economic Co-operation and Development (OECD)

International - English annual/ £52.00

Covers major developments affecting the commercial fisheries of OECD countries, including government policy and action.

01254 REVIEW OF MARITIME TRANSPORT

United Nations Economic Commission for Europe - UN/ECE

International - French annual/1968 98pp US$20.00 ISSN No. 0085-560X

Reviews the main developments in world maritime transport, assessing expected short-term forecasts of sea-borne trade for some 35 leading countries.

01255 RUBBER TRENDS

Business International Ltd

International - English quarterly/ £360.00 P.A. ISSN No. 0035-9564

Looks at industrial and market trends in the international rubber industry.

01256 SCIENCE, TECHNOLOGY AND INDUSTRY REVIEW

Organisation for Economic Co-operation and Development (OECD)

International - English twice yearly/ £21.00 P.A.

Contains reports and articles on science, technology, and related industrial policy issues which are of current interest to member countries.

01257 SETTING UP BUSINESS IN...

Department of Trade and Industry Export Publications

International - English/£20.00 per country

Basic economic and commercial information on a country mainly aimed at those intending to operate a business in that country.

01258 SHIPYARD ORDERS WEEKLY REPORT

Lloyd's Register of Shipping

International - English weekly/ £400.00 P.A.

A list of confirmed orders reported during the week, showing selected data, including cancellations and completions.

01259 SOCIAL INDICATORS OF DEVELOPMENT
World Bank Publications
International - English annual/ c400pp US$24.95

Published in association with the John Hopkins University Press. Allows readers to monitor and evaluate social and economic progress in the world with 170 plus national economic profiles.

01260 STAFF STUDIES FOR THE WORLD ECONOMIC OUTLOOK
International Monetary Fund
International - English annual/ US$20.00

Prepared by the staff of the IMF, these studies comprise supporting material for the "World Economic Outlook".

01261 STAINLESS STEEL DATABOOK 2ND EDITION
Metal Bulletin Books Ltd.
International - English 1991/£72.00

World list of stainless steel refining plants and international directory of producers.

01262 STATE OF TRADE REPORTS
International Textile Manufacturers Federation
International - English quarterly/ SF100.00

Compiled from reports from ITMF member associations and shows country-by-country changes in the spinning and weaving sectors (short-staple) for production, outstanding orders and stock. Highlighted are the regional and global trends for production and stocks for yarns and fabrics.

01263 STATISTICAL REPORT ON ROAD ACCIDENTS (ECMT SERIES)
Organisation for Economic Co-operation and Development (OECD)
International - English annual/ £12.50

Usually published two years in arrears.

01264 STATISTICAL YEARBOOK
United Nations Publications
International - English/French annual/US$100.00

Compilation of international statistics relating to population, manpower, national accounts, wages, prices, consumption, balance of payments, finance, health, education, culture, forestry and fishing, industrial production, energy, external trade, communication, transport, and international tourism.

01265 STATISTICS EUROPE: SOURCES FOR SOCIAL, ECONOMIC AND MARKET RESEARCH
CBD Research Ltd
Jarvey, J. M./International - English irregular/320pp £47.50

Latest edition available is the 6th edition 1994.

01266 STATISTICS OF FOREIGN TRADE SERIES A: MONTHLY STATISTICS OF FOREIGN TRADE
Organisation for Economic Co-operation and Development (OECD)
International - English monthly/ £78.00 P.A. ISSN No. 0474-5388

01267 STATISTICS OF FOREIGN TRADE SERIES C: FOREIGN TRADE BY COMMODITIES
Organisation for Economic Co-operation and Development (OECD)
International - English/French annual/469pp per volume/price varies ISSN No. 0474-540X

Each year several volumes are produced, each volume dedicated to a group of OECD countries. Data relates to the previous year.

01268 STATISTICS OF WORLD TRADE IN STEEL
United Nations Economic Commission for Europe - UN/ECE
International - English annual/ US$25.00

Refers to exports of semi-finished and finished steel products by exporting country, commodity, region, destination, and to total trade between countries in ECE region.

01269 THE STEEL MARKET IN 199- AND THE OUTLOOK FOR 199-
Organisation for Economic Co-operation and Development (OECD)
International - English annual/ £14.00

Annual analysis of the steel market for the OECD and the world.

01270 STUDIES OF ECONOMIES IN TRANSFORMATION
World Bank Publications
International - English varies/price varies with paper

Published by the World Bank. Informal documents that present the results of policy analysis and research on the states of the former Soviet Union.

01271 SUPPLEMENT TO THE STATISTICAL INDICATORS OF SHORT-TERM ECONOMIC CHANGES IN ECE COUNTRIES: SOURCES AND DEFINITIONS
United Nations Publications
International - English/415pp US$65.00

Current data on demographic and economic trends in Europe, North America, and Japan. Latest edition available is for 1992.

01272 SYSTEM OF NATIONAL ACCOUNTS - STATISTICAL PAPERS, SERIES F
United Nations Publications
International - English annual/ 246pp US$17.00

01273 TIMBER BULLETIN
United Nations Economic Commission for Europe - UN/ECE
International - English 7 issues per year/US$95.00, single issue US $15.00

Seven issues, which for 1992 covered monthly prices, forest product statistics, market review, trade flow data, future prospects and long term roundwood prices.

01274 TOURISM POLICY AND INTERNATIONAL TOURISM IN OECD COUNTRIES
Organisation for Economic Co-operation and Development (OECD)
International - English/French irregular/208pp £24.00

Latest edition available is for November 1990.

01275 TRANSPORT ECONOMY AND POLICY - ECMT SERIES
Organisation for Economic Co-operation and Development (OECD)
International - English irregular/ various prices

A series devoted to transport policy and related issues. Typical titles include "Freight transport and the environment 1991".

01276 TRAVEL AND TOURISM ANALYST
Business International Ltd
International - English 6 issues per year/£495.00 P.A. ISSN No. 0269-3755

Contains reports on transport, hotels and accomodation, markets, financial services and leisure industries.

01277 TRENDS IN DEVELOPING ECONOMIES
World Bank Publications

International - English annual/ 600pp US$25.95 ISSN No. 1014-7004

Provides brief reports and economic indicators for most World Bank borrowing countries.

01278 TRENDS IN PRIVATE INVESTMENT IN DEVELOPING COUNTRIES
World Bank Publications

International - English irregular/ 55pp US$25.95 series ISSN No. 1012-8069

International Finance Corporation Discussion Paper No. 14. Statistical data on private investment trends in 40 developing countries.

01279 TRENDS IN PRIVATE INVESTMENT IN DEVELOPING COUNTRIES 1993: STATISTICS FOR 1970-91
World Bank Publications

Pfeffermann, G.P., and Madarassy, A./International - English irregular/ 66pp US$6.95 ISBN No. 0821323113

International Finance Corporation Discussion Paper No. 16. Statistical data on private investment trends in developing countries.

01280 TRENDS IN PRODUCTION OF LEAD AND ZINC
World Bank Publications

International - English irregular/ 1990 75pp £60.00

Reviews the development of world mine and metal production of lead and zinc since 1960 and the principal factors which have influenced their rate of growth.

01281 TUNGSTEN STATISTICS: ANNUAL REPORT OF THE UNCTAD COMMITTEE ON TUNGSTEN
United Nations Publications

International - English annual/70pp US$25.00

Statistical report on the prices, production, consumption, trade and stocks of tungsten worldwide. Latest edition available is for 1991.

01282 UN/ECE DISCUSSION PAPERS
OECD - Organisation for Economic Co-operation and Development

International - English quarterly/ 1991 prices on application

A series covering economic, environmental, social and political topics for Europe and North America. "Critical examination of economic theories and their confrontation with empirical evidence" is the aim of the publication.

01283 UN/ECE ECONOMIC STUDIES
OECD - Organisation for Economic Co-operation and Development

International - English/price on application

A monographic series which covers economic topics for Europe and North America.

01284 UNCTAD COMMODITY YEARBOOK
United Nations Publications

International - English/French annual/403pp US$68.00

Statistics at the world, regional and country levels for trade and consumption in selected agricultural primary commodities and minerals, ores and metals.

01285 UNESCO STATISTICAL YEARBOOK 1992
World Bank Publications

International - English/French/ Spanish annual/1963 1,064pp FR375.00

Data on population, education, science, technology, culture and communication in more than 200 countries.

01286 UNCTAD STATISTICAL POCKETBOOK
United nations Publications

International - English irregular/ 122pp US$5.00

Second edition on the evolution of the world economy in the past five years with a summary on topics such as growth, trade and development. Latest edition available is for 1989.

01287 URBAN MANAGEMENT PROGRAM PAPERS
World Bank Publications

International - English irregular/ price varies with paper

Informal documents discussing research findings, best practice and promising options for strengthening the contribution that cities make in developing countries to economic and social growth.

01288 WEEKLY EXCHANGE & INTEREST RATE OUTLOOK
Barclays Bank Plc

International - English weekly/ £150.00 P.A. or £30.00 per report

Weekly guide to major exchange and interest rates.

01289 WORLD BANK AND THE ENVIRONMENT
World Bank Publications

International - English annual/ 160pp US$8.95 ISSN No. 1014-2132

Reviews progress in integrating environmental concerns and environmental economics into the operations of the World Bank.

01290 WORLD BANK ANNUAL REPORT
World Bank Publications

International - English annual/ 264pp free ISSN No. 0252-2942

01291 WORLD BANK ATLAS
World Bank Publications

International - English/French/ Spanish annual/36pp US$7.95 ISSN No. 0512-2457

Basic economic and demographic data.

01292 WORLD BANK CGIAR STUDY PAPERS
World Bank Publications

International - English/French irregular/price varies with paper

Informal documents prepared under the auspices of the Secretariat of the Consultative Group on International Agricultural Research. Includes papers on particular research issues and on the work of the international research centres in selected countries.

01293 WORLD BANK COMMODITY WORKING PAPERS
World Bank Publications

International - English irregular/ price varies with paper

Informal documents that provide information on commodities that are important in economic development.

01294 WORLD BANK COUNTRY STUDIES
World Bank Publications

International - English irregular/ US$11.95

Working documents that give material originally presented to provide country information and analysis needed by the bank in planning and lending technical assistance.

01295 WORLD BANK DEVELOPMENT COMMITTEE REPORTS
World Bank Publications

International - English irregular/ price varies with report

Informal documents relating to the development committee which advises and reports to the Board of Governors of the World Bank and IMF on broad questions of the transfer of real resources to developing countries.

01296 WORLD BANK DISCUSSION PAPERS
World Bank Publications

International - English irregular/ price varies with paper

Informal documents that present unpolished research or country analysis. They are prepared to encourage discussion and comment. Papers for which rapid publication is particularly important are often issued in this series.

01297 WORLD BANK ECONOMIC REVIEW
World Bank Publications

International - Engish 3 times per year/1986 c200pp $25.00 P.A. ISSN No. 0258-6770

01298 WORLD BANK POLICY PAPERS
World Bank Publications

International - English irregular/ price varies with paper

Informal documents that present the results of policy analysis and research to encourage discussion and comment.

01299 WORLD BANK REGIONAL AND SECTORAL SERIES
World Bank Publications

International - English irregular/ price varies with paper

A paperback book series derived from World Bank experience in specific geographic areas or economic sectors.

01300 WORLD BANK RESEARCH OBSERVER
World Bank Publications

International - English twice a year/ c100pp US$20.00 P.A. ISSN No. 0257-3032

01301 WORLD BANK RESEARCH PROGRAM: ABSTRACTS OF CURRENT STUDIES
World Bank Publications

International - English annual/ c130pp US$8.95

A compilation of current research programmes funded by the World Bank's research programme.

01302 WORLD BANK STAFF OCCASIONAL PAPERS
World Bank Publications

International - English irregular/ price varies with paper ISSN No. 0074-199X

Present the research work of World Bank staff.

01303 WORLD BANK SYMPOSIUM SERIES
World Bank Publications

International - English irregular/ price varies with paper

Edited collections of papers presented at World Bank symposia and conferences on development related topics. Other international agencies often co-sponsor the meetings.

01304 WORLD BANK TECHNICAL PAPERS
World Bank Publications

International - English irregular/ price varies with paper

Informal documents that present knowledge acquired through the Bank's operational experience. They contain material that is practical rather than theoretical and include state-of-the-art reports. They can also concern matters that cut across sectoral lines, such as technology.

01305 WORLD COMMODITY FORECASTS
Business International Ltd

International - English 6 issues per year/£290.00 P.A.

Monitors trends in the supply-demand balances of 27 commodities and uses the data to generate forecasts of spot prices for the next 18 months.

01306 WORLD COMPETITIVENESS REPORT
IMD

International - English annual/1980 700pp SF700.00

Covers 38 countries and ranks them by 330 criteria.

01307 WORLD COPPER DATABOOK 2ND EDITION
Metal Bulletin Books Ltd.

International - English 1992/£91.00

Copper, mines, traders, and detailed lists of stockholders, warehouses, assayers etc. Includes contract specifications and associations.

01308 WORLD DEBT TABLES: COUNTRY TABLES 1970-79
World Bank Publications

International - English/1988 c264pp US$24.95

Historical reference work.

01309 WORLD DEBT TABLES: EXTERNAL FINANCE FOR DEVELOPING COUNTRIES
World Bank Publications

International - English annual/ 744pp US $125.00 ISSN No. 0253-2859

In two volumes with supplements. Volume 1 analysis and summary tables, volume 2 country tables. Also available on diskette.

01310 WORLD DEPARTMENT STORE INDEX
Hadleigh Marketing Services

International - English annual/ c400pp £500.00

Market research type report covering the top world department store groups.

01311 WORLD DEVELOPMENT REPORT
World Bank Publications

International - English annual/ 304pp US$70.00 ISSN No. 0163-5085

Gives detailed statistics on 120 developing and developed countries. 1992 topic was "Development and the Environment" includes "World development indicators".

01312 WORLD DIRECTORY: LEAD AND ZINC MINES AND PRIMARY METALLURGICAL WORKS
International Lead and Zinc Study Group

International - English/1991 100pp £120.00

Directory all principal lead and zinc mines and metallurgical plants indicating annual mine production rates for lead, zinc, copper, silver, and annual smelter and refinery capacities.

01313 WORLD DIRECTORY: SECONDARY LEAD PLANTS
International Lead and Zinc Study Group

International - English/1991 39pp £90.00

A full listing of lead smelters and refineries in over 40 countries processing secondary materials showing types of plant operated and current capacities.

01314 WORLD DIRECTORY: SECONDARY ZINC PLANTS
International Lead and Zinc Study Group

International - English/1991 26pp £50.00

A listing of zinc plants producing zinc metal and alloys from secondary materials showing processes used and current capacities. Also includes details of plants recovering zinc from steel plant flue dusts.

01315 WORLD ECONOMIC FACTBOOK
Euromonitor Plc

International - English/1992 400pp £145.00 ISBN No. 0863384714

Statistical and economic information on 200 countries.

01316 WORLD ECONOMIC OUTLOOK; A SURVEY BY THE STAFF OF THE INTERNATIONAL MONETARY FUND

International Monetary Fund

International - English/French/Spanish/Arabic twice a year/US$30.00 ISSN No. 0256-6877

Consists of several chapters with statistical appendices covering balance of payments, policy options, inflation and interest rates, debt and capital flows. Also features scenarios based on alternative policy measures.

01317 WORLD ECONOMIC PROSPECTS

Euromonitor Plc

International - English/1992 300pp £95.00 ISBN No. 08633845125

Gives strategic country overviews, political stability and risk analysis, economic situation and prospects, and evaluation of market potential.

01318 WORLD ECONOMIC SURVEY: CURRENT TRENDS AND POLICIES IN THE WORLD ECONOMY

United Nations Publications

International - English annual/US$55.00

An annual report on the salient economic developments of the world. Quantitative forecasts are used to assess general prospects.

01319 THE WORLD ECONOMY

Blackwell Publishers

International - English every 2 months/£144.50 P.A. ISSN No. 0378-5920

01320 WORLD ELECTRONICS COMPANIES FILE

Elsevier Science Publishers Ltd

International - English monthly/£645.00 P.A. ISSN No. 0951-5747

Corporate reports and company profiles in the international electronics industry.

01321 WORLD GRAIN STATISTICS

International Wheat Council

International - English annual/£45.00

Includes long term data series on grain supply and demand for wheat, flour, consumption, trade, prices, export and import, national support prices, and bread prices, coarse grains production and trade. Also has data on shipments and freight rates.

01322 WORLD HEALTH STATISTICS ANNUAL

World Health Organisation

International - English/French annual/350pp US$90.00

Presents global statistical information designed to provide both country and global overviews of changing trends in health status and causes of death.

01323 WORLD HEALTH STATISTICS QUARTERLY

World Health Organisation

International - English quarterly/US$80.00 ISSN No. 0043-8510

Statistical analysis of health problems. Each issue focuses on a particular topic.

01324 WORLD METAL STATISTICS

Metal Bulletin Books Ltd.

International - English monthly/£850.00 P.A. ISSN No. 0043-8758

Published by the World Bureau of Metal Statistics, marketed by Metal Bulletin. Covers country by country trade, production and consumption statistics. Includes stock and price data, for aluminium, antimony, cadmium, copper, lead, molybdenum, nickel, silver, tin and zinc.

01325 WORLD MINERAL STATISTICS 1985-89

British Geological Survey

International - English annual/£50.00

World production and trade statistics for 65 mineral commodities from 1985 to 1989.

01326 WORLD NICKEL STATISTICS

Metal Bulletin Books Ltd.

International - English monthly/£112.00 P.A. ISSN No. 0965-0830

Published by the World Bureau of Metal Statistics, marketed by Metal Bulletin. Covers mine production, smelter/refineries, world refined production/consumption, nickel stocks, and East/West trade.

01327 WORLD OUTLOOK

Business International Ltd

International - English annual/£195.00 ISSN No. 0424-3331

Covers main political and economic events in 165 countries. Includes a six year series of macroeconomic indicators, and numerical GDP forecasts for selected countries.

01328 WORLD POPULATION PROJECTIONS, 199-: ESTIMATES AND PROJECTION WITH RELATED DEMOGRAPHIC STATISTICS

World Bank Publications

Bos, E. et al/International - English

annual/528pp US$34.95

Contains demographic profiles and projections for 202 countries, economies and territories, including the former Soviet Union, and projections for the world as a whole through the year 2150.

01329 WORLD POPULATION PROSPECTS - POPULATION STUDIES

United Nations Publications

International - English irregular/1992 607pp US$85.00

Global data based on the twelfth round of demographic estimates and projections undertaken by the UN Secretariat. Data covers major areas, regions and countries. Data is published 2-3 years in arrears.

01330 WORLD RETAIL DIRECTORY AND SOURCEBOOK

Euromonitor Plc

International - English/1993 £195.00 ISBN No. 086338515X

Has in-depth company profiles, directory of major retailers, further information sources, summary of retailing statistics, and analysis of retail legislation.

01331 WORLD SALES REPORT - ANTI-PIRACY REPORT

International Federation of the Phonographic Industry

Mark Kingston/International - English annual/1990 36pp

Gives data on world sales of phonographic products and piracy in the music industry.

01332 WORLD STAINLESS STEEL STATISTICS 1992

Metal Bulletin Books Ltd.

International - English annual/£225.00 ISSN No. 0141-0806

Data given over five or more years. Covers world production with the exception of Eastern Europe.

01333 WORLD STATISTICS IN BRIEF

United Nations Publications

International - English annual/118pp US$7.50

Annual compilation of international statistics which presents information on 169 countries giving popular statistical indicators. Contains demographic, economic and social statistics for the world as a whole.

01334 WORLD TABLES
World Bank Publications

International - English annual/
702pp US$35.95 ISSN No. 1043-
5573

Published in association with John
Hopkins University Press. Gives
up-to-date economic and social
data for 146 countries, regions,
territories, and income groups.

01335 WORLD TIN STATISTICS
Metal Bulletin Books Ltd.

International - English monthly/
£125.00 P.A. ISSN No. 0965-0822

Published by the World Bureau of
Metals Statistics, marketed by
Metal Bulletin. Includes production
and consumption statistics, mine
and refined production, stocks and
prices.

01336 WORLD TRADE -
STAINLESS, HIGH SPEED
AND OTHER ALLOY STEEL
International Steel
Statistics Bureau

International - English quarterly/
63pp £300.00 P.A.

A quarterly cumulative report
detailing the export trade of major
steel producing countries in
selected alloy products. Includes an
annual companion volume covering
value of trade.

01337 WORLD TRADE:
STAINLESS, HIGH SPEED
AND OTHER ALLOY STEEL
United Kingdom Iron and
Steel Statistics Bureau

International - English quarterly/
£250.00 ISSN No. 0952-5742

Looks at the export of major steel
producing countries in selective
alloy products. An annual
cumulation is included.

01338 WORLD TRADE STEEL
International Steel
Statistics Bureau

International - English quarterly/
71pp £300.00 P.A. ISSN No. 0952-
5734

Details the export trade of 16 major
steel producing countries
collectively accounting for over 85%
of world exports. Available six

months after the quarter ends. An
annual companion volume covering
value of trade is included.

01339 WORLD TRADE STEEL
United Kingdom Iron and
Steel Statistics Bureau

International - English quarterly/
£250.00 ISSN No. 0952-5734

Looks at the export trade of 16
major steel producing countries
collectively accounting for 85% of
world trade, covering 28 product
groups and 100 export markets. An
annual cumulation is included.

01340 YEARBOOK OF TOURISM
STATISTICS
United Nations
Publications

International - English annual/
929pp US$80.00

Comprehensive compendium of
significant internationally
comparable data for the analysis of
tourism development in the world,
at regional and national levels. In
two volumes.

EUROPE

02001 AGRA SPECIAL REPORTS AND MARKET STUDIES

Agra Europe (London) Ltd

Europe - English/price varies with report

A series of market research type reports on European agricultural and food products and markets. Price varies with each report.

02002 AGRICULTURAL REVIEW FOR EUROPE

United Nations Economic Commission for Europe - UN/ECE

Europe - English/price on application

Six volumes cover agricultural trade, grain, livestock and meat, milk and dairy products, and eggs. Data is given on holdings, production, various agricultural products, and prices. Publication dates vary from volume to volume in the range of 1989 to 1991.

02003 ANNUAL BULLETIN OF COAL STATISTICS FOR EUROPE

United Nations Economic Commission for Europe - UN/ECE

Europe - English annual/US$25.00

Data on production, consumption, stocks, and trade in various types of solid fuel, employment and labour productivity.

02004 ANNUAL BULLETIN OF ELECTRIC ENERGY STATISTICS FOR EUROPE

United Nations Economic Commission for Europe - UN/ECE

Europe - English annual/US$25.00

Contains data on capacity of plants, production, consumption, supplies to consumers, fuel consumption in power plants and corresponding production of energy and international exchanges of energy.

02005 ANNUAL BULLETIN OF GAS STATISTICS FOR EUROPE

United Nations Economic Commission for Europe - UN/ECE

Europe - English annual/US$25.00

Data covers production, stocks, consumption, trade of natural gas, light petroleum gases, other petroleum and manufactured gases, usage in terajoules, and number of household consumers.

02006 ANNUAL BULLETIN OF GENERAL ENERGY STATISTICS FOR EUROPE

United Nations Economic Commission for Europe - UN/ECE

Europe - English annual/US$30.00

Covers production of energy by form, an overall energy balance in a common unit (terajoule) and deliveries of petroleum for inland consumption.

02007 ANNUAL BULLETIN OF HOUSING AND BUILDING STATISTICS FOR EUROPE

United Nations Economic Commission for Europe - UN/ECE

Europe - English annual/US$25.00

Covers dwellings, rents and prices, employment, construction and construction prices.

02008 ANNUAL BULLETIN OF STEEL STATISTICS FOR EUROPE

United Nations Economic Commission for Europe - UN/ECE

Europe - English annual/94pp US$25.00

Provides basic data on steel production and trade, consumption and trade of raw materials, movements of scrap, consumption of energy in the industry, and steel deliveries.

02009 ANNUAL BULLETIN OF TRANSPORT STATISTICS FOR EUROPE

United Nations Economic Commission for Europe - UN/ECE

Europe - English annual/285pp US$55.00

Data on transport in Europe, Canada and the USA. Road, rail, inland waterways, containers, sea port goods, oil pipelines, and goods by commodity type are included.

02010 ANNUAL ECONOMIC REVIEW

European Bank for Reconstruction and Development

Europe - English annual/131pp

Covers Central and Eastern Europe and the former Soviet Union, giving details of economic progress.

02011 ANNUAL REVIEW OF ENGINEERING INDUSTRIES AND AUTOMATION

United Nations Economic Commission for Europe - UN/ECE

Europe - English annual/245pp US$50.00 ISSN No. 0255-9293

Covers general developments of engineering industries in North America and Europe. Analysis of trends in given sectors, such as telecommunications, are included, with comprehensive data on trade. Latest edition available is for 1990.

02012 ANNUAL REVIEW OF THE CHEMICAL INDUSTRY

United Nations Economic Commission for Europe - UN/ECE

Europe - English annual/US$50.00 ISSN No. 0255-4291

Analysis of developments and trends in the production, consumption, prices and trade in chemical products in Europe and North America. Latest edition available is for 1990.

02013 ARL RESEARCH REPORTS: PHARMACEUTICALS AND HEALTHCARE

Answers Research Ltd

Europe - English annual/£200.00 - £12,000 per report

Market research reports which look at the European pharmaceutical and healthcare industries.

02014 BIS REPORTS

BIS Group

Europe - English

Market reports on the office and home electronics industries.

02015 BLOC

Bloc

Stuart Anderson/Europe - English 6 per year/40pp ISSN No. 1051-2675

Business journal covering Eastern Europe.

02016 THE BOOK OF EUROPEAN FORECASTS

Euromonitor Plc

Europe - English/1992 311pp £145.00 ISBN No. 0863384013

Forecast data from a variety of sources presented in a statistical format.

02017 THE BOOK OF EUROPEAN REGIONS

Euromonitor Plc

Europe - English/1992 473pp £145.00 ISBN No. 0863384285

Covers the economy, population, urban development and national comparison.

02018 BSRIA EUROPEAN REPORTS

Building Services Reasearch and Information Association

Europe - English/£2,500 per report

A series of market reports which look at the European markets for heating and ventilation products.

02019 BUSINESS EUROPE

Business International Ltd

Europe - English 50 issues per year/£650.00 P.A.

Newsletter giving key short term updates on business and economic issues.

02020 CENTRAL EUROPEAN
Central European

Gavin Gray/Europe - English/ca44

Covers finance and business in Central and Eastern Europe.

02021 COMECON DATA
Macmillan Distribution Ltd

The Vienna Institute for Comparative Economic Studies/ Europe - English annual/£50.00

Handbook of statistical data on the economies of E. Europe, includes Yugoslavia, and the former Soviet Union. Published in alternate years with "Comecon Foreign Trade Data".

02022 CONSUMER EUROPE 199-
Euromonitor Plc

Europe - English annual/650pp £415.00

Statistical sourcebook covering 250 consumer products sold within 17 European markets.

02023 CONSUMER SOUTHERN EUROPE 199-
Euromonitor Plc

Europe - English annual/1993 350pp £275.00

Statistical sourcebook on selected products and markets.

02024 COUNTRY FORECASTS: REGIONAL OVERVIEW EUROPE
Business International Ltd

Europe - English quarterly/£360.00 P.A.

Contains an executive summary, sections on political and economic outlooks and the business environment, a fact sheet, and key economic indicators. Includes a quarterly Global Outlook.

02025 COUNTRY REPORT: CENTRAL AND EASTERN EUROPE 1991: BULGARIA, POLAND, ROMANIA, SOVIET UNION, CZECHOSLOVAKIA, HUNGARY
Office for Offical Publications of the European Communities

Europe - Multi-lingual/1991 184pp ECU17.50

Basic economic and business analysis.

02026 CROSS BORDER RETAILING IN EUROPE
Corporate Intelligence Research Publications

Europe - English/1991 220pp £250.00

Covers retailing by country and sector of cross border retailers and foreign retailers presence.

02027 DATAMONITOR REPORTS: EUROPEAN CONSUMER DURABLES
Datamonitor

Europe - English annual/£600.00 each

Market research type reports which look at the European market for consumer durables and consumer electronics.

02028 DATAMONITOR REPORTS: EUROPEAN COSMETICS
Datamonitor

Europe - English annual/£600.00 each

Market research type reports which look at the European cosmetics industry and products.

02029 DATAMONITOR REPORTS: EUROPEAN DRINKS INDUSTRY
Datamonitor

Europe - English annual/£600.00 each

Market research type reports which look at the European market for alcoholic and soft drinks.

02030 DATAMONITOR REPORTS: EUROPEAN FOOD INDUSTRY
Datamonitor

Europe - English annual/£600.00 each

Market research type reports which look at the European market for food products.

02031 DATAMONITOR REPORTS: EUROPEAN OIL AND PETROCHEMICAL INDUSTRIES
Datamonitor

Europe - English annual/£600.00 each

Market research type reports which look at the European oil and petrochemical industries.

02032 DATAMONITOR REPORTS: EUROPEAN PAPER INDUSTRY
Datamonitor

Europe - English annual/£600.00 each

Market research type reports which look at the European paper and pulp industry.

02033 DATAMONITOR REPORTS: EUROPEAN TRANSPORT INDUSTRY
Datamonitor

Europe - English annual/£600.00 each

Market research type reports which look at the European transport industry including road and air transport.

02034 DOING BUSINESS IN...
Price Waterhouse

International - English/free

Gives a description of the business and economic situation in a particular country. All developed countries are covered.

02035 DOING BUSINESS IN...
Ernst & Young

International - English 1-2 years/ free

An individual report on each country giving broad details of the business and economic climate. Payment issues, credit terms and business practice are also included. Mainly intended for those doing business with the country concerned.

02036 DOING BUSINESS WITH EASTERN EUROPE
Business International Ltd

Europe - English quarterly/400pp £540.00 per volume

Ten volume set, updated every three months which gives a broad picture of Eastern European markets and business practice.

02037 EAST EUROPE BUSINESS FOCUS
East Europe Business Focus

Giselle Jones/Europe - English ten issues per year/c16pp £350.00 P.A. ISSN No. 0959-9010

Newsletter covering all aspects of business in Eastern Europe.

02038 THE EAST EUROPEAN CHEMICAL INDUSTRY: RESTRUCTURING FOR EAST EUROPE
Business International Ltd

T. Baker/Europe - English/1993 £495.00

Looks at the chemical industry in East European countries in the light of recent political and economic developments and forecasts the future for the industry.

02039 THE EAST EUROPEAN CHEMICAL MONITORING SERVICE
Business International Ltd

Europe - English monthly/£480.00 P.A.

Looks at the chemical industry in E. European countries and includes news, features, and statistics.

02040 EAST EUROPEAN FINANCE UPDATE
Business International Ltd

Europe - English quarterly/£695.00 P.A.

Provides information on government and institutional sources of finance, the banking system and other financial policies and practices.

02041 EAST EUROPEAN INDUSTRIAL MONITORING SERVICE

Business International Ltd

Europe - English monthly/£495.00 P.A.

Includes developments in 12 chemical and industrial sectors using local press - media sources.

02042 EAST/WEST BUSINESS AND TRADE

Welt Publishing Co

Walter B. Smith II/Europe - English bi-weekly/1972 4pp US$249 P.A. ISSN No. 0731-7727

Newsletter covering all aspects of business, formerly entitled Soviet Business and Trade.

02043 EASTERN EUROPE ANALYST

World Reports Ltd

Europe - English quarterly/£200.00 P.A. ISSN No. 0965-0350

Covers the (former) Eastern European financial markets.

02044 EASTERN EUROPE AND THE COMMONWEALTH OF INDEPENDENT SATES 1992

Europa Publications

Europe - English 5th edition/583pp ISSN No. 0962-1040

Detailed guide covering economics geography and social conditions in Eastern Europe.

02045 EASTERN EUROPE AND THE USSR: THE CHALLENGE OF FREEDOM

Office for Offical Publications of the European Communities

Europe - Multi-lingual/1991 253pp ECU14.30

Broad analysis of the relationship between Europe and the new C.I.S. States.

02046 EAST-WEST JOINT VENTURES AND INVESTMENT NEWS

United Nations Publications

Europe - English quarterly/ US$80.00 P.A.

Legal and statistical news on East-West joint ventures.

02047 ECONOMIC BULLETIN FOR EUROPE

United Nations Economic Commission for Europe - UN/ECE

Europe - English annual/125pp US$60.00

Examines world trade and the economies in Europe and North America based on the latest data. Also covers east-west trade with special chapters focusing on specialized topics in east-west trade. Volumes 33-40 were published quarterly, from Volume 41 1991 it will be published annually.

02048 ECONOMIC SURVEY OF EUROPE

United Nations Economic Commission for Europe - UN/ECE

Europe - English annual/US$80.00 ISSN No. 0070-8712

Review of economic developments in Europe and North America. In addition, it gives an analysis of Eastern Europe and the former Soviet Union.

02049 ECONOMIST INTELLIGENCE UNIT: RESEARCH REPORTS: AUTOMOTIVE INDUSTRY

Business International Ltd

Europe - English/price varies with report

Market research reports which look at the international automotive industry. Prices range from £150.00 to £500.00 per report.

02050 ECONOMIST INTELLIGENCE UNIT: RESEARCH REPORTS: RETAIL AND CONSUMER GOODS

Business International Ltd

Europe - English/price varies with report

Market research reports which look at European retailing and consumer goods. Prices range from £150.00 to £500.00 per report.

02051 THE EEA AGREEMENT

European Free Trade Association

Europe - English/French/German/ Norwegian/Swedish/Finnish irregular/15pp free

Contents of the European Economic Area agreement.

02052 EEA - EFTA FACT SHEETS

European Free Trade Association

Europe - English/French/German/ 1991 5pp free

Eight leaflets which deal with issues related to the European Economic Area and EFTA. Issues such as free circulation of goods and capital are dealt with.

02053 EFTA

European Free Trade Association

Europe - English annual/free

A pocket leaflet with statistics on EFTA.

02054 EFTA ANNUAL REPORT

European Free Trade Association

Europe - English/French/German annual/1967 free

Looks at the work of EFTA for the preceding year.

02055 EFTA TRADE

European Free Trade Association

Europe - English/French/German annual/1112pp free

Specialist economic publication containing details of member countries trade.

02056 EFTA - WHAT IT IS AND WHAT IT DOES

European Free Trade Association

Europe - English/French/German irregular/free

Information pack type material which gives general information on the structure and functions of EFTA.

02057 ELSEVIER ADVANCED TECHNOLOGY REPORTS

Elsevier Science Publishers Ltd

Europe - English irregular/price varies with report

A series of market research reports covering different products in the European electronics and telecommunications industry.

02058 ENERGY STATISTICS AND BALANCES IN NON-OECD COUNTRIES

Organisation for Economic Co-operation and Development (OECD)

Europe - English/French annual/ 80pp £45.00

Data is published 2-3 years in arrears.

02059 ERC EUROPEAN MARKET REPORTS: CONSUMER DURABLES AND HOUSEHOLD GOODS

ERC Statistics International

Europe - English annual/£4,500-£6,000 per report

Market research reports on consumer durables (such as microware ovens), household goods (such as cookware), covering 7-8 eight main European countries.

02060 ERC EUROPEAN MARKET REPORTS: FOOD
ERC Statistics International

Europe - English annual/£4,500-£6,000 per report

Market research reports on food products covering 7-8 main European countries.

02061 ERC EUROPEAN MARKET REPORTS: LEISURE PRODUCTS
ERC Statistics International

Europe - English annual/£4,500-£6,000 per report

Market research reports on the toy industry and leisure products, covering 7-8 main European countries.

02062 ERC EUROPEAN MARKET REPORTS: PHARMACEUTICALS AND HEALTHCARE PRODUCTS
ERC Statistics International

Europe - English annual/£4,500-£6,000 per report

Market research reports on personal care, pharmaceutical and cosmetic products covering seven to eight main European countries.

02063 EUROPE RETAIL
Business International Ltd

Europe - English 24 issues per year/£365.00 P.A. ISSN No. 0960-0191

Regular news, analysis and surveys of the European retail sector.

02064 EUROPE'S 15,000 LARGEST COMPANIES
Ekonomisk Litteratur AB

Europe - English annual

Includes contact details, sales, profit, parent company, name of managing director, number of employees, capital structure.

02065 EUROPE'S TOP RETAILERS - VOLUME 1
Corporate Intelligence Research Publications

Europe - English annual/325pp £495.00

Covers 40 plus of Europe's top retailers in a market research type format.

02066 EUROPEAN ADVERTISING, MARKETING AND MEDIA DATA
Euromonitor Plc

Europe - English annual/770pp £195.00

Key statistical data and directory information.

02067 EUROPEAN AUTOMOTIVE COMPONENTS BUSINESS
Business International Ltd

Europe - English quarterly/£425.00 P.A.

Covers the European motor components business and its markets.

02068 EUROPEAN BUSINESS PLANNING FACTORS: KEY ISSUES FOR CORPORATE STRATEGY IN THE 1990'S
Euromonitor Plc

Europe - English/1992 450pp £95.00 ISBN No. 086338442

Socio-economic developments and a broad range of business factors are covered.

02069 THE EUROPEAN COMMUNITY AND ITS EASTERN NEIGHBOURS
Office for Offical Publications of the European Communities

Europe - English/1991

Part of the "Europe on the Move Series". Looks at the economic relationship between the Community and Europe.

02070 THE EUROPEAN COMMUNITY AND THE MEDITERRANEAN COUNTRIES
Office for Official Publications of the European Communities

Europe - English/1991

Part of the "Europe on the Move Series". Looks at the economic prospects for the enlargement of the Community.

02071 EUROPEAN COMPENDIUM OF MARKETING INFORMATION
Euromonitor Plc

Europe - English/1992 421pp £160.00 ISBN No. 0863384404

Covers industry and market trends for a broad range of consumer products, with statistical analysis.

02072 EUROPEAN CONSUMER PACKAGING MARKETING DIRECTORY
Euromonitor Plc

Europe - English/1993 500pp £160.00 ISBN No. 0863384498

02073 EUROPEAN DRINKS INDEX
Hadleigh Marketing Services

Europe - English annual/c400pp £500.00

Market research type report covering over 175 companies in 16 countries.

02074 EUROPEAN FOOD INDEX
Hadleigh Marketing Services

Europe - English annual/c600pp £500.00

Market research type report covering over 200 companies in 16 countries.

02075 EUROPEAN FUND DIRECTORY
Professional and Business Information Plc

Europe - English annual/£595.00

Directory which gives detailed financial data on European fund management companies.

02076 EUROPEAN INVESTMENT REGION SERIES: RESEARCH REPORTS
Business International Ltd

Europe - English/£195.00 per report

Looks at Europe's most dynamic regions and assesses their economic prospects.

02077 EUROPEAN MARKET SHARE REPORTER
Gale Research International

Europe - English annual/1993 £95.00

Information on 2000 plus companies, 1,500 products and commodities arranged by SIC code.

02078 EUROPEAN MARKETING DATA AND STATISTICS
Euromonitor Plc

Europe - English annual/£150.00

Statistical yearbook containing data on key economic and marketing parameters for 32 European countries.

02079 EUROPEAN MOTOR BUSINESS
Business International Ltd

Europe - English quarterly/£450.00 P.A. ISSN No. 0267-8233

Covers the West European motor business and its markets.

02080 EUROPEAN PACKAGING
Packaging Industry Research Association (PIRA)

Europe - English irregular/1992 £600.00

A two volume set covering the packaging industry in Europe, giving details of markets and companies.

02081 EUROPEAN PRINTING
Packaging Industry Research Association (PIRA)

Europe - English annual/£450.00

Details the European printing industry.

02082 EUROPEAN STEEL STOCKHOLDING DATABOOK
Metal Bulletin Books Ltd

Europe - English every 4 years/

£80.00

02083 EUROPEAN SUPERMARKET INDEX

Hadleigh Marketing Services

Europe - English annual/c400pp £500.00

Market research type report covering the top groups in 16 countries.

02084 EUROPEAN SUPPLIES BULLETIN

Sea Fish Industry Authority

Europe - English quarterly/£85.00 P.A. ISSN No. 0142-937X

Gives the quantity and value of imports and exports of fish and fish products. Covers Europe and North America.

02085 FACTS THROUGH FIGURES: A STATISTICAL PORTRAIT OF EFTA IN THE EUROPEAN ECONOMIC AREA

Office for Official Publications of the European Communities

Europe - English/1991 35pp free

Basic statistical picture of the EFTA in the European Economic Area.

02086 FOOD DISTRIBUTION IN EUROPE IN THE 1990'S

Corporate Intelligence Research Publications

Europe - English/1990 650pp £250.00

Covers country food markets, producers, and retailers.

02087 THE FREE TRADE AGREEMENTS OF THE EFTA COUNTRIES AND THE EUROPEAN COMMUNITIES

European Free Trade Association

Europe - English/French/German annual/1967 free

02088 FROST & SULLIVAN REPORTS: AUTOMOTIVE INDUSTRIES

Frost & Sullivan Ltd

Europe - English irregular/price varies with report

Market research type reports. Prices range from £250.00 to £2,500.00.

02089 FROST & SULLIVAN REPORTS: BIOTECHNOLOGY, ENERGY AND THE ENVIRONMENT

Frost & Sullivan Ltd

Europe - English irregular/price varies with report

Market research type reports. Prices range from £250.00 to £2,500.00.

02090 FROST & SULLIVAN REPORTS: CONSUMER GOODS

Frost & Sullivan Ltd

Europe - English irregular/price varies with report

Market research type reports. Prices range from £250.00 to £2,500.00.

02091 FROST & SULLIVAN REPORTS: DEFENCE AND SECURITY INDUSTRIES

Frost & Sullivan Ltd

Europe - English irregular/price varies with report

Market research type reports. Prices range from £250.00 to £2,500.00.

02092 FROST & SULLIVAN REPORTS: ELECTRONICS

Frost & Sullivan Ltd

Europe - English irregular/price varies with report

Market research type reports. Prices range from £250.00 to £2,500.00.

02093 FROST & SULLIVAN REPORTS: FOOD AND DRINKS INDUSTRIES

Frost & Sullivan Ltd

Europe - English irregular/price varies with report

Market research type reports. Prices range from £250.00 to £2,500.00.

02094 FROST & SULLIVAN REPORTS: HEALTHCARE, PHARMACEUTICALS AND CHEMICALS

Frost & Sullivan Ltd

Europe - English irregular/price varies with report

Market research type reports. Prices range from £250.00 to £2,500.00.

02095 FROST & SULLIVAN REPORTS: LEISURE GOODS

Frost & Sullivan Ltd

Europe - English irregular/price varies with report

Market research type reports. Prices range from £250.00 to £2,500.00.

02096 FROST & SULLIVAN REPORTS: PAPER AND PACKAGING INDUSTRIES

Frost & Sullivan Ltd

Europe - English irregular/price varies with report

Market research type reports. Prices range from £250.00 to £2,500.00.

02097 FT MANAGEMENT REPORTS

FT Management Reports

Europe - English irregular/price varies with report

A series of market research type reports which look at industrial, consumer and service sectors. Prices for European reports range from £150.00 to £400.00.

02098 HOUSING FINANCE IN EUROPE

Building Societies Association/Council of Mortgage Lenders

Europe - English irregular/£20.00

A survey of the housing finance industry in France, Germany, Italy, Spain and the UK.

02099 INDUSTRIAL POLICY IN OECD COUNTRIES: ANNUAL REVIEW

Organisation for Economic Co-operation and Development (OECD)

Europe - English annual/£20.00

Examines recent government initatives and reviews trends in industrial performance in OECD countries. Also contains information on Central and Eastern European industrial policies.

02100 INSIGHT: EAST EUROPEAN BUSINESS REPORT

Insight International Publishing Ltd

Europe - English monthly/£200.00 P.A.

Newsletter covering all aspects of business.

02101 INVESTING, LICENSING AND TRADING CONDITIONS ABROAD: EUROPE

Business International Ltd

Europe - English annual/£825.00 for complete European series

Loose-leaf format covering the state role in industry, competition rules, price controls, remittability of funds, taxation, incentives and capital sources, labour and foreign trade.

02102 KEY NOTE EUROVIEWS

ICC Information Group Ltd

Europe - English annual/price varies with report

A series of market research reports on consumer and some industrial markets. Typical titles include "European drinking habits".

02103 MARKET RESEARCH EUROPE

Euromonitor Plc

Europe - English monthly/£445.00 P.A. ISSN No. 0308-3446

Covers consumer markets and products in the major economies of the region.

02104 MARKETING IN EUROPE
Business International Ltd

Europe - English monthly/£560.00 P.A. ISSN No. 0025-3723

A monthly bulletin on consumer goods markets.

02105 MARKETPOWER MARKET RESEARCH REPORTS: FOOD INDUSTRY
Marketpower

Europe - English/£350.00-£9,000.00 per report

Market research reports on the food service and food industries. Typical titles include "Food production in Europe".

02106 MARKETPOWER MARKET RESEARCH REPORTS: PACKAGING INDUSTRY
Marketpower

Europe - English/£350.00-£9,000.00 per report

Market research reports on the packaging industry. Typical titles include "European canning prospects".

02107 MSI REPORTS: FOOD INDUSTRY IN EUROPE
Marketing Strategies for Industry

Europe - English irregular/price varies with report

Market research type reports which look at the European food sector, including such topics as "Frozen food in Germany". Reports cost about £90-£450 each depending on depth of report.

02108 MSI REPORTS: RETAILING IN EUROPE
Marketing Strategies for Industry

Europe - English irregular/price varies with report

Market research type reports which look at the European retailing sector, including such topics as supermarkets. Reports cost about £90-£450 each depending on depth of report.

02109 NHBC EURO-GUIDE OVERVIEW
National House Building Council

Europe - English irregular/1990 300pp £25.00 ISBN No. 0907257151

Statistics on private house-building in the EEC and EFTA.

02110 OVERALL ECONOMIC PERSPECTIVE TO THE YEAR 2000
United Nations Economic Commission for Europe - UN/ECE

Europe - English/1990 224pp US$48.00 ISBN No. 9211164666

Data for long term economic planning. Over 100 charts and graphs are included.

02111 OVUM REPORTS: ELECTRONICS INDUSTRIES
Ovum Ltd

Europe - English/price varies with report

Market research type reports which look at the European markets for software and hardware industries. Reports cost £250-£1500 each.

02112 PIRA REPORTS
Packaging Industry Research Association (PIRA)

Europe - English/price varies with report

A series of market research reports covering the packaging industry, giving details of markets and companies. Prices range from £350.00 to £650.00.

02113 PROSPECTS FOR EAST-WEST EUROPEAN TRANSPORT (ECMT SERIES)
Organisation for Economic Co-operation and Development (OECD)

Europe - English irregular/various prices

A series devoted to transport policy and related issues. Other titles include "Freight transport and the environment 1991".

02114 RETAIL TRADE EUROPE
Euromonitor Plc

Europe - English/1992 900pp £350.00 ISBN No. 0863384072

Retail market analysis on 27 European countries, including statistical information.

02115 RETAILING IN EASTERN EUROPE
Corporate Intelligence Research Publications

Europe - English/1992 205pp £345.00

Retail market backgrounds and Western entrants in 14 countries.

02116 RETAILING IN EUROPE
Corporate Intelligence Research Publications

Europe - English irregular/£695.00 for complete set or £120.00 per country

Market research type reports which look at the retailing sector in 15 European countries.

02117 STATISTICS OF ROAD TRAFFIC ACCIDENTS IN EUROPE
United Nations Economic Commission for Europe - UN/ECE

Europe - English annual/US$25.00 ISSN No. 0081-5160

Comprehensive data on accidents and casualties covering Europe and the USA. Also contains data on road vehicles in use, registration of new vehicles, kilometres run by road vehicles, and estimates of population and distribution by age group.

02118 THE STEEL MARKET
United Nations Economic Commission for Europe - UN/ECE

Europe - English annual/US$27.00 ISSN No. 0497-9478

Covers developments and trends in the international steel market, with data on foreign trade, supply and demand, export prices and production capacities. A separate section covers iron and steel-making raw materials. Part II comprises monographs prepared by nations authorities.

02119 TIMBER BULLETIN FOR EUROPE
United Nations Economic Commission for Europe - UN/ECE

Europe - English annual/US$95.00

An annual bulletin that provides an overview of the sector and its related trends and activities. It includes statistics on production, exports and imports, trade flow, and developemnts of forest product markets.

02120 WESTERN EUROPE 1993
Europa Publications

Europe - English every 3 years/1989 560pp £165.00

Covers politics, economics and statistics.

02121 ENERGY - STATISTICAL YEARBOOK
Eurostat

EEC - English/French/German annual/116pp ECU23.50 ISSN No. 0081-489X

Energy economics, energy supplied balance sheets, and aggregates, are given for the EC economies.

02122 A COMMUNITY OF TWELVE: KEY FIGURES
Office for Official Publications of the European Communities

EEC - English 1991/32pp free

Basic statistical picture of the European Community.

02123 A SOCIAL PORTRAIT OF EUROPE
Office for Official Publications of the European Communities

EEC - Multi-lingual/1991 142pp ECU10.00

A collection of social statistics, with graphs and charts.

02124 AGRICULTURAL INCOME
Eurostat

EEC - English/French/German annual/154pp ECU26.00

Covers agricultural income for each member state and in aggregate and looks at trends in income.

02125 AGRICULTURAL MARKETS: PRICES
Office for Official Publications of the European Communities

EEC - Multi-lingual 4 issues per year/£66.80 P.A. ISSN No. 0250-9601

EEC wide agricultural prices and comparisons.

02126 AGRICULTURAL PRICES - PRICE INDICES & ABSOLUTE PRICES
Eurostat

EEC - English/French quarterly/ 210pp ECU25.00 P.A.

Trends in producer prices and means of production prices.

02127 AGRICULTURAL PRICES - PRICES INDICES AND ABSOLUTE PRICES
Eurostat

EEC - English/French/German annual/1991 370pp ECU20.50

Trends in producer prices and means of production prices in a cumulative format. Latest edition available is for 1981-1990.

02128 THE AGRICULTURAL SITUATION IN THE COMMUNITY
Office for Official Publications of the European Communities

EEC - Multi-lingual annual/421pp £19.90

Covers the state of agriculture in the Community as a whole.

02129 AGRICULTURE, FORESTRY AND FISHERIES
Eurostat

EEC - English/French/German 10pp ECU192.00 P.A.

Rapid report type publications which summarize the main results of larger statistical studies. Available as a separate subscription.

02130 AGRICULTURE IN EUROPE: DEVELOPMENT, CONSTRAINTS, AND PERSPECTIVES
Office for Official Publications of the European Communities

EEC - English/German/1992 97pp ECU7.00

02131 AGRICULTURE - STATISTICAL YEARBOOK
Eurostat

EEC - English/French/German annual/260pp ECU27.50

Contains all the principal statistics collected for the agricultural, forestry and fishery industries.

02132 ANALYTICAL TABLES OF EXTERNAL TRADE: NIMEXE, EXPORTS
Office for Official Publications of the European Communities

EEC - Multi-lingual annual/ ECU420.00, single issue ECU 42.00

Multi-volume work which covers imports and exports of different products. Organized in volumes A-L and M-Z. Individual product sectors available.

02133 ANIMAL PRODUCTION - QUARTERLY STATISTICS
Eurostat

EEC - English/French/German quarterly 110pp ECU74.00 P.A.

Meat, eggs and poultry, milk and milk products and external trade are covered.

02134 ANNUAL REPORT OF THE ENVIRONMENT INSTITUTE
Office for Official Publications of the European Communities

EEC - English annual/78pp ECU10.00

Last published in 1990, it gives an overview of the environment and quality of life in the Community.

02135 ANNUAL REPORT OF THE EUROPEAN FOUNDATION FOR THE IMPROVEMENT OF LIVING AND WORKING CONDITIONS
Office for Official Publications of the European Communities

EEC - Multi-lingual annual/80pp ECU8.75

02136 ANNUAL REPORT OF THE EUROPEAN INVESTMENT BANK
Office for Official Publications of the European Communities

EEC - Multi-lingual annual/108pp free ISSN No. 0071-2868

02137 BASIC STATISTICS OF THE COMMUNITY
Eurostat

EEC - English annual/ECU10.00

A pocket-sized book which gives all the main statistics of the European Community. Now in the 29th edition.

02138 BULLETIN OF ENERGY PRICES
Office for Official Publications of the European Communities

EEC - English/French twice yearly/ ECU7.60

A survey of import and consumer prices for oil, coal, gas and electricity in the Community. Also included in the global subscription "Energy".

02139 BULLETIN OF THE ECONOMIC AND SOCIAL COMMITTEE
Office for Official Publications of the European Communities

EEC - Multi-lingual monthly/ ECU54.00 ISSN No. 0010-2423

Describes the work of the Economic and Social Committee of the European Community.

02140 BULLETIN OF THE EUROPEAN COMMUNITIES
Office for Official Publications of the European Communities

EEC - English 10 issues per year/ £465.00 P.A. ISSN No. 0007-5116

The C series covers the complete range of Community information apart from legislation which was covered in the L series. The Supplement covers tenders and contracts.

02141 BUSINESS AND RESEARCH ASSOCIATES REPORTS: FURNITURE INDUSTRIES
Business and Research Associates

EEC - English/price varies with report

Market research reports which look at the EEC office and domestic furniture industries.

02142 BUSINESS EUROPE
Business International Ltd

EEC - English quarterly/£650.00 P.A.

In-depth review articles on issues and developments within the European Community. Incorporates "European trends" and "Management Europe" formerly published by the EIU.

02143 CARRIAGE OF GOODS - INLAND WATERWAYS
Eurostat

EEC - English/French/German annual/184pp ECU19.00

Statistics on the carriage of goods by inland waterway irrespective of country of origin of the vessel.

02144 CARRIAGE OF GOODS - RAILWAYS
Eurostat

EEC - English/French/German annual/155pp ECU20.50

Statistics on the carriage of goods on the main railway networks which are open to public traffic.

02145 CARRIAGE OF GOODS - ROADS
Eurostat

EEC - English/French/German annual/146pp ECU16.00

Statistics on the carriage of goods by road, carried on vehicles registered in the EEC.

02146 CEDEFOP NEWS - VOCATIONAL TRAINING IN EUROPE
Office for Official Publications of the European Communities

EEC - English irregular/free

Published by the European Centre for the Development of Vocational Training.

02147 COAL AND THE INTERNATIONAL ENERGY MARKET
Office for Official Publications of the European Communities

EEC - English/French/German/ Spanish/1991 135pp ECU17.50

02148 COMMUNICATIONS
Office for Official Publications of the European Communities

EEC - Multi-lingual irregular/1992 12pp free

Published as part of the European Investment Bank briefing series. Looks at the communications infrastructure in a wider Europe and its investment needs.

02149 THE COMMUNITY BUDGET: THE FACTS IN FIGURES
Office for Official Publications of the European Communities

EEC - English/French/German irregular/c100pp ECU11.00

Annotated summary of the results of Eurobarometer public opinion surveys of the European Communities.

02150 COMPARATIVE TABLES OF THE SOCIAL SECURITY SCHEMES IN THE MEMBER STATES OF THE EUROPEAN COMMUNITIES. GENERAL SCHEME (EMPLOYEES INDUSTRY AND COMMERCE)
Office for Official Publications of the European Communities

EEC - English/French/German irregular/127pp ECU12.00

The latest edition available is the 15th edition, published in 1989.

02151 COMPLETING THE INTERNAL MARKET: AN AREA WITHOUT INTERNAL FRONTIERS
Office for Official Publications of the European Communities

EEC - English irregular/127pp ECU31.00

Sixth report of the European Commission concerning the implementation of the Internal Market. Published in 1991.

02152 COMPLETING THE INTERNAL MARKET: CURRENT STATUS 1ST JANUARY 1992
Office for Official Publications of the European Communities

EEC - Multi-lingual/1992 ECU100.00

Published in 6 volumes, this publication considers progress towards the single market in various sectors.

02153 CONSUMER POLICY IN THE SINGLE MARKET
Office for Official Publications of the European Communities

EEC - Multi-lingual/1991 42pp free

Reviews progress towards standardisation of consumer policies in the single market.

02154 CONSUMER PRICE INDEX
Eurostat

EEC - English monthly/ECU61.00 P.A. ISSN No. 0258-0861

Gives the general price index for European Community countries. Four quarterly issues cover price trends in 8 main groups of consumption, and four supplements give trends in purchasing power parities and price level indices. Austria, USA, Canada and Japan are included for comparison.

02155 CONSUMER PRICES IN THE EEC 1990
Eurostat

EEC - English/French/1991 131pp ECU15.00

Results of an 1989 survey. It lists average prices for the capital city of each country and average national prices with price level indices.

02156 THE COURIER
Office for Official Publications of the European Communities

EEC - English/French bi-monthly/ free

Looks at development and aid issues for the EEC and its relations with Third World countries. Compliments "Europe on the Move".

02157 CROP PRODUCTION - QUARTERLY STATISTICS
Eurostat

EEC - English/French/German quarterly/120pp ECU74.00 P.A.

Covers land use, arable crops, and fruit and vegetable production. Includes data on weather conditions, supply balance sheets and external trade.

02158 DEMOGRAPHIC STATISTICS
Eurostat

EEC - English/French/German irregular/191pp ECU11.00

All the principal series of demographic statistics are given in a format which allows comparison across EC countries. Latest edition published is for 1992.

02159 DIGEST OF STATISTICS ON SOCIAL PROTECTION IN EUROPE - VOLUME 1: OLD AGE
Eurostat

EEC - English/French/208pp ECU17.00

Data given from 1980 to 1988 shows expenditure on benefits as well as the corresponding number of beneficiaries.

02160 DISABLED PERSONS: STATISTICAL DATA
Office for Official Publications of the European Communities

EEC - Multi-lingual/1991 190pp ECU17.00

Population and social conditions.

02161 DOING BUSINESS IN...
Price Waterhouse

International - English/free

Gives a description of the business and economic situation in a particular country. All developed countries are covered.

02162 DOING BUSINESS IN...
Ernst & Young

International - English 1-2 years/
free

An individual report on each country giving broad details of the business and economic climate. Payment issues, credit terms and business practice are also included. Mainly intended for those doing business with the country concerned.

02163 EARNINGS - INDUSTRY AND SERVICES
Eurostat

EEC - Multi-lingual half yearly/
ECU77.00 P.A.

Gives details of the harmonized statistics on the earnings of manual and non-manual workers in industry and some parts of the service sector. Also gives data on purchasing power standards for workers across the Community.

02164 EC-NICS TRADE
Eurostat

EEC - English irregular/170pp
ECU9.50

Statistical analysis of trade between the EC and newly industrialised countries (NICs). Latest data published 1991.

02165 ECONOMIC ACCOUNTS FOR AGRICULTURE AND FORESTRY
Eurostat

EEC - English/French/German
annual/231pp ECU13.00

Contains the most recent data on agricultural and forestry accounts at the national and regional level. Latest edition published is for 1984-1989.

02166 ECONOMY AND FINANCE
Eurostat

EEC - English/French/German/
10pp ECU192.00 P.A.

These are small, rapid results type publications which summarize the main results of larger surveys. Annual subscriptions are available.

02167 THE ECU AND ITS ROLE IN THE PROCESS TOWARDS MONETARY UNION
Office for Official Publications of the European Communities

EEC - Multi-lingual/1991 39pp free

Published as a monograph issue of European Economy No 48 1991.

02168 ECU - EMS INFORMATION
Eurostat

EEC - English/French/German/
10pp ECU67.00

Regular publication on the private and official use of the ECU and the daily trends in the EMS.

02169 THE ECU REPORT
Office for Official Publications of the European Communities

EEC - Multi-lingual/1991 180pp
ECU10.00

Overview and analysis of the ECU to date.

02170 THE EEA AGREEMENT
European Free Trade Association

EEC - English/French/German/
Norwegian/Swedish/Finnish
irregular/15pp free

Contents of the European Economic Area agreement.

02171 EIB - INFORMATION
Office for Official Publications of the European Communities

EEC - Multi-lingual quarterly/free
ISSN No. 0250-3891

News about the activities of the European Investment Bank.

02172 ELECTRICITY PRICES 1985-91
Eurostat

EEC - English/French/German/
1992 169pp ECU12.00

Contains electricity prices in national currencies.

02173 EMPLOYMENT AND INDUSTRIAL RELATIONS GLOSSARIES
Office for Official Publications of the European Communities

EEC - Multi-lingual/1991 422pp
ECU24.00 per country

Provides an introduction to the key employment issues, industrial relations and labour law system for each EC country. A separate volume is published for each country.

02174 EMPLOYMENT IN EUROPE
Office for Official Publications of the European Communities

EEC - English annual/ECU11.25
ISSN No. 1016-5444

Looks at the present employment situation in the Community and future prospects. Published about one year in arrears.

02175 ENERGY 1960-1988
Eurostat

EEC - English/French/German/
1991 260pp ECU9.00

Data extracted from the Sirene database, used for the definition of EC energy policy.

02176 ENERGY AND INDUSTRY - PANORAMA OF THE EEC INDUSTRY - STATISTICAL SUMMARY 1992
Eurostat

EEC - English/French/German/
1992 547pp ECU25.00

Brings together the two editions of Panorama of EC industry.

02177 ENERGY BALANCE SHEETS
Office for Official Publications of the European Communities

EEC - English/French/German
annual/233pp ECU24.00

Data is published 1-2 years in arrears.

02178 ENERGY - COMPLETE SUBSCRIPTION
Office for Official Publications of the European Communities

EEC - Multi-lingual/ECU172.00
P.A.

Comprises all the main statistical publications on energy in one subscription. Includes monthly statistics, rapid reports, annual statistics, energy in Europe and bulletin of energy prices.

02179 ENERGY IN EUROPE: ANNUAL ENERGY REVIEW
Office for Official Publications of the European Communities

EEC - English annual/133pp

Published about one year in arrears.

02180 ENERGY IN THE EUROPEAN COMMUNITY
Office for Official Publications of the European Communities

EEC - Multi-lingual/45pp free

Gives an overview of the energy situation in the Community, the 14th edition was published in 1990.

02181 ENERGY - MONTHLY STATISTICS
Eurostat

EEC - English/French/German
monthly/71pp ECU77.00 P.A.

Short-term trends in the energy economy.

02182 ENGINEERING CONSULTANCY IN THE EUROPEAN COMMUNITY
Association of Consulting Engineers

EEC - English/1990 price on application

A series of market research type reports which look at the engineering consultancy market in the European Community.

02183 ENVIRONMENT STATISTICS

Eurostat

EEC - English/French/German/ 10pp ECU192.00

Small publications which appear regularly in a rapid results format. They summarize the main findings of larger statistical studies on the environment. Annual subscriptions are available.

02184 ENVIRONMENT STATISTICS

Office for Official Publications of the European Communities

EEC - Multi-lingual irregular/162pp ECU3.70

Data is published one year in arrears.

02185 EUROATOM SUPPLY AGENCY: ANNUAL REPORT

Office for Official Publications of the European Communities

EEC - English/French/German annual/30pp ECU6.00

Latest data available is for 1990.

02186 EUROBAROMETER: PUBLIC OPINION IN THE EUROPEAN COMMUNITY

Office for Official Publications of the European Communities

EEC - Multi-lingual 2 issues per year/ECU10.00 postage charge

02187 EUROPA TRANSPORT - OBSERVATION OF THE TRANSPORT MARKETS

Office for Official Publications of the European Communities

EEC - English/French/German quarterly/ECU50.00

Subscription includes annual report.

02188 EUROPE 2000: OUTLOOK FOR THE DEVELOPMENT OF THE COMMUNITY'S TERRITORY

Office for Official Publications of the European Communities

EEC - Multi-lingual/1992 208pp ECU15.00

Reviews future expansion of the community and the potential inclusion of Nordic/Central European countries.

02189 EUROPE IN FIGURES 3RD EDITION

Eurostat

EEC - English/1992 ECU16.50

A basic socio-economic picture of the European Community.

02190 EUROPE ON THE MOVE

Office for Official Publications of the European Communities

EEC - English irregular/8pp free

A series of leaflets giving basic information on the institutions and activities of the European Community. Features such topics as the EMS and monetary union.

02191 EUROPEAN COMMUNITIES ENCYCLOPEDIA AND DIRECTORY

Europa Publications

EEC - English every 3 years/1992 416pp £155.00

Basic data on EC countries.

02192 EUROPEAN COMMUNITIES: FINANCIAL REPORT

Office for Official Publications of the European Communities

EEC - Multi-lingual annual/62pp

02193 EUROPEAN COMMUNITY: 1992 AND BEYOND

Office for Official Publications of the European Communities

EEC - English/1991 34pp free

Describes the Community in general terms and the economic/ business challenges currently facing it.

02194 THE EUROPEAN COMMUNITY AND ITS EASTERN NEIGHBOURS

Office for Official Publications of the European Communities

EEC - English/1991

Part of the "Europe on the Move Series". Looks at the economic relationship between the Community and Eastern Europe.

02195 THE EUROPEAN COMMUNITY AND THE MEDITERRANEAN COUNTRIES

Office for Official Publications of the European Communities

EEC - English/1991

Part of the "Europe on the Move Series". Looks at the economic prospects for the enlargement of the Community.

02196 THE EUROPEAN COMMUNITY AND THE THIRD WORLD

Office for Official Publications of the European Communities

EEC - English/1991

Part of the "Europe on the Move Series". Looks at the Community's economic relations with the Third World.

02197 EUROPEAN COMMUNITY DIRECT INVESTMENT

Office for Official Publications of the European Communities

EEC - English/French irregular/ 204pp

Data is published 2-3 years in arrears.

02198 EUROPEAN COMMUNITY FOREST HEALTH REPORT

Office for Official Publications of the European Communities

EEC - Multi-lingual/1991 26pp free

02199 EUROPEAN DATABANKS: A GUIDE TO OFFICIAL STATISTICS

Eurostat

EEC - English/1992 91pp ECU3.00

Provides an overview of the databanks available, divided by field covered.

02200 EUROPEAN ECONOMY

Office for Official Publications of the European Communities

EEC - Multi-lingual quarterly/ ECU84.00 P.A. ISSN No. 0379-217X

Includes analysis and studies, with guidelines for Member States.

02201 EUROPEAN ECONOMY - SUPPLEMENT - SERIES A: ECONOMIC TRENDS

Office for Official Publications of the European Communities

EEC - English/French/German/ Italian 11 issues per year/ECU68.00 P.A.

Covers main economic indicators.

02202 EUROPEAN ECONOMY - SUPPLEMENT - SERIES B: BUSINESS AND CONSUMER SURVEY RESULTS

Office for Official Publications of the European Communities

EEC - English/French/German/ Italian 11 issues per year/ECU68.00 P.A.

02203 EUROPEAN FILE

Office for Official Publications of the European Communities

EEC - Multi-lingual monthly/ ECU10.00 postage charge ISSN No. 0379-3133

Consists of short reviews looking at current Community topics.

02204 EUROPEAN INSURANCE AND THE SINGLE MARKET 1992

Eurostat

EEC - English/French/German/ 1991 117pp ECU15.00

Statistics on the European insurance market especially as affected by the single market. Industry structure, financial investments, operating results, employment and future trends are covered.

02205 EUROPEAN INVESTMENT BANK PAPERS

European Investment Bank

EEC - English/French irregular

Occasional studies on the bank's activities.

02206 EUROPEAN INVESTMENT REGION SERIES

Business International Ltd

EEC - English irregular/£195.00 per report

Covers the economic prospects for a single investment region, with a comprehensive regional profile included. The reports appear as part of the EIU Special Report Series.

02207 EUROPEAN PARLIAMENT: EP NEWS

Office for Official Publications of the European Communities

EEC - Multi-lingual monthly/free ISSN No. 0250-5754

Useful for forthcoming legislation etc on industry.

02208 EUROPEAN REGIONAL DEVELOPMENT FUND: ANNUAL REPORT FROM THE COMMISSION TO THE COUNCIL, THE EUROPEAN PARLIAMENT AND THE ECONOMIC AND SOCIAL COMMITTEE

Office for Official Publications of the European Communities

EEC - Multi-lingual annual/124pp ECU14.00

Data published 1-2 years in arrears.

02209 EUROPEAN RETAIL

Business International Ltd

EEC - English 24 issues per year/ £375.00 P.A. ISSN No. 0960-0191

Covers retailers, suppliers, the single market, country profiles and European legislation.

02210 EUROPEAN RETAILING IN THE 1990S

Corporate Intelligence Research Publications

EEC - English/1991 £195.00

Market research type report which looks at the retailing sector in the EEC and considers the implications, for various retail sectors, of the single market.

02211 EUROPEAN STATISTICS - A GUIDE TO OFFICIAL SOURCES

Eurostat

EEC - English/1991 44pp ECU1.00

Lists EC, national, regional and international statistical offices and organizations.

02212 EUROPEAN UPDATE

Business International Ltd

EEC - English monthly/£195.00 P.A.

Highlights economic developments and their implications in the EC. Covers the 12 EC economies. A quarterly World Trade forecast is included.

02213 EUROSTAT

Office for Official Publications of the European Communities

EEC - Multi-lingual/1992 8pp free

Describes the statistical service Eurostat, and is a useful guide to what is available.

02214 EUROSTATISTICS: DATA FOR SHORT-TERM ECONOMIC ANALYSIS

Eurostat

EEC - English 11 issues per year/ ECU94.00 P.A.

Covers general economic and social trends in the European Community, the USA and Japan. Exchange and interest rates, balance of payments, unemployment, industrial production, steel, agriculture and foreign trade are included.

02215 EXTERNAL TRADE - ANALYTICAL TABLES - IMPORTS AND EXPORTS

Eurostat

EEC - English/French/German annual/5,650pp ECU630.00 for full series

External trade statistics according to the combined nomenclature. There are 26 volumes in total; 13 volumes deal with imports and 13 with exports. Volumes 1-12 with "products by country" and volume 13 with "countries by product". Data is published 1-2 years in arrears.

02216 EXTERNAL TRADE AND BALANCE OF PAYMENTS

Eurostat

EEC - English/French/German/ 10pp ECU192.00 P.A.

Rapid results type publications which summarize the main statistical findings of larger publications. A separate annual subscription is available.

02217 EXTERNAL TRADE AND BALANCE OF PAYMENTS - MONTHLY STATISTICS

Eurostat

EEC - English/French/German monthly/150pp ECU160.00 P.A.

General summary of foreign trade in the EC and trends in trade by country and product. Also features trade by main non-EC countries and gives balance of payment figures.

02218 EXTERNAL TRADE AND BALANCE OF PAYMENTS - STATISTICAL YEARBOOK 1991. RECAPITULATION 1958-90

Eurostat

EEC - English/French/German annual/158pp ECU11.55

Contains all the statistics on external trade, broken down by country and product. Includes the main statistical series from 1958-1990.

02219 EXTERNAL TRADE BY MODE OF TRANSPORT

Eurostat

EEC - English/French/German irregular/1991 177pp ECU15.00

Covers both external trade and trade between EC countries.

02220 EXTERNAL TRADE - GENERALIZED SYSTEM FOR TARIFF PREFERENCES (GSP)

Eurostat

EEC - English/French/German irregular/473pp ECU65.00

Covers imports in two volumes for 1990 under the generalized system of tariff preference (GSP).

02221 FACT SHEETS ON THE EUROPEAN PARLIAMENT AND THE ACTIVITIES OF THE EUROPEAN COMMUNITY

Office for Official Publications of the European Communities

EEC - Multi-lingual/1991 360pp ECU17.50

02222 FACTS THROUGH FIGURES: A STATISTICAL PORTRAIT OF THE EUROPEAN COMMUNITY IN THE EUROPEAN ECONOMIC AREA

Office for Official Publications of the European Communities

EEC - English 1991/31pp free

Basic statistical picture of the European Community in the extended European Economic Area.

02223 FAMILY BUDGETS - COMPARATIVE TABLES 1988

Eurostat

EEC - English/1993

Contains comparative data for six member countries; Denmark, Greece, France, Ireland, Luxembourg, and the Netherlands. Household consumption and expenditure, income, accommodation and holidays are included. The six other member states are included from 1993.

02224 THE FINANCES OF EUROPE

Office for Official Publications of the European Communities

EEC - English/French/German/ 1992 439pp ECU18.50 ISBN No. 9282623068

Describes how the EEC is financed.

02225 FINANCIAL REPORTS 199-: EUROPEAN COAL AND STEEL COMMUNITY

Office for Official Publications of the European Communities

EEC - Multi-lingual annual/96pp ISSN No. 1011-4637

Reports are published 1-2 years in arrears.

02226 FISHERIES - STATISTICAL YEARBOOK

Eurostat

EEC - English/French/German annual/167pp ECU15.00

Contains the principal statistics for the fisheries industry for EC states and other major fishing nations.

02227 FORECAST OF EMISSIONS FROM ROAD TRAFFIC IN THE EUROPEAN COMMUNITIES

Office for Official Publications of the European Communities

EEC - English/1992 286pp ECU23.75

02228 FOREST HEALTH REPORT 1991: TECHNICAL REPORT ON THE 1990 SURVEY

Office for Official Publications of the European Communities

EEC - English/French/1992 146pp ECU16.00

02229 THE FREE TRADE AGREEMENTS OF THE EFTA COUNTRIES AND THE EUROPEAN COMMUNITIES

European Free Trade Association

EEC - English/French/German annual/1967 free

02230 GAMBLING IN THE SINGLE MARKET: A STUDY OF THE CURRENT LEGAL AND MARKET SITUATION

Office for Official Publications of the European Communities

EEC - English 1991/412pp ECU300.00

Three volumes which consider the gambling market in each of the member states.

02231 GAS PRICES 1985-91

Eurostat

EEC - English/French/German annual/1992 165pp ECU12.00

Updates the annual surveys on gas prices in Community countries.

02232 GENERAL GOVERNMENT ACCOUNTS AND STATISTICS

Office for Official Publications of the European Communities

EEC - English/French irregular/ c500pp £12.10

National accounts, finance and balance of payments. Covers longer time periods - 1991 edition covered 1970-1988.

02233 GENERAL REPORT ON THE ACTIVITIES OF THE EUROPEAN COMMUNITIES

Office for Official Publications of the European Communities

EEC - English annual/464pp ECU15.00 ISSN No. 0069-6749

Provides a general picture of the Community activities over the past year.

02234 GOVERNMENT FINANCING OF RESEARCH AND DEVELOPMENT 1980-1990

Eurostat

EEC - English/French/German/ 1991 108pp ECU6.50

Overall analysis of the public financing of R&D from 1980 to 1990 with detailed data on objectives.

02235 GREEN EUROPE

Office for Official Publications of the European Communities

EEC - English/ECU10.00 postage charge ISSN No. 0250-5886

Describes agriculture in the Community in general terms and the challenges currently facing it.

02236 GUIDE TO THE COUNCIL OF THE EUROPEAN COMMUNITIES

Office for Official Publications of the European Communities

EEC - Multi-lingual/1992 181pp ECU10.00

Useful background information.

02237 IMPLEMENTATION OF THE REFORM OF THE STRUCTURAL FUNDS ANNUAL REPORT

Office for Official Publications of the European Communities

EEC - Multi-lingual annual/335pp ECU10.00

Data on level and destination of structural (infrastructure etc) funds.

02238 IN DEPTH COUNTRY APPRAISALS ACROSS THE EC - CONSTRUCTION AND BUILDING SERVICES

Building Services Research and Information Association

UK - English 1991/1,000pp £895.00 ISBN No. 0860222721

Study covering 11 European countries with general country profiles and specific profiles of the construction industry. Available as a full set or as individual country reports.

02239 INDUSTRIAL PRODUCTION - QUARTERLY STATISTICS

Office for Official Publications of the European Communities

EEC - English/French/German quarterly/ECU54.00 P.A. ISSN No. 0254-0649

Includes a subscription to "Rapid reports: Energy and industry".

02240 INDUSTRIAL TRENDS - MONTHLY

Eurostat

EEC - English/French/German monthly/90pp ECU77.00 P.A. ISSN No. 0258-1922

Short-term trends in EC industry. Production, turnover, prices, and employees are included. Figures for the USA and Japan are given for comparison.

02241 INSTITUTE OF GROCERY DISTRIBUTION REPORTS

Institute of Grocery Distribution

EEC - English irregular/£200.00 per report

A series of market research reports on retailing and distribution in the Europe. Large retailers and local consumer profiles are featured. Prices may vary with each report.

02242 INTEGRATED TARIFF OF THE EUROPEAN COMMUNITIES

Office for Official Publications of the European Communities

EEC - English irregular/1,999pp ECU185.00

Used for the purpose of applying Community measures to imports, exports and trade between member states. Known as "TARIC", it also serves as a basis for the working tariffs and tariff files in the member states. Published as part of the Official Journal No C 170 dated 6.7.92.

02243 THE INTERNAL MARKET: A CHALLENGE FOR THE WHOLESALE TRADE
Office for Official Publications of the European Communities

EEC - English/French/Dutch/1991 162pp ECU17.00

Discussion document with some data on the future of the wholesale trade within the single market.

02244 INTERNATIONAL TRADE IN SERVICES
Eurostat

EEC - English/French/irregular 202pp ECU16.00

Presents trade in services between the 12 EC countries. Data for the USA and Japan are given for comparison.

02245 INVENTORY OF TAXES LEVIED IN THE MEMBER STATES OF THE EUROPEAN COMMUNITIES BY THE STATE AND LOCAL AUTHORITIES (LANDER, DEPARTMENTS, REGIONS, DISTRICTS, PROVINCES, COMMUNES).
Office for Official Publications of the European Communities

EEC - English/French annual/ 726pp ECU80.00

The 14th edition was published in 1991.

02246 IRON AND STEEL - MONTHLY STATISTICS
Eurostat

EEC - English/French/German monthly/22pp ECU68.00 P.A. ISSN No. 0378-7559

Short-term economic statistics on production of pig iron, crude steel, rolled finished products, production indices, orders, deliveries, scrap, workers and consumption in ECSC countries.

02247 IRON AND STEEL - STATISTICAL YEARBOOK
Eurostat

EEC - English/French/German annual/135pp ECU24.00

Annual statistics on the structure and the economic situation of the Community's iron and steel industries.

02248 JOURNAL OF COMMON MARKET STUDIES
Blackwell Publishers

EEC - English quarterly/£42.50 P.A. ISSN No. 0021-9886

Looks at economic integration.

02249 L'EUROPE DE L'ENERGIE: OBJECTIF 1992 ET PERSPECTIVES 2010
Office for Official Publications of the European Communities

EEC - French/1991 171pp ECU18.00

Looks at the objectives of current energy policy and forecasts the energy situation to 2010.

02250 LABOUR COSTS: SURVEY 199-: INITIAL RESULTS
Office for Official Publications of the European Communities

EEC - English/French/German/ Italian annual/333pp ECU15.00

Data published 2-3 years in arrears.

02251 LABOUR FORCE SURVEY 1983-89
Eurostat

EEC - English/French/German/ 1991 97pp ECU10.00

Cumulative statistics from the annual labour force surveys, further years will be added as they become available.

02252 LABOUR FORCE SURVEY - RESULTS 1990
Eurostat

EEC - English/French/German annual/211pp ECU22.00

Covers population, employment, hours and unemployment. Results are published 1-2 years in arrears.

02253 THE LARGE MARKET OF 1993 AND THE OPENING-UP OF PUBLIC PROCUREMENT
Office for Official Publications of the European Communities

EEC - Multi-lingual/1991 41pp ECU6.00

Information and data on requirements for tending of public/ government contracts.

02254 LOANS TO BUILD THE EUROPEAN COMMUNITY
Office for Official Publications of the European Communities

EEC - Multi-lingual/1991 20pp free

Data on regional/sectoral assistance provided by the EEC.

02255 MONEY AND FINANCE
Eurostat

EEC - English quarterly/ECU15.00

P.A. ISSN No. 0255-6510

Gives the financial statistics for European Community countries, USA and Japan. Financial accounts, money supply, public finance, interest rates, and the EMS are included.

02256 NATIONAL ACCOUNTS ESA - AGGREGATES 1970-90
Eurostat

EEC - English/French/German/ 1991 235pp ECU18.00

Results of the main national accounts aggregates compiled in accordance with the European System of Integrated Economic Accounts (ESA). Trends and comparisons for the Community as a whole, the USA, and Japan.

02257 NATIONAL ACCOUNTS ESA - DETAILED TABLES BY BRANCH 1984-89
Eurostat

EEC - English/French/German/ 1991 219pp ECU20.00

Data is provided for transactions in goods and services, value added, renumeration of employees, gross capital formation, and final consumption of households as well as break-down by branch of employment.

02258 NATIONAL ACCOUNTS ESA - DETAILED TABLES BY SECTOR 1970-88
Eurostat

EEC - English/French/1991 423pp ECU50.00

Contains detailed results of the national accounts of EEC member countries. Presents data relating to transactions on goods and services as well as the division of work by sector.

02259 NATIONAL ACCOUNTS ESA - INPUT-OUTPUT TABLES
Eurostat

EEC - English/French/irregular ECU10.00

Detailed by country, these volumes contain input-output tables for Denmark, Spain, and Italy. Production activity, goods and services, interindustry transactions, inputs and foreign trade. Latest data available is for 1985.

02260 NET EARNINGS OF MANUAL WORKERS IN MANUFACTURING INDUSTRY IN THE COMMUNITY
Office for Official Publications of the European Communities

EEC - English/French/German/ 41pp ECU5.00

General statistics on levels of earnings. Generally published one year in arrears.

02261 NEWS FROM THE FOUNDATION
Office for Official Publications of the European Communities

EEC - Multi-lingual 5 issues per year/8pp free ISSN No. 0258-1965

Describes the work of the European Foundation for the Improvement of Living and Working Conditions and gives notice of reports published.

02262 OFFICIAL JOURNAL OF THE EUROPEAN COMMUNITIES: SERIES C
Office for Official Publications of the European Communities

EEC - English daily/£465.00 P.A.

The C series covers the complete range of Community information apart from legislation which is covered in the L series. The Supplement covers tenders and contracts.

02263 OPERATION OF NUCLEAR POWER STATIONS
Eurostat

EEC - English/French/German annual/142pp ECU11.50

Presents the main operating statistics for the past year. Includes monthly operating results for each power station.

02264 OPINIONS AND REPORTS OF THE ECONOMIC AND SOCIAL COMMITTEE
Office for Official Publications of the European Communities

EEC - English irregular/ECU360.00

Reports of the Economic and Social Committee of the European Parliament.

02265 PANORAMA OF EC INDUSTRY 1991-1992
Office for Official Publications of the European Communities

EEC - English/1991 422pp ECU110.00

Describes EEC industry down to sector level. A good source for cross country analysis. Further editions planned.

02266 POPULATION AND SOCIAL CONDITIONS
Eurostat

EEC - English/French/German irregular/ECU192.00

Small publications which summarize the main results of surveys, studies or publications. Supplied with a subscription to "Population and social conditions" publications.

02267 PORTRAIT OF THE REGIONS
Office for Official Publications of the European Communities

EEC - English/French/German/ 1992 ECU250.00

Three volume set, produced from Eurostat data, which covers over 200 regions of the EC, giving statistical data on each region. Includes basic socio-economic data.

02268 PRODUCTION, PRICES AND INCOMES IN EC AGRICULTURE
Eurostat

EEC - English/1991 227pp ECU17.00

An analysis of the economic accounts for agriculture 1973-1988.

02269 QUARTERLY NATIONAL ACCOUNTS ESA
Eurostat

EEC - English quarterly/ECU27.00 P.A. ISSN No. 1010-1764

Lists the principal national accounts aggregates on a quarterly basis, with comparative figures for the USA and Japan.

02270 RAPID REPORTS
Office for Official Publications of the European Communities

EEC - English/French/German irregular

Irregular bulletins which cover economy and finance, population and social conditions, regions, energy and industry, agriculture, forestry and fisheries, foreign trade, and the environment.

02271 THE REGIONAL IMPACT OF COMMUNITY POLICIES
Office for Official Publications of the European Communities

EEC - Multi-lingual/1991 162pp ECU10.00

Part of the EP Research and documentation series.

02272 REGIONS
Eurostat

EEC - English/French/German/ 10pp ECU192.00

Regular bulletin showing the main aspects of social and economic life of EEC regions presented in a rapid results type format. Data is taken from a variety of publications and sources.

02273 REGIONS - STATISTICAL YEARBOOK
Eurostat

EEC - English/French/German annual/235pp ECU23.20 ISSN No. 0081-4997

Shows the main aspects of social and economic life of EEC regions.

02274 REPORT ON COMPETITION POLICY
Office for Official Publications of the European Communities

EEC - English annual/ECU29.00

Describes competition policy established between the EC, Eastern European countries, and EFTA countries.

02275 RESEARCH ON THE "COST OF NON-EUROPE" - BASIC FINDINGS
Office for Official Publications of the European Communities

EEC - English/French/1991 ECU360.00

16 volume set in either French or English. Describes the cost of the absence of a true common market in various sectors and activities and points to the benefits of a common market. Much quoted in the UK press.

02276 RESULTS OF THE BUSINESS SURVEY CARRIED OUT AMONG MANAGEMENTS IN THE COMMUNITY
Office for Official Publications of the European Communities

EEC - Multi-lingual monthly/ ECU77.00 P.A.

Surveys such issues as business confidence and future expectations. Regularly quoted in the UK press.

02277 SECOND SURVEY ON STATE AIDS IN THE EUROPEAN COMMUNITY IN THE MANUFACTURING SECTOR AND CERTAIN OTHER SECTORS
Office for Official Publications of the European Communities

EEC - English/French/German irregular/102pp ECU12.00

02278 SERVICES AND TRANSPORT - MONTHLY STATISTICS
Eurostat

EEC - English/French/German monthly/22pp ECU67.00 P.A.

Monthly statistics designed to measure the flow of transport and services in the post 1992 environment.

02279 SHORT-TERM ENERGY OUTLOOK FOR THE EUROPEAN COMMUNITY
Office for Official Publications of the European Communities

EEC - English irregular/30pp

Published as a supplement to "Energy in Europe, May 1991". It gives an overview of the energy outlook for the Community.

02280 SIGMA
Eurostat

EEC - English/French/German bi-monthly/30pp free

The bulletin of European statistics from Eurostat.

02281 THE SINGLE FINANCIAL MARKET - 2ND EDITION
Office for Official Publications of the European Communities

EEC - Multi-lingual/61pp ECU8.00

Second revised and updated edition, published 1991.

02282 SOCIAL EUROPE - GENERAL REVIEW
Office for Official Publications of the European Communities

EEC - English/French/German 3 issues per year/ECU100.00 P.A. ISSN No. 0255-0776

Subscription includes supplements. Broad coverage of socio-economic issues.

02283 SOCIAL PROTECTION EXPENDITURE AND RECEIPTS 1980-89
Eurostat

EEC - English/French/German/ 69pp ECU8.00

Presents social protection expenditure and receipts for the EEC according to a system of integrated statistics.

02284 SOME STATISTICS ON SERVICES 1988
Office for Official Publications of the European Communities

EEC - English/French/333pp ECU14.00

Published in 1991, this concerns statistics on services and transport.

02285 THE STATISTICAL CONCEPT OF THE TOWN IN EUROPE
Office for Official Publications of the European Communities

EEC - English/French/1992

ECU12.00

Miscellaneous statistics on urban matters.

02286 STATISTIQUES DE L'ENVIRONEMENT
Eurostat

EEC - French irregular/1991 108pp ECU13.50

02287 STRUCTURE AND ACTIVITY OF INDUSTRY - ANNUAL SURVEY - MAIN RESULTS
Eurostat

EEC - English/French/German annual/301pp ECU25.00

Annual statistics on industrial structure and activity. Latest results are for 1987-88.

02288 STRUCTURE AND ACTIVITY OF INDUSTRY - DATA BY REGIONS
Eurostat

EEC - English/French annual/ 123pp ECU11.00

Statistics on industrial structure and activity broken down by region. Latest results are for 1987-88.

02289 STRUCTURE AND ACTIVITY OF INDUSTRY - DATA BY SIZE OF ENTERPRISE
Eurostat

EEC - English/French/German annual/301pp ECU10.50

Statistics on industrial structure and activity broken down by size of enterprise. Latest results are for 1986/87/88.

02290 TARGET 1992
Office for Official Publications of the European Communities

EEC - English monthly/8pp free

Describes progress towards the single internal market.

02291 TOURISM - ANNUAL STATISTICS
Eurostat

EEC - English/French/German annual/304pp ECU25.00

Accommodation, guest flows, capacity, employment, tourist expenditure, border arrivals, the balance of payments items and travel and passenger transport.

02292 TOURISM IN EUROPE: SUMMARIZING THE WORK OF THE EUROPEAN PARLIAMENT IN THE CONTEXT OF THE EUROPEAN YEAR OF TOURISM
Office for Official Publications of the European Communities

EEC - English/1992 11pp/free

02293 TRANSPORT - ANNUAL STATISTICS 1970-1989
Eurostat

EEC - English/French/German/ 1991 247pp ECU18.00

Statistics on the infrastructure, equipment, and operations of the different modes of transport and on traffic accidents.

02294 UNEMPLOYMENT
Eurostat

EEC - English/French/German monthly/10pp ECU50.00 P.A.

Gives harmonized data on umemployment, using the ILO definition, for European Community countries.

02295 UPLAND AREA: BASIC DATA AND STATISTICS
Office for Official Publications of the European Communities

EEC - English/French/Italian/1991 200pp ECU17.25

02296 VOCATIONAL TRAINING - INFORMATION BULLETIN
Office for Official Publications of the European Communities

EEC - Multi-lingual half-yearly/ ECU10.00 ISSN No. 0378-5068

Includes decision-making, program planning and administration, with some data on training levels.

02297 XIII MAGAZINE
Office for Official Publications of the European Communities

EEC - English quarterly/free

Covers information technology, telecommunications and related industries.

EUROPEAN UNION

Area profile:

European Union

Comprises Belgium, Denmark, France, Germany, Greece, Italy, Ireland, Luxembourg, Netherlands, Portugal, Spain, and the UK.

Area:

2,368,000 sq.km

Population:

344,924,700 (1991)

Administrative centre:

Brussels

Currency:

ECU - European Currency Unit

Institutional bank:

European Investment Bank

Government and Political structure:

Directly elected European Parliament with 158 members, an Economic & Social Committee with 189 members nominated by member governments, a Commission with 17 members nominated by member governments, a Council made up of representatives of member governments, and the Court of Justice.
Source; Eurostat 1992

03001 ENERGY - STATISTICAL YEARBOOK
Eurostat

EEC - English/French/German annual/116pp ECU23.50 ISSN No. 0081-489X

Energy economics, energy supplied balance sheets, and aggregates, are given for the EC economies.

03002 A COMMUNITY OF TWELVE: KEY FIGURES
Office for Official Publications of the European Communities

EEC - English 1991/32pp free

Basic statistical picture of the European Community.

03003 A SOCIAL PORTRAIT OF EUROPE
Office for Official Publications of the European Communities

EEC - Multi-lingual/1991 142pp ECU10.00

A collection of social statistics, with graphs and charts.

03004 AGRICULTURAL INCOME
Eurostat

EEC - English/French/German annual/154pp ECU26.00

Covers agricultural income for each member state and in aggregate and looks at trends in income.

03005 AGRICULTURAL MARKETS: PRICES
Office for Official Publications of the European Communities

EEC - Multi-lingual 4 issues per year/£66.80 P.A. ISSN No. 0250-9601

EEC wide agricultural prices and comparisons.

03006 AGRICULTURAL PRICES - PRICE INDICES & ABSOLUTE PRICES
Eurostat

EEC - English/French quarterly/ 210pp ECU25.00 P.A.

Trends in producer prices and means of production prices.

03007 AGRICULTURAL PRICES - PRICES INDICES AND ABSOLUTE PRICES
Eurostat

EEC - English/French/German annual/1991 370pp ECU20.50

Trends in producer prices and means of production prices in a cumulative format. Latest edition available is for 1981-1990.

03008 THE AGRICULTURAL SITUATION IN THE COMMUNITY
Office for Official Publications of the European Communities

EEC - Multi-lingual annual/421pp £19.90

Covers the state of agriculture in the Community as a whole.

03009 AGRICULTURE, FORESTRY AND FISHERIES
Eurostat

EEC - English/French/German 10pp ECU192.00 P.A.

Rapid report type publications which summarize the main results of larger statistical studies. Available as a separate subscription.

03010 AGRICULTURE IN EUROPE: DEVELOPMENT, CONSTRAINTS, AND PERSPECTIVES
Office for Official Publications of the European Communities

EEC - English/German/1992 97pp ECU7.00

03011 AGRICULTURE - STATISTICAL YEARBOOK
Eurostat

EEC - English/French/German annual/260pp ECU27.50

Contains all the principal statistics collected for the agricultural, forestry and fishery industries.

03012 ANALYTICAL TABLES OF EXTERNAL TRADE: NIMEXE, EXPORTS
Office for Official Publications of the European Communities

EEC - Multi-lingual annual/ ECU420.00, single issue ECU 42.00

Multi-volume work which covers imports and exports of different products. Organised in volumes A-L and M-Z. Individual product sectors available.

03013 ANIMAL PRODUCTION - QUARTERLY STATISTICS
Eurostat

EEC - English/French/German quarterly 110pp ECU74.00 P.A.

Meat, eggs and poultry, milk and milk products and external trade are covered.

03014 ANNUAL REPORT OF THE ENVIRONMENT INSTITUTE
Office for Official Publications of the European Communities

EEC - English annual/78pp ECU10.00

Last published in 1990, it gives an overview of the environment and quality of life in the Community.

03015 ANNUAL REPORT OF THE EUROPEAN FOUNDATION FOR THE IMPROVEMENT OF LIVING AND WORKING CONDITIONS
Office for Official Publications of the European Communities

EEC - Multi-lingual annual/80pp ECU8.75

03016 ANNUAL REPORT OF THE EUROPEAN INVESTMENT BANK
Office for Official Publications of the European Communities

EEC - Multi-lingual annual/108pp free ISSN No. 0071-2868

03017 BASIC STATISTICS OF THE COMMUNITY
Eurostat

EEC - English annual/ECU10.00

A pocket-sized book which gives all the main statistics of the European Community. Now in the 29th edition.

03018 BULLETIN OF ENERGY PRICES
Office for Official Publications of the European Communities

EEC - English/French twice yearly/ ECU7.60

A survey of import and consumer prices for oil, coal, gas and electricity in the Community. Also included in the global subscription "Energy".

03019 BULLETIN OF THE ECONOMIC AND SOCIAL COMMITTEE
Office for Official Publications of the European Communities

EEC - Multi-lingual monthly/ ECU54.00 ISSN No. 0010-2423

Describes the work of the Economic and Social Committee of the European Community.

03020 BULLETIN OF THE EUROPEAN COMMUNITIES
Office for Official Publications of the European Communities

EEC - English 10 issues per year/ £465.00 P.A. ISSN No. 0007-5116

The C series covers the complete range of Community information apart from legislation which was covered in the L series. The Supplement covers tenders and contracts.

03021 BUSINESS AND RESEARCH ASSOCIATES REPORTS: FURNITURE INDUSTRIES
Business and Research Associates
EEC - English/price varies with report
Market research reports which look at the EEC office and domestic furniture industries.

03022 BUSINESS EUROPE
Business International Ltd
EEC - English quarterly/£650.00 P.A.
In depth review articles on issues and developments within the European Community. Incorporates "European trends" and "Management Europe" formerly published by the EIU.

03023 CARRIAGE OF GOODS - INLAND WATERWAYS
Eurostat
EEC - English/French/German annual/184pp ECU19.00
Statistics on the carriage of goods by inland waterway irrespective of country of origin of the vessel.

03024 CARRIAGE OF GOODS - RAILWAYS
Eurostat
EEC - English/French/German annual/155pp ECU20.50
Statistics on the carriage of goods on the main railway networks which are open to public traffic.

03025 CARRIAGE OF GOODS - ROADS
Eurostat
EEC - English/French/German annual/146pp ECU16.00
Statistics on the carriage of goods by road, carried on vehicles registered in the EEC.

03026 CEDEFOP NEWS - VOCATIONAL TRAINING IN EUROPE
Office for Official Publications of the European Communities
EEC - English irregular/free
Published by the European Centre for the Development of Vocational Training.

03027 COAL AND THE INTERNATIONAL ENERGY MARKET
Office for Official Publications of the European Communities
EEC - English/French/German/Spanish/1991 135pp ECU17.50

03028 COMMUNICATIONS
Office for Official Publications of the European Communities
EEC - Multi-lingual irregular/1992 12pp free
Published as part of the European Investment Bank briefing series. Looks at the communications infrastructure in a wider Europe and its investment needs.

03029 THE COMMUNITY BUDGET: THE FACTS IN FIGURES
Office for Official Publications of the European Communities
EEC - English/French/German irregular/c100pp ECU11.00
Annotated summary of the results of Eurobarometer public opinion surveys of the European Communities.

03030 COMPARATIVE TABLES OF THE SOCIAL SECURITY SCHEMES IN THE MEMBER STATES OF THE EUROPEAN COMMUNITIES. GENERAL SCHEME (EMPLOYEES INDUSTRY AND COMMERCE)
Office for Official Publications of the European Communities
EEC - English/French/German irregular/127pp ECU12.00
The latest edition available is the 15th edition, published in 1989.

03031 COMPLETING THE INTERNAL MARKET: AN AREA WITHOUT INTERNAL FRONTIERS
Office for Official Publications of the European Communities
EEC - English irregular/127pp ECU31.00
Sixth report of the European Commission concerning the implementation of the Internal Market. Published in 1991.

03032 COMPLETING THE INTERNAL MARKET: CURRENT STATUS 1ST JANUARY 1992
Office for Official Publications of the European Communities
EEC - Multi-lingual/1992 ECU100.00
Published in 6 volumes, this publication considers progress towards the single market in various sectors.

03033 CONSUMER POLICY IN THE SINGLE MARKET
Office for Official Publications of the European Communities
EEC - Multi-lingual/1991 42pp free
Reviews progress towards standardization of consumer policies in the single market.

03034 CONSUMER PRICE INDEX
Eurostat
EEC - English monthly/ECU61.00 P.A. ISSN No. 0258-0861
Gives the general price index for European Community countries. Four quarterly issues cover price trends in 8 main groups of consumption, and four supplements give trends in purchasing power parities and price level indices. Austria, USA, Canada and Japan are included for comparison.

03035 CONSUMER PRICES IN THE EEC 1990
Eurostat
EEC - English/French/1991 131pp ECU15.00
Results of an 1989 survey. It lists average prices for the capital city of each country and average national prices with price level indices.

03036 THE COURIER
Office for Official Publications of the European Communities
EEC - English/French bi-monthly/free
Looks at development and aid issues for the EEC and its relations with Third World countries. Compliments "Europe on the Move".

03037 CROP PRODUCTION - QUARTERLY STATISTICS
Eurostat
EEC - English/French/German quarterly/120pp ECU74.00 P.A.
Covers land use, arable crops, and fruit and vegetable production. Includes data on weather conditions, supply balance sheets and external trade.

03038 DEMOGRAPHIC STATISTICS
Eurostat
EEC - English/French/German irregular/191pp ECU11.00
All the principal series of demographic statistics are given in a format which allows comparison across EC countries. Latest edition published is for 1992.

03039 DIGEST OF STATISTICS ON SOCIAL PROTECTION IN EUROPE - VOLUME 1: OLD AGE
Eurostat
EEC - English/French/208pp ECU17.00

Data given from 1980 to 1988 shows expenditure on benefits as well as the corresponding number of beneficiaries.

03040 DISABLED PERSONS: STATISTICAL DATA
Office for Official Publications of the European Communities
EEC - Multi-lingual/1991 190pp ECU17.00
Population and social conditions.

03041 DOING BUSINESS IN...
Price Waterhouse
International - English/free
Gives a description of the business and economic situation in a particular country. All developed countries are covered.

03042 DOING BUSINESS IN...
Ernst & Young
International - English 1-2 years/free
An individual report on each country giving broad details of the business and economic climate. Payment issues, credit terms and business practice are also included. Mainly intended for those doing business with the country concerned.

03043 EARNINGS - INDUSTRY AND SERVICES
Eurostat
EEC - Multi-lingual half-yearly/ECU77.00 P.A.
Gives details of the harmonized statistics on the earnings of manual and non-manual workers in industry and some parts of the service sector. Also gives data on purchasing power standards for workers across the Community.

03044 EC-NICS TRADE
Eurostat
EEC - English irregular/170pp ECU9.50
Statistical analysis of trade between the EC and newly industrialized countries (NICs). Latest data published 1991.

03045 ECONOMIC ACCOUNTS FOR AGRICULTURE AND FORESTRY
Eurostat
EEC - English/French/German annual/231pp ECU13.00
Contains the most recent data on agricultural and forestry accounts at the national and regional level. Latest edition published is for 1984-1989.

03046 ECONOMY AND FINANCE
Eurostat
EEC - English/French/German/10pp ECU192.00 P.A.

These are small, rapid results type publications which summarize the main results of larger surveys. Annual subscriptions are available.

03047 THE ECU AND ITS ROLE IN THE PROCESS TOWARDS MONETARY UNION
Office for Official Publications of the European Communities
EEC - Multi-lingual/1991 39pp free
Published as a monograph issue of European Economy No 48 1991.

03048 ECU - EMS INFORMATION
Eurostat
EEC - English/French/German/10pp ECU67.00
Regular publication on the private and official use of the ECU and the daily trends in the EMS.

03049 THE ECU REPORT
Office for Official Publications of the European Communities
EEC - Multi-lingual/1991 180pp ECU10.00
Overview and analysis of the ECU to date.

03050 THE EEA AGREEMENT
European Free Trade Association
EEC - English/French/German/Norwegian/Swedish/Finnish irregular/15pp free
Contents of the European Economic Area agreement.

03051 EIB - INFORMATION
Office for Official Publications of the European Communities
EEC - Multi-lingual quarterly/free ISSN No. 0250-3891
News about the activities of the European Investment Bank.

03052 ELECTRICITY PRICES 1985-91
Eurostat
EEC - English/French/German/1992 169pp ECU12.00
Contains electricity prices in national currencies.

03053 EMPLOYMENT AND INDUSTRIAL RELATIONS GLOSSARIES
Office for Official Publications of the European Communities
EEC - Multi-lingual/1991 422pp ECU24.00 per country
Provides an introduction to the key employment issues, industrial relations and labour law system for each EC country. A separate volume is published for each country.

03054 EMPLOYMENT IN EUROPE
Office for Official Publications of the European Communities
EEC - English annual/ECU11.25 ISSN No. 1016-5444
Looks at the present employment situation in the Community and future prospects. Published about one year in arrears.

03055 ENERGY 1960-1988
Eurostat
EEC - English/French/German/1991 260pp ECU9.00
Data extracted from the Sirene database, used for the definition of EC energy policy.

03056 ENERGY AND INDUSTRY - PANORAMA OF THE EEC INDUSTRY - STATISTICAL SUMMARY 1992
Eurostat
EEC - English/French/German/1992 547pp ECU25.00
Brings together the two editions of Panorama of EC industry.

03057 ENERGY BALANCE SHEETS
Office for Official Publications of the European Communities
EEC - English/French/German annual/233pp ECU24.00
Data is published 1-2 years in arrears.

03058 ENERGY - COMPLETE SUBSCRIPTION
Office for Official Publications of the European Communities
EEC - Multi-lingual/ECU172.00 P.A.
Comprises all the main statistical publications on energy in one subscription. Includes monthly statistics, rapid reports, annual statistics, energy in Europe and bulletin of energy prices.

03059 ENERGY IN EUROPE: ANNUAL ENERGY REVIEW
Office for Official Publications of the European Communities
EEC - English annual/133pp
Published about one year in arrears.

03060 ENERGY IN THE EUROPEAN COMMUNITY
Office for Official Publications of the European Communities
EEC - Multi-lingual/45pp free
Gives an overview of the energy situation in the Community, the 14th edition was published in 1990.

03061 ENERGY - MONTHLY STATISTICS
Eurostat

EEC - English/French/German monthly/71pp ECU77.00 P.A.

Short- term trends in the energy economy.

03062 ENGINEERING CONSULTANCY IN THE EUROPEAN COMMUNITY
Association of Consulting Engineers

EEC - English/1990 price on application

A series of market research type reports which look at the engineering consultancy market in the European Community.

03063 ENVIRONMENT STATISTICS
Eurostat

EEC - English/French/German/ 10pp ECU192.00

Small publications which appear regularly in a rapid results format. They summarize the main findings of larger statistical studies on the environment. Annual subscriptions are available.

03064 ENVIRONMENT STATISTICS
Office for Official Publications of the European Communities

EEC - Multi-lingual irregular/162pp ECU3.70

Data is published one year in arrears.

03065 EUROATOM SUPPLY AGENCY: ANNUAL REPORT
Office for Official Publications of the European Communities

EEC - English/French/German annual/30pp ECU6.00

Latest data available is for 1990.

03066 EUROBAROMETER: PUBLIC OPINION IN THE EUROPEAN COMMUNITY
Office for Official Publications of the European Communities

EEC - Multi-lingual 2 issues per year/ECU10.00 postage charge

03067 EUROPA TRANSPORT - OBSERVATION OF THE TRANSPORT MARKETS
Office for Official Publications of the European Communities

EEC - English/French/German quarterly/ECU50.00

Subscription includes annual report.

03068 EUROPE 2000: OUTLOOK FOR THE DEVELOPMENT OF THE COMMUNITY'S TERRITORY
Office for Official Publications of the European Communities

EEC - Multi-lingual/1992 208pp ECU15.00

Reviews future expansion of the community and the potential inclusion of Nordic/Central European countries.

03069 EUROPE IN FIGURES 3RD EDITION
Eurostat

EEC - English/1992 ECU16.50

A basic socio-economic picture of the European Community.

03070 EUROPE ON THE MOVE
Office for Official Publications of the European Communities

EEC - English irregular/8pp free

A series of leaflets giving basic information on the institutions and activities of the European Community. Features such topics as the EMS and monetary union.

03071 EUROPEAN COMMUNITIES ENCYCLOPEDIA AND DIRECTORY
Europa Publications

EEC - English every 3 years/1992 416pp £155.00

Basic data on EC countries.

03072 EUROPEAN COMMUNITIES: FINANCIAL REPORT
Office for Official Publications of the European Communities

EEC - Multi-lingual annual/62pp

03073 EUROPEAN COMMUNITY: 1992 AND BEYOND
Office for Official Publications of the European Communities

EEC - English/1991 34pp free

Describes the Community in general terms and the economic/ business challenges currently facing it.

03074 THE EUROPEAN COMMUNITY AND ITS EASTERN NEIGHBOURS
Office for Official Publications of the European Communities

EEC - English/1991

Part of the "Europe on the Move Series". Looks at the economic relationship between the Community and Eastern Europe.

03075 THE EUROPEAN COMMUNITY AND THE MEDITERRANEAN COUNTRIES
Office for Official Publications of the European Communities

EEC - English/1991

Part of the "Europe on the Move Series". Looks at the economic prospects for the enlargement of the Community.

03076 THE EUROPEAN COMMUNITY AND THE THIRD WORLD
Office for Official Publications of the European Communities

EEC - English/1991

Part of the "Europe on the Move Series". Looks at the Community's economic relations with the Third World.

03077 EUROPEAN COMMUNITY DIRECT INVESTMENT
Office for Official Publications of the European Communities

EEC - English/French irregular/ 204pp

Data is published 2-3 years in arrears.

03078 EUROPEAN COMMUNITY FOREST HEALTH REPORT
Office for Official Publications of the European Communities

EEC - Multi-lingual/1991 26pp free

03079 EUROPEAN DATABANKS: A GUIDE TO OFFICIAL STATISTICS
Eurostat

EEC - English/1992 91pp ECU3.00

Provides an overview of the databanks available, divided by field covered.

03080 EUROPEAN ECONOMY
Office for Official Publications of the European Communities

EEC - Multi-lingual quarterly/ ECU84.00 P.A. ISSN No. 0379-217X

Includes analysis and studies, with guidelines for Member States.

03081 EUROPEAN ECONOMY - SUPPLEMENT - SERIES A: ECONOMIC TRENDS
Office for Official Publications of the European Communities

EEC - English/French/German/ Italian 11 issues per year/ECU68.00 P.A.

Covers main economic indicators.

03082 EUROPEAN ECONOMY - SUPPLEMENT - SERIES B: BUSINESS AND CONSUMER SURVEY RESULTS

Office for Official Publications of the European Communities

EEC - English/French/German/ Italian 11 issues per year/ ECU68.00 P.A.

03083 EUROPEAN FILE

Office for Official Publications of the European Communities

EEC - Multi-lingual monthly/ ECU10.00 postage charge ISSN No. 0379-3133

Consists of short reviews looking at current Community topics.

03084 EUROPEAN INSURANCE AND THE SINGLE MARKET 1992

Eurostat

EEC - English/French/German/ 1991 117pp ECU15.00

Statistics on the European insurance market especially as affected by the single market. Industry structure, financial investments, operating results, employment and future trends are covered.

03085 EUROPEAN INVESTMENT BANK PAPERS

European Investment Bank

EEC - English/French irregular

Occasional studies on the bank's activities.

03086 EUROPEAN INVESTMENT REGION SERIES

Business International Ltd

EEC - English irregular/£195.00 per report

Covers the economic prospects for a single investment region, with a comprehensive regional profile included. The reports appear as part of the EIU Special Report Series.

03087 EUROPEAN PARLIAMENT: EP NEWS

Office for Official Publications of the European Communities

EEC - Multi-lingual monthly/free ISSN No. 0250-5754

Useful for forthcoming legislation etc on industry.

03088 EUROPEAN REGIONAL DEVELOPMENT FUND: ANNUAL REPORT FROM THE COMMISSION TO THE COUNCIL, THE EUROPEAN PARLIAMENT AND THE ECONOMIC AND SOCIAL COMMITTEE

Office for Official Publications of the European Communities

EEC - Multi-lingual annual/124pp ECU14.00

Data published 1-2 years in arrears.

03089 EUROPEAN RETAIL

Business International Ltd

EEC - English 24 issues per year/ £375.00 P.A. ISSN No. 0960-0191

Covers retailers, suppliers, the single market, country profiles and European legislation.

03090 EUROPEAN RETAILING IN THE 1990'S

Corporate Intelligence Research Publications

EEC - English/1991 £195.00

Market research type report which looks at the retailing sector in the EEC and considers the implications, for various retail sectors, of the single market.

03091 EUROPEAN STATISTICS - A GUIDE TO OFFICIAL SOURCES

Eurostat

EEC - English/1991 44pp ECU1.00

Lists EC, national, regional and international statistical offices and organizations.

03092 EUROPEAN UPDATE

Business International Ltd

EEC - English monthly/£195.00 P.A.

Highlights economic developments and their implications in the EC. Covers the 12 EC economies. A quarterly World Trade forecast is included.

03093 EUROSTAT

Office for Official Publications of the European Communities

EEC - Multi-lingual/1992 8pp free

Describes the statistical service Eurostat, and is a useful guide to what is available.

03094 EUROSTATISTICS: DATA FOR SHORT-TERM ECONOMIC ANALYSIS

Eurostat

EEC - English 11 issues per year/ ECU94.00 P.A.

Covers general economic and social trends in the European Community, the USA and Japan. Exchange and interest rates, balance of payments, unemployment, industrial production, steel, agriculture and foreign trade are included.

03095 EXTERNAL TRADE - ANALYTICAL TABLES - IMPORTS AND EXPORTS

Eurostat

EEC - English/French/German annual/5,650pp ECU630.00 for full series

External trade statistics according to the combined nomenclature. There are 26 volumes in total; 13 volumes deal with imports and 13 with exports. Volumes 1-12 with "products by country" and volume 13 with "countries by product". Data is published 1-2 years in arrears.

03096 EXTERNAL TRADE AND BALANCE OF PAYMENTS

Eurostat

EEC - English/French/German/ 10pp ECU192.00 P.A.

Rapid results type publications which summarize the main statistical findings of larger publications. A separate annual subscription is available.

03097 EXTERNAL TRADE AND BALANCE OF PAYMENTS - MONTHLY STATISTICS

Eurostat

EEC - English/French/German monthly/150pp ECU160.00 P.A.

General summary of foreign trade in the EC and trends in trade by country and product. Also features trade by main non-EC countries and gives balance of payment figures.

03098 EXTERNAL TRADE AND BALANCE OF PAYMENTS - STATISTICAL YEARBOOK 1991. RECAPITULATION 1958-90

Eurostat

EEC - English/French/German annual/158pp ECU11.55

Contains all the statistics on external trade, broken down by country and product. Includes the main statistical series from 1958-1990.

03099 EXTERNAL TRADE BY MODE OF TRANSPORT

Eurostat

EEC - English/French/German irregular/1991 177pp ECU15.00

Covers both external trade and trade between EC countries.

03100 EXTERNAL TRADE - GENERALIZED SYSTEM FOR TARIFF PREFERENCES (GSP)

Eurostat

EEC - English/French/German irregular/473pp ECU65.00

Covers imports in two volumes for 1990 under the generalized system of tariff preference (GSP).

03101 FACT SHEETS ON THE EUROPEAN PARLIAMENT AND THE ACTIVITIES OF THE EUROPEAN COMMUNITY

Office for Official Publications of the European Communities

EEC - Multi-lingual/1991 360pp ECU17.50

03102 FACTS THROUGH FIGURES: A STATISTICAL PORTRAIT OF THE EUROPEAN COMMUNITY IN THE EUROPEAN ECONOMIC AREA

Office for Official Publications of the European Communities

EEC - English 1991/31pp free

Basic statistical picture of the European Community in the extended European Economic Area.

03103 FAMILY BUDGETS - COMPARATIVE TABLES 1988

Eurostat

EEC - English/1993

Contains comparative data for six member countries; Denmark, Greece, France, Ireland, Luxembourg, and the Netherlands. Household consumption and expenditure, income, accommodation and holidays are included. The six other member states are included from 1993.

03104 THE FINANCES OF EUROPE

Office for Official Publications of the European Communities

EEC - English/French/German/ 1992 439pp ECU18.50 ISBN No. 9282623068

Describes how the EEC is financed.

03105 FINANCIAL REPORTS 199-: EUROPEAN COAL AND STEEL COMMUNITY

Office for Official Publications of the European Communities

EEC - Multi-lingual annual/96pp ISSN No. 1011-4637

Reports are published 1-2 years in arrears.

03106 FISHERIES - STATISTICAL YEARBOOK

Eurostat

EEC - English/French/German annual/167pp ECU15.00

Contains the principal statistics for the fisheries industry for EC states and other major fishing nations.

03107 FORECAST OF EMISSIONS FROM ROAD TRAFFIC IN THE EUROPEAN COMMUNITIES

Office for Official Publications of the European Communities

EEC - English/1992 286pp ECU23.75

03108 FOREST HEALTH REPORT 1991: TECHNICAL REPORT ON THE 1990 SURVEY

Office for Official Publications of the European Communities

EEC - English/French/1992 146pp ECU16.00

03109 THE FREE TRADE AGREEMENTS OF THE EFTA COUNTRIES AND THE EUROPEAN COMMUNITIES

European Free Trade Association

EEC - English/French/German annual/1967 free

03110 GAMBLING IN THE SINGLE MARKET: A STUDY OF THE CURRENT LEGAL AND MARKET SITUATION

Office for Official Publications of the European Communities

EEC - English 1991/412pp ECU300.00

Three volumes which consider the gambling market in each of the member states.

03111 GAS PRICES 1985-91

Eurostat

EEC - English/French/German annual/1992 165pp ECU12.00

Updates the annual surveys on gas prices in Community countries.

03112 GENERAL GOVERNMENT ACCOUNTS AND STATISTICS

Office for Official Publications of the European Communities

EEC - English/French irregular/ c500pp £12.10

National accounts, finance and balance of payments. Covers longer time periods - 1991 edition covered 1970-1988.

03113 GENERAL REPORT ON THE ACTIVITIES OF THE EUROPEAN COMMUNITIES

Office for Official Publications of the European Communities

EEC - English annual/464pp ECU15.00 ISSN No. 0069-6749

Provides a general picture of the Community activities over the past year.

03114 GOVERNMENT FINANCING OF RESEARCH AND DEVELOPMENT 1980-1990

Eurostat

EEC - English/French/German/ 1991 108pp ECU6.50

Overall analysis of the public financing of R&D from 1980 to 1990 with detailed data on objectives.

03115 GREEN EUROPE

Office for Official Publications of the European Communities

EEC - English/ECU10.00 postage charge ISSN No. 0250-5886

Describes agriculture in the Community in general terms and the challenges currently facing it.

03116 GUIDE TO THE COUNCIL OF THE EUROPEAN COMMUNITIES

Office for Official Publications of the European Communities

EEC - Multi-lingual/1992 181pp ECU10.00

Useful background information.

03117 IMPLEMENTATION OF THE REFORM OF THE STRUCTURAL FUNDS ANNUAL REPORT

Office for Official Publications of the European Communities

EEC - Multi-lingual annual/335pp ECU10.00

Data on level and destination of structural (infrastructure etc) funds.

03118 IN DEPTH COUNTRY APPRAISALS ACROSS THE EC - CONSTRUCTION AND BUILDING SERVICES

Building Services Research and Information Association

UK - English 1991/1,000pp £895.00 ISBN No. 0860222721

Study covering 11 European countries with general country profiles and specific profiles of the construction industry. Available as a full set or as individual country reports.

03119 INDUSTRIAL PRODUCTION - QUARTERLY STATISTICS

Office for Official Publications of the European Communities

EEC - English/French/German quarterly/ECU54.00 P.A. ISSN No. 0254-0649

Includes a subscription to "Rapid reports: Energy and industry".

03120 INDUSTRIAL TRENDS - MONTHLY

Eurostat

EEC - English/French/German

monthly/90pp ECU77.00 P.A. ISSN No. 0258-1922

Short term trends in EC industry. Production, turnover, prices, and employees are included. Figures for the USA and Japan are given for comparison.

03121 INSTITUTE OF GROCERY DISTRIBUTION REPORTS
Institute of Grocery Distribution

EEC - English irregular/£200.00 per report

A series of market research reports on retailing and distribution in Europe. Large retailers and local consumer profiles are featured. Prices may vary with each report.

03122 INTEGRATED TARIFF OF THE EUROPEAN COMMUNITIES
Office for Official Publications of the European Communities

EEC - English irregular/1,999pp ECU185.00

Used for the purpose of applying Community measures to imports, exports and trade between member states. Known as "TARIC", it also serves as a basis for the working tariffs and tariff files in the member states. Published as part of the Official Journal No C 170 dated 6.7.92.

03123 THE INTERNAL MARKET: A CHALLENGE FOR THE WHOLESALE TRADE
Office for Official Publications of the European Communities

EEC - English/French/Dutch/1991 162pp ECU17.00

Discussion document with some data on the future of the wholesale trade within the single market.

03124 INTERNATIONAL TRADE IN SERVICES
Eurostat

EEC - English/French/irregular 202pp ECU16.00

Presents trade in services between the 12 EC countries. Data for the USA and Japan are given for comparison.

03125 INVENTORY OF TAXES LEVIED IN THE MEMBER STATES OF THE EUROPEAN COMMUNITIES BY THE STATE AND LOCAL AUTHORITIES (LANDER, DEPARTMENTS, REGIONS, DISTRICTS, PROVINCES, COMMUNES).
Office for Official Publications of the European Communities

EEC - English/French annual/ 726pp ECU80.00

The 14th edition was published in 1991.

03126 IRON AND STEEL - MONTHLY STATISTICS
Eurostat

EEC - English/French/German monthly/22pp ECU68.00 P.A. ISSN No. 0378-7559

Short-term economic statistics on production of pig iron, crude steel, rolled finished products, production indices, orders, deliveries, scrap, workers and consumption in ECSC countries.

03127 IRON AND STEEL - STATISTICAL YEARBOOK
Eurostat

EEC - English/French/German annual/135pp ECU24.00

Annual statistics on the structure and the economic situation of the Community's iron and steel industries.

03128 JOURNAL OF COMMON MARKET STUDIES
Blackwell Publishers

EEC - English quarterly/£42.50 P.A. ISSN No. 0021-9886

Looks at economic integration.

03129 L'EUROPE DE L'ENERGIE: OBJECTIF 1992 ET PERSPECTIVES 2010
Office for Official Publications of the European Communities

EEC - French/1991 171pp ECU18.00

Looks at the objectives of current energy policy and forecasts the energy situation to 2010.

03130 LABOUR COSTS: SURVEY 199-: INITIAL RESULTS
Office for Official Publications of the European Communities

EEC - English/French/German/ Italian annual/333pp ECU15.00

Data published 2-3 years in arrears.

03131 LABOUR FORCE SURVEY 1983-89
Eurostat

EEC - English/French/German/ 1991 97pp ECU10.00

Cumulative statistics from the annual labour force surveys, further years will be added as they become available.

03132 LABOUR FORCE SURVEY - RESULTS 1990
Eurostat

EEC - English/French/German annual/211pp ECU22.00

Covers population, employment, hours and unemployment. Results are published 1-2 years in arrears.

03133 THE LARGE MARKET OF 1993 AND THE OPENING-UP OF PUBLIC PROCUREMENT
Office for Official Publications of the European Communities

EEC - Multi-lingual/1991 41pp ECU6.00

Information and data on requirements for tending of public/ government contracts.

03134 LOANS TO BUILD THE EUROPEAN COMMUNITY
Office for Official Publications of the European Communities

EEC - Multi-lingual/1991 20pp free

Data on regional/sectoral assistance provided by the EEC.

03135 MONEY AND FINANCE
Eurostat

EEC - English quarterly/ECU15.00 P.A. ISSN No. 0255-6510

Gives the financial statistics for European Community countries, USA and Japan. Financial accounts, money supply, public finance, interest rates, and the EMS are included.

03136 NATIONAL ACCOUNTS ESA - AGGREGATES 1970-90
Eurostat

EEC - English/French/German/ 1991 235pp ECU18.00

Results of the main national accounts aggregates compiled in accordance with the European System of Integrated Economic Accounts (ESA). Trends and comparisons for the Community as a whole, the USA, and Japan.

03137 NATIONAL ACCOUNTS ESA - DETAILED TABLES BY BRANCH 1984-89
Eurostat

EEC - English/French/German/ 1991 219pp ECU20.00

Data is provided for transactions in goods and services, value added, renumeration of employees, gross capital formation, and final consumption of households as well as break-down by branch of employment.

03138 NATIONAL ACCOUNTS ESA - DETAILED TABLES BY SECTOR 1970-88
Eurostat

EEC - English/French/1991 423pp ECU50.00

Contains detailed results of the national accounts of EEC member countries. Presents data relating to transactions on goods and services as well as the division of work by sector.

03139 NATIONAL ACCOUNTS ESA - INPUT-OUTPUT TABLES
Eurostat

EEC - English/French/irregular
ECU10.00

Detailed by country, these volumes contain input-output tables for Denmark, Spain, and Italy. Production activity, goods and services, interindustry transactions, inputs and foreign trade. Latest data available is for 1985.

03140 NET EARNINGS OF MANUAL WORKERS IN MANUFACTURING INDUSTRY IN THE COMMUNITY
Office for Official Publications of the European Communities

EEC - English/French/German/
41pp ECU5.00

General statistics on levels of earnings. Generally published one year in arrears.

03141 NEWS FROM THE FOUNDATION
Office for Official Publications of the European Communities

EEC - Multi-lingual 5 issues per year/8pp free ISSN No. 0258-1965

Describes the work of the European Foundation for the Improvement of Living and Working Conditions and gives notice of reports published.

03142 OFFICIAL JOURNAL OF THE EUROPEAN COMMUNITIES: SERIES C
Office for Official Publications of the European Communities

EEC - English daily/£465.00 P.A.

The C series covers the complete range of Community information apart from legislation which is covered in the L series. The Supplement covers tenders and contracts.

03143 OPERATION OF NUCLEAR POWER STATIONS
Eurostat

EEC - English/French/German
annual/142pp ECU11.50

Presents the main operating statistics for the past year. Includes monthly operating results for each power station.

03144 OPINIONS AND REPORTS OF THE ECONOMIC AND SOCIAL COMMITTEE
Office for Official Publications of the European Communities

EEC - English irregular/ECU360.00

Reports of the Economic and Social Committee of the European Parliament.

03145 PANORAMA OF EC INDUSTRY 1991-1992
Office for Official Publications of the European Communities

EEC - English/1991 422pp
ECU110.00

Describes EEC industry down to sector level. A good source for cross country analysis. Further editions planned.

03146 POPULATION AND SOCIAL CONDITIONS
Eurostat

EEC - English/French/German
irregular/ECU192.00

Small publications which summarize the main results of surveys, studies or publications. Supplied with a subscription to "Population and social conditions" publications.

03147 PORTRAIT OF THE REGIONS
Office for Official Publications of the European Communities

EEC - English/French/German/
1992 ECU250.00

Three volume set, produced from Eurostat data, which covers over 200 regions of the EC, giving statistical data on each region. Includes basic socio-economic data.

03148 PRODUCTION, PRICES AND INCOMES IN EC AGRICULTURE
Eurostat

EEC - English/1991 227pp
ECU17.00

An analysis of the economic accounts for agriculture 1973-1988.

03149 QUARTERLY NATIONAL ACCOUNTS ESA
Eurostat

EEC - English quarterly/ECU27.00
P.A. ISSN No. 1010-1764

Lists the principal national accounts aggregates on a quarterly basis, with comparative figures for the USA and Japan.

03150 RAPID REPORTS
Office for Official Publications of the European Communities

EEC - English/French/German
irregular

Irregular bulletins which cover economy and finance, population and social conditions, regions, energy and industry, agriculture, forestry and fisheries, foreign trade, and the environment.

03151 THE REGIONAL IMPACT OF COMMUNITY POLICIES
Office for Official Publications of the European Communities

EEC - Multi-lingual/1991 162pp
ECU10.00

Part of the EP Research and documentation series.

03152 REGIONS
Eurostat

EEC - English/French/German/
10pp ECU192.00

Regular bulletin showing the main aspects of social and economic life of EEC regions presented in a rapid results format. Data is taken from a variety of publications and sources.

03153 REGIONS - STATISTICAL YEARBOOK
Eurostat

EEC - English/French/German
annual/235pp ECU23.20 ISSN No. 0081-4997

Shows the main aspects of social and economic life of EEC regions.

03154 REPORT ON COMPETITION POLICY
Office for Official Publications of the European Communities

EEC - English annual/ECU29.00

Describes competition policy established between the EC, Eastern European countries, and EFTA countries.

03155 RESEARCH ON THE "COST OF NON-EUROPE" - BASIC FINDINGS
Office for Official Publications of the European Communities

EEC - English/French/1991
ECU360.00

16 volume set in either French or English. Describes the cost of the absence of a true common market in various sectors and activities and points to the benefits of a common market. Much quoted in the UK press.

03156 RESULTS OF THE BUSINESS SURVEY CARRIED OUT AMONG MANAGEMENTS IN THE COMMUNITY
Office for Official Publications of the European Communities

EEC - Multi-lingual monthly/
ECU77.00 P.A.

Surveys such issues as business confidence and future expectations. Regularly quoted in the UK press.

03157 SECOND SURVEY ON STATE AIDS IN THE EUROPEAN COMMUNITY IN THE MANUFACTURING SECTOR AND CERTAIN OTHER SECTORS
Office for Official Publications of the European Communities

EEC - English/French/German irregular/102pp ECU12.00

03158 SERVICES AND TRANSPORT - MONTHLY STATISTICS
Eurostat

EEC - English/French/German monthly/22pp ECU67.00 P.A.

Monthly statistics designed to measure the flow of transport and services in the post 1992 environment.

03159 SHORT-TERM ENERGY OUTLOOK FOR THE EUROPEAN COMMUNITY
Office for Official Publications of the European Communities

EEC - English irregular/30pp

Published as a supplement to "Energy in Europe, May 1991". It gives an overview of the energy outlook for the Community.

03160 SIGMA
Eurostat

EEC - English/French/German bi-monthly/30pp free

The bulletin of European statistics from Eurostat.

03161 THE SINGLE FINANCIAL MARKET - 2ND EDITION
Office for Official Publications of the European Communities

EEC - Multi-lingual/61pp ECU8.00

Second revised and updated edition, published 1991.

03162 SOCIAL EUROPE - GENERAL REVIEW
Office for Official Publications of the European Communities

EEC - English/French/German 3 issues per year/ECU100.00 P.A. ISSN No. 0255-0776

Subscription includes supplements. Broad coverage of socio-economic issues.

03163 SOCIAL PROTECTION EXPENDITURE AND RECEIPTS 1980-89
Eurostat

EEC - English/French/German/ 69pp ECU8.00

Presents social protection expenditure and receipts for the EEC according to a system of integrated statistics.

03164 SOME STATISTICS ON SERVICES 1988
Office for Official Publications of the European Communities

EEC - English/French/333pp ECU14.00

Published in 1991, this concerns statistics on services and transport.

03165 THE STATISTICAL CONCEPT OF THE TOWN IN EUROPE
Office for Official Publications of the European Communities

EEC - English/French/1992 ECU12.00

Miscellaneous statistics on urban matters.

03166 STATISTIQUES DE L'ENVIRONEMENT
Eurostat

EEC - French irregular/1991 108pp ECU13.50

03167 STRUCTURE AND ACTIVITY OF INDUSTRY - ANNUAL SURVEY - MAIN RESULTS
Eurostat

EEC - English/French/German annual/301pp ECU25.00

Annual statistics on industrial structure and activity. Latest results are for 1987-88.

03168 STRUCTURE AND ACTIVITY OF INDUSTRY - DATA BY REGIONS
Eurostat

EEC - English/French annual/ 123pp ECU11.00

Statistics on industrial structure and activity broken down by region. Latest results are for 1987-88.

03169 STRUCTURE AND ACTIVITY OF INDUSTRY - DATA BY SIZE OF ENTERPRISE
Eurostat

EEC - English/French/German annual/301pp ECU10.50

Statistics on industrial structure and activity broken down by size of enterprise. Latest results are for 1986/87/88.

03170 TARGET 1992
Office for Official Publications of the European Communities

EEC - English monthly/8pp free

Describes progress towards the single internal market.

03171 TOURISM - ANNUAL STATISTICS
Eurostat

EEC - English/French/German annual/304pp ECU25.00

Accommodation, guest flows, capacity, employment, tourist expenditure, border arrivals, the balance of payments items and travel and passenger transport.

03172 TOURISM IN EUROPE: SUMMARIZING THE WORK OF THE EUROPEAN PARLIAMENT IN THE CONTEXT OF THE EUROPEAN YEAR OF TOURISM
Office for Official Publications of the European Communities

EEC - English/1992 11pp/free

03173 TRANSPORT - ANNUAL STATISTICS 1970-1989
Eurostat

EEC - English/French/German/ 1991 247pp ECU18.00

Statistics on the infrastructure, equipment, and operations of the different modes of transport and on traffic accidents.

03174 UNEMPLOYMENT
Eurostat

EEC - English/French/German monthly/10pp ECU50.00 P.A.

Gives harmonized data on umemployment, using the ILO definition, for European Community countries.

03175 UPLAND AREA: BASIC DATA AND STATISTICS
Office for Official Publications of the European Communities

EEC - English/French/Italian/1991 200pp ECU17.25

03176 VOCATIONAL TRAINING - INFORMATION BULLETIN
Office for Official Publications of the European Communities

EEC - Multi-lingual half-yearly/ ECU10.00 ISSN No. 0378-5068

Includes decision-making, program planning and administration, with some data on training levels.

03177 XIII MAGAZINE
Office for Official Publications of the European Communities

EEC - English quarterly/free

Covers information technology, telecommunications and related industries.

ALBANIA

Country profile:	Albania
Official name:	Republic of Albania
Area:	28,748 sq.km.
	Bordered by Yugoslavia in the north and east, in the south by Greece and west by the Adriatic Sea.
Population:	3.13m (1988)
Capital:	Tirana
Language:	Albanian
Currency:	Lek 100 qintars = 1 lek
National bank:	The Albanian State Bank
Government and Political structure:	Single chamber National Assembly of 140 deputies. The pre-1990 period was dominated by a Communist political structure. Late 1990 saw the development of multi-party democracy.

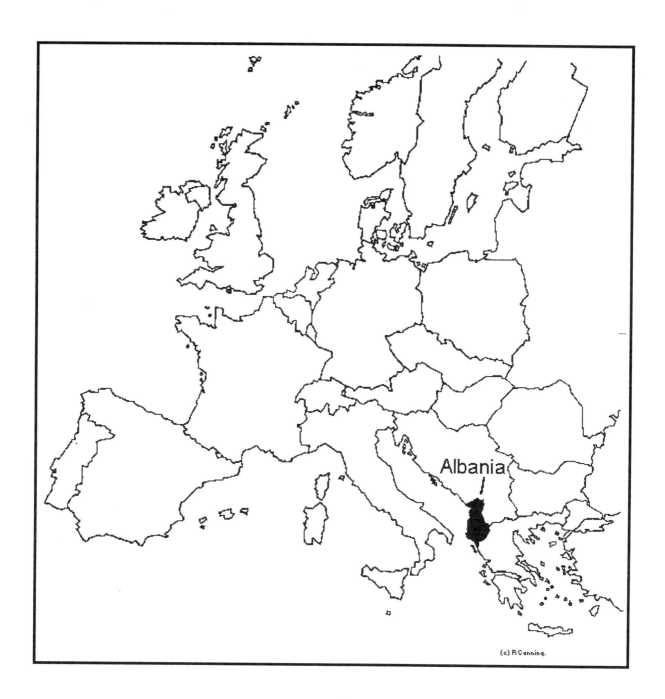

Albania

04001 ALBANIA: FROM ISOLATION TOWARDS REFORM

International Monetary Fund

Blejer, M.I. et al/Albania - English/ 1992 84pp US$15.00 ISBN No. 1557752664

Covers economic policy and has some data on economic conditions.

04002 AN AGRICULTURAL STRATEGY FOR ALBANIA

World Bank Publications

Dethier, J.J./Albania - English/1992 265pp

Development-style document prepared by a joint team from the World Bank and the European Community.

04003 COUNTRY REPORT: ROMANIA, BULGARIA, ALBANIA

Business International Ltd

Albania/Bulgaria/Romania - English quarterly/25-40pp £160.00 P.A. ISSN No. 0269-5669

Contains an executive summary, sections on political and economic structure, outlook for the next 12-18 months, statistical appendices, and a review of key political developments. Includes an annual Country Profile.

04004 DOING BUSINESS WITH EASTERN EUROPE: ALBANIA/BULGARIA

Business International Ltd

Albania/Bulgaria - English quarterly/£540.00 P.A.

Part of a ten volume set, updated every three months which gives a broad picture of the market and business pratice.

04005 ECONOMIC CHANGE IN THE BALKAN STATES: ALBANIA, BULGARIA, ROMANIA AND YUGOSLAVIA

St Martins Press Inc

Albania/Buglaria/Romania/ Yugoslavia - English/1991 173pp ISBN No. 0312057350

Review of the economic situation

and the prospects for development of a market economy. Produced by the Stockholm Institute of Soviet & East European Economics.

04006 ENGLISH, FRENCH-ALBANIAN FOREIGN TRADE

Albanian Chamber of Commerce

Albania - Albanian monthly

Business journal aimed at those trading with Albania.

04007 THE US - EASTERN EUROPEAN TRADE SOURCE BOOK

St James Press

Loiry, W.S./Bulgaria/ Czechoslovakia/Hungary/Poland/ Romania/Albania - English irregular/1991 269pp ISBN No. 1558621563

Guide to doing business with general economic/business background given.

AUSTRIA

Country profile:	Austria
Official name:	Republic of Austria
Area:	83,857 sq.km.
	Bordered by Germany and Czechoslovakia in the north, by Hungary in the east, south by Italy and Yugoslavia and west by Switzerland.
Population:	7.8m (1991)
Capital city:	Vienna
Language:	German
Currency:	Schilling 100 groschen = 1 schilling
National bank:	National Bank of Austria
Government and Political structure:	National council *(Nationalrat)* of 183 members elected for four years by proportional representation. The head of government is the Federeal Chancellor who heads a Council of Ministers.

Austria

(c) R Canning.

05001 ABFALLERHEBUNG INDUSTRIE

Bundeskammer der Gewerblichen Wirtschaft

Abteilung für Statistik/Austria - Austrian/German/1991 197pp free

Statistical coverage of the waste industry.

05002 ANNUAL FACTS SHEET

Wiener Borsekammer

Austria - English/German annual/ free

Facts about the Vienna stock exchange.

05003 ANNUAL REPORT

Wiener Borsekammer

Austria - English/German annual/ 90pp free

Annual report of the Vienna stock exchange.

05004 ARBEITSSTAETTEN-ZAEHLUNG

Oesterreichisches Statistisches Zentralamt

Austria - German every 10 years/ 1991

Results of the 1991 census of local units of employment. Ten volumes, volume Austria and 9 volumes for provinces. Classified by communities.

05005 AUFWENDUNGEN DER INDUSTRIE FÜR DEN UMWELTSCHUTZ

Bundeskammer der Gewerblichen Wirtschaft

Abteilung für Statistik/Austria - German every 3 years/1972 112pp free

Statistical coverage of the environmental production sector.

05006 AUSSENHANDEL (LÄNDER - UND WARENGLIEDERUNG)

Bundeskammer der Gewerblichen Wirtschaft

Austria - German monthly/26pp free

Looks at Austria's international trade.

05007 AUSTRIA, FACTS AND FIGURES

Bundeskanzleramt

Austria - English/German/French/ Spanish/Italian etc annual/1954 180pp free

General introduction to Austria with statistical tables.

05008 AUSTRIA IN FIGURES

Bundeskanzleramt

Austria - English/German/French/ Spanish/Italian/Portuguese annual/ 1973 c24pp free

Broad statistical picture of Austria.

05009 BANKENSTATISTICHE DATEN

Oesterreichische Nationalbank

Austria - German monthly/price varies

Banking and financial statistics from the national bank.

05010 BAUSTATISTIK

Oesterreichisches Statistisches Zentralamt

Austria - German annual/AS 480.00 & AS 510.00

Two volumes. Vol 1 gives details of output, employment, wages, energy consumption and hours of work for each sector of the building industry. Vol 2 is an analysis of the financial organization of the construction industry.

05011 BEITRÄGE ZUR REGIONALSTATISTIK - BUNDESLÄNDER-VERGLEICH

Bundeskammer der Gewerblichen Wirtschaft

Abteilung für Statistik/Austria - German every 5 years/1969 69pp

Regional statistics for Austria.

05012 BEREICHSZAELUNGEN

Oesterreichisches Statistisches Zentralamt

Austria - German every 5-7 years/ price varies

A general trade survey in 14 parts covering industry, production, transport, tourism, energy etc.

05013 BERICHTE UND STUDIEN

Oesterreichische Nationalbank

Austria - English/German quarterly free

Review-type articles dealing with Austrian banking.

05014 BESTANDSSTATISTIK DER KRAFTFAHRZEUGE IN OESTERREICH IM JAHRE

Oesterreichisches Statistisches Zentralamt

Austria - Germany annual/AS 750.00 ISSN No. 0067-6306

Statistics on motor vehicles including capacity.

05015 BODENNUTZUNGSER-HEBUNG

Oesterreichisches Statistisches Zentralamt

Austria - German every 3-4 years/ 80pp AS 260.00

Survey of land use.

05016 CA QUARTERLY: FACTS & FIGURES ON AUSTRIA'S ECONOMY

Creditanstalt - Bankverein

Austria - English quarterly/30pp

Articles on Austria's economy and selected economic indicators.

05017 CARPETS - SECTOR REPORT

Department of Trade and Industry Export Publications

Austria - English irregular/1991 42pp £30.00

Includes lists of trade figures, leading commission agents, potential and established agents.

05018 CONFECTIONERY - SECTOR REPORT

Department of Trade and Industry Export Publications

Austria - English irregular/1992 11pp £30.00

Production, import/export figures. Addresses of confectionery manufacturers; leading foreign suppliers; leading importers/ distributors; trade promotion; list of addresses for Austrian food chains and supermarkets.

05019 COUNTRY BOOKS: AUSTRIA

United Kingdom Iron and Steel Statistics Bureau

Austria - English annual/£75.00 ISSN No. 0952-584X

Annual summary tables showing production, materials consumed, apparent consumption, imports and exports of 130 products by quality and market.

05020 COUNTRY FORECASTS: AUSTRIA

Business International Ltd

Austria - English quarterly/£360.00 P.A.

Contains an executive summary, sections on political and economic outlooks and the business environment, a fact sheet, and key economic indicators. Includes a quarterly Global Outlook.

05021 COUNTRY PROFILE AUSTRIA

Department of Trade and Industry Export Publications

Austria - English irregular/1991 42pp £10.00

This publication provides a general description of the country in terms of its geography, economy and population, and gives information about trading in the country with details such as export conditions and investment covered.

05022　COUNTRY REPORT: AUSTRIA

Business International Ltd

Austria - English quarterly/25-40pp £160.00 P.A. ISSN No. 0269-5170

Contains an executive summary, sections on political and economic structure, outlook for the next 12-18 months, statistical appendices, and a review of key political developments. Includes an annual Country Profile.

05023　DEMOGRAPHISCHES JAHRBUCH OESTERREICHS

Oesterreichisches Statistisches Zentralamt

Austria - German annual/AS 360.00

Demographic data on births, deaths, marriages, divorces and internal migration. Also contains three alternative population projections.

05024　DER AUSSENHANDEL OESTERREICHS, SERIE 1A

Oesterreichisches Statistisches Zentralamt

Austria - German annual/AS 1,120.00

Statistics on imports and exports giving volume and value classified by commodity and country.

05025　DER AUSSENHANDEL OESTERREICHS, SERIE 1B

Oesterreichisches Statistisches Zentralamt

Austria - German biannual/AS 420.00

Statistics on foreign trade listing commodities according to country of origin or destination.

05026　DER AUSSENHANDEL OESTERREICHS, SERIE 2

Oesterreichisches Statistisches Zentralamt

Austria - German biannual/AS 350.00

Analysis of foreign trade according to the international SITC classification specialized for countries.

05027　DER FREMDENVERKEHR IN OESTERREICH IM JAHRE...

Oesterreichisches Statistisches Zentralamt

Austria - German annual/300pp AS 320.00 ISSN No. 0071-948X

Detailed information on foreign visitors to Austria (tourist nights, arrivals), giving figures for the month, half-year and year. Analysed by communities.

05028　DER INTENSIVOBSTBAU IN OESTERREICH

Oesterreichisches Statistisches Zentralamt

Austria - German every 3 years/AS 170.00

Statistics on the fruit farming industry.

05029　DER WEINBAU OESTERREICH

Oesterreichisches Statistisches Zentralamt

Austria - German biennial/120pp AS 170.00

Statistics on Austrian viniculture

05030　DIE ARBEITSKOSTEN IN DER INDUSTRIE ÖSTERREICHS

Bundeskammer der Gewerblichen Wirtschaft

Abteilung für Statistik/Austria - German every 3 years/1960 117pp free

Labour costs in a statistical format.

05031　DIE UNSELBSTÄNDIG BESCHAFTIGTEN DER GEWERBLICHEN WIRTSCHAFT (NACH DER KAMMERSYSTEMATIK)

Bundeskammer der Gewerblichen Wirtschaft

Abteilung für Statistik/Austria - German every 5 years/1971 36pp

Statistics on employees in Austrian industry.

05032　DOING BUSINESS IN AUSTRIA

Price Waterhouse

Austria - English irregular/1990 330-340pp free to clients

A business guide which includes sections on the investment climate, doing business, accounting and auditing, taxation and general country data. Updated by supplements.

05033　DONAU, INN UND SALZACH

Oberösterreich Touristik

Manfred Traunmüller/Austria - English/German irregular/1990 80pp AS 110.00

Statistics on tourism in upper Austria.

05034　EINKOMMENSTEUER-STATISTIK

Oesterreichisches Statistisches Zentralamt

Austria - German annual/130pp AS 290.00

Statistics on Austrian income tax.

05035　ELECTRONIC COMPONENTS - SECTOR REPORT

Department of Trade and Industry Export Publications

Austria - English/1990 27pp £30.00

Contains lists of potential agents, larger Austrian manufacturers, and import statistics on passive components and printed circuits.

05036　EMPIRICA

Nationalokondmische Gesellschaft

Austria - English/German biannual/250pp AS 200-300.00 ISSN No. 0340-8944

Austrian economic papers with an academic bias.

05037　ENERGIEVERBRAUCH DER HAUSHALTE IM JAHRE

Oesterreichisches Statistisches Zentralamt

Austria - German biennial/180pp AS 250.00

Data on energy consumption of households.

05038　ENERGIEVERSORGUNG OESTERREICHS

Oesterreichisches Statistisches Zentralamt

Austria - German monthly/175pp AS 720.00 P.A.

Statistics on energy.

05039　ERGEBNISSE DER LAND-WIRTSCHAFTLICHEN STATISTIK IM JAHRE...

Oesterreichisches Statistisches Zentralamt

Austria - German annual/128pp AS 200.00 ISSN No. 0067-2327

Statistics on agriculture giving details of land under cultivation, crop yields, size of herds and machinery used. Data analysed by provinces.

05040　FIRMENBUCH ÖSTERREICH

Jupiter Verlagsgesellschaft mbH

Austria - German annual/1949 3,392pp AS 3,600.00

Directory of Austrian industrial, trading and service firms, including names of owners, board members, directors.

05041　FORSCHUNG U DOKUMENTATION

Bundeskammer der Gewerblichen Wirtschaft

Abteilung für statistik/Austria - German every 2 years/1964 c65pp

Details of Austrian R & D expenditure.

05042 GARDENING EQUIPMENT - SECTOR REPORT

Department of Trade and Industry Export Publications

Austria - English irregular/1990 12pp £30.00

Includes lists of hardware retailers, superstores, garden centres and potential agents are also listed.

05043 GEBARUNG-SUEBERSICHTEN

Oesterreichisches Statistisches Zentralamt

Austria - German annual/300pp AS 650.00

Data on public expenditure and administration.

05044 GELD UND KREDITWESSEN, PRIVATVERSICHERUNG 1988

Oesterreichisches Statistisches Zentralamt

Austria - German every 5 years/ 406pp AS 490.00

Statistics on money, credit and insurance derived from the 1988 census.

05045 GESCHÄFTSBERICHT

Oesterreichische Nationalbank

Austria - English/German annual

Economic data from the National Bank.

05046 GEWERBESTATISTIK

Oesterreichisches Statistisches Zentralamt

Austria - German annual/2 Vols. AS 320.00 & AS 560.00

Industrial statistics covering volume & value of output of manufacturing industries. Vol 1 covers major manufacturers. Vol 2 gives figures for establishments with less than 20 employees.

05047 GEWERBESTEUER-STATISTIK

Oesterreichisches Statistisches Zentralamt

Austria - German annual/150pp AS 280.00

Statistics on the taxation of industry and trade.

05048 GROSS - UND EINZELHANDELS-STATISTIK

Oesterreichisches Statistisches Zentralamt

Austria - German annual/AS 280.00

Statistics on wholesale and retail business analysed by product and provinces.

05049 GROSSZAEHLUNG: AUSGEWAEHLTE MASSZAHLEN NACH GEMEINDEN

Oesterreichisches Statistisches Zentralamt

Austria - German every 10 years/ AS 200.00

Selected statistical parameters of the census by community, giving data on population, employment, housing etc.

05050 GRUNDERWERB

Oesterreichisches Statistisches Zentralamt

Austria - German every 3 years/AS 50.00

Data on land and property prices.

05051 HAEUSER - UND WOHNUNGSZAEHLUNG

Oesterreichisches Statistisches Zentralamt

Austria - German every 10 years/ price varies

Housing census 1991 with detailed data on buildings and housing units, classified by communities; 14 volumes, volume Austria, 9 volumes for provinces, 4 volumes with specific data on the housing census.

05052 HANDBUCH FÜR INVESTOREN-INFORMATION

Bundesministerium Für Wirtschaftliche Angelegenheiten

Austria - German/c300pp AS 500.00

Latest in the 5th ed. Investors Handbook.

05053 HINTS TO EXPORTERS VISITING AUSTRIA

Department of Trade and Industry Export Publications

Austria - English irregular/1991 60pp £5.00

This publication provides detailed practical information that is useful for those planning a business visit. Topics include currency information, economic factors, import and exchange control regulations, methods of doing business and general information.

05054 INDUSTRIE - UND GEWERBESTATISTIK

Oesterreichisches Statistisches Zentralamt

Austria - German annual/AS 320.00

Industrial and manufacturing statistics, production figures.

05055 INDUSTRIESTATISTIK

Oesterreichisches Statistisches Zentralamt

Austria - German annual/600pp AS

770.00

Industrial statistics, figures on bookkeeping records and cost accounting.

05056 INFORMATIONEN INTERNATIONALE PREISE

Bundesanstalt Für Agrarwirtschaft

Austria/Germany/Central & Western Europe - German weekly/ 1976 AS 280.00 P.A.

Covers the situation for agriculture and agricultural industries.

05057 INVESTING, LICENSING AND TRADING CONDITIONS ABROAD: AUSTRIA

Business International Ltd

Austria - English annual/£825.00 for complete European series

Loose-leaf format covering the state role in industry, competition rules, price controls, remittability of funds, taxation, incentives and capital sources, labour and foreign trade.

05058 JAHRBUCH DER WIENER BORSE

Creditanstalt - Bankverein

Austria - German annual/price varies

Yearbook of the Viennese stock exchange.

05059 JAHRESBERICHT

Fachverband der Erdölindustrie Österreichs

Austria - German annual/price varies

Review of the Austrian petroleum industry including statistics.

05060 JEWELLERY - SECTOR REPORT

Department of Trade and Industry Export Publications

Austria - English irregular/1992 10pp £30.00

Major local manufacturers and leading importers and distributors.

05061 KONSUMERHEBUNG, HAUPTERGEBNISSE

Oesterreichisches Statistisches Zentralamt

Austria - German every 10 years/ AS 120.00

Survey of household consumption.

05062 KRAFTFAHRZEUGZULAS-SUNGEN NACH ZULAS-SUNGSBEHOERDEN

Oesterreichisches Statistisches Zentralamt

Austria - German monthly/price varies

Data on motor vehicle registrations according to registration authority.

05063 LAND - UND FORSTWIRTSCHAFTLICHE ARBEITSKRAEFTE
Oesterreichisches Statistisches Zentralamt
Austria - German every 3-4 years/ AS 270.00
Data on people working in agriculture and forestry.

05064 LANDWIRTSCHAFTLICHE MASCHINENZAEHLUNG
Oesterreichisches Statistisches Zentralamt
Austria - German every 5 years/ 126pp AS 160.00
Results of a survey of agricultural machinery and equipment. There is also comparative international data.

05065 LOHNERHABUNG IN DER INDUSTRIE ÖSTERREICHS
Bundeskammer der Gewerblichen Wirtschaft
Austria - German twice a year/1962 54pp free
Details of income levels amongst blue-collar workers.

05066 LOHNSTEUERSTATISTIK
Oesterreichisches Statistisches Zentralamt
Austria - German annual/AS 350.00
Statistics on wages, taxes and national insurance.

05067 MADE IN AUSTRIA
Jupiter Verlagsgesellschaft mbH
Austria - English/German/French annual/1966 c300pp AS 900.00
Manufacturing and distribution firms engaged in export, plus banks, transportation companies.

05068 MEDICAL EQUIPMENT - SECTOR REPORT
Department of Trade and Industry Export Publications
Austria - English irregular/1991 24pp £30.00
Information on the Austrian health care market. Contains addresses of hospitals, importers and manufacturers of medical equipment.

05069 MIKROZENSUS - JAHRESERGEBNISSE
Oesterreichisches Statistisches Zentralamt
Austria - German annual/190pp AS 250.00
Data on population, employment, living conditions, demographic and socio-economic characteristics derived from the quarterly microcensus.

05070 MONATLICHE KRAFT-FAHRZEUGZULASSUNGS-STATISTIK
Oesterreichisches Statistisches Zentralamt
Austria - German monthly/AS 4,200.00 P.A.
Monthly statistics on vehicle registrations.

05071 MONATSBERICHTE ÜBER DIE ÖSTERREICHISCHE LANDWIRTSCHAFT
Bundesanstalt Für Agrarwirtschaft
Austria/Germany/Central & Western Europe - German monthly/ AS 850.00 P.A.
Agriculture, agricultural economics and rural economics are featured.

05072 MONATSBEZÜGE DER ANGESTELLTEN IN DER INDUSTRIE ÖSTERREICHS
Bundeskammer der Gewerblichen Wirtschaft
Abteilung für Statistik/Austria - German annual/1959 134pp free
Details of income levels amongst white-collar workers.

05073 MONTHLY STATISTICAL REPORTS
Wiener Borsekammer
Austria - English/German monthly
Statistics produced by the Vienna stock exchange.

05074 NUTZENERGIEANALYSE
Oesterreichisches Statistisches Zentralamt
Austria - German every 5 years/ 50pp AS 100.00
Data on energy usage and energy issues.

05075 NUTZTIERHALTUNG IN OESTERREICH
Oesterreichisches Statistisches Zentralamt
Austria - German biennial/230pp AS 360.00
A census of working animals.

05076 OESTERREICH - EG: KAUFKRAFTPARITAETEN (EIN VERGLEICH VON PRODUKTIVITAET, LEBENSSTANDARD UND PREISNIVEAUS 1990)
Oesterreichisches Statistisches Zentralamt
Austria/EEC - German every 5 years/price varies
Comparison according to international standards on economic data between Austria and the states of European Community and OECD-member countries by using purchasing power parities (PPP).

05077 OESTERREICHISCHER ZAHLENSPIEGEL
Oesterreichisches Statistisches Zentralamt
Austria - German monthly/annual/ price varies
Presents the highlights of statistical surveys in condensed tabular and graphic form. A separate annual edition in German and English covers the most important indicators for many statistical fields in time series.

05078 OESTERREICHS VOLKSEINKOMMEN
Oesterreichisches Statistisches Zentralamt
Austria - German annual/280pp AS 290.00 ISSN No. 0085-4433
Data on national income.

05079 OIL AND GAS IN AUSTRIA
Fachverband der Erdölindustrie Österreichs
Austria - English every 4 years/30pp free
Broad description of the Austrian oil and gas industry with data.

05080 PERSONEN - UND HAUSHALTSEINKOMMEN VON UNSELBSTAENDIG BESCHAEFTIGTEN UND PENSIONISTEN; ERGEBNISSE DES MIKROZENSUS
Oesterreichisches Statistisches Zentralamt
Austria - Germany every 2 years/AS 180.00
Data on employees' income for individuals and households.

05081 PFERDE - UND RINDER-RASSENERHEBUNG IM JAHRE...
Oesterreichisches Statistisches Zentralamt
Austria - German every 8-10 years/ 45pp AS 160.00
Statistics on stocks of horses and cattle.

05082 PROCUREMENT BY THE INTERNATIONAL ORGANISATIONS IN VIENNA - SECTOR REPORT
Department of Trade and Industry Export Publications
Austria - English irregular/1991 11pp £30.00
Details on procurement procedures and budgets. Major purchasing requirements of the IAEA, UNFDAC, UNWRA and UNIDO are uncluded.

05083 QUARTERLY COUNTRY TRADE REPORTS: AUSTRIA

United Kingdom Iron and Steel Statistics Bureau

Austria - English quarterly/£200.00 P.A. for 4 countries ISSN No. 0952-584X

Cumulative trade statistics of imports and exports of steel products.

05084 RATIONAL APPROACH TO LABOUR AND INDUSTRY - ECONOMIC AND SOCIAL PARTNERSHIP IN AUSTRIA

Bundeskanzleramt

Austria - English/German/French/Spanish every 2 years/1974 c20pp free

05085 REISEGEWOHNHEITEN DER OESTERREICHER IM JAHRE... (HAUPTURLAUBE, KURZURLAUBE, DIENST - UND GESCHAEFTSREISEN)

Oesterreichisches Statistisches Zentralamt

Austria - German annual/AS 270.00

Details of journeys including seasons, duration, frequency and type of traffic employed. Both inland and external journeys are covered.

05086 SATELLITE TELEVISION RECEIVING EQUIPMENT - SECTOR REPORT

Department of Trade and Industry Export Publications

Austria - English irregular/1990 12pp £30.00

Details of Austrian television and broadcasting services, satellite ground stations, cable television and Austrian manufacturers of satellite receiving equipment.

05087 SCHRIFTENREIHE DER BUNDESANSTALT FÜR AGRARWIRTSCHAFT

Bundeskammer der Gewerblichen Wirtschaft

Austria/Germany/Central & Western Europe - German/English summaries 2-3 times a year

Reviews the situation for agriculture and agricultural industries.

05088 SCHRIFTTUM DER AGRARWIRTSCHAFT

Bundesanstalt Für Agrarwirtschaft

Austria/German/Central & Western Europe - German bimonthly/1961 90pp AS 710.00 P.A. ISSN No. 0036-6986

Reviews agricultural economics and regional agriculture.

05089 SECURITY EQUIPMENT - SECTOR REPORT

Department of Trade and Industry Export Publications

Austria - English irregular/1992 17pp £30.00

05090 SOZIALSTATISTISCHE AUSWERTUNGEN AUS DER KONSUMERHEBUNG

Oesterreichisches Statistisches Zentralamt

Austria - German every 10 years/AS 100.00

Socio-economic data on household structure and expenditure.

05091 SOZIALSTATISTISCHE DATEN

Oesterreichisches Statistisches Zentralamt

Austria - German every 3 years/AS 360.00

Regular publication providing an overview of all available social and demographic statistics to describe living conditions.

05092 STATISTIK DER AKTIEN-GESELLSCHAFTEN IN OESTERREICH IM JAHRE...

Oesterreichisches Statistisches Zentralamt

Austria - German annual/135pp AS 220.00 ISSN No.0081-5233

Statistics on Austrian joint stock companies.

05093 STATISTIK DER KAMMER UND FACHGRUPPEN-MITGLIEDER

Bundeskammer der Gewerblichen Wirtschaft

Abteilung für Statistik/Austria - German annual/1955 70pp

Statistics from the Austrian chambers of commerce.

05094 STATISTISCHE NACHRICHTEN

Oesterreichisches Statistisches Zentralamt

Austria - German monthly/AS 1250.00 P.A.

Monthly statistical newsletter which contains statistical data on all aspects of the population and economy, a textual commentary and background information.

05095 STATISTISCHES HANDBUCH FUER DIE REPUBLIK OESTERREICH

Oesterreichisches Statistisches Zentralamt

Austria/International - German annual/AS 570.00 ISSN No. 0081-5314

Reports on population, economy, tourism, finance, foreign trade, transport, employment, education, social welfare and culture. There are also comparative international statistics.

05096 STATISTISCHES MONATSHEFT

Oesterreichische Nationalbank

Austria - German monthly/AS 1,000.00 P.A.

Monthly statistical bulletin from the national bank.

05097 STRASSENVERKEHRSSI-CHERHEIT IM JAHRE...

Oesterreichisches Statistisches Zentralamt

Austria - German annual/231pp AS 300.00

Details of road traffic including numbers of accidents and damage caused to both people and property. Also gives details of driving licences issued.

05098 SURVEY OF THE AUSTRIAN ECONOMY

Wirtschaftsstudio Des Österreich - Ischen Gesellschafts - und Wirtschaftsmuseums

Austria - English/German annual/70pp AS 150.00

Austrian economy and its international position in data, diagrams and tables. Covers population, labour market, national accounts, GDP, GNP, economic growth, salaries and wages, prices, standard of living, foreign trade, tourism, communications, taxation and public budgets.

05099 TOYS - SECTOR REPORT

Department of Trade and Industry Export Publications

Austria - English irregular/1991 13pp £30.00

Austrian toy market, including local manufacturers, importer/distributors and retailers. Details of prospects for British manufacturers and promotions are shown. Also included are addresses of agents and import statistics.

05100 UMSATZSTEUER-STATISTIK

Oesterreichisches Statistisches Zentralamt

Austria - German annual/AS 280.00

Statistics on value added tax.

05101 UMWELT IN OESTERREICH, DATEN UND TRENDS

Oesterreichisches Statistisches Zentralamt

Austria - German every 2-3 years/AS 150.00

Developments about the environmental media such as soil, air, water, flora and fauna and main figures on energy.

05102 UMWELTBEDINGUNGEN DES WOHNENS; ERGEBNISSE DES MIKROZENSUS
Oesterreichisches Statistisches Zentralamt

Austria - German every 3 years/ 93pp AS 110.00

Data collected during the partial census on damage to houses by external factors and including the characteristics of dwelling places, such as facilities in general.

05103 VERMOEGENSTEUER-STATISTIK
Oesterreichisches Statistisches Zentralamt

Austria - German every 3 years/AS 380.00

Statistics on property tax.

05104 VIENNA STOCK EXCHANGE YEARBOOK
Wiener Borsekammer

Austria - English/German annual/ price varies

05105 VOLKSZAEHLUNG
Oesterreichisches Statistisches Zentralamt

Austria - German every 10 years/ AS 100.00

Results of the 1991 population census, classified by communities. Main figures of phase 1: Burgenland, Kaernten, Niederoesterreich, Oberoesterreich, Salzburg, Steiermark, Tirol, Vorarlberg, Wien, Oesterreich. Main figures for all provinces of phase 2 will be available in 1993.

05106 WIRTSCHAFTSSTATISTIK DER ELEKTRIZITAETS-VERSORGUNGS - UNTERNEHMEN
Oesterreichisches Statistisches Zentralamt

Austria - German annual/AS 80.00

Information on structure and performance of the electricity industry.

05107 WOHNUNGSDATEN
Oesterreichisches Statistisches Zentralamt

Austria - German annual/256pp

This publication collates two

collections of statistics - house building statistics and the results of the partial census giving details of types of house and their facilities. Two volume work giving details for the whole country and also for the larger districts.

05108 ZIVILLUFTFAHRT IN OESTERREICH
Oesterreichisches Statistisches Zentralamt

Austria - English/German annual/ 147pp AS 170.00

Shares of both Austrian airlines and foreign airlines.

05109 ÖSTERREICHS INDUSTRIE IN ZAHLEN
Bundeskammer der Gewerblichen Wirtschaft

Austria - German/irregular 1980 24pp free

Brief description of Austrian industry.

05110 ÖSTERREICHS WIRTSCHAFT IN ZAHLEN
Creditanstalt - Bankverein

Austria - English/German/French annual/free

The Austrian economy in figures.

BALTIC STATES

Country profile(s);	Baltic States – Estonia, Latvia, Lithuania
	The Baltic States are bounded in the north by the Baltic Sea, in the west by Poland, and to the east and south by the states of the C.I.S.
Official name:	Estonia
Area:	45,100 sq.km.
Population:	1.58m (1990)
Capital:	Tallin
Language(s):	Estonian, Russian
Currency:	Soviet Rouble
National bank:	Bank of Estonia
Government and Political structure:	Separated from the former USSR in 1991. Parliamentary democracy with 80 members elected to the Supreme Soviet.
Official name:	Latvia
Area:	63,700 sq.km.
Population:	2.68m (1990)
Capital:	Riga
Language(s):	Latvian, Russian
Currency:	Soviet Rouble
Government and Political structure:	Separated from the former USSR in 1991. Parliamentary democracy with 201 members elected to the Supreme Soviet.
Official name:	Lithuania
Area:	65,200 sq.km.
Population:	3.7m (1990)
Capital:	Vilnius
Language(s):	Lithuanian, Russian
Currency:	Soviet Rouble
Government and Political structure:	Separated from the former USSR in 1991. Parliamentary democracy with 120 members elected to the Supreme Soviet.

Baltic
States

(c) R.Canning.

06001 AN ECONOMIC OUTLINE
OF THE REPUBLIC OF
LATVIA
**Latvian Chamber of
Commerce and Industry**

Henriks Silenieks/Latvia - English
twice a year/December 1991 48pp
US$5.00

Economic and statistical
information.

06003 THE BALTIC STATES:
WHAT PRICE FREEDOM
**The Economist Intelligence
Unit**

Estonia/Latvia/Lithuania - English
irregular/1990 54pp £95.00

Analyses the implications for the
Baltic States of succession from the
USSR.

06004 THE C.I.S. MARKET ATLAS
Business International Ltd

C.I.S./Baltic States - English/1993
303pp US$595.00

Demographics, micro-economics,
statistics, social and political
indicators for all the former Soviet
republics.

06005 COUNTRY RISK SERVICE:
BALTIC REPUBLICS
Business International Ltd

Baltic Republics - English quarterly/
minimum subscription £1,855.00
P.A. for 7 countries

Contains credit risk ratings, risk
appraisals, cross-country
databases, data on diskette, and 2
year projections.

06006 ESTONIA, LATVIA,
LITHUANIA
**State Committee for
Statistics of The Republic
of Latvia**

Estonia/Latvia/Lithuania - English/
1991 145pp US$13.00

Statistical data book covering the
years 1980 to 1990.

06007 LATVIA - A GUIDEBOOK
FOR A BUSINESSMAN
**State Committee for
Statistics of The Republic
of Latvia**

Latvia - English annual/US$3.00

Contains general information on
Latvia and useful organizations and
institutions. Latest edition is for
1992.

06008 LATVIA IN FIGURES
**State Committee for
Statistics of The Republic
of Latvia**

Latvia - English annual/96pp
US$4.00

Brief statistical profile of Latvia
giving the main socio-economic
figures.

06009 LATVIA TODAY
**State Committee for
Statistics of The Republic
of Latvia**

Latvia - English irregular/1992
US$2.00

Booklet providing information on
history, climate, employment,
population and economic activity.

06010 NATIONAL ECONOMY OF
LATVIA
**State Committee for
Statistics of The Republic
of Latvia**

Latvia - Latvian/Russian with
contents listing in English irregular/
340pp c.340pp US$25.00

Economic profile giving the main
indicators in statistical format.

06011 PUBLIC FINANCES OF THE
REPUBLIC OF LATVIA
**State Committee for
Statistics of The Republic
of Latvia**

Latvia - English/Latvian annual/
46pp US$2.00

A compilation of central and local
government financial data. Latest
edition is for 1991.

06012 STATISTICAL YEARBOOK
OF LATVIA
**State Committee for
Statistics of The Republic
of Latvia**

Latvia - English/Latvian irregular/
US$13.00

Broad socio-economic description
of Latvia with essential economic
indicators included.

BELGIUM

Country profile:	Belgium
Official name:	Kingdom of Belgium
Area:	30,518 sq.km. Bordered in the north by the Netherlands and the North Sea, in the south by France, and in the east by Germany and Luxembourg.
Population:	9.98m (1991)
Capital:	Brussels
Language(s):	Flemish (Dutch), French, German
Currency:	Franc 100 centimes = 1 Franc
National bank:	National Bank of Belgium
Government and Political structure:	Constitutional monarchy with a Senate and Chamber of Representatives elected by proportional representation every four years. Regional government is based in Brussels, Flanders and Wallonia.

Belgium

07001 ANNALES DES MINES DE BELGIQUE
Ministère des Affaires Economiques - Administration des Mines
Belgium - French/Dutch 2 issues per year/BF1,250 ISSN No.0003-4290
Statistics relevant to the mining industry.

07002 ANNUAIRE DE STATISTIQUES REGIONALES
Institut National de Statistique: Ministère des Affaires Economiques
Belgium - French/Dutch annual BF625 ISSN No. 0770-0369
National statistics broken down on a regional basis.

07003 ANNUAIRE STATISTIQUE
Fédération de l'Industrie du Gaz (FIGAZ)
Belgium - French/Dutch annual/ 50pp
Statistics from the trade association dealing with the gas industry.

07004 ANNUAIRE STATISTIQUE DE LA BELGIQUE
Institut National de Statistique: Ministère des Affaires Economiques
Belgium - French/Dutch annual/ BF1,560 ISSN No. 0770-0415
Demographic, economic, social, cultural and political statistics.

07005 ANNUAIRE STATISTIQUE DE POCHE
Institut National de Statistique: Ministère des Affaires Economiques
Belgium - French/Dutch annual BF185 ISSN No. 0067-5431
Statistical pocket book giving abridged figures from the 'Annuaire Statistique de la Belgique'.

07006 ANTWERP FACETS
Diamond High Council
International - English/Dutch quarterly/1989 c140pp BF800 P.A.
Articles on the Antwerp diamond trade and on the diamond industry in general.

07007 APERCU ÉCONOMIQUE TRIMESTRIEL
Ministère des Affaires Economiques
Belgium - French/Dutch quarterly/ BF850 P.A. ISSN No. 0773-9664
Economic analysis covering Belgium and its seven most important trading partners.

07008 THE AUTOMOBILE IN THE WORLD
Fabrimétal
Belgium - English annual/BF1,500

07009 BELGIUM: ECONOMIC AND COMMERCIAL INFORMATION
Belgian Foreign Trade Office
Belgium/Luxembourg - English/ French/Dutch/German/Spanish quarterly/68pp
Profile of Belgian economy and business.

07010 BILANS ENERGÉTIQUES
Ministère des Affaires Economiques: Administration de l'Energie
Belgium - French/Dutch annual

07011 BULLETIN DE LA BANQUE NATIONALE DE BELGIQUE
Banque Nationale de Belgique
Belgium - French monthly/BF750 ISSN No. 0005-5611
Financial, social and industrial statistics

07012 BULLETIN DE STATISTIQUE
Institut National de Statistique: Ministère des Affaires Economiques
Belgium - French/monthly/BF1,760 P.A. ISSN No. 0045-1703
Regular statistical bulletin with economic coverage.

07013 BULLETIN DES ASSURANCES
Union Professionnelle des Entreprises d'Assurances (UPEA)
Belgium - French/Dutch 4 issues per annum
Bulletin giving information on legal and other technical aspects of insurance.

07014 BULLETIN FINANCIER
Bank Bruxelles Lambert
Jean - Baptiste de Dorlodot Belgium + OECD countries/ French/Dutch monthly/10-15pp free ISSN No. 0771-6273
Economics.

07015 BULLETIN - KAMER VOOR HANDEL EN NIJVERHEID VAN HET: ARRONDISEMENT LEUVEN
Kamer voor Handel en Nijverheid van Het Arrondisement Leuven
Belgium - Dutch monthly/10pp BF150 P.A.

Monthly bulletin of the Louvain Chamber of Commerce, covering local economic projects, small business projects and promotion of regional industry.

07016 BULLETIN MENSUEL D'INFORMATION
Fédération de l'Industrie du Verre
Belgium - French monthly/25pp free
Monthly bulletin about the Belgian glass industry.

07017 BULLETIN MENSUEL DE L'ÉNERGIE ÉLECTRIQUE
Ministère des Affaires Economiques: Administration de l'Energie
Belgium - French/Dutch monthly
Electricity industry.

07018 BÉTON
Fédération de l'Industrie du Béton
Belgium - French/Dutch bi-monthly BF1,200 P.A.
General review of the precast concrete industry.

07019 CAUSES DE DECÈS
Institut National de Statistique
Belgium - Dutch/French annual/ BF425
Figures for causes of death in Belgium.

07020 CHRONIQUE DE CONJONCTURE BELGE
Bank Bruxelles Lambert
J. P. Hologne Belgium - French/ Dutch monthly/2pp free
Economic situation in Belgium.

07021 COMMUNIQUÉ HEBDOMADAIRE
Institut National de Statistique: Ministère des Affaires Economiques
Belgium - French/weekly/BF1,750 P.A. ISSN No. 0771-0364
Weekly economic journal.

07022 CONSTRUCTION
Confédération Nationale de la Construction
Belgium - French/Dutch weekly/ 36pp BF5,000 P.A.
Information on the building industry.

07023 COUNTRY BOOKS: BELGIUM - LUXEMBOURG
United Kingdom Iron and Steel Statistics Bureau
Belgium/Luxembourg - English annual/£75.00 ISSN No. 0952-5858

Annual summary tables showing production, material consumed, apparent consumption, imports and exports of 130 products by quality and market.

07024 COUNTRY FORECASTS: BELGIUM
Business International Ltd

Belgium - English quarterly/£360.00 P.A.

Contains an executive summary, sections on political and economic outlooks and the business environment, a fact sheet, and key economic indicators. Includes a quarterly Global Outlook.

07025 COUNTRY PROFILE BELGIUM
Department of Trade and Industry Export Publications

Belgium - English irregular/1992 78pp £10.00

This publication provides a general description of the country in terms of its geography, economy and population, and gives information about trading in the country with details such as export conditions and investment covered.

07026 COUNTRY PROFILE: BELGIUM, LUXEMBOURG
Business International Ltd

Belgium/Luxembourg - English annual/35pp £50.00 ISSN No. 0269-4352

Annual survey of political and economic background.

07027 COUNTRY REPORT: BELGIUM
Barclays Bank Plc

Belgium - English irregular/free

Economic indicators and brief economic profile.

07028 COUNTRY REPORT: BELGIUM - LUXEMBOURG
Business International Ltd

Belgium/Luxembourg - English quarterly/25-40pp £160.00 ISSN No. 0269-4158

Contains an executive summary, sections on political and economic structure, outlook for the next 12-18 months, statistical appendices, and a review of key political developments. Includes an annual Country Profile.

07029 COURRIER HEBDOMADAIRE
Centre de Recherche et d'Information Socio-Politique - CRISP

Belgium - French weekly/c40pp BF9,500 P.A.

07030 DISTRIBUTION D' AUJHOURD'HUI
Comité Belge de la Distribution

Belgium - French monthly/80pp BF318 (per issue) ISSN No. 0012-3935

Food and drinks industry.

07031 DOING BUSINESS IN BELGIUM
Ernst & Young International

Belgium - English irregular/1991 110-130pp free to clients

A business guide which includes an executive summary, the business environment, foreign investment, company structure, labour, taxation, financial reporting and auditing, and general country data.

07032 DOING BUSINESS IN BELGIUM
Price Waterhouse

Belgium - English irregular/1990 330-340pp free to clients

A business guide which includes sections on the investment climate, doing business, accounting and auditing, taxation and general country data. Updated by supplements.

07033 ENTREPRENDRE
Chambre de Commerce et d'Industrie de Bruxelles

Belgium - French monthly (free to members)

Official journal of the Chamber of Commerce with current news on all aspects of business and finance.

07034 FABRIMÉTAL MAGAZINE
Fabrimétal

Belgium - French/Dutch monthly BF1,100 P.A. ISSN No. 0377-9084

Monthly profile giving data on the Belgium metals industry.

07035 FEBELGRA - INFORMATIONS
Febelgra-Fédération Belge des Industries Graphiques A.S.B.L

Jos Rossie/Belgium - French/Dutch bi-monthly/8pp

Official bulletin of the Federation - for members only.

07036 FINANCING FOREIGN OPERATIONS: BELGIUM
Business International Ltd

Belgium - English annual/£695.00 for complete European series

Business overview, currency outlook, exchange regulations, monetary system, short/medium/long-term and equity financing techniques, capital incentives, cash management, investment trade finance and insurance.

07037 FPE
La Fédération Professionelle des Producteurs et Distributeurs d'Électricité de Belgique (FPE)

Belgium - French annual

Electricity Industry.

07038 GLOBAL FORECASTING SERVICE: BELGIUM
Business International Ltd

Belgium - English quarterly £350.00 P.A.

Economic forecasts.

07039 HINTS TO EXPORTERS VISITING BELGIUM
Department of Trade and Industry Export Publications

Belgium - English irregular/1991 87pp £5.00

This publication provides detailed practical information that is useful for those planning a business visit. Topics include currency information, economic factors, import and exchange control regulations, methods of doing business and general information.

07040 IN DEPTH COUNTRY APPRAISALS ACROSS THE EC - CONSTRUCTION AND BUILDING SERVICES - BELGIUM
Building Services Research and Information Association

Belgium - English 1991/£125.00

Part of an 11 volume survey of the construction industry in EC countries.

07041 INDUSTRIE DU VERRE - EVOLUTIE 1980-1991
Fédération de l'Industrie du Verre

Belgium - French/Dutch/1992 20pp

Statistical summary giving 10 years figures; covers production, trade, numbers employed etc.

07042 INVESTING, LICENSING AND TRADING CONDITIONS ABROAD: BELGIUM
Business International Ltd

Belgium - English annual/£825.00 for complete European series

Loose-leaf format covering the state role in industry, competition rules, price controls, remittability of funds, taxation, incentives and capital sources, labour and foreign trade.

07043 THE JOURNAL: BELGO LUXEMBOURG CHAMBER OF COMMERCE

Belgo - Luxembourg Chamber of Commerce In Great Britain

D. Partner-Macremans Belgium - English bi-monthly/16pp

One main feature plus smaller articles.

07044 L'ASSURANCE EN BELGIQUE

Union Professionnelle des Entreprises d'Assurances (UPEA)

Belgium - French/Dutch annual

Annual report of the official federation of insurance companies.

07045 L'ECONOMIE BELGE EN 19..

Ministère des Affaires Economiques: Direction Générale des Etudes et de la Documentation

Belgium - French annual/BF500 ISSN No. 0771-5641

Annual survey of the Belgian economy.

07046 LES COMPTES NATIONAUX DE LA BELGIQUE

Institut National de Statistique

Belgium - French annual/80pp

A detailed breakdown of the national accounts, giving 10 years' figures.

07047 LES GROUPES D'ENTERPRISES

Centre de Recherche et d'Information Socoi-Politique

Belgium - French annual 1990 400pp BF900

07048 LES TRANSPORTS EN BELGIQUE

Ministère des Communications et de l'Infrastructure

Belgium - French/Dutch annual/ 200pp

Detailed statistics on road, rail, air and waterborne transport.

07049 LETTRE DE CONJONCTURE

Ministère des Affaires Economiques: Direction Générale des Etudes et de la Documentation

Belgium - French monthly/BF400 P.A. ISSN No. 0772-0831

Covers production, employment and public finance.

07050 MONTHLY REVIEW AND MONTHLY NEWSLETTER - KAMER VOOR HANDEL EN NIJVERHEID VAN HET GEWEST GENT

Kamer voor Handel en Nijverheid van Het

Belgium - Dutch monthly

Review of the Ghent Chamber of Commerce and newsletter with trade purposes.

07051 MOUVEMENT DE LA POPULATION DES COMMUNES BELGES

Institut National de Statistique

Belgium - Dutch/French annual/ BF360 ISSN No. 0067-5458

Population figures.

07052 NEWSLETTER - FIGAZ

Fédération de l'industrie du Gaz (FIGAZ)

Belgium - French/Dutch monthly/ 4pp

Covers the gas industry.

07053 NOMBRE DE DÉTENTEURS D'AUTORADIOS ET DE TÉLÉVISEURS (PAR COMMUNE)

Institut National de Statistique

Belgium - French/Dutch annual/ BF135

Statistics on TV and radio ownership.

07054 NOTES SUR LA SITUATION ÉCONOMIQUE

Conseil Central de l'Economie

H. Selderslashs/Belgium - French/ Dutch monthly (except August)/ 27pp free

Economic situation in Belgium.

07055 OECD ECONOMIC SURVEYS: BELGIUM, LUXEMBOURG

Organisation for Economic Co-operation and Development (OECD)

Belgium/Luxembourg - English irregular/£102.00

Surveys on economic developments.

07056 PARC DES VÉHICULES À MOTEUR

Institut National de Statistique

Belgium - French/Dutch annual/ BF210

Statistics on car parking and numbers of vehicles.

07057 PARC DES VÉHICULES UTILITAIRES

Ministère des Communications et de l'Infrastructure

Belgium - French/Dutch annual/ 24pp free

Statistics on fleets of commercial vehicles.

07058 PERSPECTIVES DE POPULATION - LE ROYAUME

Institut National de Statistique

Belgium - French/Dutch irregular/ BF350

Population perspectives to 2025.

07059 POPULATION DES COMMUNES FUSSIONÉES

Institut National de Statistique

Belgium - French/Dutch irregular/ BF70

Population figures for communes.

07060 QUARTERLY COUNTRY TRADE REPORTS: BELGIUM - LUXEMBOURG

United Kingdom Iron and Steel Statistics Bureau

Belgium/Luxembourg - English quarterly/£200.00 P.A. for 4 countries ISSN No. 0952-5858

Cumulative trade statistics of imports and exports of steel products.

07061 RAPPORT ANNUEL

Fédération de l'Industrie Cimentière

M. J. P. Latteur/Belgium - French/ Dutch annual/1976 47pp free

Annual report of the concrete industry trade association.

07062 RAPPORT ANNUEL

Fédération des Entreprises de Métaux Non Ferreux

Belgium - French/Dutch annual/free

Annual report covering the non-ferrous metals industry.

07063 RAPPORT ANNUEL ET RAPPORT D'ACTIVITÉ

Fabrimétal

Belgium - French/Dutch annual/free

07064 RAPPORT ANNUEL - FIGAZ

Fédération de l'Industrie du Gaz (FIGAZ)

Belgium - French/Dutch annual/ 24pp

Annual report of trade association for the gas industry.

07065 RAPPORT ANNUEL
Fédération des Industries Transformatrices de Papier et Carton (FETRA)
Belgium - French/Dutch annual/ 50pp free
Annual report of the Belgian Federation of Paper and Cardboard Processors.

07066 RAPPORT - BANQUE NATIONALE DE BELGIQUE
Banque Nationale de Belgique
Belgium - English/French/Dutch annual/ISSN No. 0067-3978

07067 RECENSEMENT AGRICOLE ET HORTICOLE
Institut National de Statistique
Belgium - French/Dutch annual/ BF410
Agricultural and horticultural census.

07068 RECENSEMENT DE LA CIRCULATION ROUTIÈRE
Institut National de Statistique
Belgium - French/Dutch every 5 years/BF200
Census of road traffic.

07069 RECENSEMENT DE LA POPULATION ET DES LOGEMENTS AU 1 MARS 1981
Institut National de Statistique
Belgium - French every 10 years/ BF200-400
In various volumes, some giving general results, some on specific subjects such as housing, nationality and educational level.

07070 RECUEIL STATISTIQUE
Fédération des Entreprises de Métaux Non Ferreux
Belgium - French/Dutch annual
Statistics on production, import and export for the non-ferrous metals industry.

07071 RECUEIL STATISTIQUE
Fabrimétal
Belgium - French/Dutch annual/ BF2,000
Statistical coverage of the Belgium metals industry.

07072 RECUEIL STATISTIQUE - FETRA
Fédération des Industries Transformatrices de Papier et Carton (FETRA)
Belgium - French/Dutch annual/ 180pp free

Import/export figures and production statistics for the Belgian paper industry.

07073 REVUE DU TRAVAIL
Ministère de l'Emploi et du Travail
Christian Deneue/Belgium - French/ Dutch quarterly/1896 106pp free ISSN No. 0035-2705
Labour market and industrial relations.

07074 SÉRIE "ETUDES ÉCONOMIQUES"
Fabrimétal
Belgium - French irregular/free
Occasional papers on Belgium economics.

07075 STATISTIQUE ANNUELLE DE LA PRODUCTION
Institut National de Statistique
Belgium - French/Dutch annual/ BF675
Production statistics for industry.

07076 STATISTIQUE ANNUELLE DÉFINITIVE
Ministère des Affaires Economiques - Administration des Mines
Belgium - French/Dutch annual/free
Annual statistics for the consumption and production of coal, coke, iron and steel, and quarried minerals.

07077 STATISTIQUES
Ministère des Affaires Economiques - Administration des Mines
Belgium - French/Dutch biannual/ free
Statistics on coal mining.

07078 STATISTIQUES AGRICOLES
Institut National de Statistique: Ministère des Affaires Economiques
Belgium - French monthly/BF1,200 P.A. ISSN No. 0067-5466
Monthly bulletin covering the agricultural sector.

07079 STATISTIQUES ANNUELLE DE TRAFFIC INTERNATIONAL DES PORTS
Institut National de Statistique: Ministère des Affaires Economiques
Belgium - French/Dutch annual/ BF625
Statistics on international traffic in ports.

07080 STATISTIQUES DE LA CONSTRUCTION ET DU LOGEMENT
Institut National de Statistique
Belgium - French/Dutch irregular BF300
Statistics on building and housing.

07081 STATISTIQUES DU COMMERCE EXTÉRIEUR DE L'UNION ECONOMIQUE BELGO - LUXEMBOURGEOISE
Institut National de Statistique: Ministère des Affaires Economiques
Belgium/Luxembourg - French monthly/BF5,160 P.A. (includes annual) ISSN No. 0772-6694

07082 STATISTIQUES DU COMMERCE EXTÉRIEUR DE L'UNION ECONOMIQUE BELGO - LUXEMBOURGEOISE (ANNUAL)
Institut National de Statistique: Ministère des Affaires Economiques
Belgium/Luxembourg - French annual/BF825
Foreign trade.

07083 STATISTIQUES DU COMMERCE INTÉRIEUR ET DES TRANSPORTS
Institut National de Statistique: Ministère des Affaires Economiques
Belgium - French monthly/BF2,200 P.A. ISSN No. 0773-4255
Statistical coverage of internal trade, transport, distribution and tourism.

07084 STATISTIQUES DÉMOGRAPHIQUES
Institut National de Statistique: Ministère des Affaires Economiques
Belgium - French 4 issues per year/ BF1,475 P.A. ISSN No. 0067-5490
Demographic statistics.

07085 STATISTIQUES FINANCIÈRES
Institut National de Statistique: Ministère des Affaires Economiques
Belgium - French 3 issues per year/ BF730 P.A.
Financial statistics for Belgium.

07086 STATISTIQUES - FÉDÉRATION DE L'INDUSTRIE DU VERRE
Fédération de l'Industrie du Verre
Belgium - French/Dutch annual/ 15pp free

Official statistics about the glass industry in Belgium.

07087 STATISTIQUES INDUSTRIELLES
Institut National de Statistique: Ministère des Affaires Economiques

Belgium - French monthly/BF1,920 P.A. ISSN No. 0772-7704

Statistical coverage of industrial production and industry.

07088 STATISTIQUES SOCIALES
Institut National de Statistique: Ministère des Affaires Economiques

Belgium - French quarterly BF575 P.A. ISSN No. 0067-5563

Covers Belgian society and social conditions.

07089 THE TRADE DIRECTORY
Belgo - Luxembourg Chamber of Commerce In Great Britain

Belgium/Luxembourg - English irregular/68pp

Useful addresses, commercial and economic news and a classified list of companies.

07090 TRADE ENQUIRIES
Belgo - Luxembourg Chamber of Commerce In Great Britain

Belgium/Luxembourg/UK - English fortnightly/6pp free

Contains trade enquiries and business opportunities between the 3 countries.

07091 TRENDS: FINANCEEL EKONOMISCH MAGAZINE
N. V. Trends

Belgium - Dutch weekly/169pp BF6,190 P.A.

Financial economics are covered.

07092 VADEMECUM DES ENTREPRISES (FEUILLES MOBILES)
Ministère des Affaires Economiques: Direction Générale des Etudes et de la Documentation

Belgium - French/Dutch loose-leaf kept up to date with 2 annual supplements/BF500 (main work) BF 150 (annual subscription)

Regulations relevant to industry and commercial enterprises including details of assistance to industrial projects.

07093 VADEMECUM: LE FINANCEMENT DU COMMERCE EXTÉRIEUR
Fabrimétal

Belgium - French/Dutch every 3 years/BF7,420

Pocket book format giving data on international trade and balance of payments.

07094 VÉHICULES À MOTEUR NEUFS MIS EN CIRCULATION
Institut National de Statistique

Belgium - French/Dutch annual/ BF210 ISSN No. 0773-3070

Statistics on numbers of new cars in circulation.

BULGARIA

Country profile:	Bulgaria
Official name:	Republic of Bulgaria
Area:	110,990 sq.km.
	Bordered by Romania in the north, the Black Sea in the east, Turkey and Greece in the south and Yugoslavia in the west.
Population:	8.98m (1991)
Capital:	Sofia
Language(s):	Bulgarian
Currency:	Lev 100 stotinki = 1 Lev.
National bank:	National Bank of Bulgaria.
Government and Political structure:	Pre 1991 politics were dominated by the Communist party. Under the 1991 constitution there is a directly elected President and multi-party democracy. The National Assembly has 240 seats.

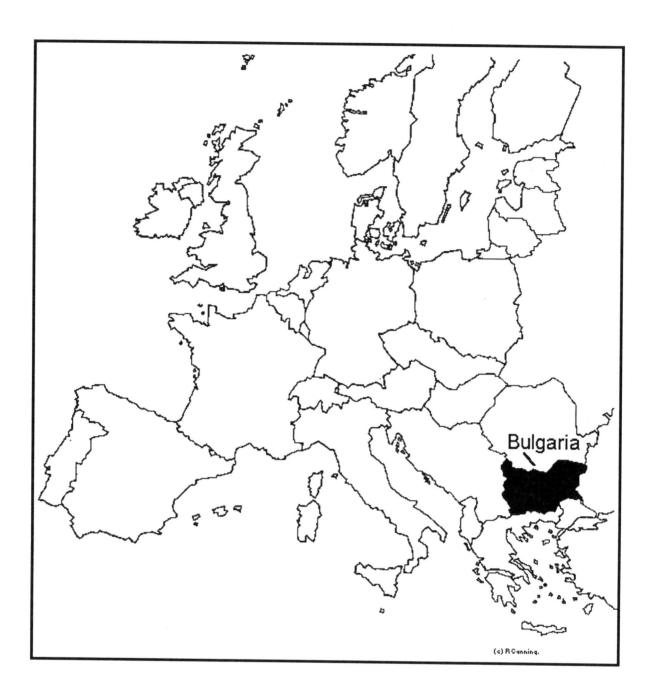

Bulgaria

08001 168 HOUNS BBN
168 Houns Ltd

Uliana Petrova/Bulgaria - English weekly/16pp US$364.00 P.A.

Source for business and trade news.

08002 AREA, YIELDS AND CROPS PRODUCTION
National Statistical Institute

Bulgaria - Bulgarian annual/65pp US$6.00

08003 BULGARIA
National Statistical Institute

Bulgaria - English/Bulgarian/French annual/50pp US$3.00

Statistical information on business and social aspects of Bulgaria.

08004 BULGARIA'S AGRICULTURE: SITUATION, TRENDS AND PROSPECTS
Office for Official Publications of the European Communities

Bulgaria - English/1991 226pp ECU24.00

08005 BULGARIAN FOREIGN TRADE
Bulgarian Chamber of Commerce and Industry

Bulgaria - English/Russian/French/ German 6 per year/50pp ISSN No. 0264-8892

08006 COMPARISON OF ENTERPRISES BY BASIC FINANCIAL ECONOMIC INDICATORS, CHARACTERIZING EFFICIENCY
National Statistical Institute

Bulgaria - Bulgarian annual/70pp US$6.00

08007 COUNTRY FORECASTS: BULGARIA
Business International Ltd

Bulgaria - English quarterly/ £360.00 P.A.

Contains an executive summary, sections on political and economic outlooks and the business environment, a fact sheet, and key economic indicators. Includes a quarterly Global Outlook.

08008 COUNTRY PROFILE - BULGARIA
Department of Trade and Industry Export Publications

Bulgaria - English annual/10pp

Economic and business overview of Bulgaria.

08009 COUNTRY PROFILE: BULGARIA 1991
Office for Official Publications of the European Communities

Bulgaria - English/French 1992 115pp ECU11.00

08010 COUNTRY REPORT - ROMANIA, BULGARIA, ALBANIA
The Economist Intelligence Unit

Bulgaria/Romania/Albania - English quarterly plus annual country profile/50pp £150.00 P.A.

08011 COUNTRY RISK SERVICE: BULGARIA
Business International Ltd

Bulgaria - English quarterly/ minimum subscription £1,855.00 P.A. for 7 countries

Contains credit risk ratings, risk appraisals, cross-country databases, data on diskette, and 2 year projections.

08012 CULTURE
National Statistical Institute

Bulgaria - Bulgarian/English preface annual/150pp US$10.00

Statistical data on cinema, museums, performing arts and broadcast media.

08013 CURRENT ECONOMIC BUSINESS
National Statistical Institute

Bulgaria - English/Bulgarian monthly/40pp US$17.00 P.A.

Brief review of the economy by sector.

08014 DOING BUSINESS IN EASTERN EUROPE: BULGARIA
Kogan Page

Bulgaria - English/1993 256pp £25.00 ISBN No. 0749406895

Part of the CBI initiative on Eastern Europe. Describes the business and economic environment.

08015 DOMESTIC TRADE IN BULGARIA
National Statistical Institute

Bulgaria - Bulgarian every 5 years/ 100pp US$5.00

Data on the domestic economy.

08016 ECONOMIC NEWS OF BULGARIA
Bulgarian Chamber of Commerce and Industry

Bulgaria - English/Russian/French/

German/Spanish monthly/ISSN No. 0205-1400

08017 ECONOMICS OF BULGARIA
National Statistical Institute

Bulgaria - Bulgarian annual/200pp US$17.00

Covers main economic indicators and primary economic factors in the Bulgarian economy.

08018 ENVIRONMENT
National Statistical Institute

Bulgaria - Bulgarian every six months/18pp US$10.00 P.A.

Statistical data on environmental protection.

08019 FOREIGN TRADE OF THE REPUBLIC OF BULGARIA
National Statistical Institute

Bulgaria - Bulgarian irregular/395pp US$30.00

Details international trade by commodity/type of trade.

08020 HEALTH SERVICES
National Statistical Institute

Bulgaria - Bulgarian annual/130pp US$20.00

Statistical data on health and health services.

08021 HINTS TO EXPORTERS VISITING BULGARIA
Department of Trade and Industry Export Publications

Bulgaria - English irregular/1990 60pp £5.00

This publication provides detailed practical information that is useful for those planning a business visit. Topics include currency information, economic factors, import and exchange control regulations, methods of doing business and general information.

08022 HOUSING
National Statistical Institute

Bulgaria - Bulgarian annual/67pp US$7.00

Reviews the living situation and housing conditions.

08023 POPULATION
National Statistical Institute

Bulgaria - Bulgarian annual/280pp US$25.00

Detailed demographic information.

08024 POPULATION
National Statistical Institute

Bulgaria - Bulgarian annual/40pp US$3.00

Demographic statistics with breakdowns by age, sex and birth rates.

08025 PUBLISHING AND PRESS
National Statistical Institute

Bulgaria - Bulgarian/contents in
Russian and French annual/87pp
US$9.00

Number of publications, circulation
etc.

**08026 REGIONS AND
MUNICIPALITIES IN THE
REGION OF BULGARIA**
National Statistical Institute

Bulgaria - Bulgarian annual/300pp
US$7.00

Statistical data on economic
conditions in towns and regions.

**08027 STANDARD OF LIFE OF
POPULATION IN THE
REPUBLIC OF BULGARIA**
National Statistical Institute

Bulgaria - Bulgarian annual/80pp
US$6.00

Looks at the standard of living for
the population using standard
criteria.

08028 STATISTICAL NEWS
National Statistical Institute

Bulgaria - Bulgarian/contents in
English and French quarterly/65pp
US$12 P.A.

**08029 STATISTICAL REFERENCE
BOOK OF THE REPUBLIC
OF BULGARIA**
National Statistical Institute

Bulgaria - Bulgarian annual/300pp
US$7

**08031 THE STATISTICAL
YEARBOOK OF THE
REPUBLIC OF BULGARIA**
National Statistical Institute

Bulgaria - Bulgarian annual/600pp
US$40.00

Gathers together the main statistics
from a range of series.

08032 TOURISM
National Statistical Institute

Bulgaria - English/Bulgarian annual/
127pp

Shows numbers, expenditure and
the contribution of the tourist
industry to the Bulgarian economy.

C.I.S.

Country profile:	Commonwealth of Independent States
Area:	22.4m sq.km.
	The twelve constituent republics are Armenia, Azerbaijan, Belorussia, Georgia, Kazakhstan, Kirghizia, Moldavia, Russia, Tajikistan, Turkmenistan, Ukraine, and Uzbekistan.
Population:	286.7m (1989)
Currency:	Rouble 100 kopeks = 1 Rouble
	Constituent republics have plans, to varying degrees, to establish national currencies.
National bank:	State Bank
	Constituent republics have plans, to varying degrees, to establish national financial institutions.
Government and Political structure:	At the point of writing these were in a state of flux. Common affairs are decided on a multilateral basis between the states. The Council of Heads of States is the formal mechanism by which decisions are taken.

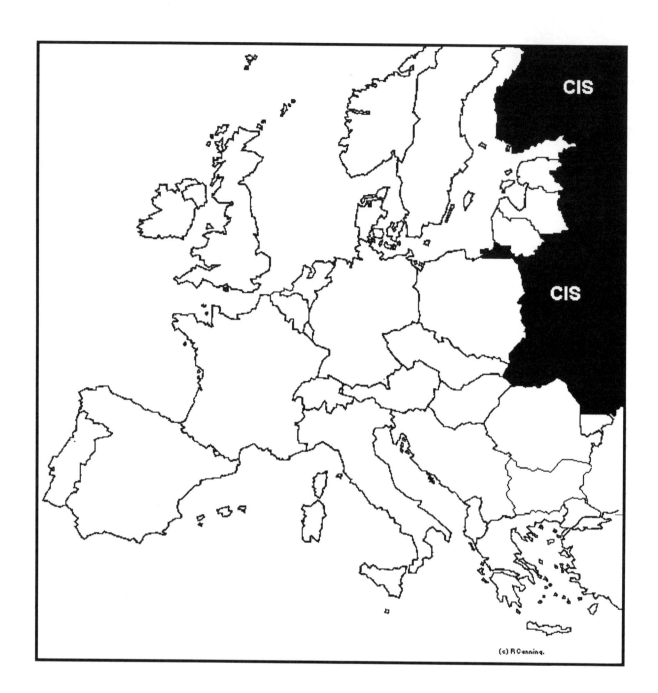

09001 ANNUAL REPORT
Moscow Narodny Bank Ltd

C.I.S. - annual/1920 20-30pp free

Annual report giving financial balances, also useful for economic background.

09002 BRITISH SOVIET BUSINESS
British Soviet Chamber of Commerce

David Winter/C.I.S./UK - English/ Russian quarterly/1991 40-60pp £72.00 P.A.

Journal covering business links between the UK and former U.S.S.R.

09003 BUSINESS MOSCOW
V/O Vneshtorgeklama

C.I.S./Russia - English/Russian every two years/1989 384pp

Directory of business organizations and foreign trade organizations in Moscow.

09004 BUSINESS WEEK: RUSSIAN LANGUAGE EDITION
Business Week International

C.I.S. - Russian monthly/1991 price varies

News selected from other "Business Week" journals which have relevance to the C.I.S.

09005 THE C.I.S. MARKET ATLAS
Business International Ltd

C.I.S./Baltic States - English/1993 303pp US$595.00

Demographics, micro-economics, statistics, social and political indicators for all the former Soviet republics.

09006 COMMONWEALTH OF INDEPENDENT STATES - FOUNDING DOCUMENTS
Ria-Novosti

C.I.S. - English translation irregular/ 1991 £25.00

Details of the creation of the C.I.S., useful for background to the political-economic structure.

09007 COMMONWEALTH OF INDEPENDENT STATES - GLOBAL FORECASTING SERVICE
The Economist Intelligence Unit

C.I.S. - English quarterly/£360.00 P.A.

Economic forecasts.

09008 COUNTRY PROFILE RUSSIA
Department of Trade and Industry Export Publications

Russia - English irregular/1992

34pp £10.00

This publication provides a general description of the country in terms of its geography, economy and population, and gives information about trading in the country with details like export conditions and investment covered.

09009 COUNTRY REPORT - COMMONWEALTH OF INDEPENDENT STATES
The Economist Intelligence Unit

C.I.S. - English quarterly with annual country profile/110pp £150.00 P.A.

This report is in the process of being sub-divided into reports on individual C.I.S. States. Check with publisher.

09010 COUNTRY REPORT: GEORGIA, ARMENIA, AZERBAIJAN, CENTRAL ASIAN REPUBLICS
Business International Ltd

Georgia/Armenia/Azerbaijan/ Central Asian Republics - English quarterly/25-40pp £160.00 P.A.

09011 COUNTRY REPORT: RUSSIA
Business International Ltd

Russia - English quarterly/25-40pp £160.00 P.A.

09012 COUNTRY REPORT: UKRAINE, BELARUS, MOLDOVA
Business International Ltd

Ukraine/Belarus/Moldova - English quarterly/25-40pp £160.00 P.A.

09013 COUNTRY RISK SERVICE: C.I.S.
Business International Ltd

C.I.S. - English quarterly/minimum subcription £1,855.00 P.A. for 7 countries

Contains credit risk ratings, risk appraisals, cross-country databases, data on diskette, and 2 year projections.

09014 COUNTRY RISK SERVICE: RUSSIA
Business International Ltd

Russia - English quarterly/minimum subscription £1,855.00 P.A. for 7 countries

09015 COUNTRY RISK SERVICE: UKRAINE
Business International Ltd

C.I.S. - English quarterly/minimum subscription £1,855.00 P.A. for 7 countries

Contains credit risk ratings, risk appraisals, cross-country databases, data on diskette, and 2 year projections.

09016 DOING BUSINESS IN EASTERN EUROPE: RUSSIA
Kogan Page

C.I.S. - English/1993 256pp £25.00 ISBN No. 074940695X

Part of the CBI initiative on Eastern Europe. Describes the business and economic environment.

09017 DOING BUSINESS IN THE USSR
Price Waterhouse

C.I.S. - English irregular/1991 330-340pp free to clients

A business guide which includes sections on the investment climate, doing business, accounting and auditing, taxation and general country data. Updated by supplements.

09018 DOING BUSINESS WITH EASTERN EUROPE: C.I.S.
Business International Ltd

C.I.S. - English quarterly/£540.00 P.A.

In two volumes; Russia and the Non-Russian republics - updated every three months which gives a broad picture of the market and business practice.

09019 ECOTASS - THE ECONOMIST AND COMMERCIAL BULLETIN OF THE NEWS AGENCY TASS
Pergamen Press Plc

C.I.S. - English/French/Italian/ German weekly/£525.00 P.A. ISSN No. 0733-5989

09020 FINANCING FOREIGN OPERATIONS: USSR
Business International Ltd

USSR - English annual/£695.00 for complete European series

Business overview, currency outlook, exchange regulations, monetary system, short/medium/ long-term and equity financing techniques, capital incentives, cash management, investment trade finance and insurance.

09021 FOOD AND AGRICULTURAL POLICY REFORMS IN THE FORMER USSR: AN AGENDA FOR TRANSITION
World Bank Publications

C.I.S. - English/1992 242pp ISSN No. 1014-997X

Part of the series "Studies of economies in transition" listed as paper number 1. A policy document with some economic data.

09022 FOREIGN DIRECT INVESTMENT IN THE STATES OF THE FORMER USSR
International Monetary Fund

Crane, K. et al/C.I.S. - English/1992 135pp ISSN No. 1014-997X

Part of the series "Studies of economies in transition", listed as paper number 5.

09023 HINTS TO EXPORTERS VISITING THE USSR
Department of Trade and Industry Export Publications

C.I.S. - English irregular/1991 £5.00

This publication provides detailed practical information that is useful for those planning a business visit. Topics include currency information, economic factors, import and exchange control regulations, methods of doing business and general information.

09024 INFORMATION MOSCOW
Information Moscow

C.I.S./Russia - English every 6 months 300pp

Directory of useful addresses including foreign representatives government departments.

09025 INVESTING, LICENSING AND TRADING CONDITIONS ABROAD: C.I.S.
Business International Ltd

C.I.S. - English annual/£825.00 for complete European series

Loose-leaf format covering the state role in industry, competition rules, price controls, remittability of funds, taxation, incentives and capital sources, labour and foreign trade.

09026 ISVESTIA
Isvestia

C.I.S. - Russian daily/£234.00 P.A.

Daily newspaper with a Tuesday business supplement covering Moscow and Leningrad.

09027 JOINT VENTURES IN THE C.I.S.
Association of Joint Ventures International Consortia and Organizations

Mr Jevgenij Minin/C.I.S. - English/ Russian bi-monthly/1990 64pp US$5.00 P.A. ISSN No. 0869-5369

Reviews the level and operations of joint ventures.

09028 MOSCOW MAGAZINE
Multi Media International

C.I.S. - English/Russian 6 times a year/US$89.00 P.A. ISSN No. 0868-8400

Useful for business environment type information and for those trading in the Moscow area.

09029 PROGRAMME FOR PRIVATISATION OF STATE AND MUNICIPAL ENTERPRISES IN RF IN 1992
Ria-Novosti

C.I.S./Russia - English translation irregular/1991 £35.00

English translation of governmental publication.

09030 REFERENCE BOOK - JOINT VENTURES IN THE C.I.S.
Association of Joint Ventures International Consortia and Organizations

Oleg Volkov/C.I.S. - Russian annual/1990 735pp US$50.00

09031 RIA-NOVOSTI REFERENCE BOOK
Ria-Novosti

C.I.S. - English translation irregular/ 1992 £20.00

Reference hand book giving basic data on the former Soviet Union.

09032 RUSSIA EXPRESS; EXECUTIVE
International Industrial Information Ltd

C.I.S. - English fornightly/28pp £295.00 P.A.

Business newsletter which includes details of joint ventures.

09033 RUSSIAN TRADE EXPRESS
I.M.A. Neva Media

C.I.S. - English/German quarterly/ c30pp US$72.00 P.A. outside C.I.S.

Business quarterly mainly targeted at those trading in the C.I.S.

09034 SOVIET ANALYST
World Reports Ltd

International - English 10 times per year/£175.00 P.A. ISSN No. 0049-1713

Covers the former Soviet Union financial markets.

09035 SOVIET INDEPENDENT BUSINESS DIRECTORY
North River Press

FYI Information Resources for a Changing World, Co-operative Reserve/C.I.S. - English/1990 564pp

Profiles of over 2100 businesses.

09036 SOVIET TRADE DIRECTORY
Flegon Press

Alec Flegon/C.I.S. - English/ Russian irregular/1988 744pp

Detailed trade directory.

09037 TRADING PARTNERS USSR
Sovinform Ltd

C.I.S. - English irregular/1990 ISSN No. 0958-9368

Company directory classified by industry.

09038 UKRAINE BUSINESS REVIEW
Ukraine Business Agency

C.I.S. - English quarterly/£145.00 P.A.

Business news format covering recent developments.

09039 USSR CALENDAR OF EVENTS ANNUAL
Academic International Press

C.I.S. - English annually/1987 519pp

09040 USSR FACTS AND FIGURES ANNUAL
Academic International Press

Alan P Pollard/C.I.S. - English annual/1977 545pp US$92.00

Gives key statistical data on the C.I.S.

CYPRUS

Country profile:	Cyprus
Official name:	Republic of Cyprus
Area:	9,251 sq.km.
Population:	702,100 (1990)
Capital:	Nicosia
Language(s):	Greek, Turkish
Currency:	Pound 100 cents = 1 Cyprus pound
National bank:	Central and Issuing Bank
Government and Political structure:	Elected President, who appoints the Council of Ministers and House of Representatives (parliament) with 80 elected members.

Cyprus

10001 AGRICULTURAL STATISTICS

Department of Statistics and Research

Cyprus - English annual/1969
C£3.00

Covers land use, crops, livestock etc.

10002 ANNUAL REPORT

Cyprus Employers and Industrialists Federation

Cyprus - Greek annual/70pp

Details activities of Cypriot industry.

10003 ANNUAL REPORT - CENTRAL BANK OF CYPRUS

Central Bank of Cyprus

Cyprus - English annual/80pp

Reviews the national banking/financial situation over the past year.

10004 ANNUAL REPORT

Cyprus Ports Authority

Nicos Nicolaou Marketing and Public Relations Officer/Cyprus - English/Greek annual/1977 35pp free ISSN No. 1013-3232

Volume of maritime traffic.

10005 ANNUAL REPORT OF THE MINES SERVICE

Mines Service

Cyprus - English annual/30pp

10006 BANK OF CYPRUS BULLETIN

Bank of Cyprus Ltd

Cyprus - English quarterly/1973 30pp free

Looks at the banking sector and related issues.

10007 BULLETIN - CENTRAL BANK OF CYPRUS

Central Bank of Cyprus

Cyprus - English quarterly/90pp

Bulletin with financial/monetary balances.

10008 BUSINESS GUIDE TO CYPRUS

Cyprus Popular Bank Ltd

Maria Zambarloukou/Cyprus - English irregular/62pp

10009 BUSINESS PROFILE SERIES - CYPRUS

Cyprus Popular Bank Ltd

Cyprus - English irregular/56pp

Brief profile intended for those doing business in Cyprus.

10010 COMMUNITY, SOCIAL AND PERSONAL SERVICES STATISTICS

Department of Statistics and Research

Cyprus - English annual/1992 C£3.00

Socio-economic data and social security type statistical coverage.

10011 CONSTRUCTION AND HOUSING STATISTICS

Department of Statistics and Research

Cyprus - English annual/1966 C£3.00 ISSN No. 0253-8725

Main construction statistical series and number of housing completions.

10012 COUNTRY PROFILE; MALTA, CYPRUS

The Economist Intelligence Unit

Cyprus/Malta - English annual/34pp £150.00 P.A.

10013 COUNTRY RISK SERVICE: CYPRUS

Business International Ltd

Cyprus - English quarterly/minimum subscription £1,855.00 P.A. for seven countries

Contains credit risk ratings, risk appraisals, cross-country databases, data on diskette, and two year projections.

10014 CRIMINAL STATISTICS

Department of Statistics and Research

Cyprus - English annual/1974 C£4.00 ISSN No. 0253-8695

10015 THE CYPRIOT EXPORTER

Cyprus Chamber of Commerce and Industry

Cyprus - English quarterly/1986 20pp free

10016 CYPRUS 1960-1990: TIME SERIES DATE

Department of Statistics and Research

Cyprus - English irregular/1992 C£10.00

Data for longer-term economic analysis.

10017 CYPRUS ECONOMY IN FIGURES

Bank of Cyprus Ltd

Cyprus - English annual/1983 5pp free ISSN No. 0256-8284

Brief look at the main features of the Cypriot economy.

10018 DEMOGRAPHIC REPORT

Department of Statistics and Research

Cyprus - English annual/1963 C£4.00 ISSN No. 0590-4846

Breakdowns by age, sex and standard criteria such as births.

10019 DOING BUSINESS IN CYPRUS

Price Waterhouse

Cyprus - English irregular/1991 330-340pp free to clients

A business guide which includes sections on the investment climate, doing business, accounting and auditing, taxation and general country data. Updated by supplements.

10020 ECONOMIC AND SOCIAL INDICATORS

Planning Bureau, Government of Cyprus

Cyprus - English/Greek annual/1989 50pp

10021 ECONOMIC NEWS

Bank of Cyprus Ltd

Cyprus - Greek quarterly/1982 8pp ISSN No. 0256-839X

10022 ECONOMIC OUTLOOK

Planning Bureau, Government of Cyprus

Cyprus - English/Greek annual/1989 130pp

Forecasts the economic situation in the short/medium-term.

10023 ECONOMIC REPORT

Department of Statistics and Research

Cyprus - English annual/1954 C£5.00 ISSN No. 0070-2412

Comprehensive information on the Cyprus economy.

10024 EDUCATION STATISTICS

Department of Statistics and Research

Cyprus - English annual/1964 C£4.00 ISSN No. 0253-8733

Covers level, and type of education including numbers participating.

CZECH AND SLOVAK REPUBLICS

Country profiles:	Czech and Slovak Republics
Official names:	Czech Republic Slovak Republic
Area:	127,900 sq.km. Czech 78,900 sq.km., Slovak 49,000 sq.km. Bordered in the north by Poland, the west by Germany, the east by the Ukraine (C.I.S.), and the south by Austria and Hungary.
Population:	15.69m Czech 10.4m, Slovak 5.28m
Capital(s):	Prague (Czech), Bratislava (Slovak)
Language(s):	Czech, Slovak, Hungarian.
Currency:	Koruna 100 haleru = 1 Koruna
National bank:	State Bank(s)
Government and Political structure:	1992 saw the division of the country into two republics, Czech and Slovak.

Czech and Slovak Republics

(c) R Canning.

11001 COUNTRY FORECASTS: CZECHOSLOVAKIA

Business International Ltd

Czechoslovakia - English annual/ £360.00 P.A.

Contains an executive summary, sections on political and economic outlooks and the business environment, a fact sheet, and key economic indicators. Includes a quarterly Global Outlook.

11002 COUNTRY PROFILE CZECH AND SLOVAK REPUBLIC

Department of Trade and Industry Export Publications

Czechoslovakia - English irregular/ 1992 42pp £10.00

This publication provides a general description of the countries in terms of their geography, economy and population, and gives information about trading in the countries with details like export conditions and investment covered.

11003 COUNTRY REPORT: CZECHOSLOVAKIA

Business International Ltd

Czechoslovakia - English quarterly/ 25-40pp £160.00 P.A.

Contains an executive summary, sections on political and economic structure, outlook for the next 12-18 months, statistical appendices, and a review of key political developments. Includes an annual Country Profile.

11004 COUNTRY RISK SERVICE: CZECHOSLOVAKIA

Business International Ltd

Czechoslovakia - English quarterly/ minimum subscription £1,855.00 P.A. for seven countries

Contains credit risk ratings, risk appraisals, cross-country databases, data on diskette, and two year projections.

11005 CZECHOSLOVAK ECONOMIC DIGEST

ORBIS Information Services /RIS/

Czechoslovakia - English bi- monthly/£5.00 ISSN No. 0045-9461

Digest of economics covering main economic factors in the economy.

11006 CZECHOSLOVAK POLICY DIGEST

Language Comprehensive Services

Czechoslovakia - English monthly/ £148.00 P.A.

Covers economic and related policy issues.

11007 CZECHOSLOVAKIA ECONOMY DIGEST

Language Comprehensive Services

Czechoslovakia - English monthly/ £185.00 P.A.

11008 CZECHOSLOVAKIA - GLOBAL FORECASTING SERVICE

The Economist Intelligence Unit

Czechoslovakia - English quarterly/ c13pp £280.00 P.A.

11009 CZECHOSLOVAKIA IN TRANSITION

The Economist Intelligence Unit

Czechoslovakia - English/1990 45pp £95.00

Review of the Czech economy during the transition to a market economy.

11010 CZECHOSLOVAKIA: MAJOR BUSINESSES

Dun and Bradstreet Ltd

Czechoslovakia - English/1992 £175.00

Comprehensive trade directory containing 4,000 short company profiles.

11011 CZECHOSLOVAKIA'S AGRICULTURE: SITUATION, TRENDS AND PROSPECTS

Office for Official Publications of the European Communities

Czechoslovakia - English/1991 206pp ECU22.00

11012 DIRECTORY OF CZECH AND SLOVAK COMPANIES

Inform. Katalog Ltd

Czechoslovakia - English/Czech annual/US$80

Directory of 17,000 companies.

11013 DOING BUSINESS IN CZECHOSLOVAKIA

Confederation of British Industry (CBI)

KPMG Peat Marwick McLintock, S.J. Berwin and Co., National Westminster Bank/Czechoslovakia - English/1991 392pp ISBN No. 074940471X

Gives economic and business background, mainly aimed at exporters and those intending to establish a business presence.

11014 DOING BUSINESS IN CZECHOSLOVAKIA

Price Waterhouse

Czechoslovakia - English irregular/ 1991 330-340pp free to clients

A business guide which includes sections on the investment climate, doing business, accounting and auditing, taxation and general country data. Updated by supplements.

11015 DOING BUSINESS IN EASTERN EUROPE: CZECHOSLOVAKIA

Kogan Page

Czechoslovakia - English/1991 320pp £25.00 ISBN No. 074940471

Part of the CBI initiative on Eastern Europe. Describes the business and economic environment.

11016 DOING BUSINESS WITH EASTERN EUROPE: CZECHOSLOVAKIA

Business International Ltd

Czechoslovakia - English quarterly/ £540.00 P.A.

Part of a ten volume set, updated every three months which gives a broad picture of the market and business practice.

11017 FINANCING FOREIGN OPERATIONS: CZECHOSLOVAKIA

Business International Ltd

Czechoslovakia - English annual/ £695.00 for complete European series

Business overview, currency outlook, exchange regulations, monetary system, short/medium/ long-term and equity financing techniques, capital incentives, cash management, investment trade finance and insurance.

11018 HINTS TO EXPORTERS VISITING CZECHOSLOVAKIA

Department of Trade and Industry Export Publications

Czechoslovakia - English irregular/ 1991 51pp £5.00

This publication provides detailed practical information that is useful for those planning a business visit. Topics include currency information, economic factors, import and exchange control regulations, methods of doing business and general information. Separate editions for the two Republics are planned.

11019 IMPORTING FROM CZECHOSLOVAKIA

Probus Europe

Czechoslovakia - English/188pp US$25.00

A guide for those wishing to import from Czechoslovakia.

11020 INTERNATIONAL CUSTOMS JOURNAL: CZECHOSLOVAKIA

Department of Trade and Industry Export Publications

Czechoslovakia - English irregular/ £25.00

This publication contains the rate of duties that apply to commodities exported into the country according to the Harmonised System (HS) of commodity coding.

11021 INVESTING, LICENSING AND TRADING CONDITIONS ABROAD: CZECHOSLOVAKIA

Business International Ltd

Czechoslovakia - English annual/ £825.00 for complete European series

Loose-leaf format covering the state role in industry, competition rules, price controls, remittability of funds, taxation, incentives and capital sources, labour and foreign trade.

11022 KOVO EXPORT

Rapid a.c.

Czechoslovakia - English quarterly/ c28pp

Details of products available for export.

11023 THE PRAGUE POST

The Prague Post

Czechoslovakia - English weekly/ US$130.00 P.A.

A weekly newspaper which includes business environment type information and news of local companies.

11024 RESOURCES CSFR

Resources CSFR

Czechoslovakia - English/Czech/ German looseleaf binder with 3 monthly update/US$80, US $28 for each 3 month update

Directory of key addresses and contacts.

11025 YOUR TRADING PARTNERS IN CZECHOSLOVAKIA

Czechoslovakia Chamber of Commerce and Industry

Czechoslovakia - English irregular/ ISBN No. 8070030739

List of companies who import and export.

DENMARK

Country profile:	Denmark
Official name:	Kingdom of Denmark
Area:	43,075 sq.km.
	Bordered by the North Sea and in the south by Germany.
Population:	5.14m (1991)
Capital:	Copenhagen
Language(s):	Danish
Currency:	Krone 100 ore = 1 Krone
National bank:	National Bank of Denmark
Government and Political structure:	Constitutional monarchy with multi-party democracy. The Folketing (Diet) has 179 members elected mostly by proportional representation.

Denmark

(c) R Canning.

12001 ANDELSBLADET
Federation of Danish Cooperatives

Denmark - Danish bi-weekly/1899 24pp DKr300 P.A. ISSN No. 0003-2913

Agricultural cooperatives.

12002 ARBEJDSMARKED: STATISTISKE EFTERRETNINGER
Danmarks Statistik

Denmark - Danish monthly/28-40pp DKr339.20 P.A. ISSN No 0108-5514

Employment, unemployment, industrial accidents, labour force.

12003 BEFOLKNING OG VALG: STATISTISKE EFTERRETNINGER
Danmarks Statistik

Denmark - Danish monthly/28-40pp DKr235.20 P.A. ISSN No 0108-5530

Vital statistics, population forecasts, housing conditions, election results, births, deaths, marriages.

12004 BYGGE - OG ANLAEGSVIRKSOHMHED: STATISTISKE EFTERRETNINGER
Danmarks Statistik

Denmark - Danish monthly/28-40pp DKr176.00 ISSN No 0108-5549

Building costs, sales of real property, construction costs indices, employment & labour costs.

12005 CENTRAL GOVERNMENT BORROWING AND DEBT
Danmarks Nationalbank

Denmark - Danish annual/184pp free ISSN No. 0902-6681

An account of domestic and foreign borrowing.

12006 CFM NEWS
Copenhagen Stock Exchange

International Denmark - English quarterly/April 1992 4pp free

Financial markets are covered, with general stock market news included.

12007 COUNTRY BOOKS: DENMARK
United Kingdom Iron and Steel Statistics Bureau

Denmark - English annual/£75.00 ISSN No. 0952-2372

Annual summary tables showing production, materials consumed, apparent consumption, imports and exports of 130 products by quality and market.

12008 COUNTRY FORECASTS: DENMARK
Business International Ltd

Denmark - English quarterly/ £360.00 P.A.

Contains an executive summary, sections on political and economic outlooks and the business environment, a fact sheet, and key economic indicators. Includes a quarterly Global Outlook.

12009 COUNTRY PROFILE DENMARK
Department of Trade and Industry Export Publications

Denmark - English irregular/1990 78pp £10.00

This publication provides a general description of the country in terms of its geography, economy and population, and gives information about trading in the country with details such as export conditions and investment covered.

12010 COUNTRY REPORT: DENMARK - ICELAND
Business International Ltd

Denmark/Iceland - English quarterly/25-40pp £160.00 P.A. ISSN No. 0269-594X

Contains an executive summary, sections on political and economic structure, outlook for the next 12-18 months, statistical appendices, and a review of key political developments. Includes an annual Country Profile.

12011 DAILY OFFICIAL LIST
Copenhagen Stock Exchange

Denmark - English/Danish daily monthly quarterly/daily DKr25.00 monthly DKr60.00 quarterly DKr60.00 P.A.

Covers trade on the Copenhagen Stock Exchange the previous trading day.

12012 DANISH ECONOMIC OUTLOOK
Unidanmark A/S

Denmark/Sweden/Norway/Finland - English 3 time per year/July 1990 approx 28pp free ISSN No. 0905-8745

Brief review of the Danish economy with an article on a current economic topic in each issue.

12013 DANSK HANDELSBLAD
Dansk Handelsblad A/S

Denmark - Danish weekly/1909 28-52pp DKr350 ISSN No. 0045-9615

Private retail grocers, supermarkets, chainstores, branded goods.

12014 DATA ON DANISH PUBLIC FOREIGN BORROWING
Danmarks Nationalbank

Denmark - Danish semi-annual/ 33pp free ISSN No. 0906-6993

Key figures for economic development and the latest details of central government borrowing abroad.

12015 DENMARKS 15,000 LARGEST COMPANIES
Ekonomisk Litteratur AB

Denmark - English/Danish annual/ DKr1550

Financial information on Danish companies.

12016 DOING BUSINESS IN DENMARK
Price Waterhouse

Denmark - English irregular/1992 330-340pp free to clients

A business guide which includes sections on the investment climate, doing business, accounting and auditing, taxation and general country data. Updated by supplements.

12017 ELFORSYNINGENS TIÅRSOVERSIGT
Danske Elvrkers Forening

Denmark - English/Danish annual/ March 1986 125pp DKr85.00 ISSN No. 0901-3369

Total energy consumption, electricity and CHP production, fuel consumption, technical systems, prices, investments and environment. The tables and figures cover the past 10 years.

12018 FACTS 199.. - MEDICINE AND HEALTH CARE
MEFA - The Association of Danish Pharmaceutical Industry

Denmark - English/Danish annual/ 118pp free ISSN No. 0107-1181

Medicine and health care.

12019 FINANS & SAMFUND
Finansrådet - Danish Bankers Association

Denmark - Danish monthly/Oct 1990 40pp DKr275.00 P.A. ISSN No. 0905-9415

Economics and finance, banks and savings banks.

12020 FRIT KØBMANDSKAB
Dansk Handelsblad A/S

Denmark - Danish 4 times per year/ 24-128pp DKr350 P.A. ISSN No. 0901-2745

Analysis of trends within grocery, shop and supermarket trade.

12021 GENERAL ERHVERVSSTATISTIK OG HANDEL: STATISTISKE EFTERRETNINGER
Danmarks Statistik

Denmark - Danish/28-40pp DKr148.80 P.A. ISSN No. 0108-5468

Accounts, vat registrations, index of retail sales.

12022 GLOBAL FORECASTING SERVICE

Business International Ltd

Denmark - English quarterly/ £350.00 P.A.

Provides a range of medium-term forecasts covering main political, economic and business trends. A quarterly Global Outlook is included.

12023 HINTS TO EXPORTERS VISITING DENMARK

Department of Trade and Industry Export Publications

Denmark - English irregular/1991 69pp £5.00

This publication provides detailed practical information that is useful for those planning a business visit. Topics include currency information, economic factors, import and exchange control regulations, methods of doing business and general information.

12024 IN DEPTH COUNTRY APPRAISALS ACROSS THE EC - CONSTRUCTION AND BUILDING SERVICES - DENMARK

Building Services Research and Information Association

Denmark - English 1991/£125.00

Part of an 11 volume survey covering the construction industry in EC countries.

12025 INDKOMST, FORBRUG OG PRISER: STATISTISKE EFTERRETNINGER

Danmarks Statistik

Denmark - Danish monthly/28-40pp DKr196.80 P.A. ISSN No. 0108-5565

Consumer surveys, consumer price index, net retail prices, wholesale prices, incomes, earnings, wealth.

12026 INDUSTRI OG ENERGI: STATISTISKE EFTERRETNINGER

Danmarks Statistik

Denmark - Danish monthly/28-40pp DKr172.80 P.A. ISSN No. 0108-5522

Energy balance sheets, energy supplies, employment and labour costs, sales and order books.

12027 THE INTERNAL MARKET AND THE SOCIAL DIMENSION

LO-Landsorganisationen i Danmark (The Danish Confederation of Trade Unions)

Denmark/EEC - English annual/ 1991 free

Denmark and the EEC labour market.

12028 INVESTING, LICENSING AND TRADING CONDITIONS ABROAD: DENMARK

Business International Ltd

Denmark - English annual/£825.00 for complete European series

Loose-leaf format covering the state role in industry, competition rules, price controls, remittability of funds, taxation, incentives and capital sources, labour and foreign trade.

12029 LABOUR RELATIONS IN DENMARK

LO-Landsorganisationen i Danmark (The Danish Confederation of Trade Unions)

Denmark - English annual/1991 free

Labour relations and labour market in Denmark.

12030 LANDBRUG: STATISTISKE EFTERRETNINGER

Danmarks Statistik

Denmark - Danish monthly/28-40pp DKr425.60 P.A. ISSN No. 0106-6680

Production and prices of crops, cereals, feedstuffs, cattle, dairy products, fertilizers.

12031 LIST OF 50 MOST ACTIVE/ LARGEST STOCKS

Copenhagen Stock Exchange

Denmark - English/Danish annual/ DKr60.00

12032 LIST OF LISTED COMPANIES

Copenhagen Stock Exchange

Denmark - English/Danish annual/ DKr350.00

Listed companies on the Copenhagen stock exchange.

12033 LIST OF STOCKBROKING COMPANIES

Copenhagen Stock Exchange

Denmark - English/Danish irregular/ DKr150.00

12034 MAGNUS, INCLUDING: DANSK ØKONOMISK BIBLIOGRAFI (DANISH ECONOMICAL BIBLIOGRAHY)

The Royal Library

Denmark - Danish MAGNUS 1991-; Dansk økon. bibl. 1950-/CD-ROM ISSN No. 0907-0834 (MAGNUS); ISSN NO. 0106-0767 (Dan. økon. Bibl).

12035 MONETARY REVIEW

Danmarks Nationalbank

Denmark - English quarterly/30pp

Review giving tables and graphs on the current financial situation.

12036 MONTHLY LIST

Copenhagen Stock Exchange

Denmark - English/Danish monthly/ DKr60.00 P.A.

Covers the trading rates every trading day and the total amount of trade in each security.

12037 NATIONALREGNSKAB, OFFENTLIGE FINANSER OG BETALINGSBALANCE: STATISTIKE EFTERRETNINGER

Danmarks Statistik

Denmark - Danish quarterly/28-40pp DKr254.00 P.A. ISSN No. 0108-5476

Local government budgets, personal taxation, corporate taxation, property taxation, custom and excise duties, balance of payments, foreign debt.

12038 THE NORDIC COUNTRIES

The Nordic Statistical Secretariat

Denmark/Sweden/Norway/Finland/ Iceland - Danish annual/40pp

Excerpts from the Yearbook of Nordic Statistics 1992.

12039 NYT FRA DANMARKS STATISTIK

Danmarks Statistik

Denmark - Danish/360 issues annually DKr1,936.00 P.A. ISSN No. 0106-9799

News from Danmarks Statistik is a rapid release service primarily aimed at news media

12040 PENGE - OG KAPITAL MARKED STATISTISKE EFTERRETNINGER

Danmarks Statistik

Denmark - Danish quarterly/28-40pp DKr243.20 P.A. ISSN No. 0108-5476

Central government's finances, bank balances, mortgages, interest rates, share issues.

12041 PER PRESS SUPERMARKEDER OG ANDRE STORE DAGLIGVAREBUTIKKER 1992

Stockmann - Gruppen A/S

Denmark - Danish annual/1992 266pp DKr895.00

Directory of 1918 supermarkets and variety stores in Denmark.

12042 QUARTERLY COUNTRY TRADE REPORTS: DENMARK

United Kingdom Iron and Steel Statistics Bureau

Denmark - English quarterly/ £200.00 P.A. for four countries ISSN No. 0960-2372

Cumulative trade statistics of imports and exports of steel products.

12043 REPORT AND ACCOUNTS
Danmarks Nationalbank

Denmark - English/Danish/130pp free

Annual report including statistical tables covering all branches of the economy.

12044 SAMFAERDSEL OG TURISME: STATISTISKE EFTERRETNINGER
Danmarks Statistik

Denmark - Danish monthly/28-40pp DKr224.80 P.A.ISSN No. 0108-5484

Holidays, nights spent in hotels, shipping, motor vehicle stock, traffic accidents.

12045 THE SCANDINAVIAN ECONOMIES
TSE Publishing Aps

Denmark/Sweden/Norway/Finland - English monthly/24pp DKr2600 P.A. ISSN NO. 0105-0848

Economic and business news in four Nordic countries.

12046 SEMI-ANNUAL REPORT
Copenhagen Stock Exchange

Denmark - English/Danish biannual/DKR40.00

Covers trade in bonds and shares that year and total amount of trade in each security.

12047 SOCIAL SIKRING OG RETSVAESEN: STATISTISKE EFTERRETNINGER
Danmarks Statistik

Denmark - Danish quarterly/28-40pp DKr232.80 P.A. ISSN No. 0108-5441

Social security benefits, housing subsidies, child benefits, sick benefits, maternity benefit, pension benefits.

12048 STATISTICS
Danish Bacon and Meat Council

Denmark/International - English annual/40pp

Pig production, consumption, prices, exports of meat etc.

12049 STATISTISK ÅRBOG
Danmarks Statistik

Denmark - Danish annual/576pp DKr180.00 ISSN No. 0070-3583

Statistical yearbook on all aspects of Danish life.

12050 STATISTISK EFTERRETNINGER
Danmarks Statistik

Denmark - Danish irregular

Statistical news is divided into 14 sub-groups with results of all periodical inquiries, surveys and censuses.

12051 STATISTISK MÅNEDSOVERSIGT
Danmarks Statistik

Denmark - Danish monthly/ DKr599.20 P.A. DKr50.40 single copy ISSN No. 0108-5603

Monthly review of essential economic statistics.

12052 STATISTISK TIÅRSOVERSIGT
Danmarks Statistiks

Denmark - Danish annual/144pp DKr78.40 ISSN No. 0070-3583

Statistical ten-year review of comparable annual statistics.

12053 UDENRIGSHANDEL: STATISTISKE EFTERRETNINGER
Danmarks Statistik

Denmark - Danish quarterly and annual/DKr695.70 P.A. ISSN No. 0109-5420

Quarterly and annual general trade statistics.

12054 UGEBREVET DANSK INDUSTRI
Dansk Industri, Confederation of Danish Industries

Denmark - Danish weekly/11th May 1992 8pp DKr1.500 P.A. ISSN No. 0906-9267

Trade and industrial matters, agreements on labour market, domestic and EEC industrial and economic policy.

12055 YEARBOOK OF NORDIC STATISTICS
The Nordic Statistical Secretariat

Denmark/Sweden/Norway/Finland - annual/DKr248 ISSN No. 0078-1088

Broad statistical review.

FINLAND

Country profile:	Finland
Official name:	Republic of Finland
Area:	304,620 sq.km.
	Bordered in the north by Norway, East by the states of the Former USSR, south by the Baltic Sea and west by the Gulf of Bothnia and Sweden.
Population:	4.99m (1991)
Capital:	Helsinki
Language(s):	Finnish, Swedish
Currency:	Finnmark 100 pennia = 1 Finnmark
National bank:	Bank of Finland
Government and Political structure:	Republic with a one chamber parliament of 200 members elected both directly and by proportional representation. A Prime Minister heads a Council of State (Cabinet).

Finland

(c) R.Canning.

13001 ANNUAL REPORT
The Finnish Bankers
Association

Finland - English annual/36pp free

Economic development, banking industry, money and foreign exchange markets, tax, capital markets, banking technology.

13002 ANNUAL REPORT
Finnair Oy

Finland - English/Finnish/Swedish annual/35pp free

Annual report of Finnair Oy, the national airline.

13003 ANNUAL REPORT
Kansallis-Osake-Pankki

Finland - English/Finnish annual

Annual report of the bank.

13004 ANNUAL REPORT
Union Bank of Finland

Finland - English/Finnish annual/ 40pp free ISSN No. 0355-0133

Details the banks activities, balances and finances.

13005 BANK OF FINLAND
ANNUAL STATEMENT
Bank of Finland

Finland - English/Finnish/Swedish/ Fmk10

Annual statement of the Bank of Finland showing balances and activities.

13006 BANK OF FINLAND
BULLETIN
Bank of Finland

Finland - English monthly/1921 60pp free ISSN No. 0784-6509

Financial markets, monetary policy, includes statistics and charts.

13007 BANK OF FINLAND
YEARBOOK
Bank of Finland

Finland - English/Finnish/Swedish annual/Fmk80 ISSN No. 0081-9468

13008 BANKRUPTCIES
Statistics Finland

Finland - Finnish monthly/1986 11pp Fmk180 ISSN No. 0784-8811

Statistical survey of Finnish bankruptcies.

13009 THE BANKS
Statistics Finland

Finland - Finnish/Swedish monthly/ 1970 25pp Fmk400 ISSN No. 0784-9303

Banks, subtitle in series "Finance".

13010 BUILDING COST INDICES
Statistics Finland

Finland - Finnish/Swedish monthly/ 1968 11pp Fmk370 ISSN No. 0784-8196

Shows costs of building inputs.

13011 BULLETIN OF STATISTICS
Statistics Finland

Finland - English/Finnish/Swedish quarterly/1924 110pp Fmk265 ISSN No. 0015-2390

General economic statistics are covered.

13012 BUSINESS FINLAND
The Central Chamber of
Commerce of Finland

Finland - English annual/October 1988 Fmk150 approx. ISSN No. 0785-5540

Finnish economy.

13013 CLOTHING INDUSTRY
YEARBOOK
Central Association of the
Finnish Clothing Industry

Finland - English/Finnish annual/ 34pp free

Statistics on production, foreign trade and labour market.

13014 CONSTRUCTION AND
HOUSING YEARBOOK
Statistics Finland

Finland - English/Finnish/Swedish annual/1989 184pp Fmk220 ISSN No. 0787-572X

Building and dwelling construction, building enterprises, housing conditions, housing loans.

13015 CONSUMER BAROMETER
Statistics Finland

Finland - Finnish twice a year/1988 20pp Fmk250 ISSN No. 0784-963X

Consumer buying expectations. (A subtitle in series "Income and Expenditure").

13016 CONSUMER PRICE INDEX
Statistics Finland

Finland - Finnish/Swedish monthly/ 1968 6pp Fmk225 ISSN No. 0784-820X

Shows the CPI for Finland and trends.

13017 CORPORATE
ENTERPRISES AND
PERSONAL BUSINESS IN
FINLAND
Statistics Finland

Finland - Finnish/Swedish annual/ 1984 200pp Fmk155 ISSN No. 0788-1738

A subtitle in series "Enterprises".

13018 COUNTRY BOOKS:
FINLAND
United Kingdom Iron and
Steel Statistics Bureau

Finland - English annual/£75.00 ISSN No. 0952-5890

Annual summary tables showing production, materials consumed, apparent consumption, imports and exports of 130 products by quality and market.

13019 COUNTRY FORECASTS:
FINLAND
Business International Ltd

Finland - English quarterly/£360.00 P.A.

Contains an executive summary, sections on political and economic outlooks and the business environment, a fact sheet, and key economic indicators. Includes a quarterly Global Outlook.

13020 COUNTRY PROFILE
FINLAND
Department of Trade and
Industry Export
Publications

Finland - English irregular/1992 46pp £10.00

This publication provides a general description of the country in terms of its geography, economy and population, and gives information about trading in the country with details like export conditions and investment covered.

13021 COUNTRY REPORT:
FINLAND
Business International Ltd

Finland - English quarterly/25-40pp £160.00 P.A. ISSN No. 0269-5901

Contains an executive summary, sections on political and economic structure, outlook for the next 12-18 months, statistical appendices, and a review of key political developments. Includes an annual Country Profile.

13022 ECONOMIC SURVEY:
FINLAND
Ministry of Finance
Economics Department

Finland - English/Finnish annual/ 121pp Fmk95 ISSN No. 0430-5227

General survey, foreign trade, production, costs and prices, incomes, investment, public finance and social security, financial markets.

13023 ENERGY REVIEW
Ministry of Trade and
Industry Energy
Department

Finland - English/Finnish/Swedish annual/48pp

A national survey of the energy market in Finland.

13024 ENERGY STATISTICS
Statistics Finland

Finland - English/Finnish/Swedish annual/135pp Fmk250 ISSN No. 0784-9354

Energy supply and consumption.

13025 ENTERPRISE AND INCOME STATISTICS OF FARM ECONOMY
Statistics Finland

Finland - Finnish/Swedish annual/ 1973 80pp Fmk100 ISSN No. 0784-9966

Statistics on farm economy - a subtitle in series "Agriculture and Forestry".

13026 ENTERPRISES
Statistics Finland

Finland - Finnish/Swedish irregular/ 1972 Fmk1600 ISSN No. 0784-8463

Statistical coverage of Finnish companies.

13027 FACTS THROUGH FIGURES: A STATISTICAL PORTRAIT OF EFTA IN THE EUROPEAN ECONOMIC AREA
Office for Official Publications of the European Communities

Finland/EFTA/EEC - English single publication/1992 31pp Fmk20 ISBN No. 9282636550

13028 FARM REGISTER
National Board of Agriculture

Finland - English/Finnish/Swedish annual/1980 110pp Fmk150 ISSN No. 0784-8404

Number of farms, farm land use, ownership of farms.

13029 FINANCE
Statistics Finland

Finland - Finnish irregular/1971 Fmk2185

Financial statistics on a national basis.

13030 FINANCIAL MARKET STATISTICS
Statistics Finland

Finland - Finnish/Swedish annual/ 1969 45pp Fmk80 ISSN No. 0784-9893

13031 FINANCIAL MARKETS
Bank of Finland

Finland - English/Finnish/Swedish monthly/Jan 92 50pp free ISSN No. 0789-9955

Monthly money market news on the financial sector. Includes statistics.

13032 FINANCIAL STATEMENTS STATISTICS
Statistics Finland

Finland - English/Finnish/Swedish annual/1990 228pp Fmk230 ISSN No. 1235-1342

Statistics, financial statements of industry, construction trade, trade and business services.

13033 FINLAND
United Kingdom Iron and Steel Statistics Bureau

Finland - English annual/£80.00 ISSN No. 0952-5890

Report on the steel industry showing production, consumption and imports and exports.

13034 FINLAND IN FIGURES
Statistics Finland

Finland - English annual/1982 30pp Fmk14 ISSN No. 0357-0371

Statistics on all aspects of society.

13035 FINLAND'S BALANCE OF PAYMENTS
Statistics Finland

Finland - English/Finnish/Swedish annual/1975 40pp Fmk75 ISSN No. 0784-9990

National accounts, imports and exports.

13036 THE FINNISH BANKING SYSTEM
The Finnish Bankers Association

Finland - English/Finnish iregular/ 56pp free

General banking outlook for Finland.

13037 FINNISH BOND ISSUES
Bank of Finland

Finland - English/Finnish/Swedish annual/1983 Fmk30

Survey of bond issues in the Finnish financial sector.

13038 FINNISH BUSINESS REPORT
Oy Novomedia Ltd

Finland - Finnish monthly/US$65 P.A. ISSN No. 0780-4652

13039 THE FINNISH ECONOMY
The Research Institute of the Finnish Economy, ETLA

Finland - English quarterly/90pp Fmk1150 ISSN No. 0785-7985

Domestic development, international developments, short-term economic forecasts, assessments of economic prospects for the next five year period and special studies of topical interest.

13040 THE FINNISH FOREST INDUSTRIES: FACTS AND FIGURES
Central Association of Finnish Forest Industries

Finland - English/Swedish/German/ French annual/27pp free

Forestry industry statistics including volume and type of product.

13041 FINNISH INSURANCE REVIEW
Federation of Finnish Insurance Companies

Finland - English quarterly/10pp free

Finnish insurance industry.

13042 FINNISH LABOUR REVIEW
Finland - Ministry of Labour

Finland - English/Finnish/Swedish quarterly/1990 120pp Fmk130 P.A. ISSN No. 0787-510X

Labour policy and labour statistics.

13043 FINNISH MERCHANT MARINE
National Board of Navigation Finland

Finland - English/Finnish/Swedish annual/1918 261pp (1991) Fmk90 ISSN No. 0355-3981

Merchant vessels that are in the Finnish ship register: Technical data, owners, names, addresses etc.

13044 FINNISH METAL, ENGINEERING AND ELECTROTECHNICAL INDUSTRY
Federation of Finnish Metal Engineering and Electrotechnical Industries

Finland - Finnish annual

Statistics on production, foreign trade and manpower.

13045 FINNISH TRADE REVIEW
Finnish Foreign Trade Association

Finland - quarterly/45pp US$60.00 P.A. ISSN No. 0015-2463

Finnish national trade is reviewed.

13046 FOREIGN TRADE
Board of Customs

Finland - English/Finnish/Swedish annual/ Vol.1 680pp, Vol.2 190pp, Vol.3 380pp Vol.1 Fmk460, Vol.2 Fmk180, Vol.3 Fmk250

Vol.1: foreign trade and the harmonized system; Vol.2: statistics relating to foreign trade and customs administration; Vol.3: foreign trade by commodities and categories of goods as well as annual statistics on the total trade by countries in terms of the SITC.

13047 FOREIGN TRADE
Board of Customs

Finland - English/Finnish/Swedish monthly/1904 76pp Fmk180 per year

Trade with countries by SITC/ by countries of origin/ imports and exports by harmonized system groups. Detailed statistics by countries are available on microfiche or photocopies.

13048 GLOBAL FORECASTING SERVICE
Business International Ltd

Finland - English quarterly/£350.00 P.A.

Provides a range of medium-term forecasts covering main political, economic and business trends. A quarterly Global Outlook is included.

13049 HINTS TO EXPORTERS VISITING FINLAND
Department of Trade and Industry Export Publications

Finland - English irregular/1991 66pp £5.00

This publication provides detailed practical information that is useful for those planning a business visit. Topics include currency information, economic factors, import and exchange control regulations, methods of doing business and general information.

13050 HORTICULTURE ENTERPRISE REGISTER
National Board of Agriculture

Finland - Finnish/Swedish annual/ 1988 106pp Fmk130 ISSN No. 0784-8404

Details the Finnish horticultural sector and includes information on enterprises.

13051 HOUSEHOLD BUDGET SURVEY
Statistics Finland

Finland - Finnish/Swedish every 5 years/1961

Household expenditure and consumption. (A subtitle in series "Income and consumption").

13052 HOUSING
Statistics Finland

Finland - Finnish/Swedish irregular/ 1986 Fmk530 ISSN No. 0784-8307

Dwellings, buildings, prices, rents, housing conditions.

13053 HOUSING CONSTRUCTION
Statistics Finland

Finland - Finnish/Swedish irregular/ 1968 Fmk1220 ISSN No. 0784-8390

Housing construction by volume and type.

13054 INCOME AND CONSUMPTION
Statistics Finland

Finland - English/Finnish/Swedish irregular/1972 Fmk1145 ISSN No. 0784-8420

Income, consumption, household budgets.

13055 INCOME DISTRIBUTION STATISTICS
Statistics Finland

Finland - English/Finnish/Swedish annual/1977 90pp Fmk150 ISSN No. 0785-9880

Income distribution in Finland also broken down on a regional basis.

13056 INSURANCE COMPANIES IN FINLAND
Federation of Finnish Insurance Companies

Finland - English/Finnish/Swedish/ German annual/6pp free

Company information and general information on companies in the insurance sector.

13057 INSURANCE IN FINLAND
Federation of Finnish Insurance Companies

Finland - Finnish twice a year free ISSN No.0356-9993

Covers the Finnish insurance sector, detailing activities and balances.

13058 INSURANCE IN FINNISH SOCIETY
Federation of Finnish Insurance Companies

Finland - English annual/30pp free

The insurance market and competition.

13059 INTEREST RATES
Statistics Finland

Finland - Finnish/Swedish quarterly/1962 23pp Fmk180 ISSN No. 0784-9776X

A subtitle in series "Finance".

13060 INVESTING, LICENSING AND TRADING CONDITIONS ABROAD: FINLAND
Business International Ltd

Finland - English annual/£825.00 for complete European series

Loose-leaf format covering the state role in industry, competition rules, price controls, remittability of funds, taxation, incentives and capital sources, labour and foreign trade.

13061 KANSALLIS-OSAKE-PANKKI ECONOMIC REVIEW
Kansallis-Osake-Pankki

Finland - English Finnish annual/ free ISSN No. 022-8449

An economic review produced by the bank.

13062 KAUPPAKAMARI
The Central Chamber of Commerce of Finland

Finland - Finnish/Swedish 10 per year/1921 50pp Fmk305 P.A. ISSN No. 0781-6375

Chamber of Commerce monthly report.

13063 LABOUR FORCE STATISTICS
Statistics Finland

Finland - English/Finnish/Swedish annual/1976 300pp Fmk125 ISSN No. 0785-0050

Employment and unemployment statistics.

13064 LABOUR MARKET
Statistics Finland

Finland - English/Finnish/Swedish irregular Fmk1290 ISSN No. 0785-0107

A survey of the Finnish labour market covering numbers employed/unemployed and employment by sector etc.

13065 LISTED COMPANIES IN FINLAND
Kansallis-Osake-Pankki

Finland - English/Finnish annual/ 192pp ISSN No. 0357-5691

Covers companies on the Finnish stock exchange.

13066 MONTHLY REVIEW OF AGRICULTURAL STATISTICS
National Board of Agriculture

Finland - English/Finnish/Swedish monthly/1960 25pp Fmk240 ISSN No. 0786-938X

Statistical survey of Finnish agriculture and agricultural production.

13067 MOTOR VEHICLES IN FINLAND
Statistics Finland

Finland - English/Finnish/Swedish annual/1969 55pp Fmk90 ISSN No. 0785-613X

Details number of vehicles, type and distribution and new registrations.

13068 NATIONAL ACCOUNTS
Statistics Finland

Finland - Finnish/Swedish irregular/ 1968 Fmk930 ISSN No. 0784-8331

National statistics and key economic indicators.

13069 NATIONAL BUDGET FOR FINLAND
Ministry of Finance
Economics Department

Finland - English/Finnish annual/ 72pp Fmk42 ISSN No. 0532-9280

13070 NAVIGATOR
Finnish Shipowners Association

Finland - English/Finnish/Swedish 11 per year/Fmk100 P.A. ISSN No. 0355-7871

Shipping industry.

13071 POPULATION
Statistics Finland

Finland - Finnish/Swedish irregular/ 1968 Fmk1060 ISSN No. 0784-8447

Vital statistics, population structure, population projections.

13072 POSTS AND TELECOMMUNICATIONS OF FINLAND: STATISTICS
Posts and Telecommunications of Finland

Finland - English/Finnish/Swedish annual/116pp

13073 PRELIMINARY DATA ON INDUSTRY
Statistics Finland

Finland - English/Finnish/Swedish annual/1968 100pp Fmk170

Industrial production (a subtitle in series "Manufacturing").

13074 PRICES
Statistics Finland

Finland - Finnish/Swedish quarterly/ 1966 Fmk145 ISSN No. 0784-8315

Consumer prices.

13075 PRODUCER PRICE INDICES
Statistics Finland

Finland - Finnish/Swedish monthly/ 1968 15pp Fmk290 ISSN No. 0784-817X

13076 PUBLIC ECONOMY
Statistics Finland

Finland - Finnish/Swedish irregular/ Fmk705 ISSN No. 0784-8323

Government finance, finance of municipalities.

13077 QUARTERLY COUNTRY TRADE REPORTS: FINLAND
United Kingdom Iron and Steel Statistics Bureau

Finland - English quarterly/£200.00 P.A. for four countries ISSN No. 0952-5890

Cumulative trade statistics of imports and exports of steel products.

13078 RAILWAY STATISTICS
Finnish State Railways

Finland - English/Finnish/Swedish annual/150pp Fmk55

Rail statistics for Finland including volume of traffic.

13079 SAASTO PANKKI
Saasto Pankkiliitto

Finland - Finnish 13 issues per year/ 32pp Fmk250 ISSN No. 0036-2123

Covers the Finnish banking sector.

13080 SERVICES
Statistics Finland

Finland - Finnish irregular/1990 Fmk245 ISSN No. 0788-5423

Covers the Finnish service sector.

13081 SPAT BANKEN
Saasto Pankkiliitto

Finland - Swedish monthly/32pp Fmk250 ISSN No. 0355-614X

13082 STATEMENT OF THE BANK OF FINLAND
Bank of Finland

Finland - English/Finnish/Swedish 4 times a month/free

Balance sheet, foreign exchange reserves, note issue.

13083 STATISTICAL REPORTS OF THE BOARD OF NAVIGATION: DOMESTIC WATERBORNE TRAFFIC
National Board of Navigation Finland

Finland - Finnish/Swedish annual/ 1976 50pp (1991) Fmk25 ISSN No. 0785-6881

Domestic waterborne traffic in Finland: Both coastal and inland waterways traffic, passenger + cargo and timber floating.

13084 STATISTICAL REPORTS OF THE BOARD OF NAVIGATION: MERCHANT FLEET
National Board of Navigation Finland

Finland/Nordic countries - English/ Finnish/Swedish annual/1918 48pp (1991) Fmk55 ISSN No. 1235-158X

Finnish merchant fleet and its structure by size, type and age + gross freight revenue and value.

13085 STATISTICAL REPORTS OF THE BOARD OF NAVIGATION: SHIPPING BETWEEN FINLAND AND FOREIGN COUNTRIES
National Board of Navigation Finland

Finland/International - English/ Finnish/Swedish annual/1856 87pp(1991) Fmk58 ISSN No. 1235-2985

Shipping between Finland and foreign countries, domestic waterborne traffic, cargo, passengers, means of transport.

13086 STATISTICAL YEARBOOK OF FINLAND
Statistics Finland

Finland - English/Finnish/Swedish annual/600pp Fmk265 ISSN No. 0081-5063

Statistical survey of general economic conditions.

13087 TALOUSELAMA/ ("BUSINESS LIFE")
Talentum

Finland - Finnish 41 issues per year/ Fmk790 P.A. ISSN No. 0356-5106

All topics of current interest in business in Finland. Once a year publishes 'Top 500' on Finland's largest companies.

13088 TEKNIKKA & TALOUS/ (TECHNOLOGY AND BUSINESS)
Oy Talentum Ab

Finland - Finnish 48 issues per year/ Fmk713 P.A. ISSN No. 0785-997X

All aspects of technology, industry and business.

13089 TELECOMMUNICATION STATISTICS
Ministry of Transport and Communications

Finland/OECD Countries - English Finnish annual/60pp

13090 TEXTILE INDUSTRY STATISTICS
Association of Finnish Textile Industries

Finland - English annual/38pp free

Production, domestic and foreign trade, labour markets.

13091 TOURISM STATISTICS
Finnish Tourist Board

Finland - English/Finnish/Swedish annual/1975 ISSN No. 0355-1113

Statistical survey of the tourist industry.

13092 TRADE
Statistics Finland

Finland - Finnish/Swedish irregular/ 1968 Fmk395 ISSN No. 0784-834X

Retail trade, wholesale trade.

13093 TRANSPORT
Statistics Finland

Finland - English/Finnish/Swedish irregular/1968 Fmk910 ISSN No. 0784-8358

Motor vehicles, road accidents, accommodation.

13094 TRANSPORT AND COMMUNICATIONS YEARBOOK
Statistics Finland

Finland - English/Finnish/Swedish annual/1958 140pp Fmk120

Railway traffic, road traffic, water traffic, air traffic, communications.

13095 TRANSPORT AND THE ENVIRONMENT IN FINLAND
Ministry of Transport and Communications

Finland - English Finnish annual/ 78pp Fmk65 ISSN No. 0784-8455

Transport and environment.

13096 UNITAS: ECONOMIC QUARTERLY REVIEW
Union Bank of Finland

Finland - English/Finnish/Swedish/ German quarterly/1929 30pp free ISSN No. 0041-7130

Economic developments in Finland and abroad, money and foreign exchange, markets, investments.

13097 VOLUME INDEX OF INDUSTRIAL PRODUCTION
Statistics Finland

Finland - Finnish/Swedish monthly/ 1968 40pp Fmk350 ISSN No. 0784-8234

Indices, industrial production - a subtitle in series "Manufacturing".

13098 WAGE AND SALARY INDICES
Statistics Finland

Finland - Finnish/Swedish quarterly/1979 24pp Fmk175 ISSN No. 0784-8218

Indices, wages and salaries for Finnish workers broken down by standard criteria.

13099 WAGES
Statistics Finland

Finland - Finnish/Swedish irregular/ 1966 Fmk2405 ISSN No. 0784-8374

Wages and salaries reviewed on a national basis.

13100 YEARBOOK OF FARM STATISTICS
National Board of Agriculture

Finland - English/Finnish/Swedish annual/1983 254pp Fmk190 ISSN No.0784-8404

Statistical survey of the agricultural sector and agricultural products.

13101 YEARBOOK OF INDUSTRIAL STATISTICS: VOLUME 1
Statistics Finland

Finland - Finnish/Swedish annual/ 1913 250pp Fmk260 ISSN No. 0786-7077

Industrial production.

13102 YEARBOOK OF INDUSTRIAL STATISTICS: VOLUME 2
Statistics Finland

Finland - Finnish/Swedish annual/ 1913 380pp Fmk260 ISSN No. 0786-7085

Industrial production.

13103 YEARBOOK OF TRANSPORT STATISTICS FOR FINLAND
Ministry of Transport and Communications

Finland - English/Finland/Swedish annual/135pp Fmk90 ISSN No. 0784-8358/0785-6172

Sea transport, railway transport, road transport, air transport.

FRANCE

Country profile:	France
Official name:	French Republic
Area:	543,965 sq.km.
	Bordered in the north by the English Channel, Belgium and Luxembourg, the south by the Mediterranean and Spain, the east by Germany, Switzerland and Italy, and the west by the Atlantic Ocean.
Population:	56.89m (1991)
Capital:	Paris
Language(s):	French
Currency:	Franc 100 centimes = 1 Franc
National bank:	Banque de France
Government and Political structure:	Republic with elected President. There are two Houses of Parliament, the National Assembly with 577 deputies and the Senate with 321 Senators.

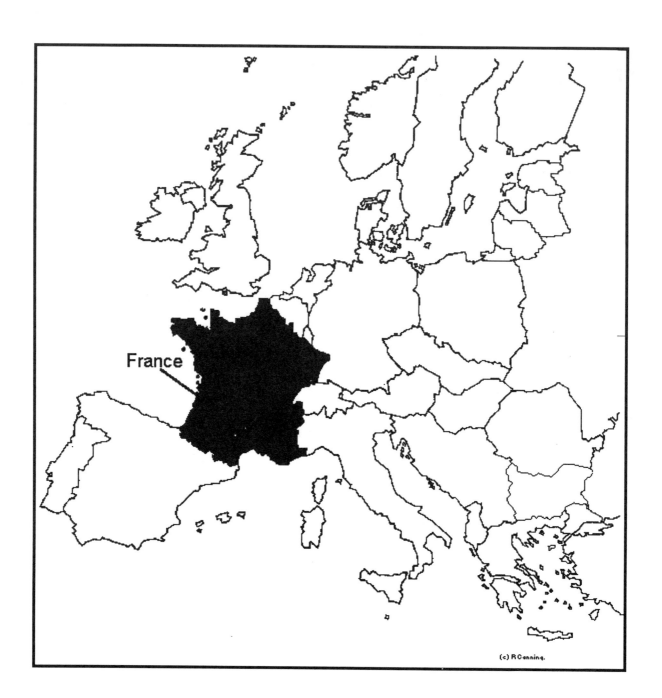

France

14001 CAHIERS: ANALYSES ET ÉTUDES
Ministère de l'Agriculture: SCEE
France - French quarterly/FrF240
ISSN No. 0998-4148
Articles on topical research or projects in the agricultural sector.

14002 ACTIONS: STATISTIQUES BOURSIÈRES MENSUELLES
Société des Bourses Françaises: Bourse de Paris
France - French monthly/FrF2,200 P.A.
Monthly stock exchange statistics.

14003 AGRESTE CONJONCTURE: COMMERCE EXTÉRIEUR AGRO-ALIMENTAIRE
Ministère de l'Agriculture: SCEE
France - French monthly/4pp FrF160 P.A.
Current statistics on agricultural and food exports.

14004 AGRESTE CONJONCTURE: FRUITS
Ministère de l'Agriculture: SCEE
France - French monthly/14pp FrF160 P.A.
Statistics on the production of certain fruits.

14005 AGRESTE CONJONCTURE: GRANDES CULTURES
Ministère de l'Agriculture: SCEE
France - French monthly/4pp FrF160 P.A.
Statistics on the production of cereals and root crops.

14006 AGRESTE CONJONCTURE: LA CONJONCTURE GÉNÉRALE (AGRICULTURE ET AGRO-ALIMENTAIRE)
Ministère de l'Agriculture: SCEE
France - French monthly/46pp FrF490 P.A.
A digest of economic information on agriculture and the food industry.

14007 AGRESTE CONJONCTURE: LAIT ET PRODUITS LAITIERS
Ministère de l'Agriculture: SCEE
France - French monthly/4pp FrF160 P.A.
Statistics on the dairy industry.

14008 AGRESTE CONJONCTURE: LÉGUMES
Ministère de l'Agriculture: SCEE
France - French monthly/20pp FrF300 P.A.
Statistics on production of various vegetables.

14009 AGRESTE CONJONCTURE: PRODUCTIONS ANIMALES
Ministère de l'Agriculture: SCEE
France - French irregular/4pp FrF110
Statistics on animal production.

14010 AGRESTE CONJONCTURE: VITICULTURE
Ministère de l'Agriculture: SCEE
France - French monthly/FrF65 P.A.
Statistics on viniculture.

14011 AGRESTE DONNÉES CHIFFRÉES: STATISTIQUES AGRICOLE ANNUELLE
Ministère de l'Agriculture: SCEE
France - French annual/155pp
Agricultural statistics.

14012 AGRESTE DONNÉES CHIFFRÉES: STATISTIQUES FORESTIÈRES
Ministère de l'Agriculture: SCEE
France - French annual/70pp
Brief economic and business guide.

14013 AGRESTE: ÉTUDES
Ministère de l'Agriculture: SCEE
France - French monthly/FrF900
ISSN No. 1150-1324
Series of reports containing statistics on all aspects of agriculture.

14014 AGRESTE SÉRIES: ANIMAUX HEBDO
Ministère de l'Agriculture: SCEE
France - French weekly/12pp FrF510 P.A.
Weekly data on slaughtering, prices etc. of cattle and pigs.

14015 AGRESTE SÉRIES: AVICULTURE
Ministère de l'Agriculture: SCEE
France - French monthly/FrF330 P.A.

Data on the poultry industry.

14016 AGRESTE SÉRIES: COMMERCE EXTÉRIEUR BOIS ET DÉRIVÉS
Ministère de l'Agriculture: SCEE
France - French quarterly/FrF140 P.A.
Covers forestry products. Import/export figures by product.

14017 AGRICULTURE: ÉCONOMIE DE L'AGRICULTURE FRANÇAISE EN EUROPE: FORCES ET FAIBLESSES
Ministère de l'Agriculture: Institut Économique Agricole (IEA)
France - French/192pp
Evolution of agriculture in France.

14018 ANNALES D'ÉCONOMIE ET DE STATISTIQUE
Institut National de la Statistique et des Études Économiques
France - French quarterly/FrF500 P.A. ISSN No. 0769-489X
Results of statistical and economic research.

14019 ANNUAIRE ABRÉGÉ
Direction Générale des Douanes et Droits Indirects
France - French annual
Foreign trade by product and trading country.

14020 ANNUAIRE DE STATISTIQUE AGRICOLE
Ministère de l'Agriculture: Service Central des Enquêtes et Études
France - French annual/4 Vols FrF260
Agricultural statistics including climate, labour, technology, training and data on production.

14021 ANNUAIRE DE STATISTIQUES INDUSTRIELLES
La Documentation Française
France - French annual/500pp FrF370
Industrial statistics.

14022 ANNUAIRE RÉTROSPECTIF DE LA FRANCE 1948-1988
Institut National de la Statistique et des Études Économiques
France - French/600pp FrF550
Statistical compilation covering 40 years of the French economy.

14023 ANNUAIRE STATISTIQUE DE LA FRANCE

Institut National de la Statistique et des Études Économiques

France - French annual/900pp FrF550 ISSN No. 0066-3654

Detailed statistics on population, economic and social trends; includes technical commentary.

14024 BILANS DE L'ÉNERGIE 1970 À 1990

La Documentation Française

France - French/68pp FrF55

Energy balances for France.

14025 BLOC-NOTES DE L'OBSERVATOIRE ÉCONOMIQUE DE PARIS

Institut National de la Statistique et des Études Économiques

France - French monthly/FrF184 P.A. ISSN No. 0180-9105

Bibliography and guide to economic information; each issue covers a specific topic.

14026 BULLETIN MENSUEL

Banque de France

France - French monthly/FrF300

Covers monetary and financial news.

14027 BULLETIN MENSUEL DE STATISTIQUE

Institut National de la Statistique et des Études Économiques

France - French monthly/FrF399 P.A. ISSN No. 0007-4713

Monthly statistical bulletin covering employment, industry etc.

14028 BULLETIN MENSUEL DE STATISTIQUE INDUSTRIELLE

Ministère de l'Industrie du Commerce Extérieur: Direction Générale des Stratégies Industrielles

France - French monthly/FrF420 P.A.

Statistical bulletin covering more than 700 products.

14029 BULLETIN TRIMESTRIEL DE LA BANQUE DE FRANCE

Banque de France

France - French quarterly/FrF145

A report on the current economic situation; technical analytic articles; statistical tables; and regulatory documentation.

14030 CAHIERS FRANÇAIS

La Documentation Française

France - French 5 issues per year/ FrF275 P.A. ISSN No. 0008-0217

Economic conditions in France.

14031 COMMERCE EXTÉRIEUR DE LA FRANCE

Centre Français du Commerce Extérieur

France - French annual/50pp FrF100

Figures for foreign trade.

14032 COMPTE RENDU ANNUEL DES OPERATIONS DE LA BANQUE DE FRANCE

Banque de France

France - French annual/free ISSN No. 0067-3927

Annual accounts of the Banque de France and analysis of French economy.

14033 COUNTRY BOOKS: FRANCE

United Kingdom Iron and Steel Statistics Bureau

France - English annual/£75.00 ISSN No. 0952-5904

Annual summary tables showing production, materials consumed, apparent consumption, imports and exports of 130 products by quality and market.

14034 COUNTRY FORECASTS: FRANCE

Business International Ltd

France - English quarterly/£360.00 P.A.

Contains an executive summary, sections on political and economic outlooks and the business environment, a fact sheet, and key economic indicators. Includes a quarterly Global Outlook.

14035 COUNTRY PROFILE: FRANCE

Business International Ltd

France - English annual/43pp £50.00 ISSN No. 026-5340

Annual survey of political and economic background. Included as part of the "Country Report" subscription.

14036 COUNTRY PROFILE: FRANCE

Department of Trade and Industry Export Publications

France - English irregular/1990 42pp £10.00

This publication provides a general description of the country in terms of its geography, economy and population, and gives information about trading in the country with details like export conditions and investment covered.

14037 COUNTRY REPORT: FRANCE

Business International Ltd

France - English quarterly/25-40pp £160.00 P.A. ISSN No. 0269-5286

Contains an executive summary, sections on political and economic structure, outlook for the next 12-18 months, statistical appendices, and a review of key political development. Includes an annual Country Profile.

14038 COURRIER DES STATISTIQUES

Institut National de la Statistique et des Études Économiques

France - French quarterly/FrF156 P.A. ISSN No. 0151-9514

Report on statistical research, publications, organizations, forthcoming events etc.

14039 CROSS CHANNEL CONNECTIONS

France - British Chamber of Commerce and Industry

France - French monthly/4pp free

Company news; cooperative ventures; import/export figures.

14040 CROSS CHANNEL TRADE

France - British Chamber of Commerce and Industry

France/UK - English monthly/ FrF250 ISSN No. 0766-4389

Each issue looks at a specific sector; also general economic news.

14041 DEUX SIÈCLES DE TRAVAIL EN FRANCE

Institut National de la Statistique et des Études Économiques

France - French/1992 200pp FrF140

Statistical examination of trends in employment over the last 200 years.

14042 DICTIONNAIRE DES SOURCES STATISTIQUES

Institut National de la Statistique et des Études Économiques

France - French irregular/3 Vols FrF120 Vols 1 & 2 or Vols 1 & 3

Offers access to principal national statistical sources by sector, keyword + originating establishments. Vol 1 - Introduction & Index, Vol 2 - Demographic & Social statistics, Vol 3 - Production, monetary and financial statistics.

14043 DOING BUSINESS IN FRANCE

Ernst & Young International

France - English irregular/1991 110-130pp free to clients

A business guide which includes an executive summary, the business environment, foreign investment, company structure, labour, taxation, financial reporting and auditing, and general country data.

14044 DOING BUSINESS IN FRANCE

Price Waterhouse

France - English irregular/1993 330-340pp free to clients

A business guide which includes sections on the investment climate, doing business, accounting and taxation and general country data. Updated by supplements.

14045 DONNÉES CHIFFRÉES: IAA

Ministère de l'Agriculture: SCEE

France - French 10 per year/ FrF560

Data on agriculture and food industries.

14046 DONNÉES ÉCONOMIQUES DE L'ENVIRONNEMENT

La Documentation Française

France - French 1991/144pp FrF90 ISBN No. 2110869569

Environmental economics.

14047 DONNÉES SOCIALES

Institut National de la Statistique et des Études Économiques

France - French every 3 years/ 464pp FrF250 ISSN No. 0758-6531

Data on social trends in France.

14048 ECOFLASH

Institut National de la Statistique et des Études Économiques

France - French 10 issues per year/ FrF112 ISSN No. 0296-4449

Summary tables and graphs on a variety of economic and social issues.

14049 ÉCONOMIE ET PRÉVISION

La Documentation Française

France - French 5 issues per year/ FrF370 P.A.

Economic forecasting for France.

14050 ÉCONOMIE ET STATISTIQUE

Institut National de la Statistique et des Études Économiques

France - French monthly/FrF513 P.A. ISSN No. 0336-1454

Economic and social review.

14051 ÉCONOMIE PROSPECTIVE INTERNATIONALE

La Documentation Française

France - French 4 issues per year/ FrF365 P.A. ISSN No. 0242-7818

14052 EMPLOI-CROISSANCE SOCIÉTÉ

La Documentation Française

France - French irregular/96pp FrF70

Employment growth in France.

14053 ENQUÊTE SUR LE COMPORTEMENT DES ENTREPRISES DE L'INDUSTRIE ET DU BATÎMENT - GÉNIE CIVIL

Banque de France

France - French annual/FrF50

Survey of 6,000 companies with data on turnover, exports, investments and forecasts.

14054 ENQUÊTE FINANCIÈRE

Banque de France

France - French quarterly/FrF120

Results of surveys carried out by Banque de France among credit institutions and industrial establishments.

14055 ENQUÊTE MENSUELLE DE CONJONCTURE

Banque de France

France - Frence 11 issues per year/ FrF350 ISSN No. 0242-5815

Economic trends in industry; production, orders, stocks, prices, employment.

14056 FASCICULES DE RÉSULTATS DE LA CENTRALE DES BILANS

Banque de France

France - French annual/FrF50

Analysis of the performance of companies and industrial sectors giving various financial ratios.

14057 FINANCING FOREIGN OPERATIONS: FRANCE

Business International Ltd

France - English annual/£695.00 for complete European series

Business overview, currency outlook, exchange regulations, monetary system, short/medium/ long-term and equity financing techniques, capital incentives, cash management, investment trade finance and insurance.

14058 FRANCE

United Kingdom Iron and Steel Statistics Bureau

France - English annual/£80 ISSN No. 0952-5904

Report on the steel industry showing production, consumption and details of imports and exports.

14059 GLOBAL FORECASTING SERVICE: FRANCE

Business International Ltd

France - English quarterly/£350.00 P.A.

Economic forecasts.

14060 GRAPH AGRI 9-

Ministère de l'Agriculture: SCEE

France - French annual/150pp FrF135

Graphs with analysis and commentary on agriculture and forestry.

14061 GRAPH AGRI 9- : LE FASCICULE RÉGIONAL

Ministère de l'Agriculture: SCEE

France - French annual/FrF140

Statistical/graphic data on agriculture at a regional level.

14062 HINTS TO EXPORTERS VISITING FRANCE

Department of Trade and Industry Export Publications

France - English irregular/1991 72pp £5.00

This publication provides detailed practical information that is useful for those planning a business visit. Topics include currency information, economic factors, import and exchange control regulations, methods of doing business and general information.

14063 IN DEPTH COUNTRY APPRAISALS ACROSS THE EC - CONSTRUCTION AND BUILDING SERVICES - FRANCE

Building Services Research and Information Association

France - English 1991/£125.00

Part of an 11 volume survey covering construction industries in EC countries.

14064 INDUSTRIES ÉLECTRIQUES ET ÉLECTRONIQUES

French Electrical and Electronics Industries Association - FIEE

France - French quarterly/8pp

Information on the electrical and electronics sectors in France.

14065 INFO

French Chamber of Commerce in Great Britain

France - English/French 7 issues per year/£35

Official publication of the French Chamber of Commerce in Great Britain.

14066   INFORMATIONS RAPIDES
Institut National de la Statistique et des Études Économiques

France - French 300 per year/ FrF1,921 P.A. ISSN No. 0151-1475

Notes on current economic developments; complete results of quarterly national accounts.

14067   INGÉNIERIE, ÉTUDES ET CONSEILS
Ministère de l'industrie (SESSI)

France - French annual/FrF100

Engineering data on size, production and exports.

14068   INSEE PREMIÈRE
Institut National de la Statistique et des Études Économiques

France - French 60 issues per year/ FrF558 ISSN No. 0997-3129

Economic and social data with commentary; each issue focuses on a specific topic.

14069   INSEE RÉSULTATS: CONSOMMATION - MODE DE VIE
Institut National de la Statistique et des Études Économiques (INSEE)

France - French 15 issues per year/ FrF1,506 P.A. ISSN No. 0998-4720

Detailed statistics on household spending and budgeting.

14070   INSEE RÉSULTATS: DÉMOGRAPHIE ET EMPLOI
Institut National de la Statistique et des Études Économiques

France - French irregular/FrF730 for 10 issues

Detailed statistics on population.

14071   INSEE RÉSULTATS: EMPLOI-REVENUS
Institut National de la Statistique et des Études Économiques (INSEE)

France - French 13 issues per year/ FrF1,302 P.A. ISSN No. 0998-4747

Detailed statistics on the employment market, salaries, domestic incomes etc.

14072   INSEE RÉSULTATS: ÉCONOMIE GÉNÉRALE
Institut National de la Statistique et des Études Économiques (INSEE)

France - French 20 issues per year/ FrF2,008

Detailed statistics on national accounts, commerce, agriculture, industry, transport, regional economics.

14073   INVESTING, LICENSING AND TRADING CONDITIONS ABROAD: FRANCE
Business International Ltd

France - English annual/£825.00 for complete European series

Loose-leaf format covering the state role in industry, competition rules, price controls, remittability of funds, taxation, incentives and capital sources, labour and foreign trade.

14074   L'ESPACE ÉCONOMIQUE FRANÇAIS
Institut National de la Statistique et des Études Économiques

France - French annual/232pp FrF150

Economic review of France.

14075   L'IMPLANTATION ÉTRANGÈRE DANS L'INDUSTRIE
Ministère de l'industrie (SESSI)

France - French annual/FrF100

Statistical report on foreign ventures in French industry. Data by sector, by origin and by region.

14076   L'INDUSTRIE PÉTROLIÈRE EN 19–
La Documentation Française

France - French annual/224pp FrF110

Covers the French petroleum industry.

14077   L'ÉCONOMIE FRANÇAISE EN 19–: LES RÉSULTATS, LES RÉFORMES
La Documentation Française

France - French annual/FrF30

Economic commentary and current issues in the economy.

14078   LA BALANCE DES PAIEMENTS DE LA FRANCE
Banque de France

France - French annual/FrF150

Detailed analysis of France's balance of payments.

14079   LA DISPERSION DES PERFORMANCES DES ENTREPRISES EN 1989: TABLEAU DE BORD DES SECTEURS INDUSTRIELS
La Documentation Française

France - French/1991 322pp FrF300

Looks at company performance in industrial sectors.

14080   LA FRANCE DES ENTREPRISES
Institut National de la Statistique et des Études Économiques

France - French annual/250pp FrF50

Comparison of French firms to those abroad; principal areas of investment; employment trends in various sectors.

14081   LA FRANCE ET SES RÉGIONS
Institut National de la Statistique et des Études Économiques

France - French every 2 years/ 150pp FrF60

14082   LA LETTRE DES INDUSTRIES ÉLECTRIQUES ET ÉLECTRONIQUES
French Electrical and Electronics Industries Association - FIEE

France - French quarterly/1990 8pp free

Economic information on electric and electronics sectors in France.

14083   LA LETTRE MENSUELLE RÉGIONALE
Banque de France

France - French monthly/free

Economic trends for each region.

14084   LA MONNAIE EN 19–
Banque de France

France - English/French annual/ 63pp FrF30 (French edition) FrF 50 (English edition)

Summarises chief statistics relevant to the financial situation.

14085   LA NOTE D'INFORMATION DE LA DIRECTION DES TRANSPORTS TERRESTRES (DTT)
Ministère de l'Equipement, du Logement et des Transports: Direction des Transports Terrestres (DTT)

France - French 6-7 issues per year/60pp FrF300 P.A. ISSN No. 0181-0103

Land transport.

14086   LA NOTE FINANCIÈRE ANNUELLE
Banque de France

France - French annual/FrF45

Analyses the financial situation in each region of France.

14087 **LA SITUATION DE L'INDUSTRIE EN 19– : RÉSULTATS DÉTAILLÉES DE L'ENQUÊTE ANNUELLE D'ENTREPRISE**
La Documentation Française
France - French annual/4 Vols
Statistics on French industry.

14088 **LA SITUATION DU SYSTÈME PRODUCTIF EN 19–**
Banque de France
France - French annual/FrF100
Annual study of production based on the results of industrial firms.

14089 **LE BULLETIN**
Ministère de l'Agriculture: SCEE
France - French monthly/FrF560 P.A.
Figures for current & preceding 2 complete years on agriculture, prices and trade.

14090 **LE COMPTE DU TOURISME EN 19– : RAPPORT DE LA COMMISSION DES COMPTES DU TOURISME**
La Documentation Française
France - French annual/144pp FrF200
Survey of tourism.

14091 **LE NOUVEL ÉCONOMISTE**
Le Nouvel Économiste SA
France - French weekly/1975
Weekly economic journal

14092 **LES 4 PAGES**
Ministère de l'Industrie du Commerce Extérieur (SESSI)
France - French monthly/4pp FrF100 P.A.
An illustrated analysis of a sector or a theme or the results of a SESSI survey.

14093 **LES CAHIERS ÉCONOMIQUES ET MONÉTAIRES**
Banque de France
France - French irregular/FrF40-80
Each issue consists of a number of articles on a specific economic or monetary topic.

14094 **LES CHIFFRES CLÉS DE L'INDUSTRIE DANS LES RÉGIONS**
Ministère de l'industrie (SESSI)
France - French annual/FrF180
Survey of industry in the regions.

14095 **LES CHIFFRES CLÉS DE L'INDUSTRIES**
Ministère de l'industrie du Commerce Extérieur: (SESSI)
France - French annual/FrF150
Essential statistics on French industry; includes tables, maps and graphs.

14096 **LES COMPTES DE L'AGRICULTURE FRANÇAISE EN 19-**
Institut National de la Statistique et des Études Économiques (INSEE)
France - French annual/200pp FrF100
Agricultural accounts giving data on supply and demand of agricultural products, revenue and expenditure and overseas trade.

14097 **LES COMPTES DE LA SÉCURITÉ SOCIALE**
La Documentation Française
France - French annual/140pp FrF90
Financial data on the social security system.

14098 **LES CONSOMMATIONS D'ÉNERGIE DANS L'INDUSTRIE EN 19– : TRAITS FONDAMENTAUX DU SYSTÈME INDUSTRIEL FRANÇAIS**
La Documentation Française
France - French annual/186pp FrF80
Use of major forms of energy in French industry.

14099 **LES DÉPENSES DES TOURISTES ÉTRANGÈRES**
La Documentation Française
France - French irregular/120pp FrF150
Covers the French tourist industry.

14100 **LES ENTREPRISES À L'ÉPREUVE DES ANNÉES 80**
Institut National de la Statistique et des Études Économiques
France - French 1991/344pp FrF130
Review of company activity in France during the 1980s.

14101 **LES ÉTUDES DE LA CENTRALE DES BILANS**
Banque de France
France - French irregular/FrF80 (sector studies) FrF 120 (thematic studies)

Sector studies offer a statistical overview of the economic sector in question. Thematic studies cover broad financial issues.

14102 **LES MOYENS DE PAIEMENT ET CIRCUITS DE RECOUVREMENT**
Banque de France
France - French annual/FrF50
Statistical yearbook covering the previous four years. Covers the financial sector.

14103 **LES PRINCIPALES BRANCHES D'ACTIVITÉ EN 19–**
Banque de France
France - French annual/FrF80
Statistical data and commentary on the major commercial sectors in France.

14104 **LES PRINCIPALES PROCÉDURES DE FINANCEMENT DES BESOINS DES ENTREPRISES ET DES MÉNAGES**
Banque de France
France - French irregular/240pp FrF90
Review of credit mechanisms and recent changes in the structure of the French financial system.

14105 **LES STATISTIQUES DE COMMERCE EXTÉRIEUR: IMPORTATIONS ET EXPORTATIONS EN NGP**
Direction Générale des Douanes et Droits Indirect
France - French annual/1300pp 4 Vols
Statistics on imports and exports classified by product, country of origin and destination.

14106 **LES STATISTIQUES MONÉTAIRES ET FINANCIÈRES TRIMESTRIELLES**
Banque de France
France - French quarterly
Detailed statistics on money supply, financial assets, interest rates, money markets etc. Provided on demand to subscribers to "Bulletin trimestriel".

14107 **LES STATISTIQUES MONÉTAIRES MENSUELLES**
Banque de France
France - French monthly
Statistics on money supply, interest rates, money markets etc. Provided on request to subscribers to "Bulletin mensuel".

14108 LETTRE MENSUELLE DE CONJONCTURE
La Chambre de Commerce et d'Industrie de Paris
France - French monthly/FrF275 P.A. ISSN No. 0479-5369
Newsletter on economic activity.

14109 MÉTHODE DES SCORES DE LA CENTRALE DES BILANS
Banque de France
France - French irregular/FrF100
Study of failed companies.

14110 MONITEUR DU COMMERCE INTERNATIONAL
Centre Français du Commerce Extérieur
France - French annual/FrF2,680
Reports on current developments in external trade.

14111 NOTE DE CONJONCTURE
Institut National de la Statistique et des Études Économiques
France - French 3 issues per year + 2 supplements/FrF191 P.A. ISSN No. 0766-6268
Short-term economic analysis.

14112 NOTES D'INFORMATION
La Chambre de commerce et d'industrie de Paris
France - French frequent/FrF23-26
Brief newsletters on a variety of commercial and financial topics.

14113 NOTES D'INFORMATIONS
Banque de France
France - French irregular/10pp free
Short studies on a variety of topics - mainly money, credit and the activities of the Banque de France.

14114 NOTES ET ÉTUDES DOCUMENTAIRES
La Documentation Française
France - French 20 issues per year/ FrF850 P.A.
Journal of economic commentary.

14115 OECD ECONOMIC SURVEY: FRANCE
Organisation for Economic Co-operation and Development (OECD)
France - English irregular/£7.50 ISSN No. 0376-6438
Survey of economic development.

14116 THE PARIS BOURSE: ORGANIZATION AND PROCEDURES
Société des Bourses Françaises: Bourse de Paris
France - English/French irregular/ FrF22
Basic structures; market safeguards; how the Paris Bourse operates.

14117 PERSPECTIVES DE FINANCEMENT DE L'ÉCONOMIE FRANÇAISE
La Documentation Française
France - French irregular/FrF200 3 Vols
Economic comment and coverage of financial sectors.

14118 POPULATION
Institut National d'Études Démographiques - INED
France/International - French 6 issues per year/1945 250pp FrF360 P.A. ISSN No. 0032-4663
Results of demographic surveys.

14119 POPULATION ET SOCIÉTÉS
Institut National d'Études Démographiques - INED
France/International - French monthly/1969 4pp FrF75 P.A. ISSN No. 0184-7783
Demography and economics.

14120 PROBLÈMES ÉCONOMIQUES
La Documentation Française
France - French 50 issues per year/ FrF510 P.A. ISSN No. 0032-9304

14121 QUARTERLY COUNTRY TRADE REPORTS: FRANCE
United Kingdom Iron and Steel Statistics Bureau
France - English quarterly/£200.00 P.A. for four countries ISSN No. 0952-5904
Cumulative trade statistics of imports and exports of steel products.

14122 RAPPORT ANNUEL - CONSEIL ÉCONOMIQUE ET SOCIAL 19–
La Documentation Française
France - French annual/228pp FrF98
Annual report of the economic and social council.

14123 RAPPORT ANNUEL DE CONSEIL NATIONAL DE CRÉDIT
Banque de France
France - French annual/FrF90
Overview of the financial situation; analysis of statistics relevant to finance and banking.

14124 RAPPORT ANNUEL DE LA COMMISSION BANCAIRE
Banque de France
France - English/French annual/ FrF210 (French version) FrF 60 (English version without statistical tables)
Reviews relevant legislation and regulatory news concerning banking establishments.

14125 RAPPORT ANNUEL DE LA ZONE FRANC
Banque de France
France & French territories - French annual/FrF190
Economic situation in French territories.

14126 RAPPORT SUR LES COMPTES DE LA NATION
Institut National de la Statistique et des Études Économiques
France - French annual/4 Vols FrF8,000
Detailed report on national accounts; French competitive position in the world.

14127 RECENSEMENT DE LA POPULATION
Institut National de la Statistique et des Études Économiques
France - French every 8 years
Results of the latest census published in a number of volumes.

14128 REGARDS SUR L'ACTUALITÉ
La Documentation Française
France - French 10 issues per year/ FrF285 ISSN No. 0337-7091
Economic commentary.

14129 REVUE ANNUELLE DE STATISTIQUE DES BOURSES FRANÇAISES DE VALEURS
Société des Bourses Françaises: Bourse de Paris
France - French annual/FrF300
Data on the activity of the market during the year.

14130 REVUE DE LA CONCURRENCE ET DE LA CONSOMMATION
La Documentation Française
France - French bi-monthly/FrF225 P.A. ISSN No. 0220-9896
Consumer economics in France.

14131 REVUE FRANÇAISE DES AFFAIRES SOCIALES
Institut National de la Statistique et des Études Économiques
France - French quarterly/1946 US$95 ISSN No. 0035-2985
Articles on a variety of subjects connected with labour and social security.

14132 RÉSULTATS ANNUELS
Direction Générale des Douanes et Droits Indirects
France - French annual/ISSN No. 0071-8645
Detailed analysis and graphical representation for previous 3 years import export data.

14133 RÉSULTATS ANNUELS DES ENQUÊTES DE BRANCHE
Ministère de l'Industrie du Commerce Extérieur (SESSI)
France - French annual/FrF75
Market information on 12 major sectors given on an annual basis.

14134 RÉSULTATS DE L'ENQUÊTE ANNUELLE D'ENTREPRISE
Ministere de l'industrie (SESSI)
France - French annual/6 Vols
Vol.1 - preliminary results; vol.2 details on 40 industrial sectors; Vol.3 - 6 detailed results on 250 sectors.

14135 RÉSULTATS MENSUELS DES ENQUÊTES DE BRANCHE
Ministère de l'industrie (SESSI)
France - French monthly/FrF300 P.A.
Series covering eight industrial sectors with data for products in those sectors on production, deliveries, stocks, imports and exports.

14136 RÉSULTATS RÉGIONALISÉS
Direction Générale des Douanes et Droits Indirect
France - French annual
Regional import/export data classified by product, region of origin and destination.

14137 RÉSULTATS TRIMESTRIELS
Direction Générale des Douanes et Droits Indirects
France - French quarterly
Statistical bulletin on foreign trade organized by product & region.

14138 RÉSULTATS TRIMESTRIELS DES ENQUÊTES DE BRANCHE
Ministère de l'Industrie du Commerce Extérieur (SESSI)
France - French quarterly/FrF200 P.A.
Market information on 12 major sectors.

14139 SITUATION FINANCIÈRE DES RÉGIONS EN 19–
Banque de France
France - French annual/2 Vols FrF40 per volume
Statistics and commentary on the financial and banking sectors in each region.

14140 STATISTIQUES DE LA CONSTRUCTION
La Documentation Française
France - French monthly/FrF325 P.A. ISSN No. 0338-4160
Construction statistics.

14141 STATISTIQUES DE LA FORMATION PROFESSIONELLE CONTINUE FINANCÉE PAR LES ENTREPRISES: TRAITEMENTS DES DÉCLARATIONS D'EMPLOYEURS NO.2483
La Documentation Française
France - French irregular/140pp FrF115
Statistics on employers financing of professional training.

14142 STATISTIQUES MONÉTAIRES ET FINANCIÈRES
Banque de France
France - French annual/FrF100
Statistical data on money supply, credit and the capital markets.

14143 STATISTIQUES QUOTIDIENNES ET MENSUELLES DU MONET
Société des Bourses Françaises: Bourse de Paris
France - French daily/FrF5,930 P.A.
Daily report and monthly summary of the options market.

14144 TABLEAUX DE BORD DES SECTEURS INDUSTRIELS
Ministère de l'industrie (SESSI)
France - French annual/FrF300
Performance of 300 sectors according to 15 ratios.

14145 TABLEAUX DE L'ÉCONOMIE FRANÇAISE
Institut National de la Statistique et des Études Économiques
France - French annual/192pp FrF70 ISSN No. 0039-8802
A survey of the French economy; includes maps, tables and international comparisons.

14146 TABLEAUX DES CONSOMMATIONS D'ÉNERGIE EN FRANCE
La Documentation Française
France - French irregular/128pp FrF90
Energy development over the last 12 years and the part played by each form of energy.

14147 TENDANCES DE LA CONJONCTURE
Institut National de la Statistique et des Études Économiques
France - French quarterly/2 Vols FrF779 P.A. ISSN No. 0497-2007
Economic indicators published in two volumes with supplements.

14148 TRAVAIL ET EMPLOI
Centre d'Études et de Recherches sur les Qualifications
France - French quarterly/ US$65.00 P.A. ISSN No. 0224-4365
Data on employment.

GERMANY

Country profile:	Germany
Official name:	Federal Republic of Germany
Area:	356,960 sq.km.
	Bordered in the north by Denmark, the North Sea and the Baltic Sea, in the south by Austria and Switzerland, in the east by Poland and Czechoslovakia and Austria, and in the west by France, Luxembourg, Belgium and the Netherlands.
Population:	79.7m (1991)
Capital:	Berlin
Language(s):	German
Currency:	Deutsche mark 100 pfennige = 1 deutsche mark
National bank:	Deutsche Bundesbank (German Federal Bank)
Government and Political structure:	Elected Bundestag (Parliament) with 662 members which elects the Federal Chancellor (Prime Minister). A Bundesrat (federal council) and President (Bundesprasident) are also elected.

(c) R Canning.

15001 THE 300 LARGEST GERMAN FIRMS

German Chamber of Industry & Commerce

Germany - English/1993 £16.00

Up-to-date information designed for the use of businessmen.

15002 ABC INDUSTRIAL INFORMATION AND REFERENCES

ABC Publishing Group

Germany - German irregular/ 7,000pp

Register of manufacturers in Germany covering industrial groups and subdivisions. Details include capital assets, number of employees, export countries.

15003 ABSATZ VON BIER

W. Kohlhammer GmbH

Germany - German monthly/ DM1.60 P.A.

Statistics on sales of beer.

15004 ABSCHLUSSE DER AK-TIENGESELLSCHAFTEN

W. Kohlhammer GmbH

Germany - German annual/ DM16.50

Figures for the balances of joint stock companies.

15005 AGRARBERICHT (AGRAR - UND ERNÄHRUNGSPOLI-TISCHER BERICHT DER BUNDESREGIERUNG ÜBER DIE LAGE DER LANDWIRTSCHAFT...)

Bundesministerium für Ernährung, Landwirtschaft und Forsten

Germany - German annual/ISSN No. 0722-8333

Review of the agricultural industry.

15006 AKTIENMARKTE

W. Kohlhammer GmbH

Germany - German monthly/ DM3.20 P.A.

Statistics on share markets.

15007 AMTLICHES KURSBLATT DER BADEN-WÜTTEM-BERGISCHEN WERTPA-PIERBÖRSE ZU STUTTGART

Baden-Wüttembergische Wertpapierbörse zu Stuttgart

Germany - German daily (except weekends)/16pp DM220.00 per quarter

List of stock market prices.

15008 ANGESTELLTENVER-DIENSTE IN INDUSTRIE UND HANDEL

W. Kohlhammer GmbH

Germany - German quarterly/ DM11.20 P.A.

Earnings of salaried employees in industry and commerce.

15009 ANNUAL REPORT

Deutsche Bundesbank

Germany - English/German annual/ free

Annual report of the bank with general economic comment.

15010 ARBEITERVERDIENSTE IN DER LANDAITSCHAFT-ARBEITNEHMERVER-DIENSTE IN INDUSTRIE UND HANDEL

W. Kohlhammer GmbH

Germany - German annual/DM1.60

Earnings of labour in agriculture: earnings of employees in industry and commerce.

15011 ARBEITS - UND SOZIALPOLITIK

Nomos Verlagsgesellschaft mbH & Co KG

Karl-Heinz Schönbach, Axel Wober/Germany - German bimonthly/DM120.00 P.A. ISSN No. 0340-8434

Covers labour and social policy.

15012 ARBEITSKOSTEN IN PRO-DUSIERENDEN GEWEBE

W. Kohlhammer GmbH

Germany - German every 4 years/ DM19.50

Labour costs in production industries.

15013 ARBEITSKRAFTE

W. Kohlhammer GmbH

Germany - German annual/ DM10.50

Labour force in Germany.

15014 ASBESTOS INDUSTRY IN WESTERN GERMANY - SECTOR REPORT

Department of Trade and Industry Export Publications

Germany - English/1990 16pp

Overview of the asbestos industry including legislation, approval standards and testing.

15015 AUSFUHRNACH VERBRAUCHS - UND KAUFERLANDERN UND WARENGRUPPEN

W. Kohlhammer GmbH

Germany - German annual/ DM21.50

Exports by countries of consumption and of purchase and by commodity groups: foreign trade in selected commodities.

15016 AUSGENAHLTE ZAHLEN FÜR DIE AGRARWIRT-SCHAFT BETRIEGSAR-BEITS - UND EINKOSMMENSVERHAL-TRUSSE BETRIEBE

W. Kohlhammer GmbH

Germany - German annual/ DM19.50

Selected figures on agriculture, structure of holdings, operating conditions and incomes, details of holdings.

15017 AUSGENAHLTE ZAHLEN FÜR DIE BAUWIRTSCHAFT

W. Kohlhammer GmbH

Germany - German monthly/ DM11.10 P.A.

Selected figures for the construction industry, including building permits, buildings completed, and construction orders.

15018 AUSGENAHLTE ZAHLEN ZUR ENERGIEWIRT-SCHAFT

W. Kohlhammer GmbH

Germany - German monthly/68pp DM9.50 P.A.

Power production.

15019 AUSSENHANDEL NACH LANDERN UND GUTER-GRUPPEN DER PRODUK-TIONSSTATISTIKEN (SPE-ZIALHANDEL)

W. Kohlhammer GmbH

Germany - German annual/ DM15.00

Foreign trade by countries and commodity groups of industry statistics.

15020 AUSSENHANDEL NACH WAREN UND LANDERN (SPEZIALHANDEL)

W. Kohlhammer GmbH

Germany - German monthly/ DM27.00 P.A.

Foreign trade by commodities and countries - special trade.

15021 BAUMOBSTFLACHEN

W. Kohlhammer GmbH

Germany - German every 5 years/ DM12.20

Areas of fruit trees in Germany.

15022 BAUTATIGKEIT

W. Kohlhammer GmbH

Germany - German annual/ DM12.00

Statistics on activities in the building industry.

15023 BERICHTE ÜBER LANDWIRTSCHAFT

Bundesministerium für Ernährung, Landwirtschaft und Forsten

Germany - German quarterly/ DM472.00 P.A. ISSN No. 0005-9080

Report on agriculture in Germany.

15024 BESCHAFTIGTE UND UMSATZ IN EINZEL-HANDEL (MESSZAHLEN)

W. Kohlhammer GmbH

Germany - German monthly/ DM4.80 P.A.

Figures for persons engaged and turnover in the retail trade.

15025 BESCHAFTIGTE UND UMSATZ IN GAST-GENERBE (MESSZAHLEN)

W. Kohlhammer GmbH

Germany - German monthly/ DM1.60 P.A.

Persons engaged and turnover in the hotel and restaurant industry.

15026 BESCHAFTIGTE UND UMSATZ IN GROSS-HANDEL (MESSZAHLEN)

W. Kohlhammer GmbH

Germany - German monthly/ DM4.80 P.A.

Statistics on employees and turnover in wholesale trade.

15027 BESCHAFTIGUNG UMSATZ, INVESTITIONEN UND KOSTENSTRUCKTUR DER UNTERNEHMEN IN DER ENERGIE - UND WASSERVER

W. Kohlhammer GmbH

Germany - German annual/ DM12.00

Cost structure in the power and water supply industries.

15028 BESCHAFTIGUNG UMSATZ UND ENERGIE-VERSORGUNG DER UNTERNEHMEN UND BE-TRIEBE IM BERGBAU UND IM VERARBEIT

W. Kohlhammer GmbH

Germany - German annual/ DM12.00

Employment, turnover and power supply of enterprises and local units in mining and manufacturing.

15029 BESCHAFTIGUNG UMSATZ UND INVESTITIONEN DER UNTERNEHMEN IM BE-RGBAU UND IM VERAR-BEITENDEN GENERBE

W. Kohlhammer GmbH

Germany - German annual/ DM14.30

Statistics for employment, turnover and investments in the mining and manufacturing industries.

15030 BESCHAFTIGUNG UMSATZ UND INVESTITIONEN DER UNTERNEHMEN IN DER ENERGIE - UND WASSERVERSORGUNG

W. Kohlhammer GmbH

Germany - German annual/ DM12.00

Statistics on employment, turnover and investment in power and water supply industries.

15031 BESCHAFTIGUNG, UMSATZ, WAREN-EINGANG, LAGER-BESTAND UND INVESTITIONEN IN EINZELHANDEL

W. Kohlhammer GmbH

Germany - German annual/ DM10.50

Purchases of goods, stocks and gross proceeds in retailing.

15032 BESITZVERHALTRISSE GRUNDSTUCKSVERKEHT FACHLICHE VORBILDUNG DER BETRIEBSLEITER

W. Kohlhammer GmbH

Germany - German biennial/ DM13.50

Statistics on land tenure, transactions of real property, and technical qualifications of farm operators.

15033 BETRIEBSGROSSEN-STRUKTUR

W. Kohlhammer GmbH

Germany - German annual/DM6.00

Size structure of property holdings.

15034 BETRIEBSSYSTEME UND STANDARDBETRIEB-SEINKOMMEN

W. Kohlhammer GmbH

Germany - German biennial/ DM15.00

Operating systems and standard income of holdings.

15035 BEVOLKERUNG GESTERN; HEUTE AND MORGEN

W. Kohlhammer GmbH

Germany - German irregular/212pp DM10.50

The population of Germany, including model computations for future development up to 2030.

15036 BEVOLKERUNGS-STRUKTUR UND WIRTSCHAFTSKRAFT DER BUNDESLANDER

W. Kohlhammer GmbH

Germany - German annual/ DM20.60

Major demographic and economic data by region.

15037 BILDUNG IN ZAHLENSPIEGEL

W. Kohlhammer GmbH

Germany - German annual/156pp DM16.50

Statistics of education.

15038 BODENNUTZUNG DER BETRIEBE

W. Kohlhammer GmbH

Germany - German biennial/ DM10.50

Land use of holdings.

15039 BODENNUTZUNG UND PFLANZLICHE

W. Kohlhammer GmbH

Germany - German annual/ DM12.00

Land use and vegetable production.

15040 BRAUWIRTSCHAFT

W. Kohlhammer GmbH

Germany - German annual/DM3.20

Finance and taxes in the brewing industry.

15041 BRITISH GERMAN TRADE

German Chamber of Industry & Commerce

Germany/UK - English bimonthly/ £50.00 P.A.

Official journal of the Chamber.

15042 BUILDING MATERIALS AND CONSTRUCTION PRODUCTS - SECTOR RE-PORT

Department of Trade and Industry Export Publications

Germany - English/1990 12pp £30.00

Overview of the construction sector with information on methods and prospects of doing business.

15043 BUNDESARBEITSBLATT

W. Kohlhammer GmbH

Germany - German monthly/135pp

Labour and social statistics.

15044 CHAMBERS OF INDUSTRY AND COMMERCE IN GERMANY

German Chamber of Industry & Commerce

Germany - English/1993 £10.00

15045 CIRET - STUDIEN

IFO - Institut für Wirtschaftsforschung e.V.

Germany - German irregular/price varies ISSN No. 0170-5679

Studies on investment and investment behaviour in Germany.

15046 COUNTRY BOOKS: GERMAN FEDERAL REPUBLIC

United Kingdom Iron and Steel Statistics Bureau

Germany - English annual/£75.00
ISSN No. 0952-5912

Annual summary tables showing production, materials consumed, apparent consumption, imports and exports of 130 products by quality and market.

15047 COUNTRY FORECASTS: GERMANY

Business International Ltd

Germany - English quarterly/ £360.00 P.A.

Contains an executive summary, sections on political and economic outlooks and the business environment, a fact sheet, and key economic indicators. Includes a quarterly Global Outlook.

15048 COUNTRY PROFILE: GERMANY

Department of Trade and Industry Export Publications

Germany - English irregular/1991 58pp £10.00

This publication provides a general description of the country in terms of its geography, economy and population, and gives information about trading in the country with details like export conditions and investment covered.

15049 COUNTRY REPORT: GERMANY

Business International Ltd

Germany - English quarterly/25-40pp £160.00 P.A.

Contains an executive summary, sections on political and economic structure, outlook for the next 12-18 months, statistical appendices, and a review of key political developments. Includes an annual Country Profile.

15050 DATEN ZUM ARBEITSMARKT

Bundesanstalt für Arbeit

Germany - German/free

Most important data from labour market and household statistics.

15051 DATENSCHUTZ IN DER EUROPÄISCHEN GEMEINSCHAFT

Nomos Verlagsgesellschaft mbH & Co KG

Spiros Simitis, Ulrich Dammann/ EEC - English/German/French/ 950pp (loose-leaf) DM148.00

Statutory provisions of data protection in the EC-member countries.

15052 DIE BANK

Bank Verlag

Germany - German monthly/1977 c56pp DM78.00

Banking politics and practice.

15053 DIE FÜNF NEUEN BUNDESLÄNDER ALS WIRTSCHAFTSPARTNER

Bundesstelle für Aussenhandelsinformation

Germany - English/German/1991 DM15.00

Doing business in the five new

15054 DIE GESETZLICHE UNFALLVERSICHERUNG IN DER BUNDESREPUBLIK DEUTSCHLAND

Bundesministerium für Arbeit und Sozialordnung

Germany - German annual/138pp DM19.00

Statistics on accident insurance.

15055 DIE MOLKEREISTRUKTUR IM BUNDESGEBIET

Bundesministerium für Ernährung, Landwirtschaft und Forsten

Germany - German every 3 years

Structure of the dairy industry.

15056 DIE RENTENBESTÄNDE IN DER GESETZLICHEN RENTENVERSICHERUNG

Bundesministerium für Arbeit und Sozialordnung

Germany - German annual/231pp DM40.00

Insurance figures.

15057 DIE WELTWIRTSCHAFT

Institut für Weltwirtschaft

Horst Siebert/Germany/ International - German quarterly/ 1950 400pp ISSN No. 0043-2652

Business and economics, economic situation and conditions, economic systems and theories.

15058 DOING BUSINESS IN GERMANY

Ernst & Young International

Germany - English irregular/1991 110-130pp free to clients

A business guide which includes an executive summary, the business environment, foreign investment, company structure, labour, taxation, financial reporting and auditing, and general country data.

15059 DOING BUSINESS IN GERMANY

Price Waterhouse

Germany - English irregular/1992 330-340pp free to clients

A business guide which includes sections on the investment climate, doing business, accounting and auditing, taxation and general country data. Updated by supplements.

15060 DOKUMENTATION ZUM BUNDESDATENSCHITZ-GESETZ

Nomos Verlagsgesellschaft mbH & Co KG

Spiros Simitis et al/International - German/c1900pp (loose-leaf) DM234.00 ISSN No. 0941-4444

Legal regulations and administrative rules and case book.

15061 DRESDNER BANK STATISTICAL SURVEY

Dresdner Bank

Economic Research Department/ Germany/OECD - English/German/ French annual/1969 8pp in each part free

Published in three parts: German economy; world economy; historical statistical survey.

15062 DUNGEMITTEL-VORSORGUNG

W. Kohlhammer GmbH

Germany - German monthly/ DM4.50 P.A.

Statistics on fertilizer supplies.

15063 ECHO AUS DEUTSCHLAND

Nomos Verlagsgesellschaft mbH & Co KG

Carl Duisburg Gesellschaft in cooperation with the Deutsche Stiftung für internationale Entwicklung DSE/Germany - German bimonthly/DM30.00 P.A. ISSN No. 0343-0405

Reports on the economy, technology, culture and development projects in Germany.

15064 ECONOMIC SURVEY INTERNATIONAL

IFO - Institut für Wirtschaftsforschung e.V.

Germany/International - German annual/DM600.00

15065 EG MAGAZIN

Nomos Verlagsgesellschaft mbH & Co KG

Volker Schmarz/EEC - German ten per year/DM40.00 P.A. ISSN No. 0343-6667

A European-wide coverage of economics, law, politics and culture.

15066 EINBANDDECKEN UND JAHRGANGSSAMMLER

Bank Verlag

Germany - German annual/ DM19.50

Review of banking politics and practice.

15067 EINFUHR NACH HORSTEL-LUNGS - UND EINKAUFS-LANDERN UND WARENGRUPPEN

W. Kohlhammer GmbH

Germany - German annual/ DM19.50

Statistics on imports by countires of production and of sale and by commodity groups.

15068 EINKOMMENSTEUER

W. Kohlhammer GmbH

Germany - German every 3 years/ DM14.40

Figures on income tax.

15069 EINNAHMEN UND AUSGA-BEN AUSGENAHLTER PRI-VATER HAUSHALTE

W. Kohlhammer GmbH

Germany - German monthly/ DM15.00 P.A.

Income and expenditure of selected private households.

15070 EISENBAHNVERKEHR, STRASSENBAHN-VERKEHR

W. Kohlhammer GmbH

Germany - German monthly

Statistics for road and rail transport.

15071 EMPLOYEES OF BRITISH FIRMS IN GERMANY: IN-COME TAX AND SOCIAL SECURITY

German Chamber of Industry & Commerce

Germany/UK - English/1992 £25.00

15072 ERZEUGUNG VON GEFLUGEL

W. Kohlhammer GmbH

Germany - German semi-annual/ DM3.20

Poultry production.

15073 EUROPARECHT

Nomos Verlagsgesellschaft mbH & Co KG

Gert Nicolaysen at al/EEC - German quarterly/DM125.00 ISSN No. 0531-2485

Deals with the analysis and development of European Community law.

15074 EUROPEAN ECONOMIES IN GRAPHS AND FIGURES

IFO - Institut für Wirtschaftsforschung e.V.

EEC - English annual/DM80.00 ISSN No. 0175-8330

IFO's annual publication contains a comprehensive overview of economic development in the form of graphs and tables for key areas of the European and World economy. Data presented

incorporates the results of the IFO business and investment surveys and include long-term economic trends in the form of time series.

15075 EXPORTIEREN IN DIE BUNDESREPUBLIK DEUTSCHLAND

Bundesstelle für Aussenhandelsinformation

Germany - German annual/1988 DM6.00

Reviews exports in the German economy.

15076 EXPORTING CONSUMER GOODS TO GERMANY - SECTOR REPORT

Department of Trade and Industry Export Publications

Germany/United Kingdom - English/ 1992 93pp £90.00

Information for exporters wishing to tap into the German consumer goods market. Contains statistics for UK exports to Germany and other annexes.

15077 FEATURES OF THE GERMAN MARKET

German Chamber of Industry & Commerce

Germany - English/1992 £10.00

Brief description of the German market.

15078 FINANCING FOREIGN OPERATIONS: GERMANY

Business International Ltd

Germany - English annual/£695.00 for complete European series

Business overview, currency outlook, exchange regulations, monetary system, short/medium/ long-term and equity financing techniques, capital incentives, cash management, investment trade finance and insurance.

15079 FOREIGN TRADE WITH SELECTED COMMODITIES

W. Kohlhammer GmbH

Germany - German annual/ DM22.50

15080 GEBIET UND BEVOLKERUNG

W. Kohlhammer GmbH

Germany - German quarterly/ DM19.50 P.A.

There are two series of figures in these publications. One gives data for the state and development of the population, the other classifies population by administrative units.

15081 GEMUSEANBAUFLACHEN

W. Kohlhammer GmbH

Germany - German annual/DM3.00

Areas of vegetable cultivation.

15082 GENERAL OUTLINE OF IMPORT CHARGES AND IMPORT REGULATIONS

Bundesstelle für Aussenhandelsinformation

Germany - English irregular/ DM10.00

15083 GERMAN AND BRITISH ECONOMY

German Chamber of Industry & Commerce

German/UK - English monthly/ £40.00 P.A.

Comparative economic statistics.

15084 GERMAN BRIEF

Frankfurter Allgemeine Zeitung GmbH Informationsdienste

Germany - English weekly/ DM895.00

Background on the German economy with news on the political situation and a regular statistical profile of the country.

15085 GERMAN DIY MARKET

Department of Trade and Industry Export Publications

Germany - English/1992 25pp £30.00

Publication reflects the increased buoyancy of this already healthy market due to the rise in the percentage of sub-standard housing since unification. Covers major German DIY chains, building cooperatives, purchasing organizations and trade associations.

15086 GERMAN SUBSIDIARY COMPANIES IN THE UNITED KINGDOM

German Chamber of Industry & Commerce

Germany/UK - English/1991 £30.00

15087 GESCHÄFTSBERICHT DER BA

Bundesanstalt für Arbeit

Germany - German annual/free

Annual report on the BA's activities for the past year.

15088 GRENZÜBERSCHREIT-ENDER GÜTERKRAFT-VERKEHR

BDF-Infoservice GmbH

Germany/EEC/Europe - German/ 1992 420pp DM98.00 ISBN No. 3574260210

Handbook of transport and travel in both EC and non-EC countries.

15089 GUTERVERKEHR DER VERKEHRSZWEIGE

W. Kohlhammer GmbH

Germany - German quarterly/ DM12.00 P.A.

Statistics on goods transport classified by branch of transport.

15090 HANDBUCH FÜR VOLKSWIRTSCHAFTLICHE BERSTUNG
Nomos Verlagsgesellschaft mbH & Co KG
Oskar Gans at al/International - Germany/c1700 (loose-leaf) DM468.00 ISBN No. 3789021385
Theory and technicalities of planning in the field of economic development.

15091 HANDBUCH FÜR INTERNATIONALE ZUSAMMENARBEIT
Nomos Verlagsgesellschaft mbH & Co KG
Vereinigung für internationale Zusammenarbeit/International - German irregular/c20,000pp (loose-leaf) DM900.00 P.A. ISSN No. 0171-8258
Digest of organizations for international cooperation.

15092 HANDEL MIT DEN OSTBLOCKLANDERN
W. Kohlhammer GmbH
Germany - German biennial/ DM9.00
Trade between Germany and East European countries.

15093 HAUSHALTSGELD - WOHER, WOHIN?
W. Kohlhammer GmbH
Germany - German irregular/1987 DM46.00
Summary results of continuous family budget surveys.

15094 HINTS TO EXPORTERS VISITING GERMANY
Department of Trade and Industry Export Publications
Germany - English irregular/1991 78pp £5.00
This publication provides detailed practical information that is useful for those planning a business visit. Topics include currency information, economic factors, import and exchange control regulation, methods of doing business and general information.

15095 HOCHSEE - UND KUSTENFISCHEREI BODENSEEFISCHEREI
W. Kohlhammer GmbH
Germany - German monthly/ DM5.70 P.A.
Statistics for sea and inland fisheries including that of Lake Constance.

15096 HOW TO APPROACH THE GERMAN MARKET
Bundesstelle für Aussenhandelsinformation
Germany - English/DM6.00
Last edition was the 5th in 1988.

15097 IFL STUDIEN ZUR ARBEITSMARKT-FORSCHUNG
IFO - Institut für Wirtschaftsforschung e.V.
Germany - German irregular/price varies ISSN No. 0175-2944
Reviews the German labour market with some statistical coverage.

15098 IFO - BRANCHENSERVICE
IFO - Institut für Wirtschaftsforschung e.V.
Germany - German monthly/ DM150.00 per sector
Key economic information from government statistics, investment surveys and associations relating to more than 20 individual sectors.

15099 IFO - DIGEST
IFO - Institut für Wirtschaftsforschung e.V.
Germany - English quarterly/40pp DM100.00 P.A. ISSN No. 0170-7663
Digest of economic trends in Germany including analysis and forecasting, business cycle indicators, industry reports and abstracts.

15100 IFO KONJUNKTURPER-SPEKTIVEN
IFO - Institut für Wirtschaftsforschung e.V.
Germany - German monthly/30pp DM150.00
Economic review type publication.

15101 IFO - SCHNELLDIENST
Dunceer & Humblot GmbH
Germany - German 3 times per month/1948 20pp DM310.00 P.A. ISSN No. 0018-974X
Current awareness journal covering industry and economic trends.

15102 IFO SCHNELLDIENST
IFO - Institut für Wirtschaftsforschung e.V.
Germany - German 3 times a month/DM310.00 P.A. ISSN No. 0018-974X
Rapid results format dealing with economic data.

15103 IFO - SPIEGEL DER WIRTSCHAFT: STRUKTUR UND KONJUNKTUR IN BILD UND ZAHL
IFO - Institut für Wirtschaftsforschung e.V.
Germany - English/German annual/ 1973 130pp DM120.00 ISSN No.

0170-3617
Statistics on the economy.

15104 IFO - STUDIEN
Dunceer & Humblot GmbH
K.H. Oppenländer/Germany - German/summaries in English quarterly/1955 320pp DM180.00 P.A. ISSN No. 0018-9731
Business and economics are covered generally with analysis of specific topics.

15105 IFO STUDIEN ZUR AGRARWIRTSCHAFT
IFO - Institut für Wirtschaftsforschung e.V.
Germany - German irregular/price varies ISSN No. 0081-7198
Reviews the German agricultural industry.

15106 IFO STUDIEN ZUR BAUWIRTSCHAFT
IFO - Institut für Wirtschaftsforschung e.V.
Germany - German irregular/price varies ISSN No. 0170-5687
The construction industry is analysed with data given.

15107 IFO STUDIEN ZUR ENERGIEWIRTSCHAFT
IFO - Institut für Wirtschaftsforschung e.V.
Germany - German irregular/price varies ISSN No. 0170-7779
Energy and its market in Germany is covered with national data included.

15108 IFO STUDIEN ZUR EUROPÄISCHEN WIRTSCHAFT
IFO - Institut für Wirtschaftsforschung e.V.
Germany/EEC - English/German irregular/price varies ISSN No. 0938-6955
Germany and the European Community economy.

15109 IFO STUDIEN ZUR FINANZPOLITIK
IFO - Institut für Wirtschaftsforschung e.V.
Germany - German irregular/price varies ISSN No. 0081-7279
Financial policy and national finances are covered.

15110 IFO STUDIEN ZUR INDUSTRIEURTSCHAFT
IFO - Institut für Wirtschaftsforschung e.V.
Germany - German irregular/price varies ISSN No. 0170-5660
Industrial analysis for Germany.

15111 IFO STUDIEN ZUR REGIONAL - UND STADTÖKONOMIE

IFO - Institut für Wirtschaftsforschung e.V.

Germany/EEC - German irregular/ price varies

Regional economic policy and economics.

15112 IFO STUDIEN ZUR STRUKTURFORSCHUNG

IFO - Institut für Wirtschaftsforschung e.V.

Germany - German irregular/price varies ISSN No. 0176-0874

Studies on the structure of German industry with data.

15113 IFO STUDIEN ZUR UMWELTÖKONMIE

IFO - Institut für Wirtschaftsforschung e.V.

Germany - German irregular/price varies ISSN No. 0175-8330

German and international economics.

15114 IFO STUDIEN ZUR VERKEHRSWIRTSCHAFT

IFO - Institut für Wirtschaftsforschung e.V.

Germany - German irregular/price varies ISSN No. 0170-5652

Data and analysis of traffic flows in Germany.

15115 IFO - WIRTSCHAFTS-KONJUKTUR

IFO - Institut für Wirtschaftsforschung e.V.

Germany - German monthly/50pp DM265.00 ISSN No. 0043-6283

Monthly bulletin dealing with economics and business.

15116 IN DEPTH COUNTRY APPRAISALS ACROSS THE EC - CONSTRUCTION AND BUILDING SERVICES - GERMANY

Building Services Research and Information Association

Germany - English 1991/£125.00

Part of an 11 volume survey covering the construction industry in EC countries.

15117 INDEX DER GROSSHANDELS - VERKAUFSPREISE

W. Kohlhammer GmbH

Germany - German monthly/ DM4.80 P.A.

Index of selling prices in wholesale trade.

15118 INDEX DER GRUNDSTOFTPREISE

W. Kohlhammer GmbH

Germany - German monthly/ DM4.30 P.A.

Price index for basic materials.

15119 INDEX DER TARIFLOHNE UND - GEHALTER

W. Kohlhammer GmbH

Germany - German quarterly/ DM48.00 P.A.

Index of agreed wages and salaries.

15120 INDIKATOREN ZUR WIRTSSCHAFTS-ENTWICKLUNG

W. Kohlhammer GmbH

Germany - German quarterly/ DM14.30 P.A.

Tables of economic indicators, seasonally adjusted time series.

15121 INDIZES DER PRODUKTION UND DER ARBEITSPRODUKTIUTAT, PRODUKTION AUSGE-NAHLTER ERZEUGUISSE IM PRODUZIE

W. Kohlhammer GmbH

Germany - German monthly/ DM9.50 P.A.

Indices of production and of labour productivity, production of selected commodities.

15122 INDIZES DER AUFTRAGS-EINGANGS DES UMSATZES UND DES AUFTRAGSBESTANDS FUR DAS VERABEITENDE BEWERBE

W. Kohlhammer GmbH

Germany - German monthly/ DM7.90 P.A.

Indices of orders received and unfilled orders for manufacturing and building industries.

15123 INFORMATIONEN ÜBER DIE FISHWIRTSCHAFT DES AUSLANDES

Bundesministerium für Ernährung, Landwirtschaft und Forsten

Germany/International - German bimonthly/DM30.00

Information on the fishing industry of foreign countries.

15124 INPUT-OUTPUT-STUDIEN

IFO - Institut für Wirtschaftsforschung e.V.

Germany - German irregular/price varies ISSN No. 0579-6415

Studies the inputs and outputs in the German economy.

15125 INTERNATIONALES GEWERBEARCHIV ZEITSCHRIFT FÜR KLEIN - UND MITTEL UNTERNEHMEN

Dunceer & Humblot GmbH

Schweizensches Institut für Gewerbliche Wirtschaft/Germany - German quarterly/1953 280pp DM72.00 P.A. ISSN No. 0020-9481

Covers trends in German small businesses.

15126 INVESTING, LICENSING AND TRADING CONDITIONS ABROAD: GERMANY

Business International Ltd

Germany - English annual/£825.00 for complete European series

Loose-leaf format covering the state role in industry, competition rules, price controls, remittability of funds, taxation, incentives and capital sources, labour and foreign trade.

15127 INVESTIONEN FUR UMWELTSCHUTZ IM PRODUZIERENDEN GEWERBE

W. Kohlhammer GmbH

Germany - German annual/ DM15.00

Investment for environmental protection in production industries.

15128 JAHRBUCH DER ABSATZ - UND VERBRAUCHS-FORSCHUNG

Dunceer & Humblot GmbH

Gesellschaft für Konsum - Markt - und Absatzforschung/Germany - German quarterly/390pp DM98.00 ISSN No. 0021-3985

Yearbook of marketing and purchasing information.

15129 JAHRBUCH FÜR FREMDENVERKEHR

Deutsches Wirtschafts-wissenschaftliches Institut für Fremdenverkehr an der Universität München e.v.

Germany - German annual/1952 DM45.00 ISSN No. 0075-2649

Covers various subjects in the field of tourism.

15130 JAHRESBERICHT - ARBEITSGEMEINSCHAFT DER DEUTSCHEN WERTPAPIERBÖRSEN

Rheinisch - Westfälische Börse zu Dusseldorf

Germany - German annual/free

Report on the German stock exchange.

15131 JAHRESBERICHT DER VDR

Verband Deutscher Reeder

Germany - German annual/64pp DM2.50

Annual report of the Merchant Shipping Federation.

15132 JAHRESBERICHT FORSCHUNG IM GESCHÄFTSBEREICH DES BUNDESMINISTERS FÜR ERNÄHRUNG, LANDWIRTSCHAFT UND FORSTEN
Bundesministerium für Ernährung, Landwirtschaft und Forsten

Germany - German annual/ DM130.00 ISSN No. 0343-7477

Annual financial report for German agriculture and forestry.

15133 JAHRESBERICHT ÜBER DIE DEUTSCHE FISCHWIRTSCHAFT
Bundesministerium für Ernährung, Landwirtschaft und Forsten

Germany - German annual/ DM59.50 ISSN No. 0075-2851

Annual report of the German fishing industry.

15134 JAHRESSCHAU DER DEUTSCHEN INDUSTRIE: DIE BEKLEIDUNGS UND WASCHE INDUSTRIE
ABC Publishing Group

Germany - German annual/price varies

Annual review of the clothing and lingerie trade and its suppliers.

15135 JAHRESSCHAU DER DEUTSCHEN INDUSTRIE: DIE CHEMISCHE INDUSTRIE
ABC Publishing Group

Germany - German annual/price varies

Chemical industry and its suppliers.

15136 JAHRESSCHAU DER DEUTSCHEN INDUSTRIE: DIE EISEN, BLECH UND METALL VERARBEITENDE INDUSTRIE
ABC Publishing Group

Germany - German annual/price varies

Iron and metal-working industries and their suppliers.

15137 JAHRESSCHAU DER DEUTSCHEN INDUSTRIE: DIE EISEN, STAHT UND NE-METALL INDUSTRIE
ABC Publishing Group

Germany - German annual/price varies

Iron, steel and non-ferrous metal industries and its suppliers.

15138 JAHRESSCHAU DER DEUTSCHEN INDUSTRIE: DIE ELEKTRO-INDUSTRIE
ABC Publishing Group

Germany - German annual/price varies

Electrical and electronics industry and its suppliers.

15139 JAHRESSCHAU DER DEUTSCHEN INDUSTRIE: DIE GIESSEREI INDUSTRIE
ABC Publishing Group

Germany - German annual/price varies

Annual review of the foundry industry and its suppliers.

15140 JAHRESSCHAU DER DEUTSCHEN INDUSTRIE: DIE KUNSTOFF-INDUSTRIE
ABC Publishing Group

Germany - German annual/price varies

Plastics industry and its suppliers.

15141 JAHRESSCHAU DER DEUTSCHEN INDUSTRIE: DIE MÖBEL INDUSTRIE
ABC Publishing Group

Germany - German annual/price varies

Annual review of the furniture industry and its suppliers.

15142 JAHRESSCHAU DER DEUTSCHEN INDUSTRIE: DIE NAHRUNGS UND GEN-USSMITTEL-INDUSTRIE
ABC Publishing Group

Germany - German annual/price varies

Annual review of the food, beverage and luxury products industry.

15143 JAHRESSCHAU DER DEUTSCHEN INDUSTRIE: DIE TEXTIL-INDUSTRIE
ABC Publishing Group

Germany - German annual/price varies

Textile industry and its suppliers.

15144 KAUFUERTE FUR BAU-LAND
W. Kohlhammer GmbH

Germany - German quarterly/ DM9.00 P.A.

Purchase values for building land.

15145 KAUFUERTE FUR LANDWIRTSCHAFTLI-CHEN GRUNDBESITZ
W. Kohlhammer GmbH

Germany - German annual/DM3.00

Purchasing values of agricultural property.

15146 KIEL REPORTS
Institut für Weltwirtschaft

Germany/International - English quarterly/1990 4pp free

Business and economics, economic situation and conditions, economic systems and theories.

15147 KIEL WORKING PAPERS
Institut für Weltwirtschaft

International - English/German irregular/DM10.00 per issue ISSN No. 0342-0787

Topics such as: cadmium in Germany, integration of Eastern Europe into the world economy.

15148 KIELER DISKUSSIONS-BEITRÄGE
Institut für Weltwirtschaft

International - English/German irregular/DM10.00 ISSN No. 0455-0420

Discussion papers on topics such as German economic policy, global recession.

15149 KIELER KURZBERICHTE
Institut für Weltwirtschaft

Germany/International - German 30 per year/1980 2-4pp free ISSN No. 0173-5241

Business and economics, economic situation and conditions, economic systems and theories.

15150 KIELER STUDIEN
Institut für Weltwirtschaft

Horst Siebert/International - English/Germany irregular/ISSN No. 0340-6989

Papers covering topics such as foreign direct investment in developing countries, international competitiveness of developing countries for risk capital.

15151 KONJUNKTUR AKTUELL
Statistisches Bundesamt

Germany - German monthly/ DM189.60

Review of economic conditions and developments.

15152 KONJUNKTUR-INDIKATOREN
IFO - Institut für Wirtschaftsforschung e.V.

Germany/International - German monthly/DM58.00 P.A. ISSN No. 0722-0227

Looseleaf sheet on key economic indicators.

15153 KONJUNKTURPOLITIK
Dunceer & Humblot GmbH

Germany - English/German bimonthly/1954 590pp DM98.00 P.A. ISSN No. 0023-3498

Business and economics. Production of goods and services in Germany.

15154 KONTEN UND STANDARDTABELLEN
W. Kohlhammer GmbH

Germany - German annual/
DM25.50

Accounts and standard tables.

15155 KORPERSCHAFTSTEUER
W. Kohlhammer GmbH

Germany - German every 3 years/
DM10.30

Figures for corporation tax.

15156 KOSTENSTRUKTUR BEI HANDELSVERTRETERN UND HANDELSMAKLERN
W. Kohlhammer GmbH

Germany - German every 4 years/
DM11.10

Cost structure of commercial
representatives and agents.

15157 KOSTENSTRUKTUR BEI RECHTSANWALTEN UND ANWALTSNOTAREN, BEI WIRTSCHAFTSPRUFERN, STEUERBERATERN
W. Kohlhammer GmbH

Germany - German every 4 years/
DM13.50

Cost structure in professions such
as lawyers and public notaries,
auditors, tax consultants etc.

15158 KOSTENSTRUKTUR DER NICHTBUNDESEIGENEN EISENBAHNEN, DES STADT SCHNELLBAHN - STRASSENBAHN - UND OMNIBUSSE
W. Kohlhammer GmbH

Germany - German every 4 year/
DM8.90

Cost structure of privately owned
transport and travel.

15159 KOSTENSTRUKTUR DER UNTERNEHMEN IM VERBRAUCHSGOTER PRODUZIEREN DEN GE- WERBE UND IN NAH- RUNGS - UND GENUSSMITTEL
W. Kohlhammer GmbH

Germany - German annual/
DM15.00

Cost structures in the consumer
goods, food, tobacco and beverage
industries and also the building
industry.

15160 KOSTENSTRUKTUR DER UNTERNEHMEN IN BANGEWERBE
W. Kohlhammer GmbH

Germany - German annual/
DM12.00

Cost structure in the building
industry.

15161 KOSTENSTRUKTUR DER UNTERNEHMEN IN BERGBAU, GRUNDSTOFF - UND PRODUKTIONSGU- TERGEWERBE
W. Kohlhammer GmbH

Germany - German annual/
DM13.50

Cost structure in mining, primary
and production industries.

15162 KOSTENSTRUKTUR DER GEWERBUEHEN GUTERKRAFTVERKEHRS, DER SPEDITIONEN UND LAGEREIEN, DER BINNEN- SEHIFTEN
W. Kohlhammer GmbH

Germany - German every 4 years/
DM9.00

Cost structure in the following
sectors: commercial goods
transport, shipping and
warehousing.

15163 KOSTENSTRUKTURIM EINZELHANDEL
W. Kohlhammer GmbH

Germany - German every 4 years/
DM13.60

Cost structure in the retail trade.

15164 KOSTENSTRUKTURIM GASTGEWERBE
W. Kohlhammer GmbH

Germany - German every 4 years/
DM7.00

Cost structure in the following
sectors: hotels and restaurants.

15165 KREDIT UND KAPITAL
Dunceer & Humblot GmbH

Germany - English/German
quarterly/1968 600pp DM112.00
P.A. ISSN No. 0023-4591

Banking and finance in Germany.

15166 KRITISCHE JUSTIZ
Nomos Verlagsgesellschaft mbH & Co KG

Thomas Blanke et al/Germany/
Europe - German quarterly/
DM42.00 P.A. ISSN No.0023-4834

Analyses questions of law in an
economic, political and social
context.

15167 LANDWIRTSCHAFTLICH GENUTZTE FLACHEN
W. Kohlhammer GmbH

Germany - German annual/DM3.00

Agriculturally used areas.

15168 LANGE REIHEN ZUR WIRTSCHAFTSENTWICK- LUNG
W. Kohlhammer GmbH

Germany - German biennial/
DM19.50

Basic data on population and
employment in addition to long-term
statistics on all aspects of the
economy.

15169 LOHNSTEUER
W. Kohlhammer GmbH

Germany - German every 3 years/
DM15.00

Figures on wage tax.

15170 LUFTVERKEHR
W. Kohlhammer GmbH

Germany - German monthly/
DM16.50 P.A.

Statistics on air transport.

15171 MARKETING MEDICAL PRODUCTS - SECTOR REPORT
Department of Trade and Industry Export Publications

Germany - English/1992 50pp
£60.00

Breakdowns on imports of medical
equipment, details of different types
of hospitals and their requirements,
information and trade fairs and
recommendations for British
manufacturers.

15172 MESSZAHLEN FUR BAULEISTUNGSPREISE UND PREISINDIZES FUR BAUWERKE
W. Kohlhammer GmbH

Germany - German quarterly/
DM7.90 P.A.

Index numbers of prices for building
services and price indices for
buildings.

15173 MILCH UND MOLKEREI- WIRTSCHAFT BUNDESREPUBLIK DEUTCHLAND, EG-MITGLIEDSTAATEN
Bundesministerium für Ernährung, Landwirtschaft und Forsten

Germany - German annual
Review of the dairy industry.

15174 MINERALOLSTEUER
W. Kohlhammer GmbH

Germany - German annual/DM4.50

Statistics on mineral oil tax.

15175 MITTEILUNGEN DES INSTITUTS FÜR ANGEWANDTE WIRTSCHAFTSFOR- SCHUNG TUEBINGEN
Institut für Angewandte Wirtschaftsforschung

Germany - German quarterly/1972
16pp DM28.00 P.A. ISSN No. 0173-
5454

Economic theory, politics, statistics,
econometrics.

15176 MONATSBERICHT
Bundesministerium für Wirtschaft

Germany - English monthly/price varies

Monthly survey of economic developments including trade and payments, prices, employment and the labour market.

15177 MONTHLY REPORT
Deutsche Bundesbank

Germany - English/German/ French/Spanish monthly/c79pp free

Issued in five series: Series 1, Banking statistics by group of banks; Series 2, Securities statistics; Series 3, Balance of payments statistics; Series 4, Seasonally adjusted economic data; Series 5, Currencies of the world.

15178 NACHRICHTEN FÜR AUSSENHANDEL
Bundesstelle für Aussenhandelsinformation

Germany/International - German 5 times weekly/DM122.20 per month

Up-to-date news on the world economy of specific interest to exporters in Germany.

15179 NORDEUROPA - FORUM
Nomos Verlagsgesellschaft mbH & Co KG

Bernd Henningsen/Northern Europe - German quarterly/ DM88.00 P.A. ISSN No. 0940-5585

Coverage of economics, law, politics and culture in the north European states.

15180 PERSONENVERKEHR DER STRASSEWERKEHRS - UNTERNEHMEN
W. Kohlhammer GmbH

Germany - German quarterly/ DM14.30 P.A.

Statistics on passenger transport by road carriers.

15181 PREISE
W. Kohlhammer GmbH

Germany - German annual/6-10pp

Prices - levels and trends.

15182 PREISE UND PREISINDIZES FUR DIE EIN - UND AUSFUHR
W. Kohlhammer GmbH

Germany - German monthly/ DM15.90 P.A.

Prices and price indices for imports and exports.

15183 PREISE UND PREISINDIZES FUR DIE LAND - UND FORSTWIRTSCHAFT
W. Kohlhammer GmbH

Germany - German monthly/

DM9.00 P.A.

Prices and price indices for agriculture and forestry.

15184 PREISE UND PREISINDIZES FUR DIE LEBENSHALTUNG
W. Kohlhammer GmbH

Germany - German monthly/ DM20.60 P.A.

Consumer prices and consumer price indices.

15185 PREISE UND PREISINDIZES FUR GEWERBUCHE PRO- DUKTE (ERZEUGPREISE)
W. Kohlhammer GmbH

Germany - German monthly/ DM12.70 P.A.

Prices and price indices for industrial products - producers prices.

15186 PREISE UND PREISINDIZES FUR VERKEHRSLEISTUNGEN
W. Kohlhammer GmbH

Germany - German annual/DM7.50

Transport prices.

15187 PRODUKTION IN PRODUZERIENDEN GEWERBE NACH WIRTSCHAFTWEIGEN UND ERZENGRISGRUPPEN
W. Kohlhammer GmbH

Germany - German annual/ DM16.50

Output of production industries by branches of economic activity and groups of products.

15188 PRODUKTION IN PRODUZIERENDEN GEWERBE DES IN - UND AUSLANDES
W. Kohlhammer GmbH

Germany - German quarterly/ DM20.60 P.A.

Statistics on the output of the production, mining and manufacturing industries, and employment and turnover in mining and manufacturing industries.

15189 QUARTERLY COUNTRY TRADE REPORTS: GERMAN FEDERAL REPUBLIC
United Kingdom Iron and Steel Statistics Bureau

Germany - English quarterly/ £200.00 P.A. for four countries ISSN No. 0952-5912

Cumulative trade statistics of imports and exports of steel products.

15190 REALSTEUERVERGLEICH GEWERBESTEUER
W. Kohlhammer GmbH

Germany - German annual/ DM10.50

Comparison of taxation on real estate, commercial tax and payroll tax, statistics on commercial tax.

15191 RECHMUNGSERGEB- NISSE DES OFFENTLICHEN GESAM- THAUSHALTS
W. Kohlhammer GmbH

Germany - German annual/ DM19.50

Accounting results of the public overall budget.

15192 RECHRUNGSERGEBNISSE DER KOMMERALEN HAUSHALTE
W. Kohlhammer GmbH

Germany - German annual/ DM22.50

Accounting results of the government budget.

15193 RECHT DER BIOTECHNOLOGIE
Nomos Verlagsgesellschaft mbH & Co KG

JÜrgen Simon/International - German/c1400pp (loose-leaf) DM248.00 ISSN No. 0941-4452

Legislative provisions with supplementary materials, case book, reports about biotechnology.

15194 RWI - KONJUNKTURBERICHTE
Dunceer & Humblot GmbH

Germany - German biannual/1950 DM100.00 P.A. ISSN No. 0023- 3447

Economic trends in the western world and Germany in particular.

15195 RWI - MITTEILUNGEN
Dunceer & Humblot GmbH

Germany - German quarterly/1950 DM200.00 P.A. ISSN No. 0035- 4465

Public sector economics.

15196 S & F - VIERTELJÄHRES- SCHRIFT FÜR SICHERHEIT UND FRIEDEN
Nomos Verlagsgesellschaft mbH & Co KG

Dieter S. Lurz et al/EEC - German quarterly/DM49.00 P.A. ISSN No. 0175-274X

New research results and political development on the sector of pacific policy and security.

15197 SCHAUMWEINSTEUER
W. Kohlhammer GmbH

Germany - German quarterly

Statistics on taxes on sparkling wine.

15198 SCHLACHTUNGEN UND FLEISCHGENINNUNG

W. Kohlhammer GmbH

Germany - German quarterly/ DM4.80 P.A.

Slaughterings and meat production.

15199 SCHRIFTENREIHE

Deutsches Wirtschafts- wissenschaftliches Institut für Fremdenverkehr an der Universität München e.V

Germany - German irregular/1953

Reports on various aspects of tourism.

15200 SCHRIFTENREIHE ... REIHE C: AGRARPOLITISCHE BE- RICHTE DER OECD

Bundesministerium für Ernährung, Landwirtschaft und Forsten

Germany - German irregular/price varies

Report on the agricultural policy of the OECD.

15201 SEESCHIFFAHRT

W. Kohlhammer GmbH

Germany - German monthly/ DM19.50 P.A.

Statistics on seaborne shipping.

15202 SONDERREIHE

Deutsches Wirtschafts- wissenschaftliches Institut für Fremdenverkehr an der Universität München e.V

Germany - German irregular/1963

Reports on various aspects of tourism.

15203 SPECIAL SERIES

Deutsche Bundesbank

Germany - English/German irregular/1971 free

Covers various topics including the Banking Act, IMF, balance of payments statistics.

15204 STAND DER REBEN UND WEINMOSTEMTE WEINERZEUGUNG UND - BESTAND

W. Kohlhammer GmbH

Germany - German irregular/ DM2.30

Figures for the state of growth of vines, the yield of wine, production and stocks of wine.

15205 STATISTICAL COMPASS

Statistisches Bundesamt

Germany - English annual/80pp DM4.50 ISSN No. 0072-4114

Pocket digest containing statistics on all aspects of economic and social life in united Germany, including employment, commerce, foreign trade, education, prices, national production and taxation. Also Germany by Länder.

15206 STATISTICHER WOCHENDIENST

Statistisches Bundesamt

Germany - German weekly/ DM130.00 ISSN No. 0177-2554

Weekly digest of statistics including economic statistics.

15207 STATISTISCHER MONATSBERICHT DES BUNDESMINISTERIUMS FÜR ERNÄHRUNG, LAND- WIRTSCHAFT UND FOR- STEN

Bundesministerium für Ernährung, Landwirtschaft und Forsten

Germany - German monthly/ DM190.00 P.A. ISSN No. 0433- 7344

Monthly statistics on food, agriculture and forestry.

15208 STATISTISCHES JAHRBUCH FUR DIE BUNDESREPUBLIK DEUTSCHLAND

Statistisches Bundesamt

Germany - German annual/746pp DM120.00

Yearbook of German statistics with broad coverage.

15209 STATISTISCHES JAHRBUCH FÜR DAS AUSLAND

Statistisches Bundesamt

Germany - German annual/376pp DM51.00

Statistical yearbook with economic coverage.

15210 STATISTISCHES JAHRBUCH ÜBER ERNÄH- RUNG, LANDWIRTSCHAFT UND FORSTEN

Bundesministerium für Ernährung, Landwirtschaft und Forsten

Germany - German annual/ DM128.00 ISSN No. 0072-1581

Statistical yearbook for food, nutrition, agriculture and forestry.

15211 STEUERHAUSHALT

W. Kohlhammer GmbH

Germany - German quarterly/ DM4.80 P.A.

Tax budget described and analysed.

15212 STRASSEN, BRUCKEN, PARKENRICHTUNGEN

W. Kohlhammer GmbH

Germany - German irregular/ DM14.60

Statistics on roads, bridges and parking facilities.

15213 SUMMARY SURVEYS OF FOREIGN TRADE

W. Kohlhammer GmbH

International - English/German monthly/DM9.50 P.A.

Broad picture of German foreign trade.

15214 TABAKGEWERBE

W. Kohlhammer GmbH

Germany - German annual/ DM30.00

Statistics on the tobacco industry.

15215 TELECOMMUNICATIONS - SECTOR REPORT

Department of Trade and Industry Export Publications

Germany - English/1991 4pp £30.00

Report covers the telecommunications market situation following reunification.

15216 TIPS FÜR IMPORTEURE

Bundesstelle für Aussenhandelsinformation

Germany - German irregular/ DM10.00

Tips for importers.

15217 TRADE ENQUIRIES FROM GERMANY

German Chamber of Industry & Commerce

Germany - English bimonthly/ £30.00 P.A.

Bulletin of trade opportunities.

15218 TRENDS AND VIEWS ON DIRECT INVESTMENT BY GERMAN-OWNED COMPANIES IN THE UK

German Chamber of Industry & Commerce

Germany/UK - English/1992 £30.00

Investment intentions, productivity, return on investment, employees, industrial and labour relations.

15219 TRENDS - DRESDNER BANK ANALYSES AND FORECASTS

Dresdner Bank

Germany/Europe/OECD - English/ German monthly/1989 c16pp free

Analyses the economic situation in Germany and puts it in a wider European context, forecasts are also given.

15220 UMSATZSTEUER
W. Kohlhammer GmbH

Germany - German biennial/
DM16.50

Statistics on turnover.

15221 UMUELT IN ZAHLEN
W. Kohlhammer GmbH

Germany - German annual/6-10pp

The environment in figures.

15222 VALUE ADDED TAX IN GERMANY
German Chamber of Industry & Commerce

Germany - English/1992 £12.00

15223 VERKEHRSWIRTSCHAFT-LICHE ZAHLEN
BDF-Infoservice GmbH

Germany - German annual/70pp
ISSN No. 0083-5021

Data on long-distance goods traffic
in Germany.

15224 VIERTELJAHRLICHE KASSENERGEBNISSE DER OFFENTLICHEN HAUSHALTE
W. Kohlhammer GmbH

Germany - German quarterly/
DM9.00 P.A.

Quarterly figures for public finance
and accounting results.

15225 VIERTELJAHRSHEFTE ZUR AUSLANDSSTATITIK
W. Kohlhammer GmbH

International - English/German
quarterly/DM15.90 P.A.

Trade statistics for Germany.

15226 VIERTELJAHRSHEFTE ZUR WIRTSCHAFTS-FORSCHUNG
Dunceer & Humblot GmbH

Deutsches Institut für
Wirtschaftsforschung/Germany -
English/German quarterly/1926
DM250.00 P.A. ISSN No. 0340-
1707

Business and economics in
Germany.

15227 WATER INDUSTRY - SECTOR REPORT
Department of Trade and Industry Export Publications

Germany - English/1991 27pp
£30.00

Overview of the German water
industry, with sections on the
organization of water supply and
sewage management. Also
contains statistics on exhibition
centres.

15228 WELTWIRTSCHAFT-LICHES ARCHIV
Institut für Weltwirtschaft

Horst Siebert/Worldwide - English/
German quarterly/1913 800pp
DM158.00 P.A. ISSN No. 0043-
2636

Business and economics,
economic situation and conditions,
economic systems and theories.
Review of world economics.

15229 WIRTSCHAFT IN ZAHLEN
Bundesministerium für Wirtschaft

Germany - German annual/186pp

Tables of economic data.

15230 WIRTSCHAFT UND STATISTIK
Statistisches Bundesamt

Germany - German monthly/
DM178.00 ISSN No. 0043-6143

Economic review.

15231 WIRTSCHAFTS-KONJUNKTUR
Dunceer & Humblot GmbH

Germany - German monthly/1949
50pp DM265.00 P.A. ISSN No.
0043-6283

Results of an ongoing survey into
industrial trends. Includes an
analysis of economic trends and
graphic representation of statistical
data.

15232 WOCHENBERICHT
Dunceer & Humblot GmbH

Germany - German weekly/1934
1016pp DM150.00 ISSN No. 0012-
1304

Economic developments and
trends in both Germany and the
western world.

15233 WSI INFORMATIONS-DIENST ARBEIT
Wirtschafts und Sozialwissen - Schaftliches Institut des Deutsche Gewerkschaftsbundes GmbH

Germany - German quarterly

Developments in work, business
and society.

15234 WSI MITTEILUNGEN
Wirtschafts und Sozialwissen - Schaftliches Institut des Deutsche Gewerkschaftsbundes GmbH

Germany - German monthly/56pp
DM20.00 P.A.

Analysis and forecast of economic
development, personal income/
investment, public expenditure,
social policy.

15235 ZAHLENKOMASS
W. Kohlhammer GmbH

Germany - English/German/
French/Spanish annual/48pp
DM3.00

Pocket-book sized summary of
comparable economic data over
preceding few years.

15236 ZEITSCHRIFT FÜR DIE GESAMTE VERSICHER-UNGSWISSENSCHAFT
Dunceer & Humblot GmbH

Deutscher Verein für
Versicherungswissenschaft/
Germany - German quarterly/1901
DM127.00 P.A. ISSN No. 0044-
2585

Insurance industry in Germany.

15237 ZEITSCHRIFT FÜR WIRT-SCHAFTS - UND SOZIAL-WISSENSCHAFTEN
Dunceer & Humblot GmbH

Artur Woll/Germany - English/
German quarterly/1880 670pp
DM148.00 P.A. ISSN No. 0342-
1783

General coverage on economic
matters and social economics.

GIBRALTAR

Country profile:	Gibraltar
Official Name:	Colony of Gibraltar
Area:	5.8 sq.km.
Population:	28,848
Capital:	Gibraltar Town
Language(s):	English, Spanish
Currency:	Gibraltar Pound (par with £ sterling) 100 pence = 1 Gibraltar pound
Government and Political structure:	Colony of UK since 1713. Governor and Gibraltar House of Assembly.

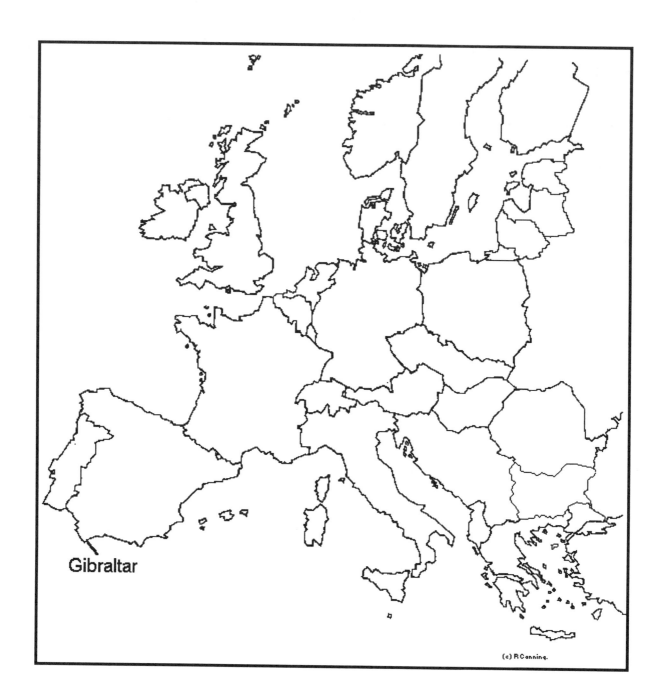

Gibraltar

(c) R.Canning.

16001 ABECOR COUNTRY REPORT - GIBRALTAR
Abecor Reports

Gibraltar - English irregular/2pp free

Brief economic report.

16002 ABSTRACT OF STATISTICS
Statistics Office

Gibraltar - English annual/1970 85pp £5.00

Covers population, housing, education, health, public safety, civil administration, employment, income and expenditure, trade, tourism, banking and finance and transport.

16004 AIR TRAFFIC SURVEY
Statistics Office

Gibraltar - English annual/1972 26pp

Data on commercial air traffic and air traffic operations to and from Gibraltar.

16005 CENSUS REPORT
Statistics Office

Gibraltar - English every ten years/ 1753 96pp £10.00

Data on demographic, economic and social characteristics of Gibraltar.

16006 EMPLOYMENT SURVEY
Statistics Office

Gibraltar - English twice yearly/ 1971 34pp

Data on employment trends, hours worked and earnings of all employees.

16007 HOTEL OCCUPANCY REPORT
Statistics Office

Gibraltar - English annual/1972 21pp

Covers hotel occupancy by type and length of stay.

16008 TOURIST SURVEY
Statistics Office

Gibraltar - English annual/1972 14pp

Analysis of tourist market, development of tourism and calculation of tourist expenditure.

GREECE

Country profile:	Greece
Official name:	Hellenic Republic
Area:	131,957 sq.km.
	Bordered in the north by Albania, Bulgaria and Yugoslavia, in the south by the Mediterranean, in the east by Turkey, and in the west by the Ionian Sea.
Population:	10.2m (1991)
Capital:	Athens
Language(s):	Greek
Currency:	Drachma 100 lepta = 1 Drachma
National bank:	Bank of Greece
Government and Political structure:	Republic with multi-party parliamentary democracy.

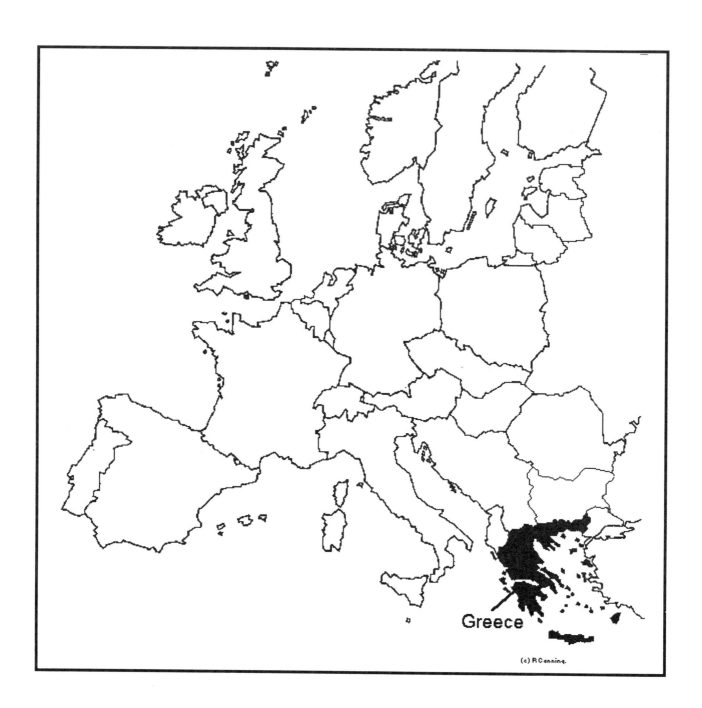

Greece

(c) R.Canning.

17001 AGRICULTURAL STATISTICS OF GREECE

National Statistical Service of Greece

Greece - English annual/US$7.00 ISSN No. 0065-4574

Covers agriculture in general including crops, livestock, land-use etc.

17002 ANNUAL INDUSTRIAL SURVEY

National Statistical Service of Greece

Greece - English annual/US$12.00 ISSN No. 0072-7393

Industrial production and statistics by type of industry.

17003 ANNUAL REPORT

National Bank of Greece S.A.

Greece - English/Greek annual/ 40pp free ISSN No. 0077-3514

17004 ANNUAL STATISTICAL SURVEY ON MINES, QUARRIES AND SALTERNS

National Statistical Service of Greece

Greece - English annual/US$4.00 ISSN No. 0072-745

Mining, quarrying and salt-works.

17005 COMMERCE EXTERIEUR DE LA GRECE

National Statistical Service of Greece

Greece - Greek/French annual/ US$26.00

External trade of Greece with levels and types of trade.

17006 CONCISE STATISTICAL YEARBOOK OF GREECE

National Statistical Service of Greece

Greece - English/Greek annual/ US$7.00 ISSN No. 0069-8245

Broad socio-economic coverage of Greek society.

17007 COUNTRY BOOKS: GREECE

United Kingdom Iron and Steel Statistics Bureau

Greece - English annual/£75.00 ISSN No. 0960-2380

Annual summary tables showing production, materials consumed, apparent consumption, imports and exports of 130 products by quality and market.

17008 COUNTRY FORECASTS: GREECE

Business International Ltd

Greece - English quarterly/£360.00 P.A.

Contains an executive summary, sections on political and economic outlooks and the business environment, a fact sheet, and key economic indicators. Includes a quarterly Global Outlook.

17009 COUNTRY PROFILE: GREECE

Department of Trade and Industry Export Publications

Greece - English irregular/1989 65pp £10.00

This publication provides a general description of the country in terms of its geography, economy and population, and gives information about trading in the country with details like export conditions and investment covered.

17010 COUNTRY REPORT: GREECE

Business International Ltd

Greece - English quarterly/25-40pp £160.00 P.A. ISSN No. 0269-591X

Contains an executive summary, sections on political and economic structure, outlook for the next 12-18 months, statistical appendices, and a review of key political developments. Includes an annual Country Profile.

17011 COUNTRY RISK SERVICE: GREECE

Business International Ltd

Greece - English quarterly/ minimum subcription £1,855.00 P.A. for seven countries

Contains credit risk ratings, risk appraisals, cross-country databases, data on diskette, and two year projections.

17012 CREDIT AND FINANCE - SECTOR REPORT

Department of Trade and Industry Export Publications

Greece - English/1992 32pp £30.00

Information on the Greek banking system and credit institutions. Greek banks and foreign banks in Greece are detailed, together with representative offices of foreign banks.

17013 DEVELOPMENT IN THE INTERNATIONAL FINANCIAL SYSTEM AND THE UNIFICATION OF THE EUROPEAN INTERNAL MARKET: EFFECTS ON THE GREEK ECONOMY

Centre of Planning and Economic Research

D.C. Maroulis/Greece/Europe - Greek/1992 544pp US$25.00

17014 ECONOMIC BULLETIN

Commercial Bank of Greece

Greece - English quarterly/c37pp

free ISSN No. 0013-0036

Covers main economic indicators and general economic developments.

17015 ECONOMIC INEQUALITIES IN GREECE: DEVELOPMENTS AND PROBABLE EFFECTS

Centre of Planning and Economic Research

L. A. Athanassiou/Greece - Greek irregular/1991

17016 EPILOGI - MONTHLY ECONOMIC REVIEW

Electra Press Publications

Stavros Papaioannoy/Greece - English/Greek monthly/1962 120pp US$150.00 P.A. ISSN No. 1105-252X

Coverage of main economic factors and developments.

17017 FINANCING FOREIGN OPERATIONS: GREECE

Business International Ltd

Greece - English annual/£695.00 for complete European series

Business overview, currency outlook, exchange regulations, monetary system, short/medium/ long-term and equity financing techniques, capital incentives, cash management, investment trade finance and insurance.

17018 GREECE IN THE 1990S: TAKING ITS PLACE IN EUROPE

Business International Ltd

Greece - English/1991 £160.00

Reviews Greek economic development and progress towards EEC stimulated convergence.

17019 GREECE: REFORM OF THE CREDIT AND FINANCIAL SECTOR

Department of Trade and Industry Export Publications

Greece - English/1992

17020 GREECE'S WEEKLY FOR BUSINESS AND FINANCE

Coronakis Press Ltd

Greece - English weekly/ US$150.00 P.A.

Economic and political comment are part of this magazine's coverage.

17021 GREEK BALANCE OF PAYMENTS: THE EFFECTS OF FULL MEMBERSHIP AND THE INTERNAL MARKET PROGRAM

Centre of Planning and Economic Research

D.C. Maroulis/Greece/EEC - Greek/1991

17022 THE GREEK ECONOMY IN FIGURES

Electra Press Publications

Evangelia Papaioannou/Greece - English/Greek annual/1980 250pp US$98.00 (Europe), US $102.75 (rest of the world) ISSN No. 1105-2503

Gives data on the economy, population, and main economic indicators.

17023 HINTS TO EXPORTERS VISITING GREECE

Department of Trade and Industry Export Publications

Greece - English irregular/1991 60pp £5.00

This publication provides detailed practical information that is useful for those planning a business visit. Topics include currency information, economic factors, import and exchange control regulations, methods of doing business and general information.

17024 IN DEPTH COUNTRY APPRAISALS ACROSS THE EC - CONSTRUCTION AND BUILDING SERVICES - GREECE

Building Services Research and Information Association

Greece - English 1991/£125.00

Part of an 11 volume survey covering the construction industry in EC countries.

17025 INVESTING, LICENSING AND TRADING CONDITIONS ABROAD: GREECE

Business International Ltd

Greece - English annual/£825.00 for complete European series

Loose-leaf format covering the state role in industry, competition rules, price controls, remittability of funds, taxation, incentives and capital sources, labour and foreign trade.

17026 LABOUR FORCE SURVEY

National Statistical Service of Greece

Greece - Greek annual/1981 90pp US$5.00 ISSN No. 0256-3576

Statistical coverage of standard labour force indicators, including numbers employed and unemployed.

17027 MONTHLY STATISTICAL BULLETIN

National Statistical Service of Greece

Greece - English/Greek monthly/ US$5.00 ISSN No. 0077-6114

Gives monthly reports on a variety of economic topics including main economic indicators.

17028 MONTHLY STATISTICAL BULLETIN

Bank of Greece

Greece - English/Greek monthly/ 110pp ISSN No. 1105-0519

17029 MOUVEMENT NATUREL DE LA POPULATION DE LA GRECE

National Statistical Service of Greece

Greece - Greek/French/ annual/ 222pp US$10.00

Covers population movements and growth in Greece.

17030 OCCUPATIONAL CHOICES OF GREEK YOUTH: AN EMPIRICAL ANALYSIS OF THE CONTRIBUTION OF INFORMATION AND SOCIO-ECONOMIC VARIABLES

Centre of Planning and Economic Research

A.G. Kostakis/Greece - English irregular/1990 US$19.00

17031 OUTPUT SUPPLY AND INPUT DEMAND IN GREEK AGRICULTURE

Centre of Planning and Economic Research

G J Margos/Greece - English/1991 US$19.00

Input-output analysis of the Greek agricultural industry.

17032 PROBLEMS AND PROSPECTS OF GREEK EXPORTS: PREREQUISITES FOR THEIR DEVELOPMENT WITHIN THE UNIFIED MARKET OF THE EEC

Centre of Planning and Economic Research

D.C. Maroulis/Greece - Greek/1992 309pp US$25.00

17033 PUBLIC FINANCE STATISTICS

National Statistical Service of Greece

Greece - English/Greek annual/ 150pp US$8.00 ISSN No. 0256-3568

Central government financial statistics, a quarterly bulletin is also available.

17034 QUARTERLY COUNTRY TRADE REPORTS: GREECE

United Kingdom Iron and Steel Statistics Bureau

Greece - English quarterly/£200.00 P.A. for four countries ISSN No. 0960-2380

Cumulative trade statistics of imports and exports of steel products.

17035 REGIONAL DEVELOPMENT INDICATORS OF GREECE

Centre of Planning and Economic Research

P.A. Kavvadias/Greece - Greek/ 1992 376pp US$25.00

Compares the relative positions of Greek regions using standard socio-economic indicators.

17036 REPORTS FOR THE FIVE-YEAR PLAN 1988-1992: EMPLOYMENT-UNEMPLOYMENT

Centre of Planning and Economic Research

Greece - Greek irregular/1990

Reviews progress in the employment area including employment generation.

17037 REPORTS FOR THE FIVE-YEAR PLAN 1988-1992: ENERGY

Centre of Planning and Economic Research

Greece - Greek irregular/1991

Reports on the current situation and progress towards national energy objectives.

17038 REPORTS FOR THE FIVE-YEAR PLAN 1988-1992: LOCAL GOVERNMENT

Centre of Planning and Economic Research

Greece - Greek irregular/1990

17039 REPORTS FOR THE FIVE-YEAR PLAN 1988-1992: PROSPECTS OF THE GREEK MONETARY SYSTEM

Centre of Planning and Economic Research

Greece - Greek irregular/1991

Reviews progress of the five year plan for financial policies and relates to the Greek situation in a wider financial Europe.

17040 REPORTS FOR THE FIVE-YEAR PLAN 1988-1992: PUBLIC ADMINISTRATION

Centre of Planning and Economic Research

Greece - Greek irregular/1991

Reviews progress in development and reform of public administration.

17041 REPORTS FOR THE FIVE-YEAR PLAN 1988-1992: REGIONAL POLICY

Centre of Planning and Economic Research

Greece - Greek irregular/1991

Reviews progress towards achieving the objectives of the official five year plan for regional development.

17042 RESULTS OF SEA FISHERY SURVEY BY MOTOR VESSELS
National Statistical Service of Greece

Greece - English/Greek annual/ US$3.00 ISSN No. 0256-3584

17043 SHIPBUILDING AND SHIPREPAIRING INDUSTRY IN GREECE
Centre of Planning and Economic Research

F G Tzamoozakis, S K Spathi/ Greece - English/1991 US$19.00

Review of the shipbuilding industry including levels of activity and trends.

17044 SHIPPING STATISTICS
National Statistical Service of Greece

Greece - English/Greek annual/ US$10.00 ISSN No. 0072-7423

Detailed data on the Greek shipping industry including tonnage, numbers etc.

17045 SOCIAL WELFARE AND HEALTH STATISTICS
National Statistical Service of Greece

Greece - English/Greek annual/ US$7.00 ISSN No. 0253-9454

Expenditure on social security and health sectors including numbers involved.

17046 STATISTICAL BULLETIN
Thessaloniki Port Authority

Greece - English annual/c45pp

Statistical data on movement of goods, maritime traffic and goods loaded and unloaded.

17047 STATISTICAL BULLETIN OF PUBLIC FINANCE
National Statistical Service of Greece

Greece - English/Greek quarterly/ US$16.00 ISSN No. 0256-3592

Quarterly bulletin of public finance which includes central government income and expenditure.

17048 STATISTICAL YEARBOOK OF GREECE
National Statistical Service of Greece

Greece - English/Greek annual/ US$20.00 ISSN No. 0031-5071

Broad survey of Greek statistics including main economic indicators.

17049 STATISTICS OF THE DECLARED INCOME OF LEGAL ENTITIES AND ITS TAXATION
National Statistical Service of Greece

Greece - Greek annual/US$4.00

Statistical coverage of corporate income and related income tax.

17050 STATISTICS OF THE DECLARED INCOME OF PHYSICAL PERSONS AND ITS TAXATION
National Statistical Service of Greece

Greece - Greek annual/US$4.00 ISSN No. 0302-1114

Statistical coverage of personal incomes and income tax.

17051 TOURISM STATISTICS
National Statistical Service of Greece

Greece - English/Greece irregular/ 134pp US$10.00

Standard indicators are used, including number of visitors, length of stay and destination.

17052 TRANSPORT AND COMMUNICATIONS STATISTICS
National Statistical Service of Greece

Greece - Greek annual/US$7.00 ISSN No. 0256-3657

Main forms of transport are covered including road and rail.

HUNGARY

Country profile:	Hungary
Official name:	Hungarian Republic
Area:	93,030 sq.km.
	Bordered in the north by Slovakia and the Ukraine, in the south by Croatia (former Yugoslavia) and Yugoslavia, to the east by Romania, and in the west by Austria.
Population:	10.45m (1990)
Capital:	Budapest
Language(s):	Hungarian
Currency:	Forint 100 filler = 1 Forint
National bank:	The National Bank
Government and Political structure:	Republican parliamentary democracy with single chamber National Assembly. 386 elected members.

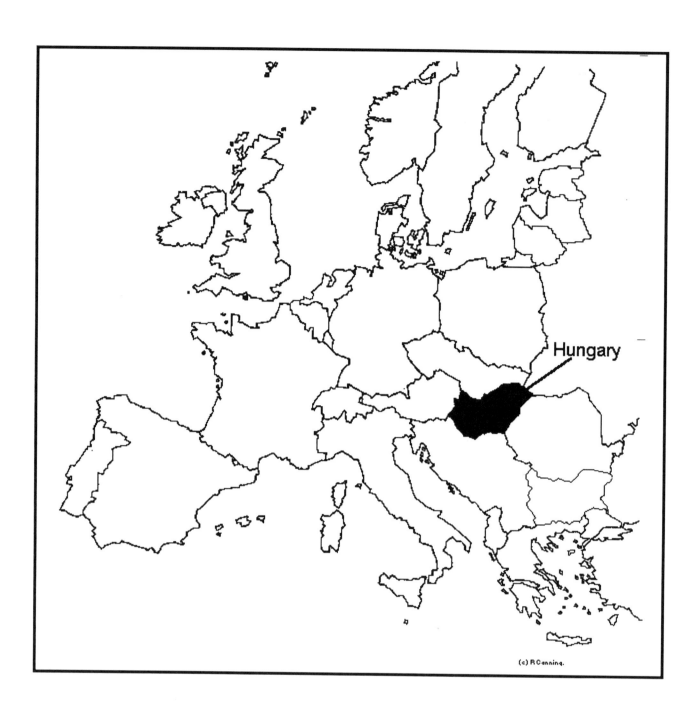

Hungary

(c) R.Canning.

18001 BULLETIN

Hungarian Central Statistical Office

Hungary/Czechoslovakia/Poland - English/Hungarian 4 times per year/ 1992 24pp

Quarterly bulletin focusing on Hungarian economy, with some coverage of Poland and Czechoslovakia.

18002 BUSINESS DIRECTORY OF THE HUNGARIAN CHAMBER OF COMMERCE

Hungarian Chamber of Commerce

Ms Judit Gyárfás/Mr Peter Forgács/ Hungary - English/German/ Russian/1990 US$40 ISSN No. 0864-8824

Hungarian business directory.

18003 BUSINESS GUIDE HUNGARY

Hungarian Chamber of Commerce

Judit Boncz et al/Hungary - English/ 1992 75pp US$25.00 ISBN No. 963785231X

Guide to doing business in Hungary.

18004 CATALOGUE OF COMPANIES INDUSTRIAL, COMMERCIAL AND BUILDING, INDUSTRIAL SECTOR

Ministry of Industry and Trade, Information Agency

Hungary - Hungarian annual/1989 5 Vols 1200pp HUF5000

Hungarian business directory.

18005 COUNTRY FORECASTS: HUNGARY

Business International Ltd

Hungary - English quarterly/ £360.00 P.A.

Contains an executive summary, sections on political and economic outlooks and the business environment, a fact sheet, and key economic indicators. Includes a quarterly Global Outlook.

18006 COUNTRY PROFILE HUNGARY

Department of Trade and Industry Export Publications

Hungary - English irregular/1992 52pp £10.00

This publication provides a general description of the country in terms of its geography, economy and population, and gives information about trading in the country with details like export conditions and investment covered.

18007 COUNTRY REPORT: HUNGARY

Business International Ltd

Hungary - English quarterly/25-40pp £160.00 P.A. ISSN No. 0269-4301

Contains an executive summary, sections on political and economic structure, outlook for the next 12-18 months, statistical appendices, and a review of key political developments. Includes an annual Country Profile.

18008 COUNTRY RISK SERVICE: HUNGARY

Business International Ltd

Hungary - English quarterly/ minimum subscription £1,855.00 P.A. for seven countries

Contains credit risk ratings, risk appraisals, cross-country databases, data on diskette, and two year projections.

18009 DOING BUSINESS IN EASTERN EUROPE: HUNGARY

Kogan Page

Hungary - English/1991 320pp £25.00 ISBN No. 0749404728

Part of the CBI initiative on Eastern Europe. Describes the business and economic environment.

18010 DOING BUSINESS IN HUNGARY

Confederation of British Industry (CBI)

KPMG Peat Marwick Mclintock, Barclays Bank, S.J. Berwin and Co., Royal Mail International/ Hungary - English/1991 346pp ISBN No. 0749404728

Guide for those wishing to visit Hungary.

18011 DOING BUSINESS IN HUNGARY

Price Waterhouse

Hungary - English irregular/1993 330-340pp free to clients ISSN No. 1059-641X

A business guide which includes sections on the investment climate, doing business, accounting and auditing, taxation and general country data. Updated by supplements.

18012 DOING BUSINESS WITH EASTERN EUROPE: HUNGARY

Business International Ltd

Hungary - English quarterly/ £540.00 P.A.

Part of a ten volume set, updated every three months which gives a broad picture of the market and business practice.

18013 EXHIBITIONS AND FAIRS IN HUNGARY 92

Hungarian Chamber of Commerce

Hungary - English/German/1992 US$5

18014 FIGYELÖ (ECONOMIC WEEKLY)

Figyelö Co

Dr Vargy György/Hungary - Hungarian weekly/February 1957 56pp HUF55

18015 GLOBAL FORECASTING SERVICE - HUNGARY

The Economist Intelligence Unit

Hungary - English quarterly/ £360.00 P.A.

Economic and political forecasting publication.

18016 HANDBOOK OF THE HUNGARIAN ECONOMY 1992

Hungarian Chamber of Commerce

Péter Balázs et al/Hungary - German irregular/1992 181pp US$10.00 ISBN No. 9637594091

Key data on the Hungarian economy which includes the main economic indicators.

18017 HETI VILAGGAZDASSY (WORLD MARKET WEEKLY)

Hungarian Chamber of Commerce

Hungary - Hungarian weekly ISSN No. 0139-1682

Weekly Hungarian business journal.

18018 HINTS TO EXPORTERS VISITING HUNGARY

Department of Trade and Industry Export Publications

Hungary - English irregular/1992 54pp £5.00

This publication provides detailed practical information that is useful for those planning a business visit. Topics include currency information, economic factors, import and exchange control regulations, methods of doing business and general information.

18019 HUNGARIAN ANNUAL STATISTICAL YEARBOOK

Hungarian Central Statistical Office

Hungary - Hungarian annual/360pp HUF563 ISSN No. 0073-4039

Gathers together the main statistical series produced.

18020 HUNGARIAN BUSINESS BOOK

International Business Club

Hungary - Hungarian/1992 471pp

Country profile type publication mainly aimed at business operating in Hungary.

18021 HUNGARIAN BUSINESS BOOK 1992

Hungarian Chamber of Commerce

Psychoinvest Training and Consulting Co./Hungary - English/German/Hungarian/1992 488pp US$10.00

A publication which looks at the business environment in Hungary.

18022 HUNGARIAN ECONOMIC REVIEW

Hungarian Chamber of Commerce

Hungary - English bi-monthly/April 1991 US$5.00 P.A. ISSN No. 1215-2439

18023 THE HUNGARIAN ECONOMY

Hungarian Foreign Trading Co

Hungary - English quarterly/24pp US$19.50 ISSN No. 0133-0365

Details the main economic influences on the economy.

18024 HUNGARIAN FINANCIAL AND STOCK EXCHANGE ALMANAC

Hungarian Chamber of Commerce

Dr Tomás Bácskai/Hungary - English annual/1991 867pp US$20.00 ISSN No. 0866-5478

A study of the Hungarian financial sector.

18025 HUNGARIAN STATISTICAL POCKET BOOK

Hungarian Central Statistical Office

Hungary - English/Hungarian annual/224pp HUF412 ISSN No. 0441-473X

Key statistical data on Hungary in an abbreviated format.

18026 HUNGARY 19–

Hungarian Central Statistical Office

Hungary - English/Hungarian/German/Spanish annual/32pp HUF111 ISSN No. 0230-5828

Annual overview of business conditions in Hungary.

18027 HUNGARY IN THE 1990S; SOWING THE SEEDS OF RECOVERY

The Economist Intelligence Unit

Christopher Mattheison/Hungary - English/1991 84pp £195.00

Overview of the Hungarian economy.

18028 IDEAS FOR JOINT VENTURE

Hungarian Chamber of Commerce

Hungary - English irregular/US$10.00

Describes potential joint ventures for prospective investors.

18029 INFORMATIONS

Hungarian Central Statistical Office

Hungary - Hungarian monthly 36pp HUF100 P.A.

Monthly journal focusing on Hungarian economy.

18030 INTERNATIONAL CUSTOMS JOURNAL: HUNGARY

Department of Trade and Industry Export Publications

Hungary - English irregular/£25.00

This publication contains the rate of duties that apply to commodities exported into the country according to the Harmonised System (HS) of commodity coding.

18031 INVESTING, LICENSING AND TRADING CONDITIONS ABROAD: HUNGARY

Business International Ltd

Hungary - English annual/£825.00 for complete European series

Loose-leaf format covering the state role in industry, competition rules, price controls, remittability of funds, taxation, incentives and capital sources, labour and foreign trade.

18032 INVESTMENT BRIEF HUNGARY 1992

Hungarian Chamber of Commerce

Press and Information Department of the Ministry of International Economic Relations/Hungary - English/German/May 1992 12pp US$8.00

A publication of interest to those wishing to invest in Hungary.

18033 INVESTORS GUIDE TO HUNGARY 199–

Hungarian Chamber of Commerce

Peter Dunai et al/Hungary - English/German every two years/1989

US$30.00

Describes the climate for investment and local regulations.

18034 MARKETING IN EUROPE, SPECIAL MARKET SURVEY; RETAILING AND WHOLESALING IN HUNGARY

The Economist Intelligence Unit

B Vink/Hungary - English/1991 49pp £110.00 single copy

Market research type review.

18035 NATIONAL ACCOUNTS, HUNGARY (1988-1990)

Hungarian Central Statistical Office

Hungary - English/Hungarian annual/1992 212pp HUF417

Covers national accounts, income, expenditure etc for local government.

18036 REAL ESTATE GUIDE HUNGARY

The Royal Institution of Chartered Surveyors

Healey and Baker Research Series/Hungary - English/1991 154pp ISBN No. 0854064877

A detailed study of the Hungarian property market.

18037 SITUATION REPORT ON THE SOCIETY AND ECONOMY OF HUNGARY

Hungarian Central Statistical Office

Hungary - English/Hungarian irregular/98pp HUF350

Socio-economic data on Hungary covering social conditions and the economic situation broadly.

18038 SITUATION, TENDANCES ET PERSPECTIVES DE L'AGRICULTURE EN HONGRIE

Office for Official Publications of the European Communities

Hungary - French/1991 158pp ECU17.00

Agricultural situation in Hungary.

18039 STATISTICAL MONTHLY REVIEW

Hungarian Central Statistical Office

Hungary - English/Hungarian monthly/110pp HUF110 P.A. ISSN No. 0018-781X

Reviews the main statistical series produced by the national statistical agency.

18040 STATISTICAL POCKET BOOK OF BUDAPEST
Hungarian Central Statistical Office

Hungary - Hungarian annual/ HUF397

Brief report type publication with key data on Budapest.

18041 YEARBOOK OF ECONOMIC STATISTICS
Hungarian Central Statistical Office

Hungary - English/Hungarian annual/440pp HUF684

Collection of the main economic statistical series.

18042 YEARBOOK OF TOURISM
Hungarian Central Statistical Office

Hungary - English/Hungarian annual/190pp HUF398 ISSN No. 0236-9524

Analyses the tourist industry in a statistical format with standard indicators such as number of visitors, travel and expenditure.

ICELAND

Country profile:	Iceland
Official name:	Republic of Iceland
Area:	103,000 sq.km.
	Surrounded by the North Atlantic and close to the Arctic circle.
Population:	255,700 (1990)
Capital:	Reykjavik
Language(s):	Icelandic
Currency:	Krona 100 aurar = 1 Krona
National bank:	Central Bank of Iceland
Government and Political structure:	Elected Alpingi (Parliament) with 63 members, divided into Upper and Lower Houses.

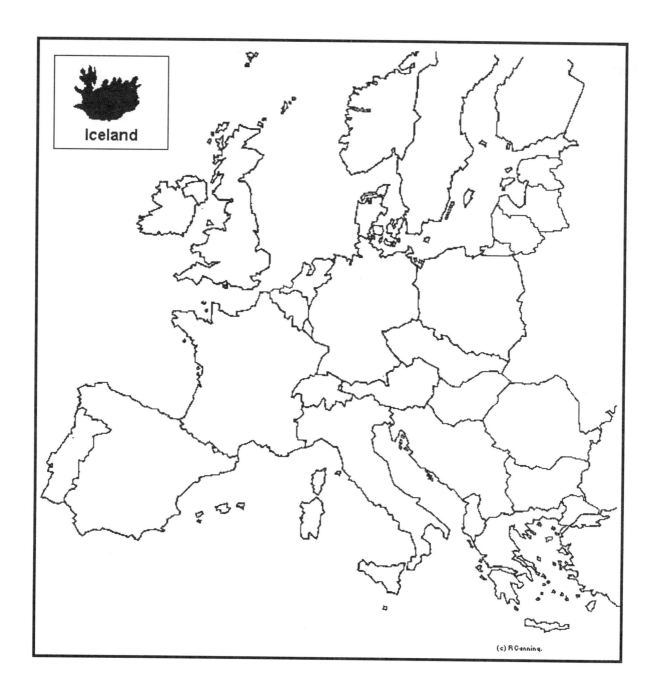

Iceland

(c) R.Canning.

19001 ANNUAL REPORT
Central Bank of Iceland

Iceland - English annual/115pp free

Economic developments; money and domestic credit; balance of payments and exchange rate policy.

19002 ANNUAL REPORT
Landsbanki Islands

Iceland - Icelandic/English summary annual/50pp free

Annual report and accounts of the Icelandic National Bank.

19003 COUNTRY PROFILE - DENMARK, ICELAND
Business International Ltd

Denmark/Iceland - English annual/ 46pp ISSN No. 026-5138

Annual survey of political and economic background. Included as part of "Country Report" subscription.

19004 COUNTRY PROFILE: ICELAND
Department of Trade and Industry Export Publications

Iceland - English irregular/1991 56pp £10.00

This publication provides a general description of the country in terms of its geography, economy and population, and gives information about trading in the country with details like export conditions and investment covered.

19005 COUNTRY REPORT: DENMARK, ICELAND
Business International Ltd

Iceland/Denmark - English quarterly/28pp £150.00 P.A. (4 issues + a "Country Profile") ISSN No. 0269-574X

Analysis of political and economic trends.

19006 COUNTRY REPORT: ICELAND
Barclays Bank Plc: Economics Dept.

Iceland - English irregular/free

Brief economic report on Iceland.

19007 ECONOMIC OUTLOOK
Landsbanki Islands

Iceland - English 4-5 times a year/ 8pp free

General economic report.

19008 ECONOMIC STATISTICS QUARTERLY
Central Bank of Iceland

Iceland - English quarterly/30pp ISSN No. 0256-193X

Domestic economic indicators, foreign trade, social indicators.

19009 THE ECONOMY OF ICELAND
Central Bank of Iceland

Iceland - English irregular/41pp

Domestic economy, foreign trade, financial markets.

19010 FOREIGN TRADE
Statistical Bureau of Iceland

Iceland - Icelandic with English headings annual/1912 460pp US$40.00 ISSN No. 1017-6365

Exports and imports classified by value, country of origin of destination, and commodity.

19011 HINTS TO EXPORTERS VISITING ICELAND
Department of Trade and Industry Export Publications

Iceland - English irregular/1990 48pp £5.00

This publication provides detailed practical information that is useful for those planning a business visit. Topics include currency information, economic factors, import and exchange control regulations, methods of doing business and general information.

19012 ICELAND REVIEW
Iceland Review

Iceland - English quarterly/1963 US$27.95 P.A. ISSN No. 0019-1094

Focuses on Icelandic culture, politics, economy, art and literature.

19013 THE ICELANDIC ECONOMY: DEVELOPMENTS 1991 AND OUTLOOK FOR 1992
National Economic Institute

Iceland - English/Icelandic 1992/ ISBN No. 9979823097

19014 ISLANDBANKI - ANNUAL REPORT
Islandbanki HF

Iceland - English annual/free

Covers both banking and related economic matters.

19015 LOCAL GOVERNMENT FINANCE
Statistical Bureau of Iceland

Iceland - Icelandic with English headings annual/1967 3 Vols $35.00 per Vol. ISSN No. 1017-6357

Statistics on local finance giving patterns of income and expenditure.

19016 MONTHLY STATISTICS
Statistical Bureau of Iceland

Iceland - Icelandic with English translation of table headings monthly/1916 500pp US$45.00

P.A. ISSN No. 0019-1078

Foreign trade, imports, exports, employment and economic indicators.

19017 NEWS FROM ICELAND
Iceland Review

Iceland - English monthly/1975 US$24.00 P.A. ISSN No. 0253-8083

General news, current events, politics, economy, trade, industry (emphasis on fishing).

19018 THE NORDIC COUNTRIES
Nordic Statistical Secretariat

Iceland/Denmark/Finland/Norway/ Sweden - English annual/1984 40pp

Snapshot from the 'Yearbook of Nordic Statistics' - pocket-size version.

19019 NORDIC LABOUR MARKET STATISTICS
Nordic Statistical Secretariat

Iceland/Denmark/Finland/Norway/ Sweden - English semi-annually

Trends in the labour market in Nordic countries.

19020 OECD ECONOMIC SURVEYS: ICELAND
Organization for Economic Co-operation and Development (OECD)

Iceland - English irregular/£7.50 ISSN No. 0376-6438

Survey on economic developments.

19021 POPULATION AND VITAL STATISTICS
Statistical Bureau of Iceland

Iceland - Icelandic with English headings every 10 years/1921 300pp US$35.00 P.A. ISSN No. 1017-6691

Population by geographic and demographic migration, life tables, population projections.

19022 STATISTICAL ABSTRACT OF ICELAND
Statistical Bureau of Iceland

Iceland - Icelandic with English headings annual/1991 260pp US$40.00 ISSN No. 1017-6683

Social, economic and industrial statistics.

19023 YEARBOOK OF NORDIC STATISTICS
Nordic Statistical Secretariat

Iceland/Denmark/Finland/Norway/ Sweden - English annual/1963 400pp

Statistics on population; exports; imports; labour market; communication; finance; social security and health-care.

IRELAND

Country profile:	Ireland
Official name:	Republic of Ireland
Area:	70,300 sq.km.
	Surrounded by the Atlantic Ocean and bordered by Northern Ireland in the north-east.
Population:	3.51m (1991)
Capital:	Dublin
Language(s):	English, Gaelic
Currency:	Punt 100 pennies = 1 Punt
National bank:	Central Bank of Ireland
Government and Political structure:	Republican parliamentary democracy. The Oireachtas (national parliament) has two houses. The Dail (Lower House) with 166 members which chooses the Taoiseach (Prime Minister).

20001 ACCOUNTANCY IRELAND
Institute of Chartered Accountants in Ireland
Ireland - English 6 issues per year/ IR£17.50

20002 ADMINISTRATION YEARBOOK AND DIARY
Institute of Public Administration
Ireland - English annual/487pp IR£32.00 ISSN No. 0073-9596
Yearbook type publication which also contains a statistical section.

20003 AGRICULTURAL INPUT PRICE INDEX
Central Statistics Office
Ireland - English monthly/IR£18.00 P.A.
Monitors the changes in the prices of agricultural inputs.

20004 AGRICULTURAL OUTPUT PRICE INDEX
Central Statistics Office
Ireland - English monthly/IR£18.00 P.A.
Monitors the changes in the prices of agricultural outputs.

20005 ANIMALS EXPORTED BY SEA AND AIR
Central Statistics Office
Ireland - English monthly/IR£18.00 P.A.
Covers cattle, sheep, pigs and horses.

20006 ANNUAL HOUSING STATISTICS BULLETIN
Government Publications
Ireland - English annual/price varies
Published by the Department of the Environment.

20007 ANNUAL REPORT OF THE CENTRAL BANK
Central Bank of Ireland
Ireland - English annual/free
Covers financial balances and international finance.

20008 ANNUAL REPORT OF THE DEPARTMENT OF LABOUR
Government Publications
Ireland - English annual/IR£4.50
Reviews the labour market and developments in employment/ unemployment.

20009 ANNUAL REPORT OF THE REVENUE COMMISSIONERS
Government Publications
Ireland - English annual/IR£4.50
Covers taxation of individuals and commercial institutions.

20010 AVERAGE EARNINGS AND HOURS WORKED
Central Statistics Office
Ireland - English quarterly/IR£6.00 P.A.
Covers each of the main employee categories in the private construction sector. Also available as part of the "Statistical Bulletin".

20011 BALANCE OF INTERNATIONAL PAYMENTS
Central Statistics Office
Ireland - English quarterly/IR£6.00 P.A.
A quarterly statistical release. Measures the balance on current and capital account. Also available as part of the "Statistical Bulletin".

20012 BANKING, INSURANCE AND BUILDING SOCIETIES - EMPLOYMENT AND EARNINGS
Central Statistics Office
Ireland - English quarterly/1988 IR£6.00
A quarterly statistical release. Measures changes in employment and average gross earnings. Also available as part of the "Statistical Bulletin".

20013 BANKING IRELAND
Fintel Publications Ltd
Ireland - English quarterly/ISSN No. 0791-1386
Covers developments in Irish banking, with some related data included.

20014 BUDGET 199-
Government Publications
Ireland - English annual
Published by the Department of Finance, this publication covers the annual central government budget provisions.

20015 BUSINESS AND FINANCE
Belenos Publications Ltd
Ireland - English weekly/1964 IR£87.00 ISSN No. 0007-6473
Irish business magazine which includes economic features and comment.

20016 BUSINESS OF ADVERTISING AGENCIES
Central Statistics Office
Ireland - English annual/IR£1.50
An annual statistical release. Measures gross amount charged by advertising agencies to clients for all services. Information is also given on employment and salary levels. Also available as part of the "Statistical Bulletin".

20017 CENSUS OF AGRICULTURE - PROVISIONAL ESTIMATES
Central Statistics Office
Ireland - English annual/IR£1.50 P.A.
Covers land usage and numbers of livestock and is a provisional estimate of the main annual Agricultural Census, published as an annual statistical release and as part of the Statistical Bulletin.

20018 CENSUS OF BUILDING AND CONSTRUCTION
Central Statistics Office
Ireland - English annual/IR£1.50
Monitors the activities of the private building sector, including civil engineering and allied trades.

20019 CENSUS OF POPULATION 1991
Central Statistics Office
Ireland - English every ten years/ IR£5.00 per county
Local population reports are available on a county-by-county basis.

20020 CENSUS OF SERVICES
Central Statistics Office
Ireland - English irregular/244pp IR£15.00
Previously called Census of Distribution, covers wholesale, retail and non-distribution services. Last census conducted 1988. In two volumes.

20021 CENTRAL BANK STATISTICS SUPPLEMENT
Central Bank of Ireland
Ireland - English monthly/free
Updates key monetary and financial data included in the Quarterly Bulletin.

20022 CII NEWSLETTER
Confederation of Irish Industry
Ireland - English fortnightly/ IR£55.00 P.A.
Information on Irish industry.

20023 CONFEDERATION OF IRISH INDUSTRY ANNUAL REPORT
Confederation of Irish Industry
Ireland - English annual
Reports on the industrial situation over the past year and includes data on industrial production and sectors.

20024 CONSTRUCTION INDUSTRY IN IRELAND REVIEW AND OUTLOOK
Government Publications
Ireland - English annual/IR£5.00

Published by the Department of the Environment this reviews the construction industry for the past year and looks at the outlook for the forthcoming year.

20025 CONSUMER PRICE INDEX
Central Statistics Office

Ireland - English quarterly/IR£6.00 P.A.

A statistical release. The official measure of inflation. Also available as part of the "Statistical Bulletin".

20026 COUNTRY BOOKS: IRISH REPUBLIC
United Kingdom Iron and Steel Statistics Bureau

Ireland - English annual/£75.00 ISSN No. 0952-5920

Annual summary tables showing production, materials consumed, apparent consumption, imports and exports of 130 products by quality and market.

20027 COUNTRY FORECASTS: IRELAND
Business International Ltd

Ireland - English quarterly/£360.00 P.A.

Contains an executive summary, sections on political and economic outlooks and the business environment, a fact sheet, and key economic indicators. Includes a quarterly Global Outlook.

20028 COUNTRY PROFILE: IRELAND
Department of Trade and Industry Export Publications

Ireland - English irregular/1991 44pp £10.00

This publication provides a general description of the country in terms of its geography, economy and population, and gives information about trading in the country with details like export conditions and investment covered.

20029 COUNTRY REPORT: IRELAND
Business International Ltd

Ireland - English quarterly/25-40pp £160.00 P.A. ISSN No. 0269-5278

Contains an executive summary, sections on political and economic structure, outlook for the next 12-18 months, statistical appendices, and a review of key political developments. Includes an annual Country Profile.

20030 DECEMBER LIVESTOCK SURVEY
Central Statistics Office

Ireland - English annual/IR£1.50

Estimates of the number of livestock with comparative data for the same month of the previous two years.

20031 DOING BUSINESS IN THE REPUBLIC OF IRELAND
Price Waterhouse

Ireland - English irregular/1990 330-340pp free to clients

A business guide which includes sections on the investment climate, doing business, accounting and auditing, taxation and general country data. Updated by supplements.

20032 ECONOMIC REVIEW AND OUTLOOK
Government Publications

Ireland - English annual/IR£4.00

Published by the Department of Finance, this publication gives a review of the year past and the outlook for the forthcoming year.

20033 ECONOMIC SERIES
Central Statistics Office

Ireland - English monthly/54pp IR£4.00 P.A. ISSN No. 0790-8407

Covers prices, industry, building, agriculture, trade and distribution, transport, finance, labour, and vital statistics.

20034 ECONOMIC TRENDS
Confederation of Irish Industry

Ireland - English monthly/IR£55.00 P.A.

Commentary on current economic trends as they affect Ireland.

20035 ESTIMATED OUTPUT, INPUT AND INCOME ARISING IN AGRICULTURE
Central Statistics Office

Ireland - English annual/IR£1.50 P.A.

Published in "advanced", "preliminary" and "final" formats.

20036 ESTIMATES FOR THE PUBLIC SERVICES
Government Publications

Ireland - English annual/IR£4.00

Published by the Department of Finance, this publication gives estimates for the cost of provision of public services. "Revised estimates" are also published during the year which amend the estimates in the light of firmer figures.

20037 ESTIMATES OF RECEIPTS AND EXPENDITURE FOR THE YEAR
Government Publications

Ireland - English annual/IR£4.00

Published by the Department of Finance this covers tax receipts and government expenditure.

20038 EXTERNAL TRADE - PROVISIONAL FIGURES
Central Statistics Office

Ireland - English monthly/IR£18.00 P.A.

A monthly statistical release. Also available as part of the "Statistical Bulletin".

20039 FAIR TRADE COMMISSION ANNUAL REPORT
Government Publications

Ireland - English annual/IR£3.00

Published by the Department of Industry and Commerce.

20040 FARM MANAGEMENT SURVEY
Teagasc - The Agricultural and Food Development

Ireland - English annual/price varies

20041 FINANCE ACCOUNTS
Government Publications

Ireland - English annual/price varies

Published by the Department of Finance, this publication covers central government financial accounts.

20042 HINTS TO EXPORTERS VISITING IRELAND
Department of Trade and Industry Export Publications

Ireland - English irregular/1991 64pp £5.00

This publication provides detailed practical information that is useful for those planning a business visit. Topics include currency information, economic factors, import and exchange control regulations, methods of doing business and general information.

20043 HIRE-PURCHASE AND CREDIT-SALES
Central Statistics Office

Ireland - English annual/IR£1.50

An annual statistical release. Measures number and value of hire-purchase and credit sale transactions. Also available as part of the "Statistical Bulletin".

20044 HOUSEHOLD BUDGET SURVEY
Central Statistics Office

Ireland - English every 7 years/ 266pp IR£9.00

Household consumption, accommodation, weekly income and expenditure. Last published 1987.

20045 IN DEPTH COUNTRY APPRAISALS ACROSS THE EC - CONSTRUCTION AND BUILDING SERVICES - IRELAND
Building Services Research and Information Association

Ireland - English 1991/£125.00

Part of an 11 volume survey covering the construction industry in EC countries.

20046 INDEX OF EMPLOYMENT

Central Statistics Office

Ireland - English monthly/IR£18.00 P.A.

Measures changes in the private construction sector. Also available as part of the "Statistical Bulletin".

20047 INDUSTRIAL DEVELOPMENT AUTHORITY ANNUAL REPORT

Industrial Development Authority

Ireland - English annual/free

Annual report of the main body responsible for the attraction of foreign investment and the development of indigenous industry. A series of leaflets aimed at potential investors/industrialists, giving brief economic background, is available from the organization.

20048 INDUSTRIAL DISPUTES

Central Statistics Office

Ireland - English quarterly/IR£6.00 P.A.

A quarterly statistical release. Details number, duration and details of industrial disputes.

20049 INDUSTRIAL EMPLOYMENT

Central Statistics Office

Ireland - English quarterly/IR£6.00 P.A.

A statistical release. Estimates total employment in industrial units with three or more employees. Also available as part of the "Statistical Bulletin".

20050 INDUSTRIAL EMPLOYMENT AND HOURS WORKED

Central Statistics Office

Ireland - English quarterly/IR£6.00 P.A.

A statistical release. Details changes in average gross earnings and average hours worked by industrial workers in 42 sectors. Also available as part of the "Statistical Bulletin".

20051 INDUSTRIAL EMPLOYMENT AND HOURS WORKED - DETAILS FOR SUPPLEMENTARY NACE SUB-SECTORS

Central Statistics Office

Ireland - English quarterly/IR£6.00 P.A.

A statistical release. Details employment, average gross earnings and hours worked by industrial workers in 20 sectors not included in the main quarterly release. Also available as part of the "Statistical Bulletin".

20052 INDUSTRIAL PRODUCTION

Central Statistics Office

Ireland - English annual/IR£7.50

Latest data available is for 1989. Also available as part of the "Statistical Bulletin".

20053 INDUSTRIAL PRODUCTION INDEX

Central Statistics Office

Ireland - English monthly/IR£18.00 P.A.

A statistical release. Measures trends in the volume of production of industrial units with three or more employees. Also available as part of the "Statistical Bulletin".

20054 INDUSTRIAL TURNOVER INDEX

Central Statistics Office

Ireland - English monthly/IR£18.00 P.A.

A statistical release. Measures changes in the level of sales of industrial products in industrial units with twenty or more employees. Also available as part of the "Statistical Bulletin".

20055 INDUSTRY AND COMMERCE

Association of Chambers of Commerce of Ireland

Ireland - English monthly/IR£25.00 P.A.

Association journal which features economic news and comment.

20056 INPUT-OUTPUT TABLES

Central Statistics Office

Ireland - English irregular/40pp IR£10.00

A description of the inputs and outputs of the different branches of the economy and their inter-relationships. Latest data available is for 1985.

20057 INSURANCE ANNUAL REPORT

Government Publications

Ireland - English annual/IR£5.00

Published by the Department of Industry and Commerce, this looks at the Irish insurance industry and includes statistics.

20058 INVESTING, LICENSING AND TRADING CONDITIONS ABROAD: IRELAND

Business International Ltd

Ireland - English annual/£825.00 for complete European series

Loose-leaf format covering the state role in industry, competition rules, price controls, remittability of funds, taxation, incentives and capital sources, labour and foreign trade.

20059 IRELAND: A DIRECTORY

Institute of Public Administration

Ireland - English annual/c500pp IR£10.35

Diary - yearbook type publication which also contains a statistical section.

20060 IRISH AGRICULTURE IN FIGURES

Teagasc - The Agricultural and Food Development Authority

Ireland - English annual/price varies

Broad statistical coverage of Irish agriculture.

20061 IRISH BANKING REVIEW

Institute of Bankers in Ireland

Ireland - English quarterly/ISSN No. 0021-1060

Covers developments in Irish banking.

20062 IRISH BROKER

Irish Broker

Ireland - English monthly/free

The official journal of the Irish Brokers Association with coverage of insurance issues.

20063 IRISH ECONOMIC STATISTICS

Central Bank of Ireland

Ireland - English annual/free

Includes in a folder details of consumer expenditure, investment, government expenditure, trade, prices, production, manpower, banking and finance.

20064 IRISH EXPORTER

Irish Exporter

Ireland - English monthly/1977 £35.00 P.A.

Exporting, importing, export finance, transport and freight.

20065 IRISH JOURNAL AGRICULTURAL AND FOOD RESEARCH

Teagasc - The Agricultural and Food Development Authority

Ireland - English 2 issues per year/ IR£44.00 P.A. ISSN No. 0791-6833

Replaces the "Irish journal of agricultural economics", published in May and December.

20066 THE IRISH MARKET: FACTS AND FIGURES 199..

Wilson Hartnell Advertising

E. Bent/Ireland - English every two years

Booklet giving brief details on the demography and economy of Ireland with particular reference to marketing.

20067 THE IRISH TIMES

Irish Times Ltd

Ireland - English daily/price varies

Main Irish quality-financial daily, features economic news.

20068 IRISH TOURIST BOARD ANNUAL REPORT

Irish Tourist Board - Bord Failte Eireann

Ireland - English annual/free

Annual report of the state sponsored organization for tourism.

20069 IRISH TRAVEL TRADE NEWS

Belgrave Group Ltd

Ireland - English monthly/£20.00 P.A. ISSN No. 0021-1419

Covers travel and tourism in Ireland with features on aspects of the trade.

20070 LABOUR COSTS SURVEY 1988

Central Statistics Office

Ireland - English irregular/50pp IR£2.65 ISSN No. 0790-9160

Labour costs. Latest edition available is for 1988.

20071 LABOUR FORCE PRELIMINARY ESTIMATE

Central Statistics Office

Ireland - English annual/IR£1.50

An annual statistical release. Details labour force by major sector in mid-April and is published in October.

20072 LABOUR FORCE SURVEY

Central Statistics Office

Ireland - English annual/88pp IR£5.00

Labour force and population estimates. Latest edition available is for 1991.

20073 LIVE REGISTER AGE-BY DURATION ANALYSIS

Central Statistics Office

Ireland - English biannual/IR£3.00

A statistical release. Details unemployment by age and duration on the live register. Also available as part of the "Statistical Bulletin".

20074 LIVE REGISTER AREA ANALYSIS

Central Statistics Office

Ireland - English monthly/IR£18.00 P.A.

A statistical release. Details unemployment by area on the live register. Also available as part of the "Statistical Bulletin".

20075 LIVE REGISTER FLOW ANALYSIS

Central Statistics Office

Ireland - English monthly/IR£18.00 P.A.

A statistical release. Details unemployment flows on and off the live register. Also available as part of the "Statistical Bulletin".

20076 LIVE REGISTER STATEMENT

Central Statistics Office

Ireland - English monthly/IR£18.00 P.A.

A statistical release. Details unemployment figures in seasonally adjusted and unadjusted formats. Also available as part of the "Statistical Bulletin".

20077 LOCAL AUTHORITY ESTIMATES

Government Publications

Ireland - English annual/IR£5.00

Published by the Department of the Environment this covers the financial estimates of local government.

20078 MANAGEMENT

Jemma Publications Ltd

Ireland - English monthly

Covers developments in Irish business and the economy, with news of businesses and management.

20079 MEDIUM TERM REVIEW

Economic and Social Research Institute

Ireland - English quarterly/ subscription price varies with size of institution ISSN No. 0790-9470

Supplements the "Quarterly Economic Commentary" with longer term analysis of the factors affecting the Irish economy.

20080 MONTHLY INDUSTRIAL SURVEY

Confederation of Irish Industry

Ireland - English monthly

Carried out using a sample of manufacturing firms.

20081 MOTOR REGISTRATIONS

Central Statistics Office

Ireland - English annual/IR£18.00

An annual statistical release. Measures number of vehicles registered and licensed and classifies by make and size.

20082 MOTOR REGISTRATIONS: FINAL, DETAILED RESULTS

Central Statistics Office

Ireland - English monthly/IR£18.00 P.A.

A monthly statistical release. Measures number of vehicles registered and licensed, and classifies by make and size. Also available as part of the "Statistical Bulletin".

20083 MOTOR REGISTRATIONS: PROVISIONAL RESULTS

Central Statistics Office

Ireland - English monthly/IR£18.00 P.A.

A monthly statistical release. Measures number of vehicles registered and licensed. Also available as part of the "Statistical Bulletin".

20084 NATIONAL ECONOMIC AND SOCIAL COUNCIL REPORTS

National Economic and Social Council

Ireland - English irregular/price varies

"The Economic and social implications of immigration" is the latest in a series of reports presented to the government and subsequently published. The council exists to provide a forum for discussion on the efficient development of the national economy and the achievement of social justice. A list of past socio-economic reports is available from the council.

20085 NATIONAL FARM SURVEY

Teagasc - The Agricultural and Food Development Authority

Ireland - English annual/IR£10.00

Latest edition available is for 1990.

20086 NATIONAL FARM SURVEY PROVISIONAL ESTIMATES

Teagasc - The Agricultural and Food Development Authority

Ireland - English annual/IR£5.00

First results of an annual survey of farm operations and economics.

20087 NATIONAL INCOME AND EXPENDITURE

Central Statistics Office

Ireland - English annual/66pp IR£5.00

National income and expenditure, capital formation and savings, transactions of the government sector. Latest edition available is for 1991.

20088 OECD ECONOMIC SURVEY

Organisation for Economic Co-operation and Development (OECD)

Ireland - English annual/£102.00

Comprehensive economic survey with forecasts.

20089 PLANNING PERMISSIONS

Central Statistics Office

Ireland - English quarterly/IR£6.00 P.A.

Monitors the number and floor area of planning permissions granted in each county.

20090 POPULATION AND LABOUR FORCE PROJECTIONS 1991-2021
Central Statistics Office

Ireland - English irregular/1988
58pp IR£5.00

20091 PRODUCTION OF MILK AND MILK PRODUCTS
Central Statistics Office

Ireland - English weekly/IR£30.00 P.A.

Detailed statistics on milk production and dairy products by product.

20092 QUANTITY SURVEYORS INQUIRY
Central Statistics Office

Ireland - English quarterly/IR£6.00 P.A.

Monitors level of progress payments on non-residential building projects.

20093 QUARTERLY BULLETIN
Central Bank of Ireland

Ireland - English/Irish quarterly/free
ISSN No. 0069-1542

Includes details of the national economy, monetary development and international monetary developments as they affect Ireland.

20094 QUARTERLY BULLETIN OF HOUSING STATISTICS
Central Statistics Office

Ireland - English quarterly

Published by the Dept. of the Environment. Covers such areas as housing completions and house building costs.

20095 QUARTERLY COUNTRY REPORTS: IRISH REPUBLIC
United Kingdom Iron and Steel Statistics Bureau

Ireland - English quarterly/£200.00 P.A. for four countries
ISSN No. 0952-5920

Cumulative trade statistics of imports and exports of steel products.

20096 QUARTERLY ECONOMIC COMMENTARY
Economic and Social Research Institute

Ireland - English quarterly/ subscription price varies with size of institution ISSN No. 0376-7191

20097 QUARTERLY REPORT ON VITAL STATISTICS
Central Statistics Office

Ireland - English quarterly/IR£3.00 P.A.

A quarterly statistical release. Measures births, marriages and deaths by county. Also available as part of the "Statistical Bulletin".

20098 QUARTERLY REVIEW
AIB Bank

Ireland - English quarterly/free

A brief economic review.

20099 REPORT OF THE COMPTROLLER AND AUDITOR GENERAL AND APPROPRIATON ACCOUNTS
Government Publications

Ireland - English annual

20100 REPORT OF THE REGISTRAR OF FRIENDLY SOCIETIES
Government Publications

Ireland - English annual/IR£5.00

Published by the Department of Industry and Commerce.

20101 REPORT ON VITAL STATISTICS 1988
Central Statistics Office

Ireland - English annual/1991
270pp IR£8.90

Data on births, marriages and deaths. Latest edition available is for 1988, published 1991.

20102 RESEARCH REPORT
Teagasc - The Agricultural and Food Development Authority

Ireland - English annual/IR£15.00

Topical coverage of the agricultural industry.

20103 RETAIL SALES INDEX
Central Statistics Office

Ireland - English monthly/IR£18.00 P.A.

A monthly statistical release. Measures changes in the value and volume of retail sales on a seasonally adjusted basis. Also available as part of the "Statistical Bulletin".

20104 REVIEW OF INDUSTRIAL PERFORMANCE
Government Publications

Ireland - English annual/IR£6.00

Reviews industrial performance for the year using government statistics.

20105 ROAD FREIGHT TRANSPORT SURVEY
Central Statistics Office

Ireland - English annual/IR£2.65

An analysis of activity for a random sample of goods vehicles. Latest edition available is for 1990.

20106 SOCIAL AND ECONOMIC REPORTS
Economic and Social Research Institute

Ireland - English frequent/price varies with report

A range of socio-economic reports published on specific topics.

20107 STATISTICAL ABSTRACT
Central Statistics Office

Ireland - English annual/416pp IR£30.00

Demography, agriculture, industry, construction, trade, social conditions, education, justice and defence, communications and prices.

20108 STATISTICAL BULLETIN
Central Statistics Office

Ireland - English quarterly/160pp IR£7.00 P.A. ISSN No. 0790-8334

Detailed results of all Central Statistical Office regular short-term inquiries.

20109 STATISTICAL REPORT OF THE REVENUE COMMISSIONERS
Central Statistics Office

Ireland - English annual

Data on government revenue.

20110 STATISTICS OF PORT TRAFFIC
Central Statistics Office

Ireland - English annual/IR£1.50

An annual statistical release. Measures number of arrivals and net register tonnage of trading and passenger vessels, details of goods handled and type of traffic. Also available as part of the "Statistical Bulletin".

20111 SUMMARY OF PUBLIC EXPENDITURE PROGRAMME
Government Publications

Ireland - English annual/IR£7.00

Published by the Department of Finance, this publication covers the central government's public expenditure programme.

20112 SUMMARY PUBLIC CAPITAL PROGRAMME
Government Publications

Ireland - English annual

Published by the Department of Finance, this publication covers the central government's public capital programme.

20113 SUPPLEMENTARY ESTIMATES
Government Publications

Ireland - English various/price varies

Published by the Department of Finance these detail financial estimates for the various activities of central government; examples would be for prisons and higher education.

20114 SURVEY OF THE EARNINGS OF PERMANENT MALE AGRICULTURAL WORKERS
Central Statistics Office

Ireland - English biennial

Latest data available is for 1988-89.

20115 TECHNOLOGY IRELAND
Eolas

Ireland - English 10 per year/1969 IR£22.50 P.A. ISSN No. 0040-1676

Information on all major industrial areas including new products, licensing and EEC news.

20116 TOURISM AND TRAVEL
Central Statistics Office

Ireland - English annual/IR£6.00

An annual statistical release. Measures number of visits, length of stay and expenditure by visitors. Also available as part of the "Statistical Bulletin".

20117 TOURISM AND TRAVEL QUARTERLY
Central Statistics Office

Ireland - English quarterly/IR£6.00 P.A.

A quarterly statistical release. Also available as part of the "Statistical Bulletin".

20118 TRADE STATISTICS
Central Statistics Office

Ireland - English monthly/200pp IR£5.00 P.A. ISSN No. 0790-9381

Monthly statistics on trade including details of imports and exports. Also available as part of the "Statistical Bulletin".

20119 TRANSPORTABLE CAPITAL GOODS PRICE INDEX FOR AGRICULTURE
Central Statistics Office

Ireland - English monthly

Published as a sub-index of the Wholesale Price Index.

20120 THE TREND OF EMPLOYMENT AND UNEMPLOYMENT 1986-1988
Central Statistics Office

Ireland - English irregular/1991 97pp IR£3.55

Data on employment levels in industry sectors and unemployment.

20121 WHOLESALE PRICE INDEX
Central Statistics Office

Ireland - English monthly/IR£18.00 P.A.

Measures the monthly change in wholesale prices (excluding VAT) of different categories of goods. Also available as part of the "Statistical Bulletin".

ITALY

Country profile:	Italy
Official name:	Republic of Italy
Area:	301,260 sq.km.
	Bordered in the north by Austria and Switzerland, in the south by the Ionian, Ligurian, Mediterranean and Tyrrhenian seas, in the east by the Adriatic sea and Yugoslavia, and in the west by France.
Population:	57.74m (1991)
Capital:	Rome
Language(s):	Italian
Currency:	Lira 100 centesimi = 1 Italian lira
National Bank:	Bank of Italy
Government and Political structure:	Republican parliamentary democracy. Elected Chamber of Deputies and Senate. The Chamber of Deputies has 630 members.

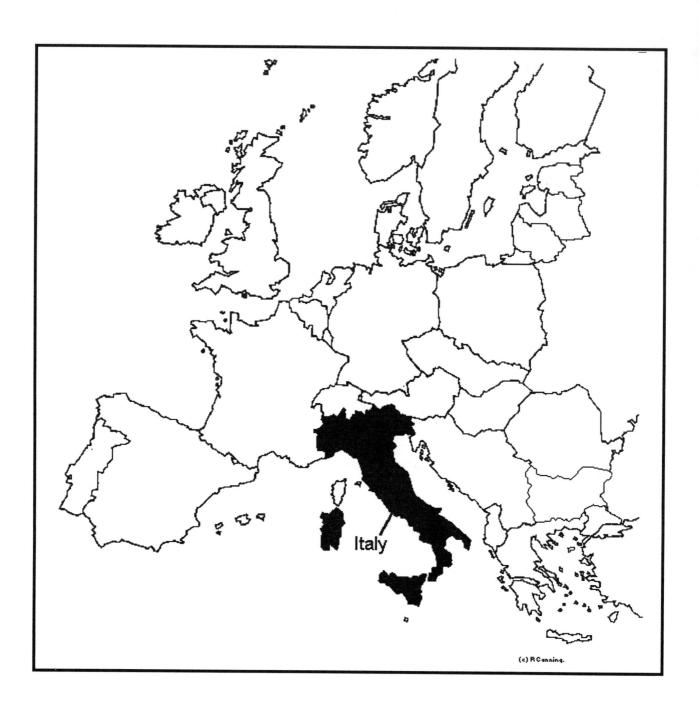

Italy

21001 ANNUARIO DI CONTABILITÀ NAZIONALE

Instituto Nazionale di Statistica

Italy - Italian annual/436pp L17,000

Detailed statistics on national accounts.

21002 ANNUARIO STATISTICO ITALIANO

Instituto Centrale di Statistica

Italy - Italian annual/L49,000 ISSN No. 0066-4545

Statistical yearbook covering all aspects of Italian life.

21003 BOLLETTINO ECONOMICO

Banca d'Italia

Italy - Italian/Abridged version in English 2 issues per year/153pp

Italian economy and balance of payments, the money and financial markets.

21004 BOLLETTINO MENSILE DI STATISTICA

Instituto Centrale di Statistica

Italy - Italian monthly/200pp L147,000 P.A. ISSN No. 0021-3136

Statistics on all aspects of social, financial and economic life including comparative statistics with other countries.

21005 BOLLETTINO STATISTICO

Banca d'Italia

Italy - Italian quarterly/free

Covers credit, financial markets, balance of payments, and public finance.

21006 BUSINESS ITALY

Business International Ltd

Italy - English 10 issues per year/ £535.00 P.A.

A managers report intended to keep managers up-to-date with this fast-changing market.

21007 COMMERCIO ESTERO

Instituto Centrale di Statistica

Italy + trading partners - Italian monthly

Brief details of imports and exports by country and product.

21008 COMPENDIO STATISTICO ITALIANO

Instituto Nazionale di Statistica

Italy - Italian/statistical abstract in English annual/616pp L24,000 ISSN No. 0069-7958

Summarized version of statistics covering all aspects of political, economic, social and cultural life.

21009 CONTI ECONOMICI TRIMESTRALI

Instituto Centrale di Statistica

Italy - Italian quarterly/20pp L12,000 P.A.

Economic and statistical indicators.

21010 CONTRIBUTI ALL'ANALISI ECONOMICA

Banca d'Italia

Italy - Italian annual/free

Covers the Italian banking sector and related economic issues.

21011 COUNTRY BOOKS: ITALY

United Kingdom Iron and Steel Statistics Bureau

Italy - English annual/£75.00 ISSN No. 0952-5939

Annual summary tables showing production, materials consumed, apparent consumption, imports and exports of 130 products by quality and market.

21012 COUNTRY FORECASTS: ITALY

Business International Ltd

Italy - English quarterly/£360.00 P.A.

Contains an executive summary, sections on politcal and economic outlooks and the business environment, a fact sheet, and key economic indicators. Includes a quarterly Global Outlook.

21013 COUNTRY PROFILE: ITALY

Department of Trade and Industry Export Publications

Italy - English irregular/1992 65pp £10.00

This publication provides a general description of the country in terms of its geography, economy and population, and gives information about trading in the country with details like export conditions and investment covered.

21014 COUNTRY REPORT: ITALY

Business International Ltd

Italy - English quarterly/25-40pp £160.00 P.A. ISSN No. 0269-5421

Contains an executive summary, sections on political and economic structure, outlook for the next 12-18 months, statistical appendices, and a review of key political developments. Includes an annual Country Profile.

21015 DOING BUSINESS IN ITALY

Price Waterhouse

Italy - English irregular/1991 330-340pp free to clients

A business guide which includes sections on the investment climate, doing business, accounting and auditing, taxation and general country data. Updated by supplements.

21016 FINANCIAL AGGREGATES FOR 1,790 ITALIAN COMPANIES (1968-1991)

Mediobanca

Italy - Italian with abstract in English annual/328pp free

Aggregate financial data on Italian companies.

21017 FINANCING FOREIGN OPERATIONS: ITALY

Business International Ltd

Italy - English annual/£695.00 for complete European series

Business overview, currency outlook, exchange regulations, monetary system, short/medium/ long-term and equity financing techniques, capital incentives, cash management, investment trade finance and insurance.

21018 GLOBAL FORECASTING SERVICE: ITALY

Business International Ltd

Italy - English quarterly/£350.00 P.A.

Economic forecasts.

21019 HINTS TO EXPORTERS VISITING ITALY

Department of Trade and Industry Export Publications

Italy - English irregular/1990 72pp £5.00

This publication provides detailed practical information that is useful for those planning a business visit. Topics include currency information, economic factors, import and exchange control regulations, methods of doing business and general information.

21020 I CONSUMI DEELE FAMIGILE

Instituto Nazionale di Statistica

Italy - Italian annual

Statistics on domestic consumption.

21021 I CONTI DEGLI ITALIANI

Instituto Centrale di Statistica

Italy - Italian annual/116pp L17,000 ISSN No. 0390-6574

Statistics on national accounts.

21022 IL CALEPINO DELL'AZIONISTA (1956-1990)

Mediobanca

Italy - Italian annual/1249pp free

Financial information on listed Italian companies.

21023 IN DEPTH COUNTRY APPRAISALS ACROSS THE EC - CONSTRUCTION AND BUILDING SERVICES - ITALY
Building Services Research and Information Association

Italy - English 1991/£125.00

Part of an 11 volume survey covering the construction industry in EC countries.

21024 INDICATORI MENSILI
Instituto Centrale di Statistica

Italy - Italian monthly/20pp L31,000 P.A. ISSN No. 0390-6620

Monthly statistical indicators on population, productivity, foreign trade, prices and salaries.

21025 INDICES AND DATA ON INVESTMENT IN LISTED ITALIAN SECURITIES (1947-1991)
Mediobanca

Italy - Italian with notes and main terms in English annual/669pp free

Indices and data on investments in listed Italian securities.

21026 INDUSTRIA COTONIERA LINIERA E DELLE FIBRE AFFINI
Associazione Cotoniera Liniera e delle Fibre Affini

Italy - Italian bi-monthly/1906 56pp L88,000

Covers the cotton and related textile industries.

21027 INTRODUCING ITALY
Instituto Centrale di Statistica

Italy - English/Italian annual/48pp free

Demographic, social and economic summary on Italy.

21028 INVESTING, LICENSING AND TRADING CONDITIONS ABROAD: ITALY
Business International Ltd

Italy - English annual/£825.00 for complete European series

Loose-leaf format covering the state role in industry, competition rules, price controls, remittability of funds, taxation, incentives and capital sources, labour and foreign trade.

21029 ITALIAN STATISTICAL ABSTRACT
Instituto Nazionale di Statistica

Italy - English annual/216pp L25,000

Statistics on all aspects of Italian life.

21030 ITALY: COUNTRY PROFILE
Business International Ltd

Italy - English annual/50pp

Annual survey of political and economic background. Included as part of "Country Report" subscription.

21031 ITALY: COUNTRY REPORT
Business International Ltd

Italy - English quarterly/28pp £150.00 P.A. (quarterly issues + a "Country Profile")

Analysis of economic and political trends every quarter.

21032 LE PRINCIPALI SOCIETA ITALIANE (1966-1990)
Mediobanca

Italy - Italian with foreward and main terms in English annual/780pp free

Financial information on Italian companies.

21033 LE REGIONI IN CIFRE
Instituto Centrale di Statistica

Italy - Italian annual/208pp free

Statistical summary on Italy.

21034 LIST OF ITALIAN STOCKS
Banca Commerciale Italiana

Italy - Italian monthly/1950 free

Review of the stock market and data on movement of stocks, prices, dividends. Also notes on current developments in major industrial companies.

21035 MONTHLY REPORT - LA BORSA VALORI DI MILANO
Borsa Valori di Milano

Italy - English/Italian monthly/8pp free

Report of the Milan stock exchange including brief profiles of some major companies.

21036 NOTIZIARI ISTAT
Instituto Nazionale di Statistica

Italy - Italian/L 17,000 per issue

Provisional results of statistical surveys published in four series: demographic and social statistics; production statistics; statistics on employment, salaries and prices; and miscellaneous.

21037 OECD ECONOMIC SURVEYS: ITALY
Organization for Economic Co-operation and Development (OECD)

Italy - English irregular/£7.50 ISSN No. 0376-6438

Economic profile and forecasts.

21038 THE PERFORMANCE OF LISTED SHARES
Borsa Valori di Milano

Italy - English/Italian annual/225pp L65,000

21039 QUARTERLY COUNTRY TRADE REPORTS: ITALY
United Kingdom Iron and Steel Statistics Bureau

Italy - English quarterly/£200.00 P.A. for four countries ISSN No. 0952-5939

Cumulative trade statistics of imports and exports of steel products.

21040 "R & S" ANNUAL DIRECTORY
Ricerche & Studi

Italy - Italian with Italian/English glossary annual/1976 4,200pp (3 Vols) US$695.00

Financial directory of leading Italian companies, encompassing the entire economic spectrum in Italy.

21041 RAPPORTO ANNUALE
Borsa Valori di Milano

Italy - Italian with English summary annual/100pp free

Annual report of the Milan Stock Exchange including a brief review of the economy.

21042 RAPPORTO ASSIN FORM SULLA SITUAZIONE DELL'INFORMATICA IN ITALIA
Associazione Costruttari: Macchine Attrezzature per l'Ufficio e per il Trattamento delle Informazion

Italy - Italian/summary in English + French annual/1981 120pp L100,000

Statistics on computer and industrial automation sectors in Italy.

21043 RASSEGNA ECONOMICA
Banco di Napoli SpA

Italy/Europe - Italian/abstract in English/quarterly/250pp 1931 free

All aspects of the economy including a specific section dealing with Southern Italy.

21044 RELAZIONE ANNUALE
Banca d'Italia

Italy - Italian/Abridged version in English annual/105pp free

Annual report of the Banca d'Italia

21045 REVIEW OF ECONOMIC CONDITIONS IN ITALY
Banco di Roma

Italy - English/every 4 months/ISSN No. 0034-6799

In depth articles on the economy and in particular financial matters.

21046 RILEVAZIONE DELLE FORZE DI LAVORO
Instituto Nazionale di Statistica

Italy - Italian irregular (several times per year)/L 12,000 per issue ISSN No. 0390-6434

Statistics on the labour force.

21047 STATISTICA ANNUALE DEL COMMERCIO CON L'ESTERO
Instituto Centrale di Statistica

Italy - Italian annual/520pp (7 Vols) Vol.1 = L41,000 Vol.2 (6 parts) = L 124,000 ISSN No. 0390-6558

Detailed statistics of foreign trade classified by product and country of origin or destination.

21048 STATISTICA DEGLI INCIDENTI STRADALI
Instituto Centrale di Statistica

Italy - Italian annual/288pp L22,000 ISSN No. 0075-188X

Statistics on road accidents covering damage to both persons and vehicles.

21049 STATISTICA DEL COMMERCIO CON L'ESTERO
Instituto Nazionale di Statistica

Italy - Italian quarterly/900pp L105,000 P.A. ISSN No. 0535-9821

Statistical bulletin on foreign trade classified by product group and by country of origin or destination.

21050 STATISTICHE DEL COMMERCIO INTERNO
Instituto Centrale di Statistica

Italy - Italian annual/116pp L12,000 ISSN No. 0075-1782

Statistics on internal trade.

21051 STATISTICHE DEL LAVORO
Instituto Centrale di Statistica

Italy - Italian annual/172pp L12,000 ISSN No. 0390-6450

Labour statistics including wages, family consumption, industrial action, and hours of work.

21052 STATISTICHE DEL TURISMO

Instituto Centrale di Statistica

Italy - Italian annual/144pp L12,000 ISSN No. 0075-1782

Statistics on tourism.

21053 STATISTICHE DELL'AGRICOLTURA, ZOOTECNIA E MEZZI DI PRODUZIONE
Instituto Nazionale di Statistica

Italy - Italian annual/532pp L41,000 ISSN No. 0075-1669

Data on all aspects of agriculture including comparative statistics covering recent years.

21054 STATISTICHE DELL'ATTIVITÀ EDILIZIÀ
Instituto Nazionale di Statistica

Italy - Italian annual/214pp L23,000 ISSN No. 0390-6558

Statistics on residential and non-residential construction.

21055 STATISTICHE DELL'INDUSTRIA COTONIERA LINIERA E DELLE FIBRE AFFINI INTERNAZIONAL: MONDO
Associazione Cotoniera Liniera e delle Fibre Affini

International - Italian annual/54pp L12,500

Statistical coverage of the cotton and related industries. Italy and the world markets are covered.

21056 STATISTICHE DELL'INDUSTRIA COTONIERA LINIERA E DELLE FIBRE AFFINI - ITALIA
Associazione Cotoniera Liniera e delle Fibre Affini

Italy - Italian annual/72pp L12,500

Covers the cotton and related textile sectors in Italy.

21057 STATISTICHE DELLA CACCIA, PESCA E COOPERAZIONE
Instituto Nazionale di Statistica

Italy - Italian annual/124pp L8,400 ISSN No. 0390 6426

Statistics on hunting, fishing and agricultural cooperatives.

21058 STATISTICHE DELLE APERE PUBBLICHE
Instituto Nazionale di Statistica

Italy - Italian annual/68pp L13,000

Statistics on public works.

21059 STATISTICHE DEMOGRAFICHE
Instituto Centrale di Statistica

Italy - Italian annual/In 2 parts Pt.1 L18,900 Pt.2 L 15,000 ISSN No. 0075-1685

Annual statistics on demography covering internal mobility, births, deaths and marriages.

21060 STATISTICHE FORESTALI
Instituto Centrale di Statistica

Italy - Italian annual/156pp L17,600 ISSN No. 0075-1707

Statistics on the forestry industry including data for preceding years, climate, production, and external trade.

21061 STATISTICHE INDUSTRIALI
Instituto Centrale di Statistica

Italy - Italian annual/584pp L41,000 ISSN No. 0075-1723

Statistics on industry, covering production, consumption, plant and machinery, prices and wages, foreign trade. The data is retrospective and there is comparative international data.

21062 SUPPLEMENTI AL BOLLETTINO STATISTICO
Banca d'Italia

Italy - Italian irregular/free

Covers credit, financial markets, balance of payments and similar topics.

21063 TAVOLE STATISTICHE
Unione Industriali Pastai Italiani (U.N.I.P.I.)

Italy - Italian annual

Statistics on the cereals-pasta industry.

21064 TEMI DI DISCUSSIONE
Banca d'Italia

Italy - English/Italian irregular/free

Covers the banking sector and related economic issues.

LIECHTENSTEIN

Country profile:	Liechtenstein
Official name:	Principality of Liechtenstein
Area:	160 sq.km.
	On the Upper Rhine, with Austria to the east and north, and Switzerland to the west and south.
Language(s):	German (Alemannish)
Capital:	Vaduz
Currency:	Swiss Franc 100 centimes = = 1 Swiss franc
Government and Political structure:	Principality, with legislative assembly of 15 members. Monetary and customs union (open border) wiith Switzerland which also provides diplomatic representation.

Liechtenstein

22001 AUSLANDERSTATISTIK
AMT FÜr Volkwirtschaft

Liechtenstein - German 3 per year

Statistics on foreign workers and immigration in Liechtenstein.

22002 BANKSTATISTIK
AMT FÜr Volkwirtschaft

Liechtenstein - German annual

Bank statistics with financial balances and related economic comment.

22003 BAUSTATISTIK
AMT FÜr Volkwirtschaft

Liechtenstein - German annual

Building and construction industry statistics.

22004 BETRIEBLICHE PERSONAL VORSORGESTATISTIK
AMT FÜr Volkwirtschaft

Liechtenstein - German annual

Domestic insurance annual statistics.

22005 COMPANIES AND TAXES IN LIECHTENSTEIN
Liechtenstein Verlag AG

Marxer, Goop, Kieber/International - English/German/French/1992 224pp SFr84 ISBN No. 3857898526

22006 DOING BUSINESS IN LIECHTENSTEIN
Price Waterhouse

Liechtenstein - English irregular/ 1991 330-340pp free to clients

A business guide which includes sections on the investment climate, doing business, accounting and auditing, taxation and general country data. Updated by supplements.

22007 EINBURGERUNGS- STATISTIK
AMT FÜr Volkwirtschaft

Liechtenstein - German annual

Demographic statistics broken down to detailed levels.

22008 ENERGIESTATISTIK
AMT FÜr Volkwirtschaft

Liechtenstein - German annual

Energy statistics with mode of generation etc.

22009 FERIENHANDBUCH FÜR DAS FÜRSTENTUM LIECHTENSTEIN
Liechtensteinische Fremdenverkehrszentrale

Berthold Konrad/Liechtenstein - English/German/French/Spanish/ Italian annual/40pp

Tourism handbook with historical background on the principality of Liechtenstein as well as useful facts and figures.

22010 FREMDENVERKEHRS- STATISTIK
AMT FÜr Volkwirtschaft

Liechtenstein - German 3 per year/ 25pp

Tourism statistics showing numbers, destinations etc.

22011 GESELLSCHAFTSFORMEN UND STEUERBELASTUNG IM FURSTENTUM LIECHTENSTEIN
Verwaltungs - Und Privat- Bank

Liechtenstein - German irregular/ 72pp Sfr5.00

Company formation data.

22012 JAHRESBERICHT
Verwaltungs - Und Privat- Bank

Liechtenstein - German annual/ 30pp

Annual report of the bank with financial balances etc.

22013 KONJUNKTURTEST
AMT FÜr Volkwirtschaft

Liechtenstein - German quarterly

Quarterly bulletin of economic indicators.

22014 KRANKENVER- SICHERUNGS- STATISTIK
AMT FÜr Volkwirtschaft

Liechtenstein - German annual

Health insurance statistics.

22015 LIECHTENSTEIN COMPANY LAW
Liechtenstein Verlag AG

Translated by Bryan Jeeves/ International - English/1992 360pp SFr96.- ISBN No. 3857899018

22016 LIECHTENSTEIN IN FIGURES
Verwaltungs - Und Privat- Bank

Liechtenstein - English irregular/ 9pp

Brief socio-economic profile of Liechtenstein.

22017 LIECHTENSTEIN IN ZAHLEN
AMT FÜr Volkwirtschaft

Liechtenstein - German annual

Contains statistical data on all basic economic indicators.

22018 LIECHTENSTEIN - PRINCIPALITY IN THE HEART OF EUROPE
Presse- und Informationsamt

Liechtenstein - English/German/ French/Italian occasional/1988 40pp free

Popular description of the main principality giving a general picture.

22019 LIECHTENSTEIN WIRTSCHAFTSFRAGEN
Verwaltungs - Und Privat- Bank

Liechtenstein - German irregular/ 20pp

General economic survey.

22020 LIECHTENSTEINISCHE GESELLSCHAFTSFORMEN
Verwaltungs - Und Privat- Bank

Liechtenstein - German loose-leaf/ Sfr40.00

22021 MOTORFAHR- ZEUGBESTAND
AMT FÜr Volkwirtschaft

Liechtenstein - German annual

Data on the Liechtenstein motor industry.

22022 STATISTIK DER INDUSTRIELLEN BETRIEBE
AMT FÜr Volkwirtschaft

Liechtenstein - German annual

Trading companies are covered in this statistical series.

22023 STATISTISCHES JAHRBUCH
AMT FÜr Volkwirtschaft

Liechtenstein - German annual/ 1977 free

Statistical yearbook of Liechtenstein with main economic series included.

LUXEMBOURG

Country profile:	Luxembourg
Official name:	Grand Duchy of Luxembourg
Area:	2,586 sq.km.
	Bordered by Belgium in west, Germany in the east, and France in the south.
Population:	384,400 (1991)
Capital:	Luxembourg
Language(s):	Luxembourgish, French, German
Currency:	Franc 100 centimes = 1 Luxembourg franc
National bank:	Monetary Institute
Government and Political structure:	Constitutional monarchy with multi-party parliamentary democracy. The Chamber of Deputies, with 60 members, chooses the Prime Minister.

Luxembourg

23001 ADMISSION TO OFFICIAL STOCK LISTING AND PUBLIC OFFER OF TRANSFERABLE SECURITIES

Société de la Bourse de Luxembourg

Luxembourg - English/French irregular/105pp LF200

Regulations on the listing of securities.

23002 ANNUAIRE STATISTIQUE DU LUXEMBOURG

Service Central de la Statistique et des Études Économiques

Luxembourg - French/German annual 500pp/LF1,100 ISSN No. 0076-1575

Statistics on population, education, employment, health, the environment as well as demographic social and economic trends.

23003 ANNUAL REPORT

Banque Internationale à Luxembourg

Luxembourg - English/French annual/70pp free

Banking and economic data.

23004 ANNUAL REPORT - SOCIÉTÉ DE LA BOURSE DE LUXEMBOURG

Société de la Bourse de Luxembourg

Luxembourg - English/French annual/1929 47pp free

Report with tables on share movements, bonds, and new issues.

23005 BELGIUM: ÉCONOMIC AND COMMERCIAL INFORMATION

Belgian Foreign Trade Office

Luxembourg/Belgium - English/French/German/Spanish/Dutch quarterly 68pp

Reports on various industrial sectors.

23006 BOURSE INFORMATIONS

Société de la Bourse de Luxembourg

Luxembourg - English/French monthly/8pp free

Contains statistics and stock exchange news.

23007 BULLETIN TRIMESTRIEL

Institut Monétaire Luxembourgeois

Luxembourg/International - French quarterly/LF1,000 P.A.

Domestic and international financial and statistical analysis.

23008 CAHIERS ÉCONOMIQUES

Banque Internationale à Luxembourg

Luxembourg - French irregular/free

Booklets covering a range of financial issues.

23009 CAHIERS ÉCONOMIQUES

Service Central de la Statistique et des Études Économiques

Luxembourg - French biennial/ LF480 ISSN No. 0070-881X

A series of publications containing detailed studies on economic themes of general interest

23010 COTE OFFICIÈLLE - BOURSE DE LUXEMBOURG

Société de la Bourse de Luxembourg

Luxembourg - French daily/1929 128pp

Bulletin of the stock exchange giving stock and share prices and price movements.

23011 COUNTRY PROFILE LUXEMBOURG

Department of Trade and Industry Export Publications

Luxembourg - English irregular/ 1992 52pp £10.00

This publication provides a general description of the country in terms of its geography, economy and population, and it gives information about trading in the country with details like export conditions and investment covered.

23012 COUNTRY REPORT: BELGIUM, LUXEMBOURG

Business International Ltd

Luxembourg/Belguim - English quarterly/22pp £130.00 (4 issues + a country profile)

Analysis of economic and political trends

23013 DE KONSUMENT

Union Luxembourgeoise des Consammateurs

Luxembourg French/German/ Luxembourg 18 issues per year/ 1963 loose-leaf BF750 P.A.

Articles on consumer topics, with some figures and tables.

23014 DOING BUSINESS IN LUXEMBOURG

Chambre de Commerce du Grand-Duché de Luxembourg

Luxembourg - English irregular/ 154pp free

Booklet describing economic and social conditions in Luxembourg intended for those considering starting a business in Luxembourg.

23015 THE ECU BOND MARKET IN 1991

Société de la Bourse de Luxembourg

Luxembourg/Europe - English/ French annual/25pp

Statistics on the ECU bond market

23016 ECU: LE MARCHÉ EURO-OBLIGITAIRE EN ECU STATISTIQUES MENSUELLES

Société de la Bourse de Luxembourg

Luxembourg - English/French monthly/13pp

Statistics on the ECU Eurobond market

23017 ÉTUDES

Institut Monétaire Luxembourgeois

Luxembourg/International - French irregular/LF250

Monetary and economic studies concerning Luxembourg and the work of the IML.

23018 HINTS TO EXPORTERS VISITING LUXEMBOURG

Department of Trade and Industry Export Publications

Luxembourg - English irregular/ £5.00

This publication provides detailed practical information that is useful for those planning a business visit. Topics include currency information, economic factors, import and exchange control regulations, methods of doing business and general information. See the Belgian edition of this series which also covers Luxembourg.

23019 INDICATEURS RAPIDES

Service Central de la Statistique et des Études Économiques

Luxembourg - French monthly/ loose-leaf (A4) LF550 P.A. ISSN No. 0019-6916

Économic indicators published loose-leaf in 14 categories

23020 INVESTING, LICENSING AND TRADING CONDITIONS ABROAD: LUXEMBOURG

Business International Ltd

Luxembourg - English annual/ £825.00 for complete European series

Loose-leaf format covering the state role in industry, competition rules, price controls, remittability of funds, taxation, incentives and capital sources, labour and foreign trade.

23021 THE JOURNAL: BELGO - LUXEMBOURG CHAMBER OF COMMERCE IN GREAT BRITAIN

Belgo - Luxembourg Chamber of Commerce in Great Britain

D. Partner-Macremans
Luxembourg/Belgium - English bi-monthly/16pp free to members

One main feature on a chosen sector and a range of shorter articles.

23022 LE BULLETIN DU STATEC

Service Central de la Statistiques et des Études Économiques

Luxembourg - French 8 issues per year/20-60pp LF100 P.A. ISSN No. 0076-1583

Studies on the Luxembourg economy and the results of statistical surveys.

23023 LE LUXEMBOURG ET SA MONNAIE

Institut Monétaire Luxembourgeois

Paul Margue/Marie-Paule Jungblut
Luxembourg - French 1990/192pp LF1,050

History of the contemporary monetary system in Luxembourg in the national and international context.

23024 THE LUXEMBOURG FRANC BOND MARKET IN 199..: STATISTICS AND REFLOW

Société de la Bourse de Luxembourg

Luxembourg - English/French annual/22pp

Statistics on the Luxembourg franc bond market.

23025 LUXEMBOURG IN FIGURES

Service Central de la Statistique et des Études Économiques

Luxembourg - English/French irregular/31pp free

General survey on production, social indicators, finance and external trade.

23026 THE LUXEMBOURG STOCK EXCHANGE: FACTS AND FIGURES

Société de la Bourse de Luxembourg

Luxembourg - English/French biannual/8pp free

Basic profile of the exchange with some statistical data.

23027 NOTE DE CONJONCTURE

Service Central de la Statistique et des Études Économiques

Luxembourg - French quarterly/appears in 4 editions LF420 P.A.

Allows analysis of short-term trends in the Luxembourg economy.

23028 OECD ÉCONOMIC SURVEYS: BELGIUM, LUXEMBOURG

Organisation for Économic Co-operation and Development (OECD)

Belgium/Luxembourg - English irregular/£102.00 ISSN No. 0376-6438

Survey on economic developments.

23029 ORGANISMES DE PLACEMENT COLLECTIF

Société de la Bourse de Luxembourg

Luxembourg - English/French/193pp LF200

Regulations governing the investment fund industry in Luxembourg.

23030 PROPOS

Société de la Bourse de Luxembourg

Luxembourg/Europe - English/French/German irregular/8-12pp

Irregular pamphlets on various economic topics.

23031 RAPPORT ANNUEL DE L'IML

Institut Monétaire Luxembourgeois

Luxembourg/International - French with English summary annual/LF250

Report of the monetary authority.

23032 RECENSEMENT DE LA POPULATION

Service Central de la Statistique et des Études Économiques

Luxembourg - French annual

Population statistics broken down in different ways.

23033 RECUEIL DE STATISTIQUES PAR COMMUNE

Service Central de la Statistique et des Études Économiques

Luxembourg - French annual/2 Vols LF200 P.A.

Regional statistics; statistical tables and maps with accompanying commentary.

23034 STATISTIQUES DU COMMERCE EXTÉRIEUR DE L'UNION ÉCONOMIQUE BELGO - LUXEMBOURGEOISE

Institut National de Statistique: Ministère des Affaires Économiques

Luxembourg/Belgium - French monthly/BF5,160 P.A. ISSN No. 0772-6694

International trade statistics for Belgium - Luxembourg.

23035 STATISTIQUES HISTORIQUES 1839-1989

Service Central de la Statistique et des Études Économiques

Luxembourg - French 1990/650pp LF1,200

Commemorates the 150th anniverary of the independence of Luxembourg; follows the same format as the 'Annuaire Statistique'.

MALTA

Country profile:	Malta
Official name:	Republic of Malta
Area:	246 sq.km.
Population:	355,900 (1990)
Capital:	Valletta
Language(s):	Maltese, English
Currency:	Pound 100 cents = 1 Maltese pound
National bank:	Central Bank of Malta
Government and Political structure:	Multi-party democracy. The House of Representatives has 65 members and chooses the Prime Minister.

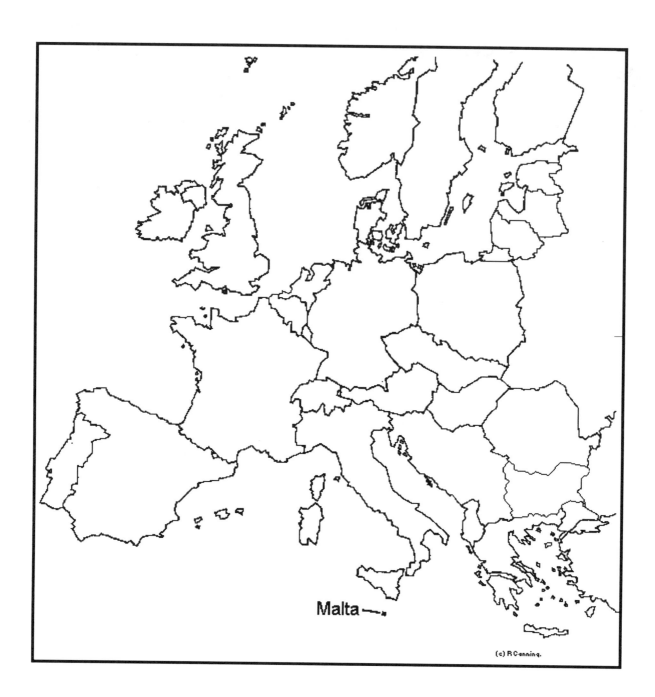

Malta ⟶

(c) R.Canning.

24001 ANNUAL ABSTRACT OF STATISTICS
Central Office of Statistics

Malta - English annual/270pp Lm1.75

Broad coverage of socio-economic data drawn from other statistical series.

24002 ANNUAL REPORT
Central Bank of Malta

Malta - English annual/82pp free

Central bank report giving financial balances and related economic data.

24003 ANNUAL REPORT
Employment & Training Corporation

Malta - English annual/free

Gives details of employment training schemes over the last year. Useful as an economic activity indicator.

24004 ANNUAL REPORT
Malta Federation of Industry

Malta - English annual/free

Reports on the activities over the past year, and matters of concern to the federation.

24005 ANNUAL REPORT
Malta International Business Authority

Malta - English annual/17pp free

Report on business developments and investment.

24006 ANNUAL REPORT
Telemalta Corporation

Malta - English annual/free

Annual report of the Maltese telecommunications authority with details of activities.

24007 BUDGET ESTIMATES
Ministry of Finance

Malta - English annual/325pp Lm1.25

Central government estimates for the forthcoming year.

24008 BUDGET SPEECH
Ministry of Finance

Malta - Maltese annual/90pp Lm1.00

Presents the budget speech and plans for forthcoming government financial year.

24009 CENSUS OF AGRICULTURE
Central Office of Statistics

Malta - English biannual/75pp Lm00.65

Details agricultural activities and outputs.

24010 CENSUS OF INDUSTRIAL PRODUCTION: REPORT AND SUMMARY TABLES
Central Office of Statistics

Malta - English annual/Lm2.00

Summarizes industrial activity by level and sector.

24011 COMMERCIAL COURIER
Malta Chamber of Commerce

Malta - English monthly/44-48pp Lm00.20 ISSN No. 0010-2938

Bulletin of the Chamber of Commerce with general business and economic news.

24012 DEMOGRAPHIC REVIEW OF THE MALTESE ISLANDS
Central Office of Statistics

Malta - English annual/70pp Lm1.00

Reviews the Maltese population by sex, age and family structure etc.

24013 DOING BUSINESS IN MALTA
Price Waterhouse

Malta - English irregular/1991 330-340pp free to clients

A business guide which includes sections on the investment climate, doing business, accounting and auditing, taxation and general country data. Updated by supplements.

24014 ECONOMIC SURVEY
Department of Information

Director Economic Planning/Malta - English annual/250pp Lm1.25

Reviews the Maltese economy over the last year.

24015 ECONOMIC TRENDS
Central Office of Statistics

Malta - English monthly/4pp free

Highlights the main economic factors in the Maltese economy.

24016 FINANCIAL REPORT
Department of Information

Accountant General/Malta - English annual/200pp Lm1.25

Reports on government finances - income, expenditure etc.

24017 INDUSTRY TODAY
Malta Federation of Industry

Malta - English quarterly/28pp Lm0.25

General news coverage of Maltese industry and industrial companies.

24018 MADE IN MALTA
Malta Export Trade Corporation

Malta - English annual/1988 c260pp free

Describes main Maltese export products, mainly aimed at the export market.

24019 MALTA YEARBOOK
De La Salle Brothers Publications

S.J.A. Clews/Malta/International - English annual/1953 c500pp £12.00

General description of Maltese life with summary of economic data.

24020 NATIONAL ACCOUNTS OF THE MALTESE ISLANDS
Central Office of Statistics

Malta - English annual/45pp Lm1.00

Covers revenue, expenditure and economic indicators such as balance of payments.

24021 QUARTERLY DIGEST OF STATISTICS
Central Office of Statistics

Malta - English quarterly/70pp Lm00.75 P.A.

First results type series covering the main statistical series.

24022 QUARTERLY REVIEW
Central Bank of Malta

Malta/International - English quarterly/114pp free

Quarterly review of the national monetary situation.

24023 SHIPPING AND AVIATION STATISTICS
Central Office of Statistics

Malta - English annual/130pp Lm1.00

24024 TRADE DIRECTORY
Malta Chamber of Commerce

Malta - English annual/Lm5.00

24025 TRADE STATISTICS
Central Office of Statistics

Malta - English three times a year/500pp Lm2.00 P.A.

Detailed breakdown of Maltese international trade.

24026 TUNISIA, MALTA - COUNTRY REPORT
Business International Ltd

Tunisia/Malta - English quarterly/£160.00 P.A.

NETHERLANDS

Country profile:	Netherlands
Official name:	Kingdom of the Netherlands
Area:	41,547 sq.km.
	Bordered in the south by Belgium, by Germany in the east and by the North Sea in the north and west.
Population:	15m (1991)
Capital:	Amsterdam
Language(s):	Dutch
Currency:	Guilder 100 cents = 1 Guilder
National bank:	Netherlands Bank
Government and Political structure:	Constitutional monarchy with multi-party parliamentary democracy. The Staten-Generaal (parliament) has two houses, the Upper House having 75 members and the Lower 150 deputies.

25001 AGRICULTURAL ECONOMIC REPORT

Landbouw - Economisch Instituut

Netherlands - Dutch annual/ISSN No. 0924-0764

Economic situation of Dutch agricultural and horticultural industry.

25002 ANNUAL PUBLICATION OF FOREIGN TRADE STATISTICS

Netherlands Central Bureau of Statistics

Netherlands - English/Dutch annual

Detailed information on imports and exports by all commodities.

25003 ANNUAL REPORT - DE NEDERLANDSCHE BANK

De Nederlandsche Bank

Netherlands - English/Dutch annual/220pp DF 26-46 ISSN No. 0167-3998

Macro-economic policy analysis and statements on the monetary policy conducted by the bank.

25004 APPENDIX TO THE MONTHLY BULLETIN OF PRICE STATISTICS

Netherlands Central Bureau of Statistics

Netherlands - Dutch/English monthly

Consumer price indices of low-income households.

25005 BANKS AND BROKERS IN THE NETHERLANDS

Nederlands Instituut voor Het Bank effectenbedrijf (NIBE)

Netherlands - English/Dutch annual/DF 42.50

25006 BRITAIN IN THE NETHERLANDS

Netherlands British Chamber of Commerce

Netherlands - English annual/ £60.00

Directory of 437 British subsidiaries based in the Netherlands.

25007 BUSINESS SURVEY

Netherlands Central Bureau of Statistics

Netherlands - English/Dutch monthly

Manufacturing industry, trends in production and turnover, orders received, order positions and stocks of production.

25008 CENTRAL ECONOMIC PLAN

Central Planning Bureau

Netherlands - English/Dutch annual/200pp DF 80 ISSN No. 0927-7226

Reviews the economic development of the past year and updates prospects for 1993.

25009 COMMERCIAL GARDENING & FARMING

Netherlands British Chamber of Commerce

Netherlands - English annual/ £15.00

Agriculture, horticulture, livestock with 1197 companies profiled.

25010 CONSUMER GOODS FOOD

Netherlands British Chamber of Commerce

Netherlands - English annual/ £20.00

Covers foodstuffs and stimulants with 984 companies profiled.

25011 CONSUMER GOODS NON-FOOD

Netherlands British Chamber of Commerce

Netherlands - English annual/ £27.00

Durable and non-durable consumer goods and related products; 1012 companies.

25012 COUNTRY BOOKS: NETHERLANDS

United Kingdom Iron and Steel Statistics Bureau

Netherlands - English annual/ £75.00 ISSN No. 0952-6005

Annual summary tables showing production, materials consumed, apparent consumption, imports and exports of 130 products by quality and market.

25013 COUNTRY FORECASTS: NETHERLANDS

Business International Ltd

Netherlands - English quarterly/ £360.00 P.A.

Contains an executive summary, sections on political and economic outlooks and the business environment, a fact sheet, and key economic indicators. Includes a quarterly Global Outlook.

25014 COUNTRY PROFILE: NETHERLANDS

Department of Trade and Industry Export Publications

Netherlands - English irregular/ 1992 67pp £10.00

This publication provides a general description of the country in terms of its geography, economy and population, and gives information about trading in the country with details like export conditions and investment covered.

25015 COUNTRY REPORT: NETHERLANDS

Business International Ltd

Netherlands - English quarterly/25-40pp £160.00 P.A. ISSN No. 0269-6134

Contains an executive summary, sections on political and economic structure, outlook for the next 12-18 months, statistical appendices, and a review of key political developments. Includes an annual Country Profile.

25016 DOING BUSINESS IN THE NETHERLANDS

Ernst & Young International

Netherlands - English irregular/ 1991 110-130pp free to clients

A business guide which includes an executive summary, the business environment, foreign investment, company structure, labour, taxation, financial reporting and auditing, and general country data.

25017 DUTCH COMPANIES AND THEIR UK AGENTS/ DISTRIBUTORS

Netherlands British Chamber of Commerce

Netherlands - annual/£15.00

25018 DUTCH COMPANIES AND THEIR UK ASSOCIATES

Netherlands British Chamber of Commerce

Netherlands/UK - English annual/ £60.00

Details on 430 plus Dutch parent companies and 630 plus associates in Great Britain.

25019 ECONOMIC INDICATORS

Netherlands Central Bureau of Statistics

Netherlands - English/Dutch irregular/DF 3.30

Covers main economic indicators including inflation and balance of payments.

25020 ECONOMIC PROSPECTS

Central Planning Bureau

Netherlands - English/Dutch annual

Preliminary forecasts of the Dutch economy in the coming year.

25021 ECONOMIC REVIEW

Bank Mees & Hope

Netherlands - English bi-monthly/ 1992 8pp

Economy, wages and prices, financial markets, government financing.

25022 ENERGY SUPPLY IN THE NETHERLANDS

Netherlands Central Bureau of Statistics

Netherlands - English/Dutch quarterly/ISSN No. 0168-5236

Quantitative and qualitative trends in energy consumption, data on manufacturing industries and monthly production figures.

25023 ENVIRONMENTAL QUARTERLY

Netherlands Central Bureau of Statistics

Netherlands - English/Dutch quarterly

Regular statistics on environmental issues, pollution levels etc.

25024 ENVIRONMENTAL STATISTICS OF THE NETHERLANDS

Netherlands Central Bureau of Statistics

Netherlands - English Dutch annual/ DF 25

Emissions, concentrations and changes in animal and plant distribution, also data on costs, standards, public health and legislation.

25025 FINANCIAL INVESTMENT COMPANIES IN THE NETHERLANDS 1991

Netherlands British Chamber of Commerce

Netherlands - annual/£20.00

List of major Dutch investment companies.

25026 FINANCING FOREIGN OPERATIONS: NETHERLANDS

Business International Ltd

Netherlands - English annual/ £695.00 for complete European series

Business overview, currency outlook, exchange regulations, monetary system, short/medium/ long-term and equity financing techniques, capital incentives, cash management, investment trade finance and insurance.

25027 GLOBAL FORECASTING SERVICE

Business International Ltd

Netherlands - English quarterly/ £150.00 P.A.

Provides a range of medium-term forecasts covering main political, economic and business trends. A quarterly Global Outlook is included.

25028 HINTS TO EXPORTERS VISITING THE NETHERLANDS

Department of Trade and Industry Export Publications

Netherlands - English irregular/ 1991 80pp £5.00

This publication provides detailed practical information that is useful for those planning a business visit. Topics include currency

information, economic factors, import and exchange control regulations, methods of doing business and general information.

25029 IN DEPTH COUNTRY APPRAISALS ACROSS THE EC - CONSTRUCTION AND BUILDING SERVICES - NETHERLANDS

Building Services Research and Information Association

Netherlands - English 1991/£125.00

Part of an 11 volume survey covering the construction industry in EC countries.

25030 INDUSTRIAL INVESTMENT PROSPECTS AND PURPOSES IN 199–

Netherlands Central Bureau of Statistics

Netherlands - English/Dutch annual/DF 3.75

25031 INDUSTRIAL PRODUCTS

Netherlands British Chamber of Commerce

Netherlands - English annual/ £30.00

Manufacturers and exporters of finished products, semi-finished articles, raw materials with 3887 companies profiled.

25032 INVESTING, LICENSING AND TRADING CONDITIONS ABROAD: NETHERLANDS

Business International Ltd

Netherlands - English annual/ £825.00 for complete European series

Loose-leaf format covering the state role in industry, competition rules, price controls, remittability of funds, taxation, incentives and capital sources, labour and foreign trade.

25033 LABOUR FORCE SURVEY

Netherlands Central Bureau of Statistics

Netherlands - English/Dutch annual/DF 42.50

Broad coverage of labour statistics including employment and unemployment.

25034 LANDBOUWEIFERS (AGRICULTURAL FIGURES)

Landbouw - Economisch Instituut

Netherlands - English/Dutch annual/1948

300 tables with data on type and size of holdings, production, income, trade, prices, costs.

25035 LIST OF GOODS FOR THE STATISTICS OF FOREIGN TRADE

Netherlands Central Bureau of Statistics

Netherlands - English/Dutch annual/DF 120.00

Useful for cross-checking the main series of foreign trade statistics.

25036 MACRO-ECONOMIC OUTLOOK

Central Planning Bureau

Netherlands - English/Dutch annual/125pp DF 40.00

Projections for the current year and first detailed forecasts for the year ahead.

25037 MAJOR COMPANIES IN THE NETHERLANDS

Netherlands British Chamber of Commerce

Netherlands - English annual/ £20.00

Details on the major companies in the Netherlands including contact points.

25038 MEN AND WOMEN SIDE BY SIDE

Netherlands Central Bureau of Statistics

Netherlands - irregular DF 15.00

Statistical observation of men and women in the Netherlands.

25039 MONTHLY BULLETIN OF AGRICULTURAL STATISTICS

Netherlands Central Bureau of Statistics

Netherlands - English/Dutch monthly/DF 137.50 P.A. ISSN No. 0024-8754

Dairy production, agriculture and horticulture products, imports and exports.

25040 MONTHLY BULLETIN OF CONSTRUCTION STATISTICS

Netherlands Central Bureau of Statistics

Netherlands - English/Dutch monthly/DF 8.50 P.A.

Construction projects, contracts, work in progress, price-indices, dwellings completed, average prices of dwellings.

25041 MONTHLY BULLETIN OF DISTRIBUTION AND SERVICE TRADE STATISTICS

Netherlands Central Bureau of Statistics

Netherlands - English/Dutch monthly/DF 9 P.A.

Turnover and stocks for wholesale and retail and services sector in the form of monthly indices.

25042 MONTHLY BULLETIN OF FINANCIAL STATISTICS
Netherlands Central Bureau of Statistics

Netherlands - English/Dutch monthly/DF 10.50 P.A.

Banking, the capital market, stock exchange, insurance & pension funds, consumer credit, balance of payments provinces and municipalities.

25043 MONTHLY BULLETIN OF HEALTH STATISTICS
Netherlands Central Bureau of Statistics

Netherlands - English/Dutch monthly/DF 10 P.A.

Data on morbidity and mortality, sickness form work, benefits paid, public health, use of health care facilities.

25044 MONTHLY BULLETIN OF POPULATION STATISTICS
Netherlands Central Bureau of Statistics

Netherlands - English/Dutch monthly/DF 6.00 P.A. ISSN No. 0024-8711

First report of population changes and development.

25045 MONTHLY BULLETIN OF PRICE STATISTICS
Netherlands Central Bureau of Statistics

Netherlands - English/Dutch monthly/DF 82 P.A.

Statistics on consumer price indices, households, industry indices, import/export indices, wages.

25046 MONTHLY BULLETIN OF SOCIO-ECONOMIC STATISTICS
Netherlands Central Bureau of Statistics

Netherlands - English/Dutch monthly/DF 04 P.A. ISSN No. 0168-549X

Labour, earnings and labour costs, income and property, social security, consumption and savings.

25047 MONTHLY BULLETIN OF THE NETHERLANDS CENTRAL BUREAU OF STATISTICS
Netherlands Central Bureau of Statistics

Netherlands - English/Dutch monthly/DF 00.00 P.A. ISSN No. 0166-0268

25048 MONTHLY BULLETIN OF TRANSPORT STATISTICS
Netherlands Central Bureau of Statistics

Netherlands - English/Dutch monthly/DF 25 P.A. ISSN No. 0024-8770

Trends and figures on goods traffic, transit, sea-going shipping, containers, road & rail, civil aviation passenger transport, car use and traffic intensity.

25049 MONTHLY STATISTICAL BULLETIN OF FOREIGN TRADE
Netherlands Central Bureau of Statistics

Netherlands - English/Dutch monthly/DF 39.50 P.A.

Imports and exports of main commodities and trade of Netherlands with other countries.

25050 MONTHLY STATISTICAL BULLETIN OF MANUFACTURING
Netherlands Central Bureau of Statistics

Netherlands - English/Dutch monthly/DF 12 P.A.

Production and sales in the major divisions of industry, specified by product.

25051 NATIONAL ACCOUNTS
Netherlands Central Bureau of Statistics

Netherlands - English/Dutch annual/DF 8.50 ISSN No. 0168-3489

Economic activities of all the peoples and institutions in the Netherlands.

25052 NETHERLANDS
United Kingdom Iron and Steel Statistics Bureau

Netherlands - English annual/ £80.00 ISSN No. 0952-6005

Report on the steel industry showing production, consumption and details of imports and exports.

25053 THE NETHERLANDS IN TRIPLO
Netherlands Central Bureau of Statistics

Netherlands - English/Dutch ISBN No. 9039902925

A scenario study of the Dutch economy 1990-2015.

25054 NETHERLANDS OFFICIAL STATISTICS
Netherlands Central Bureau of Statistics

Netherlands - English/Dutch quarterly/DF 9.50 P.A.

Official statistics in the Netherlands described.

25055 POCKET YEARBOOK
Netherlands Central Bureau of Statistics

Netherlands - English/Dutch annual/DF 9.95 ISSN No. 0168-3705

Statistical tables on the economy and society.

25056 POCKET YEARBOOK TRAFFIC AND TRANSPORT STATISTICS
Netherlands Central Bureau of Statistics

Netherlands - English/Dutch annual/DF 0.00

Draws together the main features of Dutch transport and traffic over the year.

25057 POCKETBOOK OF EDUCATIONAL STATISTICS
Netherlands Central Bureau of Statistics

Netherlands - English/Dutch annual

Full time education, teaching staff, cost of education.

25058 POPULATION PROJECTIONS FOR THE NETHERLANDS 1970-2000
Netherlands Central Bureau of Statistics

Netherlands - non-recurrent/ DF13.00

25059 PRICE ANALYSIS
Netherlands Central Bureau of Statistics

Netherlands - English/Dutch annual/DF16.25

Shows changes in level of prices over a period of one year.

25060 QUARTERLY ACCOUNTS
Netherlands Central Bureau of Statistics

Netherlands - English/Dutch quarterly/DF44.50 P.A.

Trends in the Dutch economy including gross domestic product, imports, exports, consumption, investment.

25061 QUARTERLY BULLETIN ON JUSTICE AND SECURITY
Netherlands Central Bureau of Statistics

Netherlands - English/Dutch quarterly/DF70.50

Crime, bankruptcies, fire, adoptions, prison populations, suicides, persons in custody, wills and administrative jurisdiction.

25062 QUARTERLY COUNTRY TRADE REPORTS: NETHERLANDS
United Kingdom Iron and Steel Statistics Bureau

Netherlands - English quarterly/ £200.00 P.A. for four countries ISSN No. 0952-6005

Cumulative trade statistics of imports and exports of steel products.

NORWAY

Country profile:	Norway
Official name:	Kingdom of Norway
Area:	323,878 sq.km.
	Bordered in the north by the Arctic Ocean, by the Skagerrak Straits in the south, by the North Sea in the west and by Finland, Sweden and the CIS in the east.
Population:	4.24m (1991)
Capital:	Oslo
Language(s):	Norwegian
Currency:	Kroner 100 ore = 1 Norwegian kroner
National bank:	Norges Bank
Government and Political structure:	Constitutional monarchy with multi-party parliamentary democracy. The Storting (parliament) has 165 members and chooses the Prime Minister.

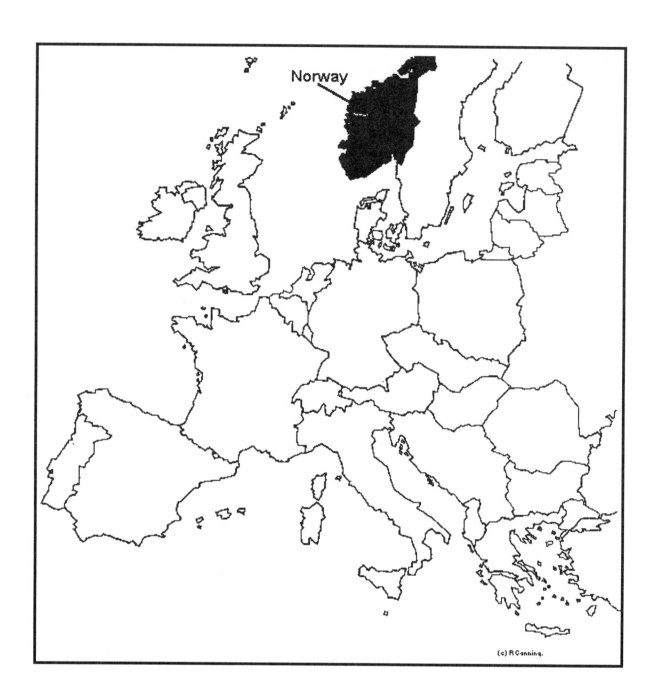

Norway

(c) R Canning.

26001 AGRICULTURAL STATISTICS

Central Bureau of Statistics

Norway - English/Norwegian
annual

Statistics on all aspects of
agriculture.

26002 ALL ABOUT THE NORWEGIAN FEDERATION OF TRADE UNIONS

The Norwegian Confederation of Trade Unions

Norway - 300pp NKr100 ISBN No.
8210027174

History, incomes policy, unions,
economic policy, industrial
relations.

26003 ANNUAL REPORT

Den Norske Bank

Norway - English/Norwegian
annual/1990 free

Annual report of Den Norske Bank.

26004 ANNUAL REPORT

Oslo Børs

Norway - English/Norwegian
annual/free ISSN No. 0085-4565

Annual report of Oslo Børs.

26005 ANNUAL STATISTICS

Oslo Børs

Norway - English/Norwegian
annual/free

Annual statistics from the Oslo
stock exchange.

26006 BØRSFERSFEKTIVER

Oslo Børs

Norway - Norwegian 1-3 times per
year/free

Articles on different topics
regarding the securities market.

26007 CONSTRUCTION STATISTICS

Central Bureau of Statistics

Norway - English/Norwegian
annual/NKr60

Details on all aspects of the
construction industry.

26008 COUNTRY BOOKS: NORWAY

United Kingdom Iron and Steel Statistics Bureau

Norway - English annual/£75.00
ISSN No. 0952-6013

Annual summary tables showing
production, materials consumed,
apparent consumption, imports and
exports of 130 products by quality
and market.

26009 COUNTRY FORECASTS: NORWAY

Business International Ltd

Norway - English quarterly/£360.00
P.A.

Contains an executive summary,
sections on political and economic
outlooks and the business
environment, a fact sheet, and key
economic indicators. Includes a
quarterly Global Outlook.

26010 COUNTRY PROFILE: NORWAY

Department of Trade and Industry Export Publications

Norway - English irregular/1991
50pp £10.00

This publication provides a general
description of the country in terms
of its geography, economy and
population, and gives information
about trading in the country with
details like export conditions and
investment covered.

26011 COUNTRY REPORT: NORWAY

Business International Ltd

Norway - English quarterly/25-40pp
£160.00 P.A. ISSN No. 0269-4182

Contains an executive summary,
sections on political and economic
structure, outlook for the next 12-18
months, statistical appendices, and
a review of key political
developments. Includes an annual
Country Profile.

26012 DOING BUSINESS IN NORWAY

Price Waterhouse

Norway - English irregular/1991
330-340pp free to clients

A business guide which includes
sections on the investment climate,
doing business, accounting and
auditing, taxation and general
country data. Updated by
supplements.

26013 ECONOMIC BULLETIN

Norges Bank

Norway - English quarterly/104pp

Quarterly report on banking and
economic conditions.

26014 ECONOMIC BULLETIN

Den Norske Bank

Norway - English/Norwegian 6-8
times per year/16-20pp US$230.00
P.A.

Norwegian and international
economy.

26015 ECONOMIC SURVEY

Central Bureau of Statistics

Norway - English/Norwegian
quarterly ISSN No. 0800-4110

Reports and analysis of current
economic trends.

26016 ELECTRICITY STATISTICS

Central Bureau of Statistics

Norway - Norwegian annual/NKr60

Statistical data covering supplies,
production, employment and
operations of installations for both
electricity and gas.

26017 ENERGY STATISTICS

Central Bureau of Statistics

Norway - English/Norwegian
annual

Statistics covering total energy
consumption, production, prices for
electricity, crude petroleum, natural
gas, coal and coke.

26018 EXTERNAL TRADE

Central Bureau of Statistics

Norway - English/Norwegian
annual

Statistics on import/exports
classified by Harmonised System
for commodities, by country of
origin or destination.

26019 FINANCIAL REVIEW

Norwegian Bankers Association

Norway - English quarterly/12pp
NKr350 P.A. ISSN No. 0800-3564

Commentary on economic and
financial subjects and review of
developments in banking.

26020 FINANCING FOREIGN OPERATIONS: NORWAY

Business International Ltd

Norway - English annual/£695.00
for complete European series

Business overview, currency
outlook, exchange regulations,
monetary system, short/medium/
long-term and equity financing
techniques, capital incentives, cash
management, investment trade
finance and insurance.

26021 FISHERY STATISTICS

Central Bureau of Statistics

Norway - English/Norwegian
annual

Statistics on all aspects of the
fishing industry.

26022 FISKETS GANG

Fiskeridirektoratet

Norway - English/Norwegian semi-
monthly/NKr330 P.A. ISSN No.
0015-3133

Current developments in the fishing
industry including statistics on
catches, exports, type of fish.

26023 FORESTRY STATISTICS

Central Bureau of Statistics

Norway - English/Norwegian
annual

Statistics on all aspects of the
forestry industry.

26024 GLOBAL FORECASTING SERVICE

Business International Ltd

Norway - English quarterly/£350.00
P.A.

Provides a range of medium-term forecasts covering main political economic and business trends. A quarterly Global Outlook is included.

26025 HINTS TO EXPORTERS VISITING NORWAY
Department of Trade and Industry Export Publications
Norway - English irregular/1991 60pp £5.00

This publication provides detailed practical information that is useful for those planning a business visit. Topics include currency information, economic factors, import and exchange control regulations, methods of doing business and general information.

26026 INCOME AND PROPERTY STATISTICS
Central Bureau of Statistics
Norway - English/Norwegian every 3 years

Statistics on personal and household incomes.

26027 INTERNATIONAL CUSTOMS JOURNAL: NORWAY
Department of Trade and Industry Export Publications
Norway - English irregular/£25.00

This publication contains the rate of duties that apply to commodities exported into the country according to the Harmonised System (HS) of commodity coding.

26028 INVESTING, LICENSING AND TRADING CONDITIONS ABROAD: NORWAY
Business International Ltd
Norway - English annual/£825.00 for complete European series

Loose-leaf format covering the state role in industry, competition rules, price controls, remittability of funds, taxation, incentives and capital sources, labour and foreign trade.

26029 KEY FIGURES
Oslo Børs
Norway/International - English/ Norwegian annual/1991 free

Stock exchange statistics.

26030 LABOUR MARKET STATISTICS
Central Bureau of Statistics
Norway - English/Norwegian annual

Current data on the labour market.

26031 LEVEL OF LIVING SURVEY
Central Bureau of Statistics
Norway - English/Norwegian irregular

Survey of persons 16-79 giving occupation, working conditions, household conditions.

26032 MANEDSINFORMASJON
Oslo Børs
Norway - Norwegian monthly/free

Statistics and articles on the securities market.

26033 MANUFACTURING STATISTICS
Central Bureau of Statistics
Norway - English/Norwegian annual/2 Vols

Figures on production, labour, wages, costs.

26034 MONTHLY BULLETIN OF EXTERNAL TRADE
Central Bureau of Statistics
Norway - English/Norwegian monthly/NKr300.00 P.A.

Detailed monthly imports/exports data.

26035 MONTHLY BULLETIN OF STATISTICS
Central Bureau of Statistics
Norway - English/Norwegian/ NKr415.00 P.A.

Detailed monthly economic statistics.

26036 NATIONAL ACCOUNTS
Central Bureau of Statistics
Norway - English/Norwegian annual

Data on domestic product, consumption, expenditure, private and public finance.

26037 NATURAL RESOURCES AND THE ENVIRONMENT
Central Bureau of Statistics
Norway - English/Norwegian annual

Comment and analytic survey derived from official statistics.

26038 NORWAY
United Kingdom Iron and Steel Statistics Bureau
Norway - English annual/£80.00 ISSN No. 0952-6013

Report on the steel industry showing production, consumption and details of imports and exports.

26039 NORWAY'S FOREIGN TRADE
Norwegian Trade Council
Norway - English/Norwegian annual/51pp free

Balance of payments, imports/ exports, investments, commodities, market shares, trading partners.

26040 NORWAY'S LARGEST COMPANIES
Ekonomisk Litteratur AB
Norway - English/Norwegian annual/NKr1,300

Financial information on Norway's largest companies.

26041 THE NORWEGIAN SECURITIES MARKET
Oslo Børs
Norway - English irregular/free

Describing the Norwegian market.

26042 OIL AND GAS ACTIVITY
Central Bureau of Statistics
Norway - English/Norwegian quarterly

Exploration, finds, production, employment, wages, taxes.

26043 OKONOMISK REVY
Norwegian Bankers Association
Norway - Norwegian 6 times per year/52pp NKr390 ISSN No. 0030-1914

Economic review.

26044 OSLO STOCK EXCHANGE INFORMATION
Oslo Børs
Norway - English monthly/free

Basic data on the stock exchange with levels of business given.

26045 PETROLEUM ACTIVITIES IN NORWAY
Den Norske Bank
Norway - English annual/1990 48pp

Petroleum activities.

26046 POPULATION STATISTICS VOL 1
Central Bureau of Statistics
Norway - English/Norwegian annual

Demographic statistics presented by municipality.

26047 POPULATION STATISTICS VOL 2
Central Bureau of Statistics
Norway - English/Norwegian annual

Population classified by age, sex, marital status.

26048 POPULATION STATISTICS VOL 3
Central Bureau of Statistics
Norway - English/Norwegian annual

Data on both external and internal migration classified by demographic characteristics.

26049 PRICE LIST
Oslo Børs
Norway - Norwegian daily/NKr2.095 P.A.

Daily price list of stocks, bonds and options.

26050 QUARTERLY COUNTRY TRADE REPORTS: NORWAY
United Kingdom Iron and Steel Statistics Bureau

Norway - English quarterly/£200.00 P.A. for four countries
ISSN No. 0952-6013

Cumulative trade statistics of imports and exports of steel products.

26051 SHIPPING INDUSTRY AND OFFSHORE ACTIVITIES
Norges Rederiforbund

Norway - English/Norwegian
4 times per year/12pp

Shipping, offshore.

26052 SOCIAL STATISTICS
Central Bureau of Statistics

Norway - English/Norwegian
annual

Main statistics on all aspects of social, political, environmental and economic life.

26053 STATISTICAL YEARBOOK OF NORWAY
Central Bureau of Statistics

Norway - English/Norwegian
annual/NKr95 ISSN No. 0377-8908

All major fields of the economy and society are covered.

26054 STATISTICS OF THE WEEK
Central Bureau of Statistics

Norway - English/Norwegian
weekly/NKr590 P.A.

Summary of current data.

26055 TRANSPORT AND COMMUNICATION STATISTICS
Central Bureau of Statistics

Norway - English/Norwegian
annual

Statistics on railways, roads, automobiles, aircraft, postal services, telecommunications, broadcasting.

26056 WAGE STATISTICS
Central Bureau of Statistics

Norway - Norwegian annual

Monthly earnings for the following groups: agriculture, manufacturing, wholesale and retail, banks, local government, hotels and restaurants.

26057 ØKONMISKE ANALYSER
Central Bureau of Statistics

Norway - Norwegian monthly

Current economic trends.

POLAND

Country profile: Poland

Official name: Polish Republic

Area: 312,680 sq.km.

Bordered by the Baltic Sea and the CIS in the north, in the south by the Czech Slovak states, in the east by Lithuania, Belorussia and Ukraine and by Germany in the west.

Population: 37.9m (1989)

Capital: Warsaw

Language(s): Polish

Currency: Zloty
100 groszy = 1 Polish zloty

National bank: The National Bank of Poland

Government and Political structure: Multi-party parliamentary democracy. The Sejm (national parliament) has 460 members.

(c) R Canning.

27001 AGRICULTURAL COMMODITY PRODUCTION AND MEANS OF PRODUCTION IN AGRICULTURE IN 199...
Central Statistical Office of Poland
Poland - Polish irregular/120pp
US$8.00

27002 AGRICULTURE (STATISTICAL YEARBOOK)
Central Statistical Office of Poland
Poland - Polish annual/300pp
US$25.00
Statistics on agriculture, with the main outputs covered.

27003 AREA AND POPULATION IN TERRITORIAL SECTION
Central Statistical Office of Poland
Poland - Polish irregular/90pp
US$8.00
Regional breakdown of demographics.

27004 THE BASIC STATISTICAL DATA ON UNITS EMPLOYING LESS THAN FIVE EMPLOYEES
Central Statistical Office of Poland
Poland - Polish irregular/20pp
US$3.00
Selected data on Polish small business units.

27005 CHANGES IN PRICES IN 199..
Central Statistical Office of Poland
Poland - Polish irregular/50pp
US$5.00
Price analysis and useful in compiling inflation levels over an extended period.

27006 CHANGES IN PRICES IN NATIONAL ECONOMY IN 1991 AND IN THE FIRST HALF-YEAR OF 1992
Central Statistical Office of Poland
Poland - Polish irregular/80pp
US$7.00
Details price levels and changes in prices which have had significant effects on the Polish economy.

27007 COMMUNITY SERVICES IN 199..
Central Statistical Office of Poland
Poland - Polish irregular/80pp
US$7.00
Statistics on social service type activities.

27008 THE CONCENTRATION OF INDUSTRIAL PRODUCTION 1990-1991
Central Statistical Office of Poland
Poland - Polish irregular/100pp
Reviews the structure of Polish industry and is useful for sectoral analysis.

27009 CONCISE STATISTICAL YEARBOOK
Central Statistical Office of Poland
Poland - Polish annual/360pp
US$16.00 ISSN No. 0079-2608
Takes the highlights from the main statistical series and presents them in an abridged format.

27010 COUNTRY FORECASTS: POLAND
Business International Ltd
Poland - English quarterly/£360.00 P.A.
Contains an executive summary, sections on political and economic outlooks and the business environment, a fact sheet, and key economic indicators. Includes a quarterly Global Outlook.

27011 COUNTRY PROFILE: POLAND
Department of Trade and Industry Export Publications
Poland - English irregular/1992 76pp £10.00
This publication provides a general description of the country in terms of its geography, economy and population, and gives information about trading in the country with details like export conditions and investment covered.

27012 COUNTRY REPORT: POLAND
Business International Ltd
Poland - English quarterly/25-40pp £160.00 P.A.
Contains an executive summary, sections on political and economic structure, outlook for the next 12-18 months, statistical appendices, and a review of key political developments. Includes an annual Country Profile.

27013 COUNTRY RISK SERVICE: POLAND
Business International Ltd
Poland - English quarterly/minimum subscription £1,855.00 P.A. for seven countries
Contains credit risk ratings, risk appraisals, cross-country databases, data on diskette, and two year projections.

27014 DEMOGRAPHY (STATISTICAL YEARBOOK)
Central Statistical Office of Poland
Poland - Polish annual/350pp
US$18.00
Statistics on demography, with standard breakdowns by age, sex and marital status.

27015 DOING BUSINESS IN EASTERN EUROPE: POLAND
Kogan Page
Poland - English/1991 320pp £25.00 ISBN No. 0749404736
Part of the CBI initiative on Eastern Europe. Describes the business and economic environment.

27016 DOING BUSINESS IN POLAND
Confederation of British Industry (CBI)
Frere Cholmeley, Polish Investment Company, Touche Ross and Co. DRT Poland/Poland - English/1991 347pp ISBN No. 0749404736
General business and economic guide mainly aimed at those exporting to, or establishing a business in Poland.

27017 DOING BUSINESS IN POLAND
Ernst & Young International
Poland - English irregular/1991 110-130pp free to clients
A business guide which includes an executive summary, the business environment, foreign investment, company structure, labour, taxation, financial reporting and auditing, and general country data.

27018 DOING BUSINESS IN POLAND
Price Waterhouse
Poland - English irregular/1992 330-340pp free to clients
A business guide which includes sections on the investment climate, doing business, accounting and auditing, taxation and general country data. Updated by supplements.

27019 DOING BUSINESS WITH EASTERN EUROPE: POLAND
Business International Ltd
Poland - English quarterly/£540.00 P.A.
Part of a ten volume set, updated every three months which gives a broad picture of the market and business practice.

27020 EMPLOYMENT, WAGES AND SALARIES
Central Statistical Office of Poland

Poland - Polish quarterly/60pp

Wage and salary levels across Polish industry.

27021 EMPLOYMENT IN NATIONAL ECONOMY 199...
Central Statistical Office of Poland

Poland - Polish irregular/95pp US$7.00

Breakdown of Polish employment by sector/type.

27022 FINANCE (STATISTICAL YEARBOOK)
Central Statistical Office of Poland

Poland - Polish irregular/300pp US$24.00 ISSN No. 0079-2640

Statistics on financial sector with standard indicators used such as inflation and money supply. Latest edition published 1987.

27023 THE FINANCES OF ECONOMIC UNITS IN 19– AND IN THE FIRST HALF-YEAR OF 19(–
Central Statistical Office of Poland

Poland - Polish irregular/80pp US$8.00

Statistical coverage of company finance in Poland.

27024 FINANCING FOREIGN OPERATIONS: POLAND
Business International Ltd

Poland - English annual/£695.00 for complete European series

Business overview, currency outlook, exchange regulations, monetary system, short/medium/long-term and equity financing techniques, capital incentives, cash management, investment trade finance and insurance.

27025 FOREIGN MARKETS
Polish Chamber of Commerce

Dr. Andrzej Zielinski/Poland/Europe/North America/Asia - English/Polish/German/Russian 3 times a week/April 1957 12-16pp Zl10,000 P.A.

Details foreign investment in Poland and has some economic coverage.

27026 FOREIGN TRADE
Polish Chamber of Commerce

Maciej Deniszczuk/Poland/International - Polish monthly/1956 24pp Zl25,000 P.A. ISSN No. 0017-7245

Deals with Polish international trade - imports, exports etc.

27027 FOREIGN TRADE (STATISTICAL YEARBOOK)
Central Statistical Office of Poland

Poland - Polish annual/250pp US$24.00

Statistics on foreign trade with details of commodities and countries.

27028 FUEL-POWER ECONOMY 199..
Central Statistical Office of Poland

Poland - Polish irregular/70pp US$7.00

Energy statistics with type of energy detailed.

27029 GENERAL TRADE INDEX AND BUSINESS GUIDE
Business Foundation Ltd

Poland - Text in English, indexes in English and Polish/1991 714pp

Comprehensive Polish trade directory.

27030 GLOBAL FORECASTING SERVICE - POLAND
The Economist Intelligence Unit

Poland - English quarterly/£360.00 P.A.

Economy and political forecasting publication.

27031 GROSS DOMESTIC PRODUCT IN 199...
Central Statistical Office of Poland

Poland - Polish irregular/120pp US$7.00

Uses standard statistical formats to detail Polish GDP.

27032 HINTS TO EXPORTERS VISITING POLAND
Department of Trade and Industry Export Publications

Poland - English irregular/1991 65pp £5.00

This publication provides detailed practical information that is useful for those planning a business visit. Topics include currency information, economic factors, import and exchange control regulations, methods of doing business and general information.

27033 HOUSING CONSTRUCTION
Central Statistical Office of Poland

Poland - Polish quarterly/15pp US$2.00 P.A.

Statistical coverage of the domestic construction sector.

27034 IMPORTING FROM POLAND
Probus Europe

Poland - English irregular/201pp US$25 ISBN No. 9627138231

Guide for those importing from Poland.

27035 INDUSTRY (STATISTICAL YEARBOOK)
Central Statistical Office of Poland

Poland - Polish annual/400pp US$26.00 ISSN No. 0079-2764

Statistics on industry broken down by sector and type of industrial production.

27036 INFORMATION ABOUT ECONOMIC SITUATION OF THE COUNTRY IN 199..
Central Statistical Office of Poland

Poland - Polish monthly/30pp US$120.00 P.A.

An overview of the Polish economy.

27037 INFORMATION BULLETIN
Narodowy Bank Polski

Poland - English/Polish monthly and annual/45-55pp monthly, 64-90pp annually ISSN No. 1/1230-0101 2/1230-0020

A banking journal which also contains general information on the Polish economy.

27038 INVESTING, LICENSING AND TRADING CONDITIONS ABROAD: POLAND
Business International Ltd

Poland - English annual/£825.00 for complete European series

Loose-leaf format covering the state role in industry, competition rules, price controls, remittability of funds, taxation, incentives and capital sources, labour and foreign trade.

27039 INVESTMENT ACTIVITY IN 199–, PART I INVESTMENT OUTLAYS
Central Statistical Office of Poland

Poland - Polish irregular/70pp US$8.00

Statistical data on investment including level and type of investment.

27040 INVESTMENT ACTIVITY IN 199–, PART II BUILDINGS INVESTMENTS IN A PRIVATE ECONOMY
Central Statistical Office of Poland

Poland - Polish irregular/70pp US$5.00

Statistical coverage of investment in the private construction sector.

27041 JOINT VENTURE COMPANIES
Central Statistical Office of Poland

Poland - Polish irregular/50pp US$5.00

Covers number and type of joint business ventures.

27042 LIVING CONDITIONS OF THE POPULATION IN 199...
Central Statistical Office of Poland

Poland - Polish irregular/100pp US$8.00

Social survey type publication detailing living conditions, housing conditions etc.

27043 MONTHLY INFORMATION ABOUT CHANGES IN PRICES IN NATIONAL ECONOMY
Central Statistical Office of Poland

Poland - Polish monthly/40pp US$3.00 P.A.

27044 NATIONAL INCOME (STATISTICAL YEARBOOK)
Central Statistical Office of Poland

Poland - Polish annual/170pp US$22.00

Details of Polish national income arranged in standard formats.

27045 OUTLAYS AND RESULTS OF INDUSTRY
Central Statistical Office of Poland

Poland - Polish quarterly/75pp US$7.00 P.A.

Data on industrial performance and end-results.

27046 PARTNER FROM POLAND
Polish Chamber of Commerce

Eugenia Dmowska/International - English/Russian quarterly/1971 48pp US$15 P.A. ISSN No. 1230-0675

Deals with foreign investment in Poland.

27047 POLAND IN NUMBERS 199–
Central Statistical Office of Poland

Poland - English/Polish/German/French/Russian irregular/14pp

Brief survey of key data on Poland including the Polish economy.

27048 POLAND: INDUSTRIAL DEVELOPMENT REVIEW
Blackwell Publishers

Poland - English irregular/1991 £50.00

Reviews the development of Polish industry and is useful when analysing structural changes in the industrial base.

27049 POLAND - SELECTED DATA
Central Statistical Office of Poland

Poland - English/Polish/French irregular/32pp US$2.50

Key statistical, socio-economic data.

27050 POLAND'S NEXT FIVE YEARS; THE DASH FOR CAPITALISM SPECIAL REPORT NO. 2110
The Economist Intelligence Unit

Poland - English/1991 £155.00

Overview of current and future state of the Polish economy.

27051 POLISH BUSINESS VOICE
Richard Cottrell and Associates

Poland - English monthly/8pp £95.00 P.A.

Business magazine with general news.

27052 POLISH FOREIGN TRADE
Polish Foreign Trade

Poland - English monthly/US$24.00 P.A. ISSN No. 0032-2881

English language magazine aimed at business people in the Polish market.

27053 POLISH TRADE MAGAZINE
Agpol Polerpostpress

Poland - English quarterly/£3.40 P.A. ISSN No. 0239-989X

English language magazine aimed at overseas business people operating in Poland.

27054 PROPERTY AND PERSONAL INSURANCE
Central Statistical Office of Poland

Poland - Polish annual/ISSN No. 0079-2853

Income distribution by demographic unit.

27055 RESULTS OF AGRICULTURAL CENSUS IN 199–
Central Statistical Office of Poland

Poland - Polish irregular/130pp US$8.00

Details land-use, outputs, livestock etc.

27056 RESULTS OF PLANT PRODUCTION IN 199–
Central Statistical Office of Poland

Poland - Polish irregular/100pp US$7.00

Covers agricultural industry.

27057 SMALL BUSINESS
Bank Gdansk

Poland - Polish monthly/ISSN No. 0137-7221

Bulletin giving general news about Polish small businesses.

27058 STATISTICAL BULLETIN
Central Statistical Office of Poland

Poland - Polish monthly/120pp US$120.00 P.A. ISSN No. 0043-518X

Monthly reports on key statistical series. Contents page and summaries in English and Russian.

27059 STATISTICAL NEWS
Central Statistical Office of Poland

Poland - Polish monthly/120pp US$60.00 P.A.

Covers statistical methods.

27060 STATISTICAL YEARBOOK
Central Statistical Office of Poland

Poland - Polish annual/700pp US$36.00 ISSN No. 0079-2780

Covers the main statistical series produced and includes the important economic series.

27061 TOURISM AND RECREATION
Central Statistical Office of Poland

Poland - Polish irregular/150pp US$12.00

Standard data Polish tourist industry and leisure in Poland.

27062 TRADE MAGAZINE - POLAND
Polish Chamber of Commerce

Barbara Bosowska/Poland - English/German quarterly/1982 US$10.00, DM 17-50 P.A. ISSN No. 0239-989X, 0239-9902

Publication covering light industry and culture.

27063 UNEMPLOYMENT IN POLAND
Central Statistical Office of Poland

Poland - Polish irregular/30pp US$3.00

Numbers, location etc. of Polish unemployed labour.

27064 YEARBOOK OF CONSTRUCTION STATISTICS
Central Statistical Office of Poland

Poland - Polish irregular/100pp
US$12.00 ISSN No. 0079-2632

Latest edition published 1987.

27065 YEARBOOK OF INVESTMENT AND FIXED ASSETS
Central Statistical Office of Poland

Poland - Polish irregular/90pp
US$7.00 ISSN No. 0079-2705

Latest edition published 1987.

27066 YEARBOOK OF LABOUR STATISTICS
Central Statistical Office of Poland

Poland - Polish irregular/50pp
US$5.00 ISSN No. 0079-2772

Employment movements across industrial/service sectors. Latest edition published 1987.

27067 YEARBOOK OF PUBLIC HEALTH STATISTICS
Central Statistical Office of Poland

Poland - Polish irregular/150pp
US$12.00

Healthcare in Poland, with a broad overview of the national health service. Latest edition published 1990.

27068 YEARBOOK OF TRANSPORT STATISTICS
Central Statistical Office of Poland

Poland - Polish irregular/30pp
US$3.00 ISSN No. 0079-2802

Information on Polish communications and the communications infrastructure. Latest edition published 1987.

PORTUGAL

Country profile:	Portugal
Official name:	Republic of Portugal
Area:	91,985 sq.km.
	Bordered by the Atlantic Ocean in the south and west and by Spain in the north and east.
Population:	9.85m (1991)
Capital:	Lisbon
Language(s):	Portugese
Currency:	Escudo 100 centavos = 1 Portugese escudo
National bank:	Banco de Portugal
Government and Political structure:	Republican parliamentary democracy. Elected President with 230 unicameral legislature, elected by proportional representation.

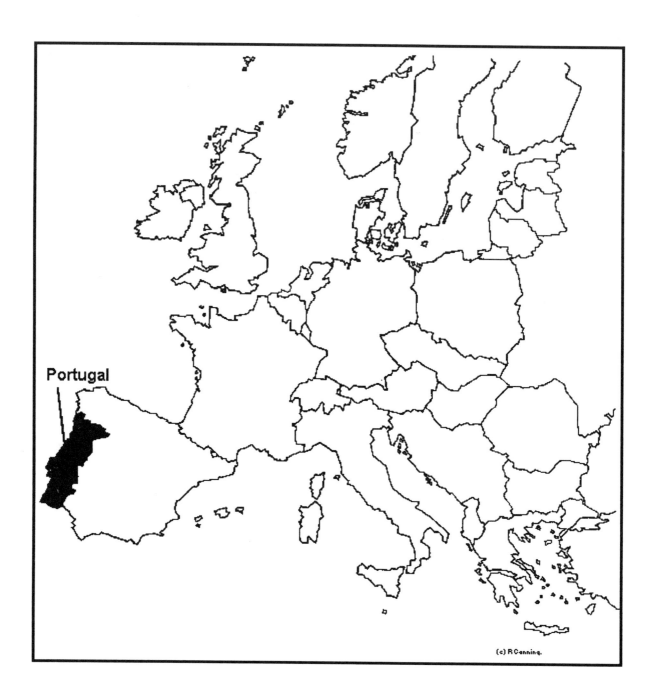

Portugal

(c) R.Canning.

28001 ANNUAL REPORT
Bolsa de Valores de Lisboa
Portugal - English/Portuguese
annual/53pp
Annual report of the Lisbon Stock
Exchange containing detailed
statistics of stock market
transactions.

28002 ANNUAL REPORT
European Free Trade
Association
Portugal - English/Portuguese/
annual/1979 free
EFTA report covering Portugal's
international trade.

28003 AS ACÇÏES NO MERCADO
DE CATAÇÏES OFICIALS
Bolsa de Valores de Lisboa
Portugal - Portuguese biannual/
1992 120pp Esc500
Individual information for each
company with listed shares on the
market with official quotations.

28004 BI-MONTHLY BULLETIN
OF THE BRITISH -
PORTUGUESE CHAMBER
OF COMMERCE
British Portuguese
Chamber of Commerce
Portugal - English/Portuguese bi-
monthly/controlled circulation free
to members ISSN No. 0892-1572
News about Portugal, company
news, Portuguese legislation,
overseas trade statistics.

28005 BOLETIN DE COTAÇÏES
Bolsa de Valores de Lisboa
Portugal - Portuguese daily/1920
Esc39,195
Official daily prices and
transactions.

28006 BOLETIN DE
OPORTUNIDADES DE
NEGÓCIO
ICEP: Investimentos
Comércio e Turismo de
Portugal
Portugal - Portuguese bi-weekly/
1987 4pp Esc5,250 P.A.
Business opportunities all over the
world.

28007 BOLETIN MENSAL DE
ESTATÍSTICA
Instituto Nacional de
Estatística
Portugal - English/Portuguese
monthly/1929 Esc2,600 P.A. ISSN
No. 0032-5082
Monthly bulletin of statistics.

28008 BUYING IN PORTUGAL -
GENERAL INFORMATION
ICEP: Investimentos
Comércio e Turismo de
Portugal
Portugal - English/Portuguese/
Spanish annual/1991 90pp
Economy, trade, investment, laws.

28009 CARACTERIZAÇÍO DAS
EMPRESAS
PORTUGUESAS
Instituto Nacional de
Estatística
Portugal - Portuguese annual/1991
Esc2,100 ISSN No. 0871-8709
Classification of companies by
district, volume of business and
numbers of staff.

28010 COMÉRCIO EXTERNO
PORTUGUÊS
ICEP: Investimentos
Comércio e Turismo de
Portugal
Portugal - English/Portuguese
annual/1979 190pp Esc1,000
Foreign trade, economic
environment.

28011 CONTAS NACIONAIS
Instituto Nacional de
Estatística
Portugal - Portuguese annual/1972
Esc1,260 ISSN No. 0870-2659
Contains the principal Portuguese
macro-economics aggregates for
the year under review.

28012 CONTAS REGIONAIS
Instituto Nacional de
Estatística
Portugal - Portuguese annual/1991
Esc2,310 ISSN No. 0871-9543
Regional accounts.

28013 COUNTRY BOOKS:
PORTUGAL
United Kingdom Iron and
Steel Statistics Bureau
Portugal - English annual/£75.00
ISSN No. 0952-4951
Annual summary tables showing
production, materials consumed,
apparent consumption, imports and
exports of 130 products by quality
and market.

28014 COUNTRY FORECASTS:
PORTUGAL
Business International Ltd
Portugal - English quarterly/
£360.00 P.A.
Contains an executive summary,
section on political and economic
outlook and the business
environment, a fact sheet, and key
economic indicators. Includes a
quarterly Global Outlook.

28015 COUNTRY PROFILE:
PORTUGAL
Department of Trade and
Industry Export
Publications
Portugal - English irregular/1990
44pp £10.00
This publication provides a general
description of the country in terms
of its geography, economy and
population, and gives information
about trading in the country with
details like export conditions and
investment covered.

28016 COUNTRY REPORT:
PORTUGAL
Business International Ltd
Portugal - English quarterly/
25-40pp £160.00 P.A. ISSN No.
0269-5456
Contains an executive summary,
sections on political and economic
structure, outlook for the next 12-18
months, statistical appendices, and
a review of key political
developments. Includes an annual
Country Profile.

28017 COUNTRY RISK SERVICE:
PORTUGAL
Business International Ltd
Portugal - English quarterly/
minimum subscription £1,855.00
P.A. for seven countries
Contains credit risk ratings, risk
appraisals, cross-country
databases, data on diskette, and
two year projections.

28018 DIGEST
Bolsa de Valores de Lisboa
Portugal - English/Portuguese
monthly/1991 4pp free
Monthly statistical information on
the securities markets.

28019 ECONOMIC INDICATORS
Banco de Portugal
Portugal - English/Portuguese
annual/1978 25pp ISSN No. 0870-
0087
Data for the past five years on
areas such as production, national
accounts, foreign trade.

28020 ESTADO DAS CULTURAS
E PREVISÍO DAS
COLHEITAS
Instituto Nacional de
Estatística
Portugal - Portuguese monthly
Major agricultural crops; forecast of
areas covered, income and
production figures.

28021 ESTATÍSTICAS
AGRÍCOLAS
Instituto Nacional de
Estatística
Portugal - Portuguese annual/1943
Esc3,360 ISSN No. 0079-4139
Trends in the agricultural sector.

28022 ESTATÍSTICAS DA CONSTRUCÇÍO ET HABITAÇÍO
Instituto Nacional de Estatística

Portugal - Portuguese annual/1970 Esc2,100 ISSN No. 0377-2225

Information on developments in the housing sector; average hourly wages in construction; describes current projects and their regional distribution.

28023 ESTATÍSTICAS DA ENERGIA
Instituto Nacional de Estatística

Portugal - Portuguese annual/1969 Esc685 ISSN No. 0377-2233

Statistics on the production, consumption, distribution and national balances of energy.

28024 ESTATÍSTICAS DA PESCA
Instituto Nacional de Estatística

Portugal - Portuguese annual/1970 Esc1,785 ISSN No. 0377-225X

Structural information on fisheries, production, foreign trade, credit and investment sectors and a characterization of fishermen and their vessels.

28025 ESTATÍSTICAS DA PROTECÇÍO SOCIAL, ASSOCIAÇÍES SINDICAIS E PATRONAIS
Instituto Nacional de Estatística

Portugal - Portuguese annual/1938 Esc2,520 ISSN No. 0870-4406

Portuguese social welfare and the annual activity of employees and employers associations.

28026 ESTATÍSTICAS DA SAUDÉ
Instituto Nacional de Estatística

Portugal - Portuguese annual/1969 Esc7,035 ISSN No. 0377-2268

Statistics on staff employed in the health sector, hospitals, births, deaths etc.

28027 ESTATÍSTICAS DAS CONTRIBUIÇÍES E IMPOSTAS
Instituto Nacional de Estatística

Portugal - Portuguese annual/1877 Esc1,155 ISSN No. 0079-4120

Provides information on taxable income and the liquidation and payment of the principal taxes.

28028 ESTATÍSTICAS DAS FINANÇAS PÚBLICAS
Instituto Nacional de Estatística

Portugal - Portuguese annual/1968 Esc4,410 ISSN No. 0377-2276

Statistical information on the public sector.

28029 ESTATÍSTICAS DAS SOCIEDADES
Instituto Nacional de Estatística

Portugal - Portuguese annual/1939-1985 (ceased publication)

Statistics on public and private companies.

28030 ESTATÍSTICAS DEMOGRÁFICAS
Instituto Nacional de Estatística

Portugal - Portuguese annual/1862 Esc5,250 ISSN No. 0037-2284

Population estimates; statistics on life expectancy; population variations and other demographic indicators.

28031 ESTATÍSTICAS DO COMÉRCIO EXTERNO
Instituto Nacional de Estatística

Portugal - Portuguese annual/1930 Esc6,090 ISSN No. 0079-4147

Provides statistical information on imports and exports by produce and country.

28032 ESTATÍSTICAS DO TURISMO
Instituto Nacional de Estatística

Portugal - Portuguese annual/1969 Esc5,460 ISSN No. 0377-2306

Statistics on tourism in Portugal and on Portuguese tourism abroad.

28033 ESTATÍSTICAS DOS SALÁRIOS
Instituto Nacional de Estatística

Portugal - Portuguese annual/1980 Esc905 ISSN No. 0870-4325

Statistics on wages and duration of work by sector, professional status and sex.

28034 ESTATÍSTICAS DOS TRANSPORTES E COMUNICAÇÍES
Instituto Nacional de Estatística

Portugal - Portuguese annual/1970 Esc5,250 ISSN No. 0377-2292

Statistics on land, sea and air transport, data on road accidents, the sale of cars and the consumption of fuel.

28035 ESTATÍSTICAS INDUSTRIAIS
Instituto Nacional de Estatística

Portugal - Portuguese annual/1943 2 Vols Esc1,474 (Vol 1) Esc7,035 (Vol 2) ISSN No. Vol 1 0377-2314 ISSN No. Vol 2 0079-418X

Vol.1 - Mining, electricity, gas and water. Vol.2 - Information on the percentage distribution of establishments with annual industry production indices and general summaries by industry.

28036 ESTATÍSTICAS MONETÁRIAS E FINANCEIRAS
Instituto Nacional de Estatística

Portugal - Portuguese annual/1930 Esc3,150 ISSN No. 0377-2322

Provides information on the following activity sectors; banking sector institutions, bank type institutions, insurance companies and mutual aid societies.

28037 ESTATÍSTICAS REGIONAIS
Instituto Nacional de Estatística

Portugal - Portuguese annual/1991 Esc4,725 ISSN No. 0871-911X

Statistics on population, social and economic trends on a regional basis.

28038 FRESH FRUITS, VEGETABLES AND FLOWERS FROM PORTUGAL
ICEP: Investimentos Comércio e Turismo de Portugal

Portugal - English/Portuguese irregular/1990 70pp

Production and foreign trade of fruits, vegetables and flowers.

28039 GLOBAL FORECASTING SERVICE: PORTUGAL
Business International Ltd

Portugal - English quarterly £350.00 P.A.

Economic forecasts.

28040 HINTS TO EXPORTERS VISITING PORTUGAL
Department of Trade and Industry Export Publications

Portugal - English irregular/1991 90pp £5.00

This publication provides detailed practical information which is useful for those planning a business visit. Topics include currency information, economic factors, import and exchange control regulations, methods of doing business and general information.

28041 IN DEPTH COUNTRY APPRAISALS ACROSS THE EC - CONSTRUCTION AND BUILDING SERVICES - PORTUGAL
Building Services Research and Information Association

Portugal - English 1991/£125.00

Part of an 11 volume survey covering the construction industry in EC countries.

28042 INDICADORES DE PRODUÇAO VEGETAL
Instituto Nacional de Estatística

Portugal - Portuguese quarterly

Analysis of vegetable production.

28043 INDÍCE DE PREÇOS NO CONSUMIDOR
Instituto Nacional de Estatística

Portugal - Portuguese monthly

Detailed statistics and analysis of consumer price indices.

28044 INDÍCE DE PRODUÇÍO INDUSTRIAL
Instituto Nacional de Estatística

Portugal - Portuguese monthly

Indices of industrial production; data allows inter-sector comparison over a period of time.

28045 INFORMAÇÍO SEMANAL
Bolsa de Valores de Lisboa

Portugal - Portugese weekly/1992 1pp Included in 'Boletin de Colaçíes'

Individual information for each company with listed shares on the market with official quotations.

28046 INQUIÉRITO AO EMPREGO: INFORMAÇAO ANTECIPADA
Instituto Nacional de Estatística

Portugal - Portuguese quarterly

Advance data on employment, unemployment and working population by occupation.

28047 INQUIÉRITO ÀS FÉRIAS DOS PORTUGUESES
Instituto Nacional de Estatística

Portugal - Portuguese annual/1988 Esc1,050 ISSN No. 0871-7192

Statistics on holidays by Portuguese citzens; period of holidays; destinations; accommodation; means of transport etc.

28048 INQUIÉRITO ÀS RECEITAS E DESPESAS FAMILIARES
Instituto Nacional de Estatística

Portugal - Portuguese every 10 years/1970 ISSN No. 0377-2365

Survey on family income and expenditure.

28049 INQUIÉRITO DE CONJUNTURA AO INVESTIMENTO
Instituto Nacional de Estatística

Portugal - Portuguese bi-annual

Estimates and forecasts on variations in investment made by companies in major sectors of activity.

28050 INQUIÉRITO MENSAL DE CONJURTURA À INDÚSTRIA TRANSFORMADORA
Instituto Nacional de Estatística

Portugal - Portuguese monthly

Data on the manufacturing sector including information on macro-economic variables: industrial supply, consumption, investment, exports and imports, stocks, prices and employment.

28051 INQUÉRITO AO TRANSPORTE RODOVIARIO DE MERCADORIAS
Instituto Nacional de Estatística

Portugal - Portuguese annual/1975 Esc2,310 ISSN No. 0870-2586

Statistics on the transport of goods by road (national and international) by Portuguese registered vehicles.

28052 INVESTING, LICENSING AND TRADING CONDITIONS ABROAD: PORTUGAL
Business International Ltd

Portugal - English annual/£825.00 for complete European series

Loose-leaf format covering the state role in industry, competition rules, price controls, remittability of funds, taxation, incentives and capital sources, labour and foreign trade.

28053 LAST TEN YEARS ON LISBON STOCK EXCHANGE
Bolsa de Valores de Lisboa

Portugal - English/Porluguese irregular/4pp free

Turnover and market capitalization figures; exchange rates and general information.

28054 MERCADOS DE VALORES MOBILIÁRIOS: ESTUDIOS E ESTATÍSTICAS
Bolsa de Valores de Lisboa

Portugal - Portuguese with English annotations quarterly/March 1992 100pp

Detailed stock market statistics.

28055 MONTHLY BULLETIN
Banco de Portugal

Portugal - English/Portuguese

monthly/1987 100pp Esc525 ISSN No. 0870-807X

Statistics on production, balance of payments and financial markets.

28056 O LIVRO BRANCO DO TURISMO
Direcçío General do Turismo

Portugal - Portuguese irregular/ 174pp Esc2,000

Present situation and organization of tourism in Portugal.

28057 O TURISMO EM
Direcçío General do Turismo

Portugal - Portuguese annual/1970 300pp Esc2,000

Statistics on tourism in Portugal covering hotels, camping, restaurants etc.

28058 O TURISMO ESTRANGEIRO EM PORTUGAL
Direcçío General do Turismo

Portugal - Portuguese annual/1979 200pp Esc1,500

Statistics on tourism in Portugal giving data on country of origin of tourists, regions visited etc.

28059 OECD ECONOMIC SURVEYS: PORTUGAL
Organisation for Economic Co-operation and Development (OECD)

Portugal - English annual/£7.50 ISSN No. 0474-5256

Survey on economic development, and current economic situtation with forecasts.

28060 PORTUGAL
United Kingdom Iron and Steel Statistics Bureau

Portugal - English 1992/£80.00

Report on the steel industry showing production, consumption and details of imports and exports.

28061 PORTUGAL: COUNTRY PROFILE
Business International Ltd

Portugal - English annual/44pp £50.00 ISSN No. 0269-5987

Annual survey of political and economic background. Included with subscription to "Country Report".

28062 PORTUGAL: COUNTRY REPORT
Business International Ltd

Portugal - English quarterly/25pp £150.00 P.A. (4 issues plus a country profile) ISSN No. 0269-5456

Analysis of economic and political trends every quarter.

28063 PORTUGAL IN FIGURES
Instituto Nacional de Estatística

Portugal - English/Portuguese annual/1969 free

A synthesis of the principle statistical information produced during the year for information to the general public.

28064 PORTUGAL SOCIAL 1992
Instituto Nacional de Estatística

Portugal - Portuguese 1992/ Esc8,925

Description of Portuguese life 1985-1989; population statistics; socio-economic conditions; family living conditions etc.

28065 PORTUGAL-IN
ICEP: Investimentas Comércio e Turismo de Portugal

Portugal - English irregular/1992 80pp Esc787.50

General and specific information about Portugal.

28066 QUARTERLY BULLETIN
Banco de Portugal

Portugal - English/Portuguese quarterly/1979 200pp Esc945 ISSN No. 0870-0095

Statistics on money, finance, exchange rates, employment, wages and balance of payments.

28067 QUARTERLY COUNTRY TRADE REPORTS: PORTUGAL
United Kingdom Iron and Steel Statistics Bureau

Portugal - English quarterly/£200.00 P.A. for four countries ISSN No. 0958-4951

Cumulative trade statistics of imports and exports of steel products.

28068 RECENSEAMENTO AGRÍCOLA
Instituto Nacional de Estatística

Portugal - Portuguese every 10 years/1952 ISSN No. 0377-2454

Statistics on Portuguese agriculture.

28069 RECENSEAMENTO DA POPULAÇÍO E DA HABITAÇÍO
Instituto Nacional de Estatística

Portugal - Portuguese every 10 years/1984 various volumes ISSN No. 0377-2454

Results of national census - data on buildings; accommodation for public and private buildings.

28070 RECENSEAMENTO DAS EMPRESAS DO SECTOR DE TRANSPORTES
Instituto Nacional de Estatística

Portugal - Portuguese every 10 years/1982 Esc2,000 ISSN No. 0870-4317

Results of a census carried out on companies in the transport sector.

28071 RECENSEAMENTO INDUSTRIAL
Instituto Nacional de Estatística

Portugal - Portuguese every 10 years/1881 ISSN No. 0870-4309

Results of a census of industrial establishments giving information on numbers, turnover, investment, expenditure, energy consumption etc.

28072 REPORT OF THE BOARD OF DIRECTORS
Banco de Portugal

Portugal - English/Portuguese annual/1867 250pp free ISSN No. 0870-0095

Annual report including a detailed economic and financial survey with statistics on output, wages, employment, prices, foreign trade, balance of payments.

28073 REVISTA EXPORTAR
ICEP: Investimentos Comércio e Turismo de Portugal

Portugal - Portuguese bi-monthly/ 1988 80pp Esc3,150 P.A.

General and specific information for the Portuguese exporter.

28074 SINTESE ANUAL DO MECADO DE VALORES MOBILIÁRIOS
Bolsa de Valores de Lisboa

Portugal - Portuguese with English annotations annual/78pp

Annual synthesis of statistics in the securities market in Portugal.

28075 STATISTICAL YEARBOOK
Instituto Nacional de Estatística

Portugal - English/Portuguese annual/1975 Esc7,035 ISSN No. 0079-4112

Overview of the social, political and economic situation in Portugal. Five years data with most detail on the immediately preceding year.

28076 SUNDRY COSTS
ICEP: Investimentas Comércio e Turismo de Portugal

Portugal - English/Portuguese annual/1991 4pp

Main investment costs in Portugal (land, offices etc.).

ROMANIA

Country profile:	Romania
Official name:	Republic of Romania
Area:	237,500 sq.km.
	Bordered by the CIS and Hungary in the north and east, in the south by Bulgaria, by the Black Sea in the east and by Yugoslavia in the west.
Population:	23m (1990)
Capital:	Bucharest
Language:	Romanian
Currency:	Leu 100 bani = 1 Romanian leu
National bank:	The National Bank of Romania
Government and Political structure:	Pre 1990 was dominated by the Communist party. Currently there is a directly elected President and a bicameral parliament with a 396 member National Assembly and a 119 member Senate.

Romania

(c) R.Canning.

29001 COUNTRY FORECASTS: ROMANIA

Business International Ltd

Romania - English quarterly/ £360.00 P.A.

Contains an executive summary, sections on political and economic outlooks and the business environment, a fact sheet, and key economic indicators. Includes a quarterly Global Outlook.

29002 COUNTRY PROFILE: ROMANIA

Department of Trade and Industry Export Publications

Romania - English irregular/1992 38pp £10.00

This publication provides a general description of the country in terms of its geography, economy and population, and gives information about trading in the country with details like export conditions and investment covered.

29003 COUNTRY REPORT - ROMANIA, BULGARIA, ALBANIA

The Economist Intelligence Unit

Romania/Bulgaria/Albania - English quarterly plus annual Country Profile/50pp £150.00 P.A. ISSN No. 0269-5669

Contains an executive summary, sections on political and economic structure, outlook for the next 12-18 months, statistical appendices, and a review of key political developments. Includes an annual Country Profile.

29004 COUNTRY RISK SERVICE: ROMANIA

Business International Ltd

Romania - English quarterly/ minimum subscription £1,855.00 P.A. for seven countries

Contains credit risk ratings, risk appraisals, cross-country databases, data on diskette, and two year projections.

29005 DOING BUSINESS IN EASTERN EUROPE: ROMANIA

Kogan Page

Romania - English/1992 256pp £25.00 ISBN No. 0749406909

Part of the CBI initiative on Eastern Europe. Describes the business and economic environment.

29006 DOING BUSINESS WITH EASTERN EUROPE: ROMANIA

Business International Ltd

Romania - English quarterly/ £540.00 P.A.

Part of a ten volume set, updated every three months which gives a broad picture of the market and business practice.

29007 EAST EUROPEAN ENERGY: ROMANIA'S ENERGY NEEDS PERSIST

US General Accounting Office

Romania - English irregular/1992 54pp free

An official government report on the energy situation in Romania by the Committee on Energy and Natural Resources of the US Senate, US General Accounting Office.

29008 FREE ROMANIA

Romania Liberia

Romania - Romanian daily

Newspaper with circulation of 400,000. Contains business and economic news.

29009 HINTS TO EXPORTERS VISITING ROMANIA

Department of Trade and Industry Export Publications

Romania - English irregular/1990 52pp £5.00

This publication provides detailed practical information that is useful for those planning a business visit. Topics include currency information, economic factors, import and exchange control regulations, methods of doing business and general information.

29010 INTERNATIONAL TRADE: ROMANIA TRADE DATA

US General Accounting Office

Romania - English irregular/1992 25pp

A report to the chairman, Committee of Finance of the US Senate, US General Accounting Office giving estimates of Romania international trade in connection with the award of "Most Favoured Nation Status" for trade purposes.

29011 NEUER WEG

Neuer Weg

Romania - German daily/Leu900 P.A.

Newspaper for the German business community in Romania, covering the business environment and news of business opportunities.

29012 PHARMACEUTICAL AND HEALTH INDUSTRIES OF EUROPE

Financial Times - Surveys Department

Bulgaria/C.I.S./Romania/ Yugoslavia - English irregular/1992 182pp ISBN No. 1853341703

Volume two of this series has sections on the pharmaceutical and health administration service in Romania, Bulgaria, the C.I.S. and Yugoslavia.

29013 ROMANIA: A COUNTRY STUDY

The Division for Sale by the Superintendant for Documents

Bachman, R.D./Romania - English irregular/1991 356pp price on application

Part of a regional handbook series which gives general socio-economic data. Produced by the Federal Research Division of the Library of Congress.

29014 ROMANIA: HUMAN RESOURCES AND THE TRANSITION TO A MARKET ECONOMY

World Bank Publications

Romania - English/1992 242pp

Covers social conditions and social policy with regard to the development of post-communist Romania.

29015 THE ROMANIAN ECONOMIC REFORM PROGRAM

International Monetary Fund

Demekas Dimitri, G./Romania - English/1991 36pp

An IMF occasional paper (No. 89) describing the transition to a market economy with some economic background data.

29016 SITUATION, TENDANCES ET PERSPECTIVES DE L'AGRICULTURE EN ROUMANIE

Office for Official Publications of the European Communities

Romania - French/1991 188pp ECU20.00

Looks at the agricultural situation in Romania and reviews its future economic prospects.

29017 STATISTICAL YEARBOOK OF THE SOCIALIST REPUBLIC OF ROMANIA

Central Statistical Board

Romania - Romanian annual

Broad coverage of main socio-economic statistical series.

SPAIN

Country profile:	Spain
Official name:	Kingdom of Spain
Area:	504,750 sq.km.
	Bordered in the north by France and the Bay of Biscay, in the south and east by the Mediterranean Sea, and in the west by Portugal and the Atlantic Ocean.
Population:	38.99m (1991)
Capital:	Madrid
Language(s):	Castilian Spanish, Catalan, Galician, Basque
Currency:	Peseta 100 centimos = 1 Spanish peseta
National Bank:	Bank of Spain
Government and Political structure:	Parliamentary monarchy with multi-party democracy elected by proportional representation. There is a bicameral Cortes (parliament) with 350-400 deputies in the Congress of Deputies and 208 senators in the Senate.

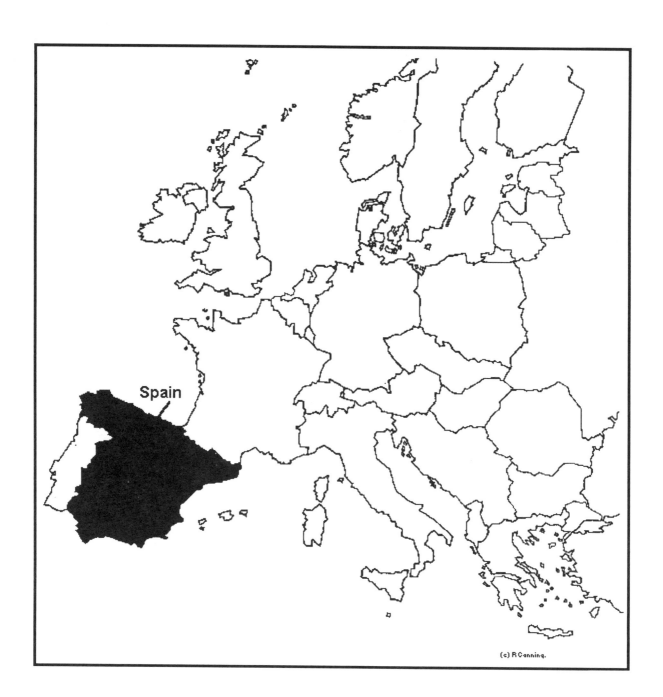

Spain

(c) R.Canning.

30001 ANUARIO ESTADÍSTICO DE ESPAÑA - AVANCE

Instituto Nacional de Estadística

Spain - Spanish annual/340pp PTA2,300

Pocket version of the statistical yearbook.

30002 ANUARIO ESTADÍSTICO DE ESPAÑA: EDICÍON EXTENSO

Instituto Nacional de Estadística

Spain - Spanish annual/930pp PTA6,200

Statistical yearbook covering demography and social, economic aspects of Spain.

30003 ANUARIO HORTOFRUTICOLA ESPAÑOL

Sucro SA

Spain/Europe - Spanish annual/ 1967 430pp PTA7,632

Annual figures in the Spanish agricultural industry and international trade.

30004 ANUARIO HORTOFRUTICOLA MERCADO INTERIOR

Sucro SA

Spain/Portugal - Spanish annual/ 1988 466pp PTA5,512

Annual statistics on the Spanish agricultural industry.

30005 AVANCE RESULTADOS

Teneo SA

Spain - Spanish annual/23pp

Report and statistics on Spanish industry.

30006 BALANCES Y CUENTAS: SEGUROS PRIVADOS

Ministerio de Economia y Hacienda: Direccíon General de Seguros

Spain - Spanish annual/1,060pp

Balances and accounts for the private insurance sector in Spain.

30007 BOLETÍN DE INFORMACIÓN TRIMESTRAL SEGUROS PRIVADOS

Ministerio de Economia y Hacienda: Direccíon General de Seguros

Spain - Spanish quarterly/1987 350pp PTA2,600 P.A. ISSN No. 1132-0052

News from the insurance sector in Spain.

30008 BOLETÍN ECONÓMICO

Banco de España

Spain - Spanish monthly/ PTA16,000 ISSN No. 0210-3737

Monthly bulletin that focuses on banking and related economic topics.

30009 BOLETÍN ESTADÍSTICO

Banco de España

Spain - Spanish monthly/ PTA36,000 ISSN No. 0005-4798

Statistical bulletin covering banking.

30010 BOLETÍN INFORMATIVO - ACCIDENTES

Ministerio del Interior: Dirección General de Trafico

Spain - Spanish annual/1960 136pp PTA1,600

Statistical series on accidents and accident victims.

30011 BOLETÍN INFORMATIVO - ANUARIO ESTADÍSTICO GENERAL

Ministerio del Interior: Dirección General de Trafico

Spain - Spanish annual/1960 222pp PTA1,600

Statistics on vehicles, driving licences etc.

30012 BOLETÍN MENSUAL DE ESTADÍSTICA

Instituto Nacional de Estadística

Spain - Spanish monthly/300pp PTA18,500 P.A.

Monthly statistical bulletin.

30013 BOLETÍN TRIMESTRAL DE COYUNTURA

Instituto Nacional de Estadística

Spain - Spanish quarterly/1980 215pp PTA5,200 P.A.

Economic commentary with tables and graphs.

30014 BUSINESS SPAIN

Business Spain

Spain - English monthly/66pp £345.00 P.A.

Monthly report on the economy, business news and consumer market trends in Spain.

30015 CENSO AGRARIO

Instituto Nacional de Estadística

Spain - Spanish irregular/7 Vols various prices

Statistical results of the agricultural census.

30016 CENSO DE EDIFICIOS - AVANCE DE RESULTADOS

Instituto Nacional de Estadística

Spain - Spanish annual/4 Vols PTA1,600

Data on the total number of buildings in each Spanish municipality classified by use.

30017 CENSO DE LOCALES - 1990

Instituto Nacional de Estadística

Spain - Spanish irregular/4 Vols

Covers location of industry.

30018 CENSO DE POBLACÍON - (1) AVANCE DE RESULTADOS (2) PROJECTO

Instituto Nacional de Estadística

Spain - Spanish every 10 years/ PTA2,100 (Avance) PTA 1,000 (Projecto)

Population census.

30019 CENSO DE VIVIENDAS 1991

Instituto Nacional de Estadística

Spain - Spanish irregular/4 Vols

Census of Spanish housing.

30020 CONTABILIDAD NACIONAL DE ESPAÑA

Instituto Nacional de Estadística

Spain - Spanish annual/1964 PTA3,700

Statistics on Spain's national accounts.

30021 CONTABILIDAD REGIONAL DE ESPAÑA

Instituto Nacional de Estadística

Spain - Spanish annual/390pp

Regional economic statistics.

30022 COUNTRY BOOKS: SPAIN

United Kingdom Iron and Steel Statistics Bureau

Spain - English annual/£75.00 ISSN No 0952-4943

Annual summary tables showing production, materials consumed, apparent consumption, imports and exports of 130 products by quality and market.

30023 COUNTRY FORECASTS: SPAIN

Business International Ltd

Spain - English quarterly/£360.00 P.A.

Contains an executive summary, sections on political and economic outlooks and the business environment, a fact sheet, and key economic indicators. Includes a quarterly Global Outlook.

30024 COUNTRY PROFILE: SPAIN

Department of Trade and Industry Export Publications

Spain - English irregular/1991 66pp £10.00

This publication provides a general description of the country in terms of its geography, economy and population, and gives information about trading in the country with details like export conditions and investment covered.

30025 COUNTRY REPORT: SPAIN

Business International Ltd

Spain - English quarterly/25-40pp £160.00 P.A. ISSN No. 0269-4263

Contains an executive summary, sections on politics and economic structure, outlook for the next 12-18 months, statistical appendices, and a review of key political developments. Includes an annual Country Profile.

30026 COUNTRY RISK SERVICE: SPAIN

Business International Ltd

Spain - English quarterly/minimum subscription £1,855.00 P.A. for seven countries

Contains credit risk ratings, risk appraisals, cross-country databases, data on diskette, and two year projections.

30027 DOING BUSINESS IN SPAIN

Ernst & Young International

Spain - English irregular/1991 110-130pp free to clients

A business guide which includes an executive summary, the business environment, foreign investment, company structure, labour, taxation, financial reporting and auditing, and general country data.

30028 DOING BUSINESS IN SPAIN

Price Waterhouse

Spain - English irregular/1990 330-340pp free to clients

A business guide which includes sections on the investment climate, doing business, accounting and auditing, taxation and general country data. Updated by supplements.

30029 EFECTOS DE COMERCIO DEVUELTOS IMPAGADOS

Instituto Nacional de Estadística

Spain - Spanish annual/168pp

Consumer credit in Spain.

30030 EL COMERCIO HISPANO-BRITANNICO

Spanish Chamber of Commerce in Great Britain

Nicolas Belmonte/Spain/UK - English/Spanish bi-monthly/1886 30pp free to members

Covers business, law, statistics and Chamber activities.

30031 EL ECONOMISTA

Economista

Spain - Spanish weekly/1886 PTA1,590 P.A. ISSN No. 0013-0656

Economic journal covering general economic issues.

30032 ENCUESTA CONTINUA DE PRESUPUESTOS FAMILIARES

Instituto Nacional de Estadística

Spain - Spanish annual/1985 PTA3,000

Statistics on domestic consumption.

30033 ENCUESTA DE POBLACIÓN ACTIVA EPA: PRINCIPALES RESULTADOS

Instituto Nacional de Estadística

Spain - Spanish irregular/1983 PTA1,200

Employment statistics on national and provincial level classified by sex and economic sector.

30034 ENCUESTA DE POBLACIÓN EPA: RESULTADOS DETALLADOS

Instituto Nacional de Estadística

Spain - Spanish quarterly/1987 PTA2,100 P.A.

Detailed statistics at national level on population.

30035 ENCUESTA DE POBLACIÓN EPA: TABLAS ANUALES

Instituto Nacional de Estadística

Spain - Spanish annual/1987 PTA1,000

Annual statistics at national level on population.

30036 ENCUESTA DE SALARIOS EN LA INDUSTRIA Y LOS SERVICIOS

Instituto Nacional de Estadística

Spain - Spanish quarterly/1989 85pp PTA1,200 P.A.

Statistics on wages in Spain.

30037 ENCUESTA INDUSTRIAL

Instituto Nacional de Estadística

Spain - Spanish annual/1978 PTA2,600

Industrial production by sector, area of industrial activity and region. Each issue covers four years.

30038 ESPAÑA EN CIFRAS

Instituto Nacional de Estadística

Spain - Spanish irregular/PTA600

Brief statistical summary of Spanish economic, social and demographic situation.

30039 ESTADÍSTICA DE HIPOTECAS

Instituto Nacional de Estadística

Spain - Spanish annual/564pp

Data on the number and type of mortgages.

30040 ESTADÍSTICA DE SOCIEDADES MERCANTILES

Instituto Nacional de Estadística

Spain - Spanish annual/1965 140pp PTA1,000

Statistical coverage of Spanish trading companies.

30041 ESTADÍSTICA DE VENTAS A PLAZOS

Instituto Nacional de Estadística

Spain - Spanish annual/212pp

Credit and consumer credit statistics.

30042 ESTADÍSTICA ESPAÑOLA

Instituto Nacional de Estadística

Spain - Spanish every four months/ PTA2,300 P.A.

Each issue contains 4-7 articles on theoretical and applied statistics.

30043 FINANCING FOREIGN OPERATIONS: SPAIN

Business International Ltd

Spain - English annual/£695.00 for complete European series

Business overview, currency outlook, exchange regulations, monetary system, short/medium/long-term and equity financing techniques, capital incentives, cash management, investment trade finance and insurance.

30044 GLOBAL FORECASTING SERVICE: SPAIN

Business International Ltd

Spain - English quarterly/11pp

Economic forecasts.

30045 HINTS TO EXPORTERS VISITING SPAIN

Department of Trade and Industry Export Publications

Spain - English irregular/1991 95pp £5.00

This publication provides detailed practical information that is useful for those planning a business visit. Topics include currency information, economic factors, import and exchange control regulations, methods of doing business and general information.

30046 IN DEPTH COUNTRY APPRAISALS ACROSS THE EC - CONSTRUCTION AND BUILDING SERVICES - SPAIN

Building Services Research and Information Association

Spain - English 1991/£125.00

Part of an 11 volume survey covering the construction industry in EC countries.

30047 INDICADORES DE COYUNTURA
Instituto Nacional de Estadística
Spain - Spanish monthly/1967-1991 (ceased publication) PTA525
Trends in the Spanish economy during previous 12 months.

30048 INDICE DE PRECIOS DE CONSUMO - BOLETÍN TRIMESTRAL
Instituto Nacional de Estadística
Spain - Spanish quarterly/30pp PTA1,000 P.A.
Bulletin on consumer price index in Spain.

30049 INDICES DE PRECIOS INDUSTRIALES
Instituto Nacional de Estadística
Spain - Spanish monthly/PTA1,400 P.A.
Industrial price indices.

30050 INFORMACIÓN AGROPECUARIA
Consorcio para el Fomento de la Riqueza Ganadera
José-Eugenio Naranje Chicharro/ Spain - Spanish weekly/1975 8pp free
Weekly prices for cereals and agricultural products - national, regional and local.

30051 INFORMACIÓN COMERCIAL ESPAÑOLA: BOLETÍN ECONÓMICO
Ministerio de Economia y Hacienda: Secretaría General de Comercio
Spain - Spanish weekly/1940 PTA12,000 P.A.
Bulletin covering economics, trade and the financial sector.

30052 INFORME ANUAL
Banco de España
Spain - English/Spanish annual/ PTA2,500 ISSN No. 0067-3315
Annual report covering banking activities.

30053 INVESTING, LICENSING AND TRADING CONDITIONS ABROAD: SPAIN
Business International Ltd
Spain - English annual/£825.00 for complete European series
Loose-leaf format covering the state role in industry, competition rules, price controls, remittability of funds, taxation, incentives and capital sources, labour and foreign trade.

30054 LA INDUSTRIA SIDERÚRGICA ESPAÑOLA EN EL AÑO 19-
Union de Empresas Siderúrgicas (UNESID)
Spain/EEC - Spanish annual/1968 104pp PTA6,000
Covers the Spanish steel industry.

30055 MEMORIA ESTADÍSTICA SEGUROS PRIVADOS
Ministerio de Economia y Hacienda: Direccíon General de Seguros
Spain - Spanish annual/913pp PTA9,100
Statistical coverage of the Spanish insurance market.

30056 MIGRACIÓNES
Instituto Nacional de Estadística
Spain - Spanish annual/1985 PTA1,600
Statistics on immigration, emigration and internal movement.

30057 MOVIMIENTO DE VIAJEROS EN ESTABLECIMIENTOS TURÍSTICOS
Instituto Nacional de Estadística
Spain - Spanish annual/240pp PTA425
Statistics both regional and national on tourism - covers both the nationality of the tourist and the category of the hotel they occupy.

30058 MOVIMIENTO NATURAL DE LA POBLACIÓN
Instituto Nacional de Estadística
Spain - Spanish irregular/1975 various prices
Statistics on population movement.

30059 NÚMEROS ÍNDICES DE LA PRODUCCIÓN INDUSTRIAL
Instituto Nacional de Estadística
Spain - Spanish monthly/PTA1,400 P.A.
Indices of Industrial production in Spain.

30060 OECD ECONOMIC SURVEYS: SPAIN
Organisation for Economic Co-operation and Development (OECD)
Spain - English irregular/£7.50 ISSN No. 0376-6438
Survey on economic developments.

30061 QUARTERLY COUNTRY TRADE REPORTS: SPAIN
United Kingdom Iron and Steel Statistics Bureau
Spain - English quarterly/£200.00 P.A. for four countries ISSN No. 0958-4943

Cumulative trade statistics of imports and exports of steel products.

30062 RAPPORT ECONÓMICO
Estructura Grupo de Estudios Económicos SA
Spain - Spanish weekly/14pp
Business and economic news.

30063 SPAIN
United Kingdom Iron and Steel Statistics Bureau
Spain - English annual/£80.00
Report on the steel industry showing production, consumption and details of imports and exports.

30064 SPAIN: A DIRECTORY AND SOURCEBOOK
Euromonitor Plc
Spain - English 1992/450pp £195.00 ISBN No. 0863384374
Contains an expert commentary on the major issues affecting Spain with a directory of the leading companies and a statistical guide to economical parameters and consumer markets.

30065 SPAIN: COUNTRY PROFILE
Business International Ltd
Spain - English annual/36pp £50.00 ISSN No. 0269-5995
Annual survey of political and economic background. Included as part of the "Country Report" subscription.

30066 THE SPANISH ECONOMY IN FIGURES
Cámara de Comercio e Industria de Madrid
Spain - English/Spanish annual/ 11pp free
Statistical summary of the Spanish economy.

30067 TRANSPORTE DE VIAJEROS INTERIOR REGULAR
Instituto Nacional de Estadística
Spain - Spanish irregular/PTA1,400
Transport statistics.

30068 VALENCIA FRUITS
Sucro SA
Fidel Pascual Tocles/Spain/Europe - Spanish weekly/1962 60pp PTA250 per issue

SWEDEN

Country profile:	Sweden
Official name:	Kingdom of Sweden
Area:	449,964 sq.km.
	Bordered in the north and north-west by Norway, in the south-east by Baltic Sea and the Kattegat, and in the east by Finland and the Gulf of Bothnia.
Population:	8.59m (1991)
Capital:	Stockholm
Language(s):	Swedish
Currency:	Krona 100 ore = 1 Swedish krona
National bank:	Riksbank
Government and Political structure:	Parliamentary multi-party democracy. The Riksdag (parliament) has 349 members elected by proportional representation.

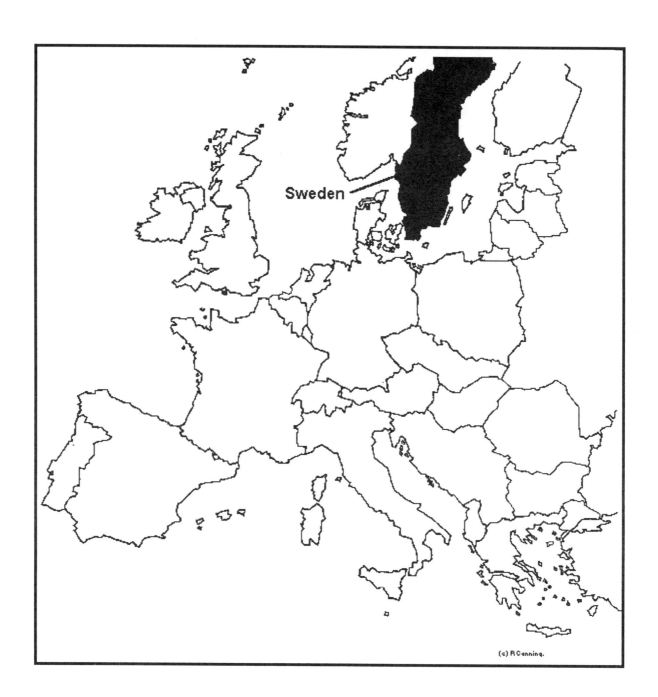

Sweden

(c) R.Canning.

31001 ACCOMMODATION STATISTICS
Statistics Sweden

Sweden - English/Swedish annual/ SKr220

Hotels and guest houses are surveyed, broken down by such factors as size and type of accommodation.

31002 AGRICULTURAL ECONOMICS
Statistics Sweden

Sweden - English/Swedish annual/ SKr400

Covers the agricultural sector from economic view point; volumes and prices for example.

31003 AGRICULTURAL ENTERPRISES AND LABOUR
Statistics Sweden

Sweden - English/Swedish annual/ SKr400

Covers numbers of enterprises and level of employment in the agricultural sector.

31004 ALKOHOLSTATISTIK
National Board of Health and Welfare

Sweden - English/Swedish annual/ 1968 39pp free

Production, trade, prices and taxation of alcohol, and alcohol-related offences and mortality.

31005 ALL REPORTS ON CULTURE AND MEDIA
Statistics Sweden

Sweden - English/Swedish annual/ SKr450

Usage of public libraries, visitors to museums.

31006 ANALYSIS SWEDEN
Swedbank

Sweden - English/Swedish monthly/Oct 1992 free

Analysis of Swedish economy and its financial markets.

31007 ANNUAL REPORT SVENSKA HANDELSBANKEN
Svenska Handelbanken

Sweden - English/Swedish annual/ 70pp free

Annual report of this major Swedish lending bank.

31008 ANNUAL REPORT SVERIGES RIKSBANK
Sveriges Riksbank

Sweden - English/Swedish annual/ 50-60pp ISSN No. 0347-5042

Sveriges Riksbank annual report detailing its activities, balances etc.

31009 ANNUAL REPORT SWEDBANK
Swedbank

Sweden - English/Swedish annual/ free

Swedbank's financial statement and details of its balances and activities in the past year.

31010 AQUACULTURE IN SWEDEN
Statistics Sweden

Sweden - English/Swedish annual/ SKr60

Coverage of the agricultural industry with details like levels and type of activity.

31011 ARBETSRAPPORT
Sveriges Riksbank

Sweden - English/Swedish irregular/November 1990 free ISSN No. 1101-6604

Central bank economic issues.

31012 ARSREDOVISNING
Postverket

Sweden - English/Swedish annual/ 50-60pp free

Sweden post annual report.

31013 ASSESSMENT OF REAL ESTATE
Statistics Sweden

Sweden - English/Swedish annual/ SKr140

Numbers and assessed values.

31014 BANKRUPTCIES
Statistics Sweden

Sweden - English/Swedish annual/ SKr220

Statistics on level and type of bankruptcy.

31015 BANKS
Financial Supervisory Authority

Sweden - English/Swedish annual/ various pagination

Annual report of the Swedish regulatory authority covering the banking sector.

31016 BANKS
Statistics Sweden

Sweden - English/Swedish annual/ SKr160

Statistical coverage of the banking sector in Sweden.

31017 BULLETIN OF THE INTERNATIONAL STATISTICAL INSTITUTE
International Statistical Institute

International - English bi-annually 4 Vols

Proceedings of the biannual session of the institute.

31018 CAUSES OF DEATH
Statistics Sweden

Sweden - English/Swedish annual/ SKr210

Mortality statistics by cause of death.

31019 COMMERCIAL BANKS
Statistics Sweden

Sweden - English/Swedish annual/ monthly/SKr580/SKr60 P.A.

Statistical coverage of the commercial banking sector.

31020 CONSUMER BUYING EXPECTATIONS
Statistics Sweden

Sweden - English/Swedish quarterly/SKr110 P.A.

Shows consumers expectations of future purchases and consumer confidence.

31021 CONSUMER PRICE INDEX
Statistics Sweden

Sweden - English/Swedish monthly/SKr600 P.A.

Details current consumer price index.

31022 CONSUMER PRICE INDEX NUMBERS 1914 ONWARDS
Statistics Sweden

Sweden - English/Swedish annual/ SKr60

Historical levels of consumer price indices.

31023 COUNTRY BOOKS: SWEDEN
United Kingdom Iron and Steel Statistics Bureau

Sweden - English annual/£75.00 ISSN No. 0952-6048

Annual summary tables showing production, materials consumed, apparent consumption, imports and exports of 130 products by quality and market.

31024 COUNTRY FORECASTS: SWEDEN
Business International Ltd

Sweden - English quarterly/£360.00 P.A.

Contains an executive summary, sections on political and economic outlooks and the business environment, a fact sheet, and key economic indicators. Includes a quarterly Global Outlook.

31025 COUNTRY PROFILE: SWEDEN
Department of Trade and Industry Export Publications

Sweden - English irregular/1988 62pp £10.00

This publication provides a general description of the country in terms of its geography, economy and population, and gives information about trading in the country with details like export conditions and investment covered.

31026 COUNTRY REPORT: SWEDEN
Business International Ltd
Sweden - English quarterly/25-40pp £160.00 P.A. ISSN No. 0269-6142

Contains an executive summary, sections on politics and economic structure, outlook for the next 12-18 months, statistical appendices, and a review of key political developments. Includes an annual Country Profile.

31027 CREDIT MARKET STATISTICS FOR THE MUNICIPALITIES AND COUNTY COUNCILS
Statistics Sweden
Sweden - English/Swedish annual/ SKr60

31028 CROP HUSBANDRY
Statistics Sweden
Sweden - English/Swedish annual/ SKr725

Statistical coverage of crop husbandry on a national basis.

31029 CULTURAL STATISTICS
Statistics Sweden
Sweden - English/Swedish annual/ SKr450

Economy, activities and cultural habits.

31030 DOING BUSINESS IN SWEDEN
Price Waterhouse
Sweden - English irregular/1991 330-340pp free to clients

A business guide which includes sections on the investment climate, doing business, accounting and auditing, taxation and general country data. Updated by supplements.

31031 EDUCATION IN SWEDEN
Statistics Sweden
Sweden - English/Swedish annual/ SKr20

Includes details of students and institutions plus statistics on the various educational sectors.

31032 ELECTRIC ENERGY SUPPLY AND DISTRICT HEATING
Statistics Sweden
Sweden - English/Swedish annual/ SKr110

31033 ELECTRONICS AND COMPUTER INDUSTRIES IN SWEDEN
Statistics Sweden
Sweden - English/Swedish annual/ 1992 SKr170

A survey detailing the level and type of activity in the above sector.

31034 EMPLOYMENT IN THE PRIVATE SECTOR
Statistics Sweden
Sweden - English/Swedish quarterly/SKr220 P.A.

Includes types and numbers of employees and employee levels.

31035 ENERGY STATISTICS FOR DWELLINGS AND PREMISES
Statistics Sweden
Sweden - English/Swedish annual/ SKr220

Consumption levels by type of energy and type of consumer.

31036 ENERGY SUPPLY
Statistics Sweden
Sweden - English/Swedish quarterly/annual/SKr250 P.A.

Statistics on the supply of energy by origin and destination.

31037 ENTERPRISES FINANCIAL ACCOUNTS
Statistics Sweden
Sweden - English/Swedish annual/ SKr280

Gives details such as profit/loss levels for private companies.

31038 ENVIRONMENTAL PROTECTION COSTS IN SWEDISH AGRICULTURE
Statistics Sweden
Sweden - English/Swedish annual/ SKr110

Surveys the economic impact of environmental protection measures in the agricultural sector.

31039 THE EXPORT/IMPORT YEAR
Statistics Sweden
Sweden - English/Swedish monthly/ SKr60 P.A.

Comments and analysis on the development of trade.

31040 FACT SHEETS ON SWEDEN
The Swedish Institute
Sweden - English/2-6pp free ISSN No. 1101-6124

Concentrated, detailed information on special topics including: Tourism in Sweden. The Swedish Economy. Taxes in Sweden. Energy and Energy Policy in Sweden. The Cooperative Movement in Sweden. Swedish Trade Policy. Sweden's Foreign Trade. The Distributive

Trade in Sweden. Swedish Consumer Policy. The Swedish Engineering Industry. The Swedish Motor Industry. Electrical Engineering and Electronics. The Swedish Steel Industry. Sweden's Chemical Industry. Forestry and the Forest Industry. The Swedish Mining Industry. The Swedish Construction Industry. Swedish Industry. Environment Protection. Labour Relations in Sweden. Mass Media in Sweden.

31041 FACTS ABOUT THE SWEDISH ECONOMY
Svenska Arbetsgivareforeningen
Sweden - English annual/1993 48pp SKr140

Contains 35 diagrams and tables together with brief explanatory comments.

31042 FAMILY FOOD EXPENDITURE SURVEY
Statistics Sweden
Sweden - English/Swedish annual/ SKr140

Gives details of families expenditure on food by type of food.

31043 FINANCIAL ACCOUNTS
Statistics Sweden
Sweden - English/Swedish quarterly/SKr220

National financial accounts given on a quarterly basis.

31044 FINANCIAL ACCOUNTS
Statistics Sweden
Sweden - English/Swedish annual/ SKr160

Annual cumulation of national accounts.

31045 FINANCIAL ENTERPRISES
Statistics Sweden
Sweden - English/Swedish annual/ SKr160

Statistical coverage of companies in the financial sector.

31046 THE FINANCIAL MARKET
Statistics Sweden
Sweden - English/Swedish annual/ SKr470

Covers money markets with factors such as levels of borrowing detailed.

31047 FINANCING FOREIGN OPERATIONS: SWEDEN
Business International Ltd
Sweden - English annual/£695.00 for complete European series

Business overview, currency outlook, exchange regulations, monetary system, short/medium/ long-term and equity financing techniques, capital incentives, cash management, investment trade finance and insurance.

31048 FISHING STATISTICS
Statistics Sweden

Sweden - English/Swedish monthly and annual/SKr800 P.A.

Statistical coverage of the fishing industry showing volumes, numbers participating etc.

31049 FOREIGN OWNED ENTERPRISES
Statistics Sweden

Sweden - English/Swedish annual/ SKr60

Statistics on the level and type of non-Swedish enterprises.

31050 FOREIGN TRADE BY BRANCH OF TRADE
Statistics Sweden

Sweden - English/Swedish annual/ SKr60

International trade is covered by industrial sector or type of activity.

31051 FOREIGN TRADE: IMPORT/ EXPORT
Statistics Sweden

Sweden - English/Swedish annual/ SKr280 ISSN No. 0281-0050

Distribution by country/commodity according to SITC.

31052 FOREIGN TRADE/ EXPORTS
Statistics Sweden

Sweden - English/Swedish quarterly/SKr1000 P.A.

Quarterly export figures by type of exports.

31053 FOREIGN TRADE/ IMPORTS
Statistics Sweden

Sweden - English/Swedish quarterly/SKr1000 P.A.

Quarterly import figures by type of import.

31054 FUELS
Statistics Sweden

Sweden - English/Swedish annual/ quarterly

Deliveries and consumption of fuels and lubricants.

31055 GLOBAL FORECASTING SERVICE
Business International Ltd

Sweden - English quarterly/£350.00 P.A.

Provides a range of medium-term forecasts covering main political, economic and business trends. A quarterly Global Outlook is included.

31056 GROSS DOMESTIC PRODUCT
Statistics Sweden

Sweden - English/Swedish quarterly/SKr600 P.A.

Gross domestic is detailed by its constituent parts.

31057 HEALTH IN FIGURES
Statistics Sweden

Sweden - English/Swedish annual/ SKr240

General coverage of health in Sweden.

31058 HINTS TO EXPORTERS VISITING SWEDEN
Department of Trade and Industry Export Publications

Sweden - English irregular/1991 73pp £5.00

This publication provides detailed practical information that is useful for those planning a business visit. Topics include currency information, economic factors, import and exchange control regulations, methods of doing business and general information.

31059 HOUSING CONSTRUCTION
Statistics Sweden

Sweden - English/Swedish quarterly/SKr220 P.A.

New construction and restoration of housing stock.

31060 HOW MUCH DO LOCAL PUBLIC SERVICES COST IN SWEDEN
Statistics Sweden

Sweden - English/Swedish annual/ SKr300

Breaks down individual services by cost to the economy and citizens.

31061 INCOME AND WEALTH DISTRIBUTION
Statistics Sweden

Sweden - English/Swedish annual/ SKr240

Households, occupation and region are some of the criteria used in this analysis.

31062 INDUSTRIAL STATISTICS - CONSUMPTION OF PURCHASED ENERGY
Statistics Sweden

Sweden - English/Swedish annual/ SKr140

Industrial energy consumption by type of industrial unit is among the statistics included.

31063 INDUSTRIAL STATISTICS - REGIONAL STATISTICS
Statistics Sweden

Sweden - English/Swedish annual/ SKr160

Swedish industry covered on a regional basis.

31064 INFORMATION TECHNOLOGY IN SWEDEN
Statistics Sweden

Sweden - English/Swedish annual/ 1992 SKr290

A broad survey giving details of the operation and penetration of information technology in Sweden.

31065 INSURANCE COMPANIES
Financial Supervisory Authority

Sweden - English/Swedish annual/ 1912 250pp SKr225

Insurance companies reported on by the national financial regulatory authority.

31066 THE INSURANCE FIRM - A BIBLIOGRAPHY
Svenska Forsakringsforeningen

Sweden - English Oct 1992/80pp SKr140 ISBN No. 919710261X

General insurance issues.

31067 INSURANCE SCHEMES ON THE SWEDISH LABOUR MARKET: STATUTORY SCHEMES: SCHEMES UNDER COLLECTIVE AGREEMENTS: INSURANCE SCHEMES
Svenska Arbetsgivareforeningen

Sweden - English irregular/15pp free

Report by the Swedish Employers' Confederation.

31068 INVESTING, LICENSING AND TRADING CONDITIONS ABROAD: SWEDEN
Business International Ltd

Sweden - English annual/£825.00 for complete European series

Loose-leaf format covering the state role in industry, competition rules, price controls, remittability of funds, taxation, incentives and capital sources, labour and foreign trade.

31069 INVESTMENTS
Statistics Sweden

Sweden - English/Swedish annual/ SKr140

A survey of investment levels in Sweden.

31070 JOURNAL OF OFFICIAL STATISTICS
Statistics Sweden

Sweden - English/Swedish four issues annually/SKr300 P.A.

The official journal of the Swedish national statistical body.

31071 THE LABOUR FORCE SURVEY
Statistics Sweden

Sweden - English/Swedish annual/ quarterly/monthly/SKr470 P.A.

Includes age, location, distribution, sector and size.

31072 THE LABOUR MARKET IN FIGURES
Statistics Sweden

Sweden - English/Swedish annual/ SKr150

Employment statistics by criteria such as sex and type of employment.

31073 THE LABOUR MARKET TENDENCY
Statistics Sweden

Sweden - English/Swedish annual/ SKr90

Employment and unemployment trends.

31074 LABOUR RELATIONS IN 18 COUNTRIES
Svenska Arbetsgivareforeningen

Australia/Austria/Belgium/Canada/ Denmark/Finland/France/Germany/ Italy/Japan/Netherlands/Norway/ Switzerland/USSR/UK/USA/ Sweden/Spain - English annual/ 148pp SKr75

Relations between employer and employees and the structure of the labour market.

31075 LARGER MANUFACTURING ENTERPRISES
Statistics Sweden

Sweden - English/Swedish annual/ SKr60

Statistical coverage of larger manufacturing companies.

31076 LIVESTOCK
Statistics Sweden

Sweden - English/Swedish monthly and annual/SKr550

National statistical coverage of livestock - numbers, type etc.

31077 LOCAL GOVERNMENT FINANCE
Statistics Sweden

Sweden - English/Swedish annual/ SKr180

Details local government finance, including income and expenditure by local service.

31078 LOCAL TAXES
Statistics Sweden

Sweden - English/Swedish annual/ SKr60

Statistical coverage of local taxation levels and type of taxation.

31079 MANUFACTURING
Statistics Sweden

Sweden - English/Swedish annual/ SKr160

Preliminary data on manufacturing industry in Sweden.

31080 MANUFACTURING ENTERPRISES
Statistics Sweden

Sweden - English/Swedish annual/ SKr140

Manufacturing companies are described and analysed by several criteria.

31081 MANUFACTURING - PART ONE
Statistics Sweden

Sweden - English/Swedish annual/ SKr280

Grouped according to Swedish standard industrial classification of all economic activities.

31082 MANUFACTURING - PART TWO
Statistics Sweden

Sweden - English/Swedish annual/ SKr280

Production of commodities and services grouped according to the Harmonized System.

31083 MONTHLY DIGEST OF SWEDISH STATISTICS
Statistics Sweden

Sweden - English/Swedish monthly/ SKr580 P.A. ISSN No. 0039-7253

Broad coverage of the Swedish economy.

31084 MONTHLY FIRST FIGURES FOR EXPORT OF GOODS
Statistics Sweden

Sweden - English/Swedish monthly/ SKr60 P.A.

Preliminary figures for exports, issued shortly after the event.

31085 MONTHLY FIRST FIGURES FOR IMPORT OF GOODS
Statistics Sweden

Sweden - English/Swedish monthly/ SKr60 P.A.

Preliminary figures for imports issued shortly after the event.

31086 MONTHLY FIRST FIGURES ON THE BALANCE OF TRADE
Statistics Sweden

Sweden - English/Swedish monthly/ SKr60 P.A.

First reports on the balance of trade later cumulated in repeated statistical publications.

31087 MOTOR VECHICLE STATISTICS
Statistics Sweden

Sweden - English/Swedish monthly and annual/SKr780 P.A.

New registrations of all vehicles.

31088 NATIONAL ACCOUNTS
Statistics Sweden

Sweden - English/Swedish annual/ SKr350

Balance of payments, imports/ exports, production, GNP and GDP are among the economic indicators included.

31089 NATURAL ENVIRONMENT IN FIGURES
Statistics Sweden

Sweden - English/Swedish annual/ SKr210

Surveys environmental matters, such as water and wildlife, nationally.

31090 NEW FIRMS IN SWEDEN
Statistics Sweden

Sweden - English/Swedish annual/ SKr60

Statistics on the growth of private companies in all sectors.

31091 NEWSLETTER OF THE SWEDISH STATISTICAL ASSOCIATION
Swedish Statistical Association

Sweden - Swedish 4 per year

31092 OCCUPATIONAL DISEASES AND OCCUPATIONAL ACCIDENTS
Statistics Sweden

Sweden - English/Swedish annual/ SKr180

Gives levels and type of disease and accidents.

31093 OTHER HEALTH AND MEDICAL CARE STATISTICS
Statistics Sweden

Sweden - English/Swedish annual/ SKr110

Statistical coverage of health care activities not covered in the main sequence of health statistics.

31094 THE OUTCOME OF THE CENTRAL GOVERNMENT BUDGET FOR THE FISCAL YEAR
Statistics Sweden

Sweden - English/Swedish annual/ SKr60

Looks at actual effects of the national budget and its operation.

31095 PESTICIDES IN SWEDISH AGRICULTURE
Statistics Sweden
Sweden - English/Swedish annual/ SKr60
Usage levels of different pesticides.

31096 POPULATION AND HOUSING CENSUS 199–
Statistics Sweden
Sweden - English/Swedish irregular in 4 parts
Population & cohabiting, dwellings, households and families.

31097 POPULATION AND VITAL STATISTICS BY MUNICIPALITY AND PARISH
Statistics Sweden
Sweden - English/Swedish annual

31098 POPULATION BY SEX, AGE AND CITIZENSHIP
Statistics Sweden
Sweden - English/Swedish annual

31099 POSTENS STATISTIKA ARSBOK, STATISTICAL YEARBOOK
Postverket
Sweden - Swedish annual/35pp SKr30
Statistical information on postal items, staff, cash transactions, organization.

31100 PRICES OF REAL ESTATE
Statistics Sweden
Sweden - English/Swedish quarterly/SKr220 P.A.
Surveys trends in real estate prices nationally.

31101 QUARTERLY COUNTRY TRADE REPORTS: SWEDEN
United Kingdom Iron and Steel Statistics Bureau
Sweden - English quarterly/£200.00 P.A. for four countries ISSN No. 0952-6048
Cumulative trade statistics of imports and exports of steel products.

31102 QUARTERLY FOREIGN TRADE STATISTICS SITC/ COUNTRY
Statistics Sweden
Sweden - English/Swedish quarterly/SKr60 P.A.
A complement to the 'first figures' report.

31103 QUARTERLY REVIEW
Sveriges Riksbank
Sweden - English/Swedish 1979 80-90pp ISSN NO. 0319-658

Banking and general economic review.

31104 REAL ESTATE AND HOUSING ECONOMY
Statistics Sweden
Sweden - English/Swedish annual/ SKr470
The property market including the domestic property market.

31105 RESEARCH STATISTICS
Statistics Sweden
Sweden - English/Swedish irregular
R & D in Sweden 1989-91, funds for R & D across the board.

31106 SAF TIDNINGEN
Svenska Arbetsgivareforeningen, SAF (Swedish Confederation of Employers)
Sweden - Swedish weekly/1953 16-32pp SKr385 P.A. ISSN No. 0349-6740
Labour market and business economics are covered.

31107 SAF'S KALENDER
Svenska Arbetsgivareforeningen
Sweden - English/Swedish/French annual/SKr156
Swedish Employers' Federation Board, delegates, general annual meeting, administration, constitution and the organizations where SAF is represented.

31108 SAF'S, LO'S, TCO'S, OCH SAVO/SR'S UPPBYGGNAD
Svenska Arbetsgivareforeningen
Sweden - English/Swedish/French/ German annual/free
The structure and organization of: The Swedish Employers' Confederation (SAF), The Trade Union Congress (LO), The Swedish Central Organization of Salaried Employees (TCO), The Swedish Confederation of Professional Associations (SACO).

31109 SAF'S VERKSAMHETS-BERATTELSE
Svenska Arbetsgivareforeningen
Sweden - Swedish annual/free
The annual report from the Swedish Employers' Confederation covering SAF's administrative work, wage negotiations, cooperation with other organizations.

31110 SAMPLE SURVEY OF INCOME TAX RETURNS FOR INDIVIDUALS AT THE ASSESSMENT
Statistics Sweden
Sweden - English/Swedish annual/ SKr140

Gives a snapshot of individual levels and types of taxation.

31111 SCB ECONOMIC INDICATORS PART ONE: ANALYSIS AND COMMENTS
Statistics Sweden
Sweden - English/Swedish annual/ SKr700
Analyses, and comments on standard economic indicators such as inflation levels.

31112 SCB ECONOMIC INDICATORS PART TWO: TABLES
Statistics Sweden
Sweden - English/Swedish annual/ SKr1900
Gives detailed figures for standard economic indicators over the past year.

31113 SKOGSSTATISTIK ARSBOK
National Board of Forestry
Sweden - Swedish annual/SKr160
Forestry statistics.

31114 SOCIAL INSURANCE STATISTICS - FACTS
Riksforsakringsverket (National Insurance Board)
Sweden - English/Swedish annual/ 80pp ISSN No. 0282-4966
Statistical coverage of the operation and economic of the Swedish social security system.

31115 STATISTICAL YEARBOOK
Sveriges Riksbank
Sweden - English/Swedish annual/ 98pp ISSN No. 0348-7342
Banks, commercial banks, savings banks, cooperative banks, finance companies, credit market, money supply, interest rates, balance of payments, exchange markets.

31116 STATISTICAL YEARBOOK OF ADMINISTRATIVE DISTRICTS OF SWEDEN
Statistics Sweden
Sweden - English/Swedish annual/ SKr270
Broad statistical survey covering economy snd society by local government unit.

31117 STATISTICAL YEARBOOK OF SWEDEN
Statistics Sweden
Sweden - English/Swedish annual/ SKr410
A broad picture of Swedish society and economy.

31118 STATISTICS FROM INDIVIDUAL COUNTRIES, NATIONAL STATISTICS FROM SWEDEN AND OTHER COUNTRIES
Statistics Sweden

International/Sweden - English/Swedish annual/SKr140 ISSN No. 0280-7610

A comparison of basic statistics from Sweden with those of other selected countries.

31119 STATISTICS FROM INTERNATIONAL ORGANISATIONS
Statistics Sweden

International/Sweden - English/Swedish annual/SKr60 ISSN No. 0280-7629

International statistics drawn from a variety of institutions.

31120 STATISTICS OF OWNERS OF QUOTED SHARES
Statistics Sweden

Sweden - English/Swedish annual/SKr60

Shows level and type and ownership.

31121 STATISTICS ON REGIONAL EMPLOYMENT
Statistics Sweden

Sweden - English/Swedish annual/SKr60

Surveys employment on a regional basis by criteria such as type of employment and sex.

31122 STATISTICS ON WORKING HOURS AND ABSENTEEISM
Svenska Arbetsgivareforeningen, SAF (Swedish Confederation of Employers)

Sweden - English annual/12pp

Statistics based on 210,000 employees concerning absenteeism and working hours.

31123 SVENSKA FORSAKRINGSARBOK
Svenska Forsakringsforeningen

Sweden - Swedish annual/380pp SKr240 ISSN No. 1102-1330

Private and public insurance, facts and figures.

31124 SWEDEN BUSINESS INCORPORATING SWEDEN BUSINESS REPORT, THE SWEDISH STOCK MARKET
Affärsvärlden and the Swedish Association for Share Promotion

Sweden - English monthly/8pp SKr1600 ISSN No. 1101-5772

31125 SWEDEN IN FIGURES
Statistics Sweden

Sweden - Swedish/German/French annual/SKr15

Statistics on all aspects of Swedish society.

31126 SWEDEN INTERNATIONAL
Erlandsson Ltd

Sweden - English quarterly/50pp ISSN No. 0349-3326

Investment, environment, the market place, pulp + paper, stockmarket, culture.

31127 SWEDEN TODAY
Anglo-Swedish Trade Co-operation

Sweden - English quarterly/38pp

Publication for Anglo-Swedish trade cooperation.

31128 SWEDEN'S LARGEST COMPANIES
Ekonomisk Litteratur AB

Sweden - English/Swedish annual/800pp SKr1180

Financial information on 8,500 Swedish companies.

31129 THE SWEDISH BUDGET
CE Fritzes

Sweden - English annual/128pp Skr96

Facts and figures on the Swedish government's budget.

31130 THE SWEDISH ECONOMY
CE Fritzes

Sweden - English annual/160pp SKr192

Production, foreign trade, labour market, investment, public sector, household finances.

31131 SWEDISH INSURANCE COMPANIES
Statistics Sweden

Sweden - English/Swedish quarterly/SKr250 P.A.

Insurance company statistics.

31132 THE SWEDISH PUBLIC SECTOR
Statistics Sweden

Sweden - English/Swedish annual/SKr200

Gives broad picture of the public sector with such things as level and type of employment and finances covered.

31133 SWEDISH SEA FISHERIES
Statistics Sweden

Sweden - English/Swedish annual/SKr140

A survey of sea fisheries with levels of activity and catches included.

31134 SWEDISH TRADE
Federation of Swedish Commerce and Trade

Sweden - Swedish 10 per year/1923 12pp ISSN No. 0349-9464

Trade imports & wholesale, transport and data communications.

31135 TRAFFIC INJURIES
Statistics Sweden

Sweden - English/Swedish annual/SKr160

Statistics on road accidents.

31136 TRAVEL AGENCIES
Statistics Sweden

Sweden - English/Swedish annual/SKr60

Covers commercial travel agencies and their levels of business.

31137 TRENDS AND FORECASTS POPULATION, EDUCATION AND LABOUR MARKET IN SWEDEN
Statistics Sweden

Sweden - Swedish annual/SKr140

Forecasts population levels, educational levels and future employment prospects.

31138 UNDERSTODS-FORENINGAR
Financial Supervisory Authority

Sweden - Swedish annual/SKr110

Friendly societies.

31139 WAGES AND EMPLOYMENT IN THE PRIVATE SECTOR
Statistics Sweden

Sweden - English/Swedish annual/SKr210

Wage levels and employment by sector/type.

31140 WAGES AND SALARIES IN SWEDEN
Statistics Sweden

Sweden - English/Swedish annual/SKr240

National pay levels by type of employee and employment type are among criteria included. Data covers 6-7 years.

31141 WAGES AND TOTAL LABOUR COSTS FOR WORKERS
Svenska Arbetsgivareforeningen

Belgium/Canada/Denmark/Finland/France/Ireland/Italy/Japan/Netherlands/Norway/Switzerland/Sweden/UK/USA/Germany/Austria - English/Swedish annual/46pp SKr200 ISSN No. 0280-4743

International survey latest issues covers 1980-1990

31142 WAGES, SALARIES AND EMPLOYMENT IN THE PUBLIC SECTOR. PART ONE GOVERNMENT EMPLOYEES
Statistics Sweden
Sweden - English/Swedish annual/ SKr160

31143 WAGES, SALARIES AND EMPLOYMENT IN PUBLIC SECTOR. PART TWO EMPLOYEES IN MUNICIPALITIES
Statistics Sweden
Sweden - English/Swedish annual/ SKr160

31144 WOMEN AND MEN IN SWEDEN - EQUALITY OF THE SEXES
Statistics Sweden
Sweden - English/Swedish annual/ SKr20
Trends and statistics on equal opportunities - areas such as employment are covered.

31145 WORK PATTERN SURVEY
Statistics Sweden
Sweden - English/Swedish annual/ SKr60
Statistics and analysis of work patterns in industry.

31146 THE WORLDS WOMEN 1970-1990
Statistics Sweden
International/Sweden - English/ 1991 SKr150
Trends and statistics on the position of women in various countries.

31147 YEARBOOK
Swedish Civil Aviation Administration
Sweden - Swedish annual/SKr42 ISSN No. 0348-2251
Air traffic, operations, financial statements.

31148 YEARBOOK OF AGRICULTURAL STATISTICS
Statistics Sweden
Sweden - English/Swedish annual/ SKr280 ISSN No. 0082-0199
Annual cumulation of agricultural statistics.

31149 YEARBOOK OF EDUCATION STATISTICS
Statistics Sweden
Sweden - English/Swedish annual/ ISSN No. 0348-6397
National education analysed by type of institution and type of student plus standard criteria.

31150 YEARBOOK OF HOUSING AND BUILDING STATISTICS
Statistics Sweden
Sweden - English/Swedish annual/ SKr280
Housing completions and other construction (new-housing) projects.

31151 YEARLY STATISTICS ON REGIONAL EMPLOYMENT
Statistics Sweden
Sweden - English/Swedish annual/ SKr160
Employment broken down by region and type of employment.

SWITZERLAND

Country profile:	Switzerland
Official name:	Confederation of Switzerland
Area:	41,293 sq.km.
	Bordered in the north by Germany, in the south by Italy, in the west and north-west by France, and in the east by Austria.
Population:	6.75m (1991)
Capital:	Berne
Language(s):	German, French, Italian
Currency:	Franc 100 centimes = 1 Swiss franc
National bank:	Swiss National Bank
Government and Political structure:	Republic with multi-party parliamentary democracy. The parliament is bicameral with 46 members in the Council of States and 200 members in the National Council. The Bundesrat (Federal Council) has a President and seven members and serves as the chief executive authority.

Switzerland

32001 A GUIDE TO FOREIGN EXCHANGE AND THE MONEY MARKETS
Credit Suisse

Switzerland - English/French/German/1992 103pp SF15.00

Gives basic data on the operation of the foreign exchange/money markets in Switzerland.

32002 ANNUAIRE STATISTIQUE DE LA SUISSE
Office Fédérale de la Suisse

Switzerland - German/French annual/400pp

Official statistics on Swiss life.

32003 ANNUAIRE SUISSE DE L'ÉCONOMIE FORESTIÈRE ET DE L'INDUSTRIE DU BOIS
Office Fédérale de la Statistique

Switzerland - French/German annual/1985 134pp

Yearbook of the forestry and forest products industry.

32004 ANNUAL REPORT
Credit Suisse

Switzerland - English/French/German annual/110pp free

A brief survey of the Swiss economy followed by details of the Bank's and the Group's accounts.

32005 ANNUAL REPORT
Geneva Stock Exchange

Switzerland - English/French annual/60pp free

Annual accounts and statistical summary.

32006 ANNUAL REPORT
Swiss Bank Corporation

Switzerland - English/French/German annual/143pp free

Annual report of bank's activities with a review of the Swiss economy in relation to world economy.

32007 ANNUAL REPORT
Schweizerische Nationalbank

Switzerland - French/German/Abridged versions in English & Italian annual/free

32008 ANNUAL REPORT
Union Bank of Switzerland

Switzerland - English/French/German annual/free

Annual report covering the banks' activities.

32009 BASEL STOCK EXCHANGE
Basel Stock Exchange

Switzerland - English/German annual/70pp free

Includes the organizational structure of the stock exchange, the annual report, a statistical survey of the exchange and a directory of members.

32010 BI-MONTHLY BULLETIN
British-Swiss Chamber of Commerce

Switzerland - English bi-monthly

Bi-monthly bulletin of the British-Swiss Chamber of Commerce.

32011 THE BRITISH-SWISS BULLETIN
British-Swiss Chamber of Commerce (in Switzerland)

Switzerland - English bi-monthly/16pp free to members

Covers trade opportunities, commercial news on British and Swiss trade, exhibitions, new products, Chamber notices and news from members.

32012 BUDGETS DE MÉNAGE
Office Fédérale de la Statistique

Switzerland - French/German annual/68pp

Covers household budgets in Switzerland.

32013 BULLETIN: CREDIT SUISSE
Credit Suisse

Switzerland/International - English/Italian/German/French quarterly (in English + Italian) 6 issues per year (in German + French)/36pp 1895

General economic information with emphasis on financial matters.

32014 CHARGE FISCALE EN SUISSE 19.. PERSONNES PHYSIQUES PAR COMMUNES
Office Fédérale de la Statistique

Switzerland - French/German annual/56pp

Local taxation in Switzerland.

32015 COMMENTAIRE ANNUEL
Direction Générale des Douanes: Section Statistique

Switzerland - French/German annual/SF33.00

External trade by branches of the economy and country; annual commentary and analysis.

32016 COMPTAGE SUISSE DE LA CIRCULATION ROUTIÈRE 19..
Office Fédérale de la Statistique

Switzerland - French/German every 5 years/83pp

Road transport statistics.

32017 COMPTES NATIONAUX DE LA SUISSE
Office Fédérale de la Statistique

Switzerland - French/German annual/1984 67pp

Swiss national accounts.

32018 CONSTRUCTIONS EXÉCUTÉES EN 199– ET CONSTRUCTIONS PROJETÉES POUR 199– EN SUISSE
Office Fédérale de la Statistique

Switzerland - French/German annual/170pp

Annual survey of construction both private and public sector by canton.

32019 COUNTRY BOOKS: SWITZERLAND
United Kingdom Iron and Steel Statistics Bureau

Switzerland - English annual/£75.00 ISSN No. 0952-6099

Annual summary tables showing production, materials consumed, apparent consumption, imports and exports of 130 products by quality and market.

32020 COUNTRY FORECASTS: SWITZERLAND
Business International Ltd

Switzerland - English quarterly/£360.00 P.A.

Contains an executive summary, sections on political and economic outlooks and the business environment, a fact sheet, and key economic indicators. Includes a quarterly Global Outlook.

32021 COUNTRY PROFILE: SWITZERLAND
Business International Ltd

Switzerland - English annual/35pp £50.00 ISSN No. 0269-6010

Annual survey of political and economic background. Included with subscription to "Country Report".

32022 COUNTRY PROFILE: SWITZERLAND
Department of Trade and Industry Export Publications

Switzerland - English irregular/1989 83pp £10.00

This publication provides a general description of the country in terms of its geography, economy and population, and gives information about trading in the country with details like export conditions and investment covered.

32023 COUNTRY REPORT: SWITZERLAND
Business International Ltd

Switzerland - English quarterly/25-40pp £160.00 P.A. (4 issues + a Country Profile) ISSN No. 0269-6169

Analysis of political and economic trends every quarter.

32024 DOCUMENTARY CREDITS, DOCUMENTARY COLLECTIONS, BANK GUARANTEES
Credit Suisse

Switzerland/International - English/French/German/Italian irregular (next ed. ca 1994)/1982 144pp SF15.00

32025 DOING BUSINESS IN SWITZERLAND
Ernst & Young International

Switzerland - English irregular/1991 110-130pp free to clients

A business guide which includes an executive summary, the business environment, foreign investment, company structure, labour, taxation, financial reporting and auditing, and general country data.

32026 DOING BUSINESS IN SWITZERLAND
Price Waterhouse

Switzerland - English irregular/1991 330-340pp free to clients

A business guide which includes sections on the investment climate, doing business, accounting and auditing, taxation and general country data. Updated by supplements.

32027 ECONOMIC TRENDS IN SWITZERLAND
Union Bank of Switzerland

Switzerland - English/French/German/Italian annual/34pp

Survey of main economic sectors in Switzerland.

32028 EFFECTIF DES VÉHICULES À MOTEUR EN SUISSE AU 30 SEPTEMBRE 19..
Office Fédérale de la Statistique

Switzerland - French/German annual/285pp

Statistics on motor vehicles by type, weight and canton.

32029 ENQUÊTE CONJONCTURELLE
Chambre de Commerce et d'Industrie de Genève

Switzerland - French annual/17pp

Statistical results of a survey of companies in Geneva.

32030 FEDERAL, CANTONAL AND COMMUNAL TAXES: AN OUTLINE OF THE SWISS SYSTEM OF TAXATION
Administration Fédérale des Contributions: Division Statistique et Documentation

Switzerland - English/French/German annual/60pp free

32031 FIL DE LA BOURSE
Geneva Stock Exchange

Véronique Sieber/Switzerland -

French/German/1991 16pp SF10.00

Covers Swiss stock markets.

32032 FINANCES PUBLIQUES EN SUISSE 19..
Office Fédérale de la Statistique

Switzerland - French/German annual/157pp

Public finance & spending.

32033 FINANCING FOREIGN OPERATIONS: SWITZERLAND
Business International Ltd

Switzerland - English annual/£695.00 for complete European series

Business overview, currency outlook, exchange regulations, monetary system, short/medium/long-term and equity financing techniques, capital incentives, cash management, investment trade finance and insurance.

32034 FOREIGN EXCHANGE AND MONEY MARKET OPERATIONS
Swiss Bank Corporation

Switzerland - English irregular/129pp free

Describes the operations of the foreign exchange and money market; also focuses on currency and interest rate options.

32035 GENEVA STOCK EXCHANGE
Geneva Stock Exchange

Switzerland - English/French/German irregular (1986 latest)/215pp SF49.00

History of the Geneva Stock Exchange and a description of the Exchange today; includes a statistical survey of the money markets.

32036 GENÈVE ET LA SUISSE: UN MARIAGE D'AMOUR ET DE RAISON
Geneva Stock Exchange

Marian Srepezynski et al/Switzerland - French/1992 185pp SF57.00

History of the union between Geneva and Switzerland from an economic and political viewpoint.

32037 GLOBAL FORECASTING SERVICE: SWITZERLAND
Business International Ltd

Switzerland - English quarterly £350.00 P.A.

Economic forecasts for Switzerland.

32038 HINTS TO EXPORTERS VISITING SWITZERLAND
Department of Trade and Industry Export Publications

Switzerland - English irregular/1991 90pp £5.00

This publication provides detailed practical information that is useful for those planning a business visit. Topics include currency information, economic factors, import and exchange control regulations, methods of doing business and general information.

32039 IMPT FÉDÉRALE DIRECT: STATISIQUE
Administration Fédérale des Contributions: Division Statistique de Documentation

Switzerland - French/German every 2 years/60pp SF14.00

Statistics on direct federal taxation in Switzerland.

32040 INFORMATIONS STATISTIQUES
Chambre de Commerce et d'Industrie de Genève

Switzerland - French monthly/4pp

Statistics on the Swiss economy and detailed statistics on Geneva.

32041 INTERNATIONAL ECONOMIC OUTLOOK
Union Bank of Switzerland

Switzerland/International - EnglishFrench/German quarterly/26pp free

Swiss and international economic comment and outlook.

32042 INVESTING, LICENSING AND TRADING CONDITIONS ABROAD: SWITZERLAND
Business International Ltd

Switzerland - English annual/£825.00 for complete European series

Loose-leaf format covering the state role in industry, competition rules, price controls, remittability of funds, taxation, incentives and capital sources, labour and foreign trade.

32043 INVESTMENT NEWSLETTER
Union Bank of Switzerland

Switzerland/International - English/French/German bi-monthly/7pp free

Economic and investment news.

32044 JAHRESBERICHT
Schweizerischer Bauernverband: Abteilung Dokumentation

Switzerland - German/French annual/75pp free

Annual report of the Swiss Farmer's Union.

32045 L'AGRICULTURE SUISSE - FORMAT DE POCHE
Office Fédérale de la Statistique

Switzerland - French/German annual/7pp

Summary of agricultural statistics by canton.

32046 L'AGRICULTURE SUISSE: IMAGES, NOMBRES, COMMENTAIRES
Office Fédérale de la Statistique

Switzerland - English/French/German irregular/50pp

Analysis of agricultural statistics, conditions of production and quantities produced.

32047 L'INDICE DES PRIX EN GROS
Office Fédérale de la Statistique

Switzerland - French/German monthly/1988 40pp

Price indices and variations for various services and groups of merchandise.

32048 L'INDICE SUISSE DES PRIX À LA CONSOMMATION
Office Fédérale de la Statistique

Switzerland - French/German monthly/1988 50pp

Consumer prices - average cost of various merchandise.

32049 LA BALANCE TOURISTIQUE DE LA SUISSE
Office Fédéral de la Statistique

Switzerland - French/German annual/1984 12pp

Receipts and expenditure in the tourist industry between Switzerland and other countries.

32050 LA CONSTRUCTION DES LOGEMENTS EN SUISSE 19..
Office Fédérale de la Statistique

Switzerland - French/German annual/44pp

Statistics on housing.

32051 LAGEBEURTEILUNG DER BAUWIRTSCHAFT
St Galler Zentrum fÜr Zukunftsforschung

Switzerland - German annual/1973 SF300.00

Forecast for the construction industry.

32052 LE TOURISME SUISSE EN CHIFFRES
Office Fédérale de la Statistique

Switzerland - French/German annual/1988 20pp

Principal statistics on Swiss tourism.

32053 LES COMMANDES, LA PRODUCTION, LES CHIFFRES D'AFFAIRES ET LES STOCKS DANS L'INDUSTRIE ET DANS LE SECTOR PRINCIPAL DE LA CONSTRUCTION
Office Fédérale de la Statistique

Switzerland - French/German annual/43pp

Construction industry statistics in detail.

32054 LES RÉSULTATS COMPTABLES DES ENTREPRISES SUISSES
Office Fédérale de la Statistique

Switzerland - French/German annual/1983/84 94pp

Financial accounts of Swiss companies.

32055 LES TRANSPORTS PUBLIC
Office Fédérale de la Statistique

Switzerland - French/German annual/240pp

Statistics on public transport.

32056 MONTHLY REPORT
Schweizerische Nationalbank

Switzerland - French/German monthly/SF40.00 P.A. ISSN No. 0036-7729

Details the activities of the bank on a monthly basis.

32057 MÉMENTO STATISTIQUE DE LA SUISSE
Office Fédérale de la Statistique

Switzerland - English/French/German/Italian annual/24pp

Snapshot of Swiss official statistics.

32058 OECD SURVEYS: SWITZERLAND
Organisation for Economic Co-operation and Development (OECD)

Switzerland - English irregular/£7.50 ISSN No. 0376-6438

Survey of economic developments.

32059 PRICES AND EARNINGS AROUND THE GLOBE
Union Bank of Switzerland

Switzerland/International - English/French/German/Italian every 3 years/free

Comparative prices, wages and purchasing power in 48 cities worldwide.

32060 PRODUCTS AND SERVICES OF SWITZERLAND
Swiss Office for Trade Promotion

Switzerland - English/French/German/Spanish irregular/600pp

SF50.00

Publication supplying information on 15,800 export products (Capital and consumer goods, semi-manufactured products) and services offered by 12,000 export firms.

32061 PROFILE 9-
Swiss Bank Corporation

Switzerland - English/French/German annual/48pp free

Condensed versions of the corporations annual report.

32062 QUARTERLY COUNTRY TRADE REPORTS: SWITZERLAND
United Kingdom Iron and Steel Statistics Bureau

Switzerland - English quarterly/£200.00 P.A. for four countries ISSN No. 0952-6099

Cumulative trade statistics of imports and exports of steel products.

32063 QUARTERLY REPORT - MONNAIE ET CONJONCTURE
Schweizerische Nationalbank

Switzerland/International - French/German/Summaries in English & Italian/1983 SF30.00 P.A.

A selection of articles covering finance, monetary conditions and the economic situation in Switzerland and overseas.

32064 RECENSEMENT FÉDÉRALE DES ENTREPRISES
Office Fédérale de la Statistique

Switzerland - French/German irregular/1985-1988 32pp - 340pp (6 Vols)

Federal census of companies.

32065 RECENSEMENT FÉDÉRALE DES ENTREPRISES AGRICOLES
Office Fédérale de la Statistique

Switzerland - French/German irregular/1985-1987 (8 Vols) 40pp - 160pp

Federal census of agriculture including agriculture, forestry and fisheries.

32066 REPORT AND ACCOUNTS
British-Swiss Chamber of Commerce

Switzerland - English annual

Annual report and accounts of the British-Swiss Chamber of Commerce based in Switzerland.

32067 SOCIÉTÉS ANONYMES EN SUISSE
Office Fédérale de la Statistique

Switzerland - French/German

annual/1986 47pp

Number of companies by economic class, by size of company and by canton.

32068 STATISTICAL DATA ON SWITZERLAND

Office Fédérale de la Statistique

Switzerland - English/French/German annual/24pp

Brief statistical survey of Switzerland.

32069 STATISTICAL YEARBOOK OF THE SWISS BANKING INDUSTRY

Schweizerische Nationalbank

Switzerland - French/German annual SF20.00

32070 STATISTIQUE ANNUELLE

Direction Générale des Douanes: Section Statistiques

Switzerland - French/German annual/1885
3 Vols SF209.00 (3 Vols available separately)

Annual statistics of Swiss imports and exports by commodity, country and means of transport.

32071 STATISTIQUE DE L'EMPLOI ET DE LA POPULATION ACTIVE OCCUPÉE

Office Fédérale de la Statistique

Switzerland - French/German quarterly/46pp

Employment statistics by sex, nature of employment etc.

32072 STATISTIQUE MENSUELLE

Direction Générale des Douanes: Section Statistiques

Switzerland - French/German monthly/1885 440pp SF237.00 P.A.

Swiss imports and exports by quantity and value, by commodity, by country of production and destination.

32073 STATISTIQUE SUISSE DES TRANSPORTS

Office Fédérale de la Statistique

Switzerland - French/German annual/136pp

Transport statistics (public & private).

32074 STATISTIQUES ÉCONOMIQUES

Chambre de Commerce et d'Industrie de Genève

Switzerland - French annual/32pp
Economic statistics.

32075 STATISTISCHE ERHEBUNGEN UND SCHÄTZUNGEN ÜBER LANDWIRTSCHAFT UND ERNÄHRUNG

Schweizerischer Bauernverband: Abteilung Dokumentation

Switzerland - German/French annual/200pp SF44.00

Statistics on agriculture and food in Switzerland.

32076 SWISS BUSINESS

Swiss Business

Switzerland - English bi-monthly/88pp SF7.00

Data on Swiss companies and market sectors

32077 THE SWISS CAPITAL MARKET: A GUIDE

Swiss Bank Corporation

Switzerland - English/French/German irregular/1986 80pp free

Overview of the Swiss capital market for would-be investors.

32078 SWISS FOREIGN TRADE STATISTICS - ANNUAL REPORT

Direction Générale des Douanes: Section Statistiques

Switzerland - French/German annual/1885 300pp SF33.00

Foreign trade by branches of the economy and countries; annual commentary and analysis.

32079 SWISS STOCK GUIDE

Union Bank of Switzerland

Switzerland - English/French/German annual/free

Guide to the operation of Swiss stock markets.

32080 SWISS TEXTILES

Swiss Office for Trade Promotion

Switzerland - English/French/German quarterly/SF74.00 P.A. ISSN No. 0040-5248

Journal of the fabric trade especially clothing fabrics.

32081 SWITZERLAND

United Kingdom Iron and Steel Statistics Bureau

Switzerland - English 1992 annual/£80.00 ISSN No. 0952-5904

Report on the steel industry showing production, consumption and details of imports and exports.

32082 SWITZERLAND IN FIGURES

Union Bank of Switzerland

Switzerland - English/French/German/Italian/Spanish/Portuguese/Japanese/Chinese/Korean annual/1pp free

Statistical snapshot of Switzerland.

32083 SWITZERLAND YOUR PARTNER

Swiss Office for Trade Promotion (OSEC)

Switzerland - English/French/German irregular

Auxiliary supplies from Switzerland; the Swiss packaging industry; food and drink from Switzerland; automation, electronics and information technology from Switzerland.

32084 TOURISME EN SUISSE

Office Fédérale de la Statistique

Switzerland - French/German annual/1984 134pp

Detailed commentary and tables on national and international tourism for the year under consideration, also retrospective data.

32085 TOURISME EN SUISSE: SEMESTRE D'HIVER

Office Fédérale de la Statistique

Switzerland - French/German annual/1984/85 37pp

Seasonal figures for supply and demand in the hotel industry and tourist trade at national, regional and local level.

32086 TOURISME EN SUISSE: SEMESTRE D'ÉTÉ

Office Fédérale de la Statistique

Switzerland - French/German annual/1985 39pp

Seasonal figures for supply and demand in the hotel industry and tourist trade at national, regional and local level.

32087 TOURISTES SUISSES À L'ETRANGER 19..

Office Fédérale de la Statistique

Switzerland/International - French/German annual/16pp

World tourism statistics.

32088 UBS INTERNATIONAL FINANCE

Union Bank of Switzerland

William Gasser/Switzerland/International - English quarterly/16pp free

International financial news and comment; comparative economic indicators for major economies.

32089 THE WORLD BANK GROUP AND SWITZERLAND

Union Bank of Switzerland

Switzerland - English/German irregular/1987 79pp free

IBRD, IDA, IFC-organization, tasks, financing and relations of these organizations with Switzerland.

TURKEY

Country profile:	Turkey
Official name:	Republic of Turkey
Area:	779,452 sq.km.
	Bordered in the north by Bulgaria and the Black Sea, in the south by Iraq, Syria and the Mediterranean Sea, in the east by Armenia, Georgia and Iran and the west by Greece and the Aegean Sea.
Population:	56.56m (1991)
Capital:	Ankara
Language(s):	Turkish, Kurdish
Currency:	Lira 100 kurus = 1 Turkish lira
National bank:	Central Bank of Turkey
Government and Political structure:	The Grand National Assembly (parliament) has 450 elected members in a multi-party parliamentary democracy.

(c) R.Canning.   Turkey

33001 ANNUAL MANUFACTURING INDUSTRY STATISTICS

State Institute of Statistics

Turkey - English/Turkish annual/ 1974

Manufacturing industry by type of product/industry is covered.

33002 ANNUAL REPORT

Central Bank of the Republic of Turkey

Turkey - English/Turkish annual

Looks at the national banking sector and related economic indicators.

33003 ANNUAL REPORT

Turkiye Smai Kalkmma Bankasi A.S. (TSKB)

Turkey - English/Turkish annual/ free

Annual report of a leading Turkish bank with some related economic comment.

33004 BUDGETS MUNICIPAL AND PROVINCIAL SPECIAL ADMINISTRATIONS AND VILLAGES

State Institute of Statistics

Turkey - English/Turkish annual/ 1931

Covers local government finance at budgetary levels.

33005 BULLETIN OF TOURISM STATISTICS

Ministry of Tourism

Turkey - English/Turkish annual/ 1972 c85pp

Frontier arrivals/departures by months, means of transport, nationality, by means of provinces of entry, tourism receipts/ expenditures.

33006 BULLETIN OF YACHT STATISTICS

Ministry of Tourism

Turkey - English/Turkish annual/ 1987 181pp

Distribution of number of yachts by months, by their flags and l.o.a's, by their length of stay in the marina, and coverage of OECD countries.

33007 CAUSES AND EFFECTS OF AIR POLLUTION

State Institute of Statistics

Turkey - English/Turkish annual/ 1991

Environmental statistics giving levels, emissions etc.

33008 CENSUS OF INDUSTRY AND BUSINESS ESTABLISHMENTS

State Institute of Statistics

Turkey - English/Turkish irregular/ 1976

33009 CENSUS OF POPULATION: ADMINISTRATIVE DIVISION

State Institute of Statistics

Turkey - English/Turkish irregular/ 1928

Demographic statistics by local government unit.

33010 CHARTER BULLETIN

Ministry of Tourism

Turkey - Turkish annual/1988 46pp

Landing and departing charter flights in and out of Turkish airports by countries where charter companies are registered.

33011 COMPANIES, COOPERATIVES AND FIRMS STATISTICS

State Institute of Statistics

Turkey - English/Turkish annual/ 1967

Shows level of economic activity by type of enterprise.

33012 CONSTRUCTION STATISTICS

State Institute of Statistics

Turkey - English/Turkish annual/ 1969

Statistical coverage of the construction industry including completions, employees etc.

33013 COUNTRY FORECASTS: TURKEY

Business International Ltd

Turkey - English quarterly/£360.00 P.A.

Contains an executive summary, sections on political and economic outlooks and the business environment, a fact sheet, and key economic indicators. Includes a quarterly Global Outlook.

33014 COUNTRY REPORT: TURKEY

Business International Ltd

Turkey - English quarterly/25-40pp £160.00 P.A. ISSN No. 0269-5464

Contains an executive summary, sections on political and economic structure, outlook for the next 12-18 months, statistical appendices, and a review of key political developments. Includes an annual Country Profile.

33015 COUNTRY RISK SERVICE: TURKEY

Business International Ltd

Turkey - English quarterly/minimum subscription £1,855.00 P.A. for seven countries

Contains credit risk ratings, risk appraisals, cross-country databases, data on diskette, and two year projections.

33016 DOING BUSINESS IN TURKEY

Business International Ltd

Turkey - English quarterly/500pp £795.00 P.A.

Details all aspects of doing business in Turkey with broad background data.

33017 DOING BUSINESS IN TURKEY

Price Waterhouse

Turkey - English irregular/1991 330-340pp free to clients

A business guide which includes sections on the investment climate, doing business, accounting and auditing, taxation and general country data. Updated by supplements.

33018 ECONOMIC INDICATORS OF TURKEY

TÜrkiye Is Bankasi A.S.

Turkey - English/Turkish annual/ 1954 c17pp free

Economic figures which are related to the Turkish economy based on official data.

33019 ECONOMIC REPORT

TÜrkiye Is Bankasi A.S.

Turkey - English/Turkish annual/ 1954 c50pp free

Survey of current economic developments in Turkey, mainly based on officially published data or their own intelligence in principal markets.

33020 EKONOMI

Turkish Cypriot Chamber of Commerce

North Cyprus - Turkish weekly/1978 8pp free

Weekly economic comment.

33021 FIGURES IN TURKEY

State Institute of Statistics

Turkey - English/Turkish annual/ 1975

Brief statistical survey covering main indicators.

33022 FINAL ACCOUNTS MUNICIPAL AND PROVINCIAL SPECIAL ADMINISTRATIONS

State Institute of Statistics

Turkey - English/Turkish annual/ 1969

Covers final (actual) figures of local government finance.

33023 FOREIGN TRADE STATISTICS

State Institute of Statistics

Turkey - English/Turkish annual/ 1926 free ISSN No.0082-6901

Foreign trade by sector/product.

33024 FOREIGN VISITORS AND TOURISM RECEIPTS

Ministry of Tourism

Turkey - English/Turkish annual/ 1987 102pp

Characteristics and tendencies of the foreign visitors, expenditure of the foreign visitors, number of nights spent, and coverage of OECD countries.

33025 GARANTI BANK ANNUAL REPORT

Garanti Bank

Turkey - English/Turkish annual/ free

Report of the bank with some related economic data.

33026 GAS AND WATER STATISTICS

State Institute of Statistics

Turkey - English/Turkish annual

Production and consumption statistics.

33027 GROSS NATIONAL PRODUCT RESULTS QUARTERLY

State Institute of Statistics

Turkey - English/Turkish quarterly/ 1947

33028 INDUSTRIAL PRODUCTION INDEXES QUARTERLY

State Institute of Statistics

Turkey - English/Turkish quarterly/ 1984

Production data by type of industry.

33029 INVESTING, LICENSING AND TRADING CONDITIONS ABROAD: TURKEY

Business International Ltd

Turkey - English annual/£825.00 for complete European series

Loose-leaf format covering the state role in industry, competition rules, price controls, remittability of funds, taxation, incentives and capital sources, labour and foreign trade.

33030 ISVEREN/EMPLOYER

Turkish Confederation of Employer Associations

Turkey - English/Turkish monthly/ October 1962 32pp

Labour, economic, social and industrial relations.

33031 LABOUR STATISTICS AND LABOUR COSTS

Turkish Confederation of Employer Associations

Turkey - English/Turkish annual

Employment, trade unions, labour costs.

33032 MANUFACTURING INDUSTRY QUARTERLY EMPLOYMENT PAYMENTS PRODUCTION EXPECTATION

State Institute of Statistics

Turkey - English/Turkish quarterly/ 1975

33033 MINING STATISTICS

State Institute of Statistics

Turkey - English/Turkish annual/ 1967

Statistical coverage of output, type of material mined etc.

33034 MONTHLY BULLETIN OF STATISTICS

State Institute of Statistics

Turkey - English/Turkish monthly/ free ISSN No. 0041-4263

First reports of data over a broad range of economic activities.

33035 MONTHLY ECONOMIC INDICATORS

State Institute of Statistics

Turkey - English/Turkish monthly/ 1980

Main economic indicators including such things as inflation and balance of payments.

33036 MONTHLY STATISTICAL BULLETIN

Central Bank of the Republic of Turkey

Turkey - English/Turkish monthly

Monthly review of the national finances.

33037 MONTHLY SUMMARY OF FOREIGN TRADE

State Institute of Statistics

Turkey - English/Turkish monthly/ 1964

Foreign trade by type of trade/ sector.

33038 MOTOR VEHICLE STATISTICS

State Institute of Statistics

Turkey - English/Turkish annual/ 1947

Elements such as number and type of vehicle are covered.

33039 OUTSTANDING EXPORTERS

Istanbul Chamber of Commerce

Turkey - English annual/1991 187pp

Major companies involved in exporting from Turkey.

33040 PRICES RECEIVED BY FARMERS

State Institute of Statistics

Turkey - English/Turkish annual/ 1977

Agricultural product prices by product.

33041 QUARTERLY BULLETIN

Central Bank of the Republic of Turkey

Turkey - English/Turkish quarterly

Reviews the Turkish banking sector.

33042 REPORT

TÜsiad - Turkish Industrialists' and Businessmen's Association

Turkey - English/Turkish annual/ 1976 TL150.000

Annual report of the association covering activities and the business environment.

33043 RETAIL PRICE STATISTICS

State Institute of Statistics

Turkey - English/Turkish annual/ 1972

Levels and trends in the retail price index.

33044 REVIEW OF ECONOMIC CONDITIONS

TÜrkiye Is Bankasi A.S.

Argun Bassorgun/Turkey - English quarterly/1954 20-30pp free

Survey of current economic developments in Turkey, mainly based on officially published data or their own intelligence in principal markets.

33045 SETTING UP BUSINESS IN TURKEY

Department of Trade and Industry Export Publications

Turkey - English irregular/1992 134pp £20.00

Comprehensive guide to establishing a business and broad economic and business practice background.

33046 SPORT STATISTICS

State Institute of Statistics

Turkey - English/Turkish annual/ 1972

Participation levels and relative levels are covered.

33047 STATISTICAL INDICATORS

State Institute of Statistics

Turkey - English/Turkish annual/ 1991 US$35.00

Standard economic indicators including balance of payments.

33048 STATISTICAL YEARBOOK OF TURKEY

State Institute of Statistics

Turkey - English/Turkish annual/ 1982 ISSN No. 0082-691X

General statistical survey.

33049 STATISTICS OF COASTAL AND INTERNATIONAL SEA TRANSPORTATION

State Institute of Statistics

Turkey - English/Turkish annual/ 1959

33050 STATISTICS OF VESSELS 18 GROSS TONNAGES AND OVER

State Institute of Statistics

Turkey - English/Turkish annual/ 1960

33051 TISK INFORMATION

Turkish Confederation of Employer Associations

Turkey - English/Turkish quarterly/ free

Labour, economic, social and industrial relations.

33052　TOURISM

Ministry of Tourism

Turkey - English/Turkish irregular/ 1989 96pp

Brief information about Turkish tourism sector, and coverage of OECD countries.

33053　TOURISM STATISTICS

State Institute of Statistics

Turkey - English/Turkish annual/ 1959

Covers number of visitors, expenditure by visitors and factors such as origin/destination of visitors.

33054　TRADE, SERVICE, HOTEL AND RESTAURANT

State Institute of Statistics

Turkey - English/Turkish annual/ 1991

Statistical coverage of the hotel-catering service sector.

33055　TRANSPORTATION AND ROAD TRAFFIC ACCIDENTS STATISTICS SUMMARY

State Institute of Statistics

Turkey - English/Turkish annual/ 1982

Covers number and category of traffic accidents.

33056　TURKEY - A BRIEF GUIDE: GENERAL REPORT

Department of Trade and Industry Export Publications

Turkey - English irregular/1990 8pp £10.00

General introduction to Turkey and its economy.

33057　TURKEY: BRITISH CONNECTED COMPANIES IN ANKARA

Department of Trade and Industry Export Publications

Middle East Branch, Overseas Trade Division 4/1B/Turkey - English/1992

33058　TURKEY: COMPUTER HARDWARE AND SOFTWARE MARKET

Department of Trade and Industry Export Publications

Turkey - English/1991 51pp £30.00

Sector report on the computer market in Turkey with data on leading companies.

33059　TURKEY - CONSULTANCY SECTOR: SECTOR REPORT

Department of Trade and Industry Export Publications

Turkey - English irregular/1990 13pp £30.00

Brief account and background for this sector and details possible approaches to the market.

33060　TURKEY - COUNTRY PROFILE

Business International Ltd

Turkey - English annual/£50.00

Annual profile of economic and political developments.

33061　TURKEY - COUNTRY REPORT

Business International Ltd

Turkey - English quarterly/£150.00 P.A. ISSN No. 0269-5464

Gives latest political and economic indicators. Subscription includes an annual "Country profile".

33062　TURKEY DEMOGRAPHY SURVEY

State Institute of Statistics

Turkey - English/Turkish irregular/ 1992 US$40.00

Population is analysed by standard factors such as sex, age and occupation.

33063　TURKEY: HEALTH REPORT

Department of Trade and Industry Export Publications

Middle East Branch, Overseas Trade Division 4/1B/Turkey - English

Survey of the medical industry in Turkey, including lists of hospitals and medical suppliers.

33064　TURKEY - MACHINE TOOLS: SECTOR REPORT

Department of Trade and Industry Export Publications

Turkey - English irregular/1990 55pp £30.00

General introduction to the market, including imports, local manufacture competition and future trends.

33065　TURKEY - MINING EQUIPMENT: SECTOR REPORT

Department of Trade and Industry Export Publications

Turkey - English irregular/1991 25pp £30.00

General introduction to the market aimed at British exporters.

33066　TURKEY MONITOR

Business International Ltd

Turkey - English 10 issues per year/

£650.00 P.A.

Reports identify and track business issues, developments and opportunities in Turkey.

33067　TURKEY: NATURAL GAS

Department of Trade and Industry Export Publications

Turkey - English/1992 38pp £30.00

Sector report on the gas industry in Turkey.

33068　TURKEY: ON THE ROAD TO PROGRESS

Middle East Economic Digest

Jim Bodgener/Turkey - English/ 1989 ISBN No. 0946510415

33069　TURKEY - RAILWAYS: SECTOR REPORT

Department of Trade and Industry Export Publications

Turkey - English irregular/1991 23pp £30.00

General background to the existing rail network including light rail, aimed at British exporters.

33070　TURKEY - TELECOMMUNICATIONS: SECTOR REPORT

Department of Trade and Industry Export Publications

Turkey - English irregular/1990 51pp £30.00

Recent developments and current situation in this sector aimed at British exporters.

33071　TURKEY: THE CONSTRUCTION SECTOR

Department of Trade and Industry Export Publications

Turkey - English/1987

33072　TURKEY - WATER AND SEWERAGE: SECTOR REPORT

Department of Trade and Industry Export Publications

Turkey - English irregular/1992 38pp £30.00

Information on the sector and government plans for it, mainly aimed at British exporters.

33073　TÜSIAD MEMBERS COMPANY PROFILES

TÜsiad - Turkish Industrialists and businessmen's Association

Turkey - English

Covers leading Turkish companies and their performance.

33074   **WHOLESALE AND CONSUMER PRICE INDEXES: MONTHLY BULLETIN**
State Institute of Statistics
Turkey - English/Turkish monthly/ 1963
Price indices and some historical data/trends.

33075   **WHOLESALE PRICE STATISTICS**
State Institute of Statistics
Turkey - English/Turkish annual/ 1974

33076   **WOMEN IN STATISTICS**
State Institute of Statistics
Turkey - English/Turkish irregular/ 1992

UNITED KINGDOM

Country profile:	United Kingdom
Official name:	United Kingdom of Great Britain and Northern Ireland
Area:	229,880 sq.km. (GB)
Population:	57.48m (1991)
Capital:	London
Language(s):	English
Currency:	Pound sterling 100 pence = 1 pound sterling
National bank:	The Bank of England
Government and Political structure:	Multi-party parliamentary democracy with a bicameral parliament, the House of Commons having 651 members, and the House of Lords an indefinite number currently standing at 1,914.

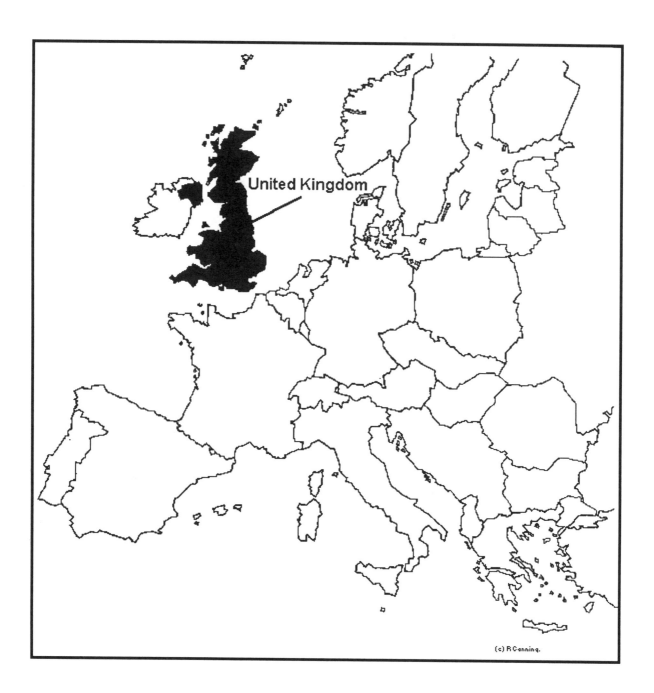

United Kingdom

(c) R Canning.

34001 1991 CENSUS AGE, SEX AND MARITAL STATUS

Office of Population Censuses and Surveys

UK - English every 10 years/price on application

National summary volume giving data for the resident population of Great Britian and its constituent parts down to county level (regional level in Scotland) by sex, age and marital status.

34002 1991 CENSUS COUNTY MONITORS

Office of Population Censuses and Surveys

UK - English every 10 years/£80.00 for complete set, £2.00 single issues

Summary statistics for each county in England and Wales. 12 tables include, population broken down by age, sex, ethnic group, tenure, household composition and economic characteristics.

34003 1991 CENSUS COUNTY REPORTS

Office of Population Censuses and Surveys

UK - English every 10 years/ individually priced for each country

A two part county report for England and Wales. Similar reports are made for each UK region and the Islands area of Scotland. Each report contains statistical tables on all topics covered in the 1991 census.

34004 50 CENTRES - A GUIDE TO OFFICE, INDUSTRIAL AND RETAIL RENTAL TRENDS IN ENGLAND, SCOTLAND AND WALES

Jones Lang Wootton

UK - English 2 per year/price on application

Shows trends in property rental prices.

34005 A BRIEF REVIEW OF THE UK DOMESTIC FURNITURE MARKET

Furniture Industry Research Association

UK - English/1992 30pp £120.00 (FIRA members £60.00)

Economic indicators, manufacturers deliveries, overseas trade, product and distribution analysis.

34006 A PICTURE OF THE BRITISH VOLUNTARY SECTOR 1979/89

Charities Aid Foundation

UK - English 1990/£5.00

General picture of the voluntary sector with supporting data.

34007 ABSTRACT OF STATISTICS FOR INDEX OF RETAIL PRICES, AVERAGE EARNINGS, SOCIAL SECURITY BENEFITS AND CONTRIBUTIONS

Lord Chancellor's Department

UK - English annual/£15.00

34008 ADVERTISING FORECAST

NTC Publications Ltd

UK - English quarterly/145pp £825.00 P.A.

Advertising expenditure forecasted by media and product.

34009 ADVERTISING STATISTICS YEARBOOK

NTC Publications Ltd

UK - English annual/145pp £42.00

Advertising expenditure for the preceding year and up to 20 years historical data for comparison purposes.

34010 AGRICULTURAL STATISTICS UNITED KINGDOM

HMSO - Her Majesty's Stationery Office

UK - English annual/£12.50 ISSN No. 0065-4590

Covers the UK agricultural industry in depth.

34011 AGRICULTURE IN SCOTLAND

Scottish Office Library

UK - English annual/£5.50

Broad review of the agricultural sector in Scotland.

34012 AGRICULTURE IN THE UNITED KINGDOM

HMSO - Her Majesty's Stationery Office

UK - English annual/£12.00

Information on the economic conditions and the prospects for the future of agriculture in the UK. Facts and figures relating to all aspects of agriculture.

34013 AN OVERVIEW OF THE UK BED MARKET

Furniture Industry Research Association

UK - English/1992 £60.00 (FIRA members £30.00)

34014 AN OVERVIEW OF THE UK OFFICE FURNITURE MARKET

Furniture Industry Research Association

UK - English/1992 30pp £120.00 (FIRA members £60.00)

Manufacturers deliveries, foreign trade, total market, future trends and the UDT directive.

34015 ANNUAL ABSTRACT OF STATISTICS

HMSO - Her Majesty's Stationery Office

UK - English annual/357pp £21.00 ISSN No. 0072-5730

Contains a wide range of statistical series with a range of years presented.

34016 ANNUAL REPORT - ADVERTISING ASSOCIATION

Advertising Association

UK - English annual/free

Useful for indications of levels of activity in the industry.

34017 ANNUAL REVIEW OF GOVERNMENT FUNDED R&D

HMSO - Her Majesty's Stationery Office

UK - English annual/£25.00

Presents government expenditure on R&D, both outurn and forward plans by department, R&D in industry, and international comparisons.

34018 APPROPRIATION ACCOUNTS

HMSO - Her Majesty's Stationery Office

UK - English annual/price on application

Compares in 12 volumes the actual expenditure of government departments during the previous financial year, and the provision in the Supply Estimates for that year. Includes the report of the Controller and Auditor General upon the figures.

34019 ARL RESEARCH REPORTS: FOOD INDUSTRIES

Answers Research Ltd

UK - English annual/about £250.00 per report

Market research reports which look at the UK food industries.

34020 ARL RESEARCH REPORTS: PHARMACEUTICAL AND HEALTHCARE

Answers Research Ltd

UK - English annual/about £250.00 per report

Market research reports which look at the UK pharmaceutical and healthcare industries.

34021 ARL RESEARCH REPORTS: VETERINARY PRODUCTS

Answers Research Ltd

UK - English annual/about £250.00 per report

Market research reports which look at the UK veterinary products industries.

34022 BANK OF ENGLAND QUARTERLY BULLETIN

Bank of England

UK - English quarterly/£27.00 P.A. ISSN No. 0005-5166

Looks at financial economics of the UK economy. Includes articles and speeches on economic topics.

34023 BANK OF ENGLAND STATISTICAL ABSTRACT

Bank of England

UK - English annual/£15.00 P.A.

Annual cumulation of statistics covering financial and general economics, it appears with the November edition of the Bank of England quarterly bulletin.

34024 BANK OF ENGLAND WORKING PAPERS

Bank of England

UK - England occasional/price varies

These replace the Discussion papers and Technical series. They cover economic and financial topics written in the Bank.

34025 THE BANKER

Financial Times Magazines Ltd

UK - English monthly/£48.00 P.A. ISSN No. 0005-5395

Looks at UK and international banking with features on selected topics, including some data.

34026 BASIC FACTS

Machine Tool Technologies Association

UK - English annual/8pp free

Small booklet giving the main statistical figures for the machine tools industry.

34027 BASIC HORTICULTURAL STATISTICS FOR THE UNITED KINGDOM 1981-90

HMSO - Her Majesty's Stationery Office

UK - English annual/free

34028 BENEFITS

The Reward Group

UK - English annual/£160.00

Trends and details of benefits and conditions of service.

34029 BEST OF BRITISH - THE TOP 20,000 COMPANIES

Jordan & Sons Ltd

UK - English annual/£225.00

In four volumes. Ranks companies by turnover, and includes contact details, number of employees, activities and geographical area.

34030 BRITAIN: AN OFFICIAL HANDBOOK

HMSO - Her Majesty's Stationery Office

UK - English annual/£80.00 ISSN No. 0068-1075

Produced by the Central Office of Information this provides a snapshot of Britain and British industry.

34031 BRITAIN'S TOP EUROPEAN OWNED COMPANIES

Jordan & Sons Ltd

UK - English annual/£150.00

Ranks companies by turnover, gives contact details, activities and brief financial details for each company.

34032 BRITAIN'S TOP FOREIGN OWNED COMPANIES

Jordan & Sons Ltd

UK - English annual/£190.00

Ranks companies by turnover, gives contact details, activities and brief financial details for each company.

34033 BRITAIN'S TOP JAPANESE OWNED COMPANIES

Jordan & Sons Ltd

UK - English annual/£150.00

Ranks companies by turnover, gives contact details, activities and brief financial details for each company.

34034 BRITAIN'S TOP PRIVATELY OWNED COMPANIES

Jordan & Sons Ltd

UK - English annual/£395.00

Ranks companies by turnover, give contact details, activities and brief financial details for each company.

34035 BRITAIN'S TOP US OWNED COMPANIES

Jordan & Sons Ltd

UK - English annual/£150.00

Ranks companies by turnover, gives contact details, activities and brief financial details for each company.

34036 BRITISH AID STATISTICS

Overseas Development Administration

UK - English annual 1966/£8.00 ISSN No. 0068-1210

Details levels and destination of British overseas aid.

34037 BRITISH COACH OPERATORS SURVEY

English Tourist Board

UK - English annual/1990 £30.00

Detailed information on the nature and extent of involvement of the coaching industry in tourism in Britain.

34038 BRITISH COAL ANNUAL REPORT

British Coal Corporation

UK - English annual/free

Includes a statistical survey of the UK coal industry.

34039 BRITISH DRINKS INDEX

Hadleigh Marketing Services

UK - English annual/c400 £400.00

Market research type report covering the top 200 + drinks manufacturers in the UK.

34040 BRITISH FOOD INDEX

Hadleigh Marketing Services

UK - English annual/c500 £400.00

Market research type report covering the top 200 + food manufacturers in the UK.

34041 BRITISH HOLIDAY INTENTIONS

English Tourist Board

UK - English annual/£15.00

Report of an annual survey to measure intentions in the forthcoming season. A short-term forecast is published in May, based on the findings.

34042 BRITISH HORECA INDEX

Hadleigh Marketing Services

UK - English annual/c350 £300.00

Market research type report covering over the top hotel and catering companies in the UK.

34043 THE BRITISH ON HOLIDAY

British Tourist Authority

UK - English annual/£12.50

Summary of results taken from the British National Travel Survey (BNIS), the BTA's annual survey of holidays taken by the British, together with some comparative results for earlier years.

34044 THE BRITISH SHOPPER

NTC Publications Ltd

UK - English annual/128pp £16.95

Data on consumer purchasing behaviour and shopping activity.

34045 BRITISH SUPERMARKET INDEX

Hadleigh Marketing Services

UK - English annual/c300 £400.00

Market research type report covering over the top supermarkets and wholesalers in the UK.

34046 BRITISH TOURIST AUTHORITY ANNUAL REPORT
British Tourist Authority

UK - English annual/£10.00 ISSN No. 0068-2667

A report on the Authority's activities during the year which looks at tourism in Britain today and the outlook for the future.

34047 BSRIA MARKET STUDIES
Building Services Research and Information Association

UK - English annual/£25.00 - £500.00

Regular market studies on the heating and ventilation market.

34048 BSRIA STATISTICS BULLETIN
Building Services Research and Information Association

UK - English quarterly/£150.00 P.A. ISSN No. 0388-6244

Statistics on the heating and ventilation market.

34049 BUILDING SOCIETIES ASSOCIATION ANNUAL REPORT
Building Societies Association/Council of Mortgage Lenders

UK - English annual/£5.00

Basic data on building societies, and levels of business.

34050 BUILDING SOCIETIES YEARBOOK
Building Societies Association/Council of Mortgage Lenders

UK - English annual/900pp £55.00 ISSN No. 0068-3566

Includes key statistics on societies and housing finance generally and balance sheets of member societies.

34051 BUILDING SOCIETIES YEARBOOK - UPDATING SERVICE
Building Societies Association/Council of Mortgage Lenders

UK - English annual/£40.00

Updates the BSA Yearbook on a regular basis. The Yearbook includes key statistics on societies and housing finance generally and balance sheets of member societies.

34052 BUILDING SOCIETY ANNUAL ACCOUNTS DATA
Building Societies Association/Council of Mortgage Lenders

UK - English annual/£40.00

12 tables based on data drawn from individual building societies' published reports and accounts.

34053 BUS AND COACH STATISTICS GREAT BRITIAN
HMSO - Her Majesty's Stationery Office

UK - English annual/£11.20

34054 BUSINESS AND RESEARCH ASSOCIATES REPORTS: FURNITURE INDUSTRIES
Business and Research Associates

UK - English/price varies with report

Market research reports which look at the UK office and domestic furniture industries.

34055 BUSINESS MONITORS: ANNUAL CENSUS OF PRODUCTION REPORTS: ANNUAL AND OCCASIONAL MONITORS COVERING MORE THAN INDUSTRY
HMSO - Her Majesty's Stationery Office

UK - English/various

Designed for those undertaking market research. They provide statistics on manufacturing, energy, mining, service and distributive industries. Changes in format and frequency are expected in 1994-1995.

Key to Series:
P = Production monitors
SD = Service & Distributive monitors
M = Miscellaneous
Frequency:
M = Monthly, Q = Quarterly
A = Annual O = Occasional

Annual Monitors covering more than one industry

PA1001	Annual census of production introductory notes
PA1002	Annual census of production summary tables
PA1003	Size analysis of United Kingdom businesses
PA1007	Minerals

Occasional Monitors covering more than one industry

PO1000	Index of commodities 1991
PO1007	UK Directory of manufacturing businesses 1989 (issued in six parts)
PO1007a	UK Directory of manufacturing businesses 1990 supplement
PO1008	Purchases inquiry 1989

34056 BUSINESS MONITORS: ANNUAL CENSUS OF PRODUCTION REPORTS: CONSTRUCTION INDUSTRY
HMSO - Her Majesty's Stationery Office

UK - English/various

Designed for those undertaking market research. They provide statistics on manufacturing, energy, mining, service and distributive industries. Changes in format and frequency are expected in 1994-1995.

Construction
PA500   Construction industry

34057 BUSINESS MONITORS: ANNUAL CENSUS OF PRODUCTION REPORTS: ENERGY AND WATER SUPPLY INDUSTRIES
HMSO - Her Majesty's Stationery Office

UK - English/various

Designed for those undertaking market research. They provide statistics on manufacturing, energy, mining, service and distributive industries. changes in format and frequency are expected in 1994-1995.

Key to Series:
P = Production monitors
SD = Service & Distributive monitors
M = Miscellaneous monitors
Frequency:
M = Monthly Q = Quarterly
A = Annual O = Occasional

Annual Census of Production reports

Energy and water supply industries

PA111	Coal extraction and manufacture of solid fuels
PA120	Coke ovens
PA130	Extraction of mineral oil and natural gas
PA140	Mineral oil processing
PA161	Production and distribution of electricity
PA162	Public gas supply
PA170	Water supply industry

34058 BUSINESS MONITORS: ANNUAL CENSUS OF PRODUCTION REPORTS: EXTRACTION OF MINERAL AND ORES OTHER THAN FUELS: MANUFACTURE OF METALS, MINERAL PRODUCTS AND CHEMICALS
HMSO - Her Majesty's Stationery Office

UK - English/various

Designed for those undertaking market research. They provide statistics on manufacturing, energy, mining, service and distributive industries. Changes to format and frequency are expected in 1994-1995.

Key to Series:

P = Production monitors
SD = Service & Distributive monitors
M = Miscellaneous monitors
Frequency:
M = Monthly Q = Quarterly
A = Annual O = Occasional

PA210	Extraction and preparation of metalliferous ores
PA221	Iron and steel industry
PA222	Steel tubes
PA223	Drawing, cold rolling and cold forming of steel
PA224	Non-ferrous metals industry
PA231	Extraction of stone, clay, sand and gravel
PA239	Extraction of miscellaneous minerals (including salt)
PA241	Structural clay products
PA242	Cement, lime and plaster
PA243	Building products of concrete, cement or plaster
PA244	Asbestos goods
PA245	Working of stone and other non-metallic minerals
PA246	Abrasive products
PA247	Glass and glassware
PA248	Refractory and ceramic goods
PA251	Basic industrial chemicals
PA255	Paints, varnishes and printing ink
PA256	Specialized chemical products mainly for industrial and agricultural purposes
PA257	Pharmaceutical products
PA258	Soap and toilet preparations
PA259	Specialized chemical products mainly for household and office use
PA260	Production of manmade fibres

34059 BUSINESS MONITORS: ANNUAL CENSUS OF PRODUCTION REPORTS: METAL GOODS, ENGINEERING AND VEHICLE INDUSTRIES
HMSO - Her Majesty's Stationery Office

UK - English/various

Designed for those undertaking market research. They provide statistics on manufacturing, energy, mining, service and distributive industries. Changes in format and frequency are expected in 1994-1995.
Key to Series: P = Production monitors
SD = Service & Distributive monitors
M = Miscellaneous monitors
Frequency:
M = Monthly, Q = Quarterly
A = Annual, O = Occasional

Monthly Production Monitor

PM3510	Car and commercial vehicle production
PA311	Foundries
PA312	Forging, pressing and stamping
PA313	Bolts, nuts, washers, rivets, springs, non-precision chains, metal treatment
PA314	Metal doors, windows etc
PA316	Hand tools and finished metal goods
PA320	Industrial plant steelwork
PA321	Agricultural machinery and tractors
PA322	Metal-working machine tools and engineers' tools

PA323	Textile machinery
PA324	Machinery for the food, chemical and related industries; process engineering contractors
PA325	Mining machinery, constructional; and mechanical handling equipment
PA326	Mechanical power transmission equipment
PA327	Machinery for the printing, paper, wood, leather, rubber, glass and related industries, laundry and dry-cleaning machinery
PA328	Miscellaneous machinery and mechanical equipment
PA329	Ordnance, small arms and ammunition
PA330	Manufacture of office machinery and data processing equipment
PA341	Insulated wires and cables
PA342	Basic electrical equipment
PA343	Electrical equipment for industrial use, and batteries and accumulators
PA344	Telecommunications equipment, electronic measuring equipment, electronic capital goods and passive electronic components
PA345	Miscellaneous electronic equipment
PA346	Domestic-type electric appliances
PA347	Electric lamps and other electrical lighting equipment
PA351	Motor vehicles and their engines
PA352	Motor vehicle bodies, trailers and caravans
PA353	Motor vehicle parts
PA361	Shipbuilding and repairing
PA362	Railway and tramway vehicles
PA363	Cycles and motorcycles
PA364	Aerospace equipment manufacturing and repairing
PA365	Miscellaneous vehicles
PA371	Measuring, checking and precision instruments and apparatus
PA372	Medical and surgical equipment and orthopaedic appliances
PA373	Optical precision instruments and photographic equipment
PA374	Clock, watches and other timing devices

34060 BUSINESS MONITORS: ANNUAL CENSUS OF PRODUCTION REPORTS: MISCELLANEOUS MONITORS
HMSO - Her Majesty's Stationery Office

UK - English/various

Designed for those undertaking market research. They provide statistics on manufacturing, energy, mining, service and distributive industries. Changes in format and frequency are expected in 1994-1995.
Key to Series:
P = Production monitors
SD = Service & Distributive monitors
M = Miscellaneous monitors
Frequency:
M = Monthly Q = Quarterly
A = Annual O = Occasional

Miscellaneous Monitors Monthly

MM1	Motor vehicle registrations
MM17	Price index numbers for current cost accounting (monthly supplement)

MM19	Aerospace and electronics cost indices (1980 = 100 and 1985 = 100)
MM20	Overseas trade statistics of the UK
MM22	Producer price indices

Quarterly

MQ5	Insurance companies' and private pension funds' investments
MQ6	Overseas travel and tourism
MQ10	Overseas trade analysed in terms of industries

Annual

MA3	Company finance
MA4	Overseas transactions
MA6	Overseas travel and tourism
MA20	Overseas trade statistics of the United Kingdom (annual revision)
MA21	Guide to the classification of the overseas trade statistics

Occasional

MO4	Census of overseas assets 1987
MO14	Survey of scientific research and development carried out within the UK (1985)
MO18	Price index numbers for current cost accounting, summary volume 1983-1987

34061 BUSINESS MONITORS: ANNUAL CENSUS OF PRODUCTION REPORTS: OTHER MANUFACTURING INDUSTRIES
HMSO - Her Majesty's Stationery Office

UK - English/various

Designed for those undertaking market research. They provide statistics on manufacturing, energy, mining, service and distributive industries. Changes in format and frequency are expected in 1994-1995.
Key to Series:
P = Production monitors
SD = Service & Distributive monitors
M = Miscellaneous monitors
Frequency:
M = Monthly, Q = Quarterly
A = Annual' O = Occasional

PA411	Organic oils and fats
PA412	Slaughtering of animals and production of meat and by-products
PA413	Preparation of milk and milk products
PA414	Processing of fruit and vegetables
PA415	Fish processing
PA416	Grain milling
PA419	Bread, biscuits and flour confectionery
PA420	Sugar and sugar by-products
PA421	Ice-cream, cocoa, chocolate and suger confectionery
PA422	Animal feeding stuffs
PA423	Starch and miscellanous foods
PA424	Spirit distilling and compounding
PA426	Wines, cider and perry
PA427	Brewing and malting
PA428	Soft drinks
PA429	Tobacco industry
PA431	Woollen and worsted industry
PA432	Cotton and silk industries

PA433	Throwing, texturing, etc, of continuous filament yarn
PA434	Spinning and weaving of flax, hemp and ramie
PA435	Jute and polypropylene yarns and fabrics
PA436	Hosiery and other knitted goods
PA437	Textile finishing
PA438	Carpets and other textile floor coverings
PA439	Miscellaneous textiles
PA441	Leather (tanning and dressing) and fellmongery
PA442	Leather goods
PA451	Footwear
PA453	Clothing, hats, gloves and furs
PA455	Household textiles and other made-up textiles
PA461	Sawmilling, planing, etc of wood
PA462	Manufacture of semi-finished wood products and further processing and treatment of wood
PA463	Builders' carpentry and joinery
PA464	Wooden containers
PA465	Miscellaneous wooden articles
PA466	Articles of cork and plaiting materials, brushes and brooms
PA467	Wooden and upholstered furniture and shop and office fittings
PA471	Pulp, paper and board
PA472	Conversion of paper and board
PA475	Printing and publishing
PA481	Rubber products (including retreating and specialist repairing of tyres)
PA483	Processing of plastics
PA491	Jewellery and coins
PA492	Musical instruments
PA493	Photographic and cinematographic processing laboratories
PA494	Toys and sports goods
PA495	Miscellaneous manufacturing industries

34062 BUSINESS MONITORS: PRODUCTION MONITORS: ENERGY INDUSTRIES
HMSO - Her Majesty's Stationery Office

UK - English/various

Designed for those undertaking market research. They provide statistics on manufacturing, energy, mining, service and distributive industries.
Key to Series:
P = Production monitors
SD = Service & Distributive monitors
M = Miscellaneous monitors
Frequency:
M = Monthly, Q = Quarterly
A = Annual, O = Occasional

Energy industries

PQ1113	Coal mining
PQ1300	Extraction of mineral oil and natural gas
PAS1401	Mineral oil refining, lubricating oils / 1402 and greases

34063 BUSINESS MONITORS: PRODUCTION MONITORS: EXTRACTION OF MINERALS AND ORES OTHER THAN FUELS: MANUFACTURE OF METALS, MINERAL PRODUCTS AND CHEMICALS
HMSO - Her Majesty's Stationery Office

UK - English/various pages price varies

Designed for those undertaking market research. They provide statistics on manufacturing, energy, mining, service and distributive industries. Changes to format and frequency are expected in 1994-1995.
Key to Series:
P = Production monitors
SD = Service & Distributive monitors
M = Miscellaneous monitors
Frequency:
M = Monthly, Q = Quarterly
A = Annual, O = Occasional

PAS2234	Steel wire and steel wire products
PAS2245	Aluminium and aluminium alloys
PAS2246	Copper, brass and other copper alloys
PAS2247	Miscellaneous non-ferrous metals and their alloys
PQ2310	Extraction of stone, clay, sand and gravel, working of /2450 stone and miscellaneous non-metallic minerals
PAS2330	Salt and miscellaneous minerals / 2396
PAS2410	Structural clay products
PAS2420	Cement, lime and plaster
PQ2437	Miscellaneous building products of concrete, cement or plaster
PAS2440	Asbestos
PAS2460	Abrasives
PAS2471	Flat glass
PAS2478	Glass containers and miscellaneous /2479 glass products
PAS2481	Refractory goods
PAS2489	Ceramic goods
PQ2511	Inorganic chemicals (except industrial gases)
PQ2512	Basic organic chemicals (except specialized pharmaceutical chemicals)
PAS2513	Fertilizers
PAS2514	Synthetic resins, plastics materials / 2515 and synthetic rubber
PAS2516	Dyestuffs and pigments
PQ2551	Paints, varnishes and painters' fillings
PAS2552	Printing ink
PAS2562	Formulated adhesives, sealants and /2569 adhesive film, cloth and coil
PAS2563	Chemical treatment of oils, fats and /2564 essential oils and flavouring materials
PQ2567	Miscellaneous chemical products for industrial use
PAS2568	Formulated pesticides
PQ2570	Pharmaceutical products
PAS2581	Soap and synthetic detergents
PAS2582	Perfumes, cosmetics and toilet preparations

PAS2591	Photographic materials, chemicals and polishes and /2599 miscellaneous specialized products mainly for household and office use
PAS2600	Production of man-made fibres

34064 BUSINESS MONITORS: PRODUCTION MONITORS: METAL GOODS, ENGINEERING AND VEHICLE INDUSTRIES
HMSO - Her Majesty's Stationery Office

UK - English/various pages price varies

Designed for those undertaking market research. They provide statistics on manufacturing, energy, mining, service and distributive industries. Changes to format and frequency are expected in 1994-1995.
Key to Series:
P = Production monitors
SD = Service & Distributive monitors
M = Miscellaneous
Frequency:
M = Monthly, Q = Quarterly
A = Annual, O = Occasional

PAS3111	Iron castings
PAS3112	Non-ferrous metal foundries
PAS3120	Forging, pressing and stamping
PAS3137	Bolts, nuts, washers, rivets, springs, and non-precision chains
PAS3138	Heat and surface treatment of metals (including sintering)
PAS3142	Metal doors, windows etc
PAS3161	Hand tools and implements
PAS3162	Cutlery, spoons, forks, and similar tableware; razors
PAS3163	Metal storage vessels (mainly non industrial) /3164 and packaging products of metal
PAS3165	Domestic heating and cooking appliances (non-electric)
PAS3166	Metal furniture and safes
PAS3167	Domestic utensils of metal
PAS3169	Miscellaneous finished metal products
PAS3204	Fabricated constructional steelwork
PAS3205	Boilers and process plant fabrications
PAS3211	Agricultural machinery
PAS3212	Wheeled tractors
PAS3221	Metal-working machine tools
PAS3222	Engineers' small tools
PAS3230	Textile machinery
PAS3244	Food, drink and tobacco processing machinery; packaging and bottling machinery
PAS3245	Chemical industry machinery; furnaces and kilns; gas, water and waste treatment plant
PAS3251	Mining machinery
PAS3254	Construction and earth-moving equipment
PAS3255	Mechanical lifting and handling equipment
PAS3261	Precision chains, other mechanical power /3262 transmission equipment, ball, needle and roller bearings PAS3275 Machinery for working wood, rubber, plastics, leather, and making paper, glass, bricks and similar materials; laundry and dry-cleaning machinery

PAS3276 Printing, bookbinding and paper goods machinery
PAS3281 Industrial (including marine) engines
PAS3283 Compressors and fluid power equipment
PAS3284 Refrigerating, space heating, ventilating and air-conditioning equipment
PAS3285 Scales, weighing machinery and portable power tools
PAS3286 Miscellaneous industrial and commercial machinery
PAS3287 Pumps
PAS3288 Industrial valves
PAS3289 Miscellaneous mechanical, marine and precision engineering
PAS3290 Ordnance, small arms and ammunition
PAS3301 Office machinery
PAS3302 Electronic data processing equipment
PAS3410 Insulated wires and cables
PAS3420 Basic electrical equipment
PAS3432 Batteries
PAS3433 Alarms and signalling equipment
PQ3434 Electrical equipment for motor vehicles, cycles and aircraft
PAS3435 Miscellaneous electrical equipment for industrial use
PAS3441 Telegraph and telephone equipment
PQ3442 Electrical instruments and control systems
PAS3443 Radio and electronic capital goods
PQ3444 Components other than active components, mainly for electronic equipment
PAS3452 Gramophone records and pre-recorded tapes
PQ3453 Electronic sub-assemblies and active components
PAS3454 Electronic consumer goods and miscellaneous equipment
PAS3460 Domestic electrical appliances
PAS3470 Electric lamps and lighting equipment
PQ3510 Motor vehicle and their engines
PAS3521 Motor vehicle bodies
PAS3522 Trailers, semi-trailers and caravans /3523 PQ3530 Motor vehicle parts
PQ3610 Shipbuilding and repairing
PAS3620 Railway and tramway vehicles
PAS3633 Motorcycles, cycles and miscellaneous vehicles /3634 /3650 PAS3640 Aerospace equipment manufacturing and repairing
PAS3710 Measuring, checking and precision instruments and apparatus
PAS3720 Medical and surgical equipment and orthopaedic appliances
PAS3731 Spectacles, unmounted lenses and optical /3732 precision instruments
PAS3733 Photographic and cinematographic equipment
PAS3740 Clocks, watches and other timing devices

34065   BUSINESS MONITORS: PRODUCTION MONITORS: OTHER MANUFACTURING INDUSTRIES
HMSO - Her Majesty's Stationery Office

UK - English/various pages price varies

Designed for those undertaking market research. They provide statistics on manfacturing, energy, mining, service and distributive industries. Changes to format and frequency are expected in 1994-1995.
Key to Series:
P = Production monitors
SD = Service & Distributive monitors
M = Miscellaneous monitors
Frequency:
M = Monthly, Q = Quarterly
A = Annual, O = Occasional

Other manufacturing industries
PAS4115 Margarine and compound cooking fats
PAS4116 Organic oils and fats
PQ4122 Bacon curing and meat products
PAS4123 Poultry and poultry products
PAS4126 Animal by-products
PAS4147 Fruit and vegetable products
PAS4150 Fish products
PAS4180 Starch
PAS4196 Bread and flour confectionery
PAS4197 Biscuits and crispbread
PQ4213 Ice-cream
PAS4239 Miscellaneous foods
PAS4240 Spirit distilling and compounding
PAS4261 Wines, cider and perry
PAS4270 Brewing and Malting
PQ4283 Soft drinks
PAS4290 Tobacco
PAS4310 Woollen and worsted
PAS4321 Spinning and doubling of cotton, / 4336 throwing, texturing etc of continuous filament yarn
PAS4322 Weaving of cotton, silk and manmade fibres
PAS4340 Spinning and weaving of flax, hemp and ramie
PAS4350 Jute and polypropylene yarns and fabrics
PQ4363 Hosiery and other weft knitted goods
PQ4364 Warp knitted goods
PAS4370 Textile finishing
PQ4384 Carpets and other textile floor, / 4385 coverings
PAS4395 Lace
PAS4396 Rope, twine and net
PQ4398 Narrow fabrics
PQ4399 Miscellaneous textiles
PAS4410 Leather and fellmongery and /4420 leather goods
PQ4510 Footwear
PQ4531 Weatherproof outerwear
PQ4532 Men's and boys' tailored outerwear
PQ4533 Women's and girls' tailored outerwear
PQ4534 Work clothing and men's and boys' jeans
PQ4535 Men's and boys' shirts, underwear and nightwear
PQ4536 Women's and girls' light outerwear, lingerie and infants' wear
PAS4537 Hats, caps and millinery
PQ4538 Gloves
PQ4539 Miscellaneous dress industries
PAS4555 Soft furnishings
PQ4556 Canvas goods, sacks and miscellaneous made-up textiles
PQ4557 Household textiles
PAS4630 Builders' carpentry and joinery
PAS4640 Wooden containers, miscellaneous /4650 wooden articles

PAS4663 Articles of cork and plaiting materials, /4664 brushes and brooms
PAS4671 Wooden and upholstered furniture
PAS4672 Shop and office fitting
PAS4710 Pulp, paper and board
PAS4721 Wallcoverings
PAS4722 Household and personal hygiene paper products
PAS4723 Stationery
PAS4724 Packaging products of paper and pulp
PAS4725 Packaging products of board and / 4728 miscellaneous paper and board products
PAS4751 Printing and publishing of newspapers
PAS4752 Printing and publishing of periodicals
PAS4753 Printing and publishing of books
PAS4754 Miscellaneous printing and publishing
PAS4811 Rubber tyres and inner tubes, retreading and /4820 specialists repairing of rubber tyres
PAS4812 Miscellaneous rubber products
PAS4813 Plastic-coated textile fabrics and plastics /4833 floor coverings
PAS4832 Plastics semi-manufactures
PAS4834 Plastics building products
PAS4835 Plastics packaging products
PAS4836 Miscellaneous plastics products
PAS4910 Jewellery and coins
PAS4920 Musical instruments
PAS4941 Toys and Games
PAS4942 Sports goods
PAS4954 Miscellaneous stationers' goods
PAS4959 Miscellaneous manufacturing industries

34066   BUSINESS MONITORS: SERVICE AND DISTRIBUTIVE MONITORS
HMSO - Her Majesty's Stationery Office

UK - English/various pages price varies

Designed for those undertaking market research. They provide statistics on manufacturing, energy, mining, service and distributive industries. Changes to format and frequency are expected in 1994-1995.
Key to Series:
P = Production monitors
SD = Service & Distributive monitors
M = Miscellaneous monitors
Frequency:
M = Monthly, Q = Quarterly
A = Annual, O = Occasional

SERVICE AND DISTRIBUTIVE MONITORS
Monthly
SDM28   Retail sales

Quarterly
SDQ7   Assets and liabilities of finance houses and other consumer credit companies
Annual
SDA25   Retailing
SDA26   Wholesaling
SDA27   Motor trades
SDA28   Catering and allied trades
SDA29   Service trades

34067 BUSINESS RATIO REPORTS

ICC Information Group Ltd

UK - English annual/£225.00 per report

About 200 reports, each covering a different industrial sector. Each report ranks about 100 companies by specific financial performance ratios, with aggregate ratios given for the industry as a whole.

34068 CATALOGUE OF STATISTICAL SOURCES

Association of British Insurers

UK - English irregular/£100.00 (price includes a series of titles)

Details of all known sources of statistical information about UK insurers and insurance. Series of appendices giving statistical information mentioned in the catalogue.

34069 CBI/BSL REGIONAL TRENDS SURVEY

Confederation of British Industry (CBI)

UK - English quarterly/£195.00 (£ 120.00 to CBI members) P.A.

Sample figures for the 11 standard UK regions derived from the CBI Industrial Trends survey. Detailed regional analysis and guide to key regional characteristics is included.

34070 CBI/COOPERS DELOITTE FINANCIAL SERVICES SURVEY

Confederation of British Industry (CBI)

UK - English quarterly/£195.00 (£ 120.00 to CBI members) P.A.

Covers volume of business, charges, costs, profitability, employment, training, capital expenditure, and business optimism. Survey sample covers 50% of the total employment in this sector.

34071 CHARITY TRENDS

Charities Aid Foundation

UK - English annual/£12.95

Covers the voluntary sector in the UK with some international data.

34072 CIVIL SERVICE STATISTICS

HMSO - Her Majesty's Stationery Office

UK - English annual/£10.80

Facts and figures about civil servants including numbers by grade, department and location.

34073 CLERICAL AND OPERATIVE REWARDS

The Reward Group

UK - English twice a year/£255.00 P.A. single issue £150.00

Analysis of basic pay and average earnings for all main clerical and operative positions, by job, company size, location and industry.

34074 CML QUARTERLY LENDING FIGURES

Building Societies Association/Council of Mortgage Lenders

UK - English quarterly/£16.00 P.A.

Covers the lending activities of banks, building societies, centralized lenders, and other lending institutions.

34075 CONSUMER SPENDING FORECASTS

Staniland Hall Associates

UK - English quarterly/£420.00 P.A.

General forecast for overall spending and more detailed forecasts for spending by sector.

34076 CO-OPERATIVE STATISTICS

Co-operative Union Ltd

UK - English annual/44pp £79.00

Economic and statistical information on the cooperative movement in the UK.

34077 COST OF LIVING REPORTS - REGIONAL COMPARISONS

The Reward Group

UK - English twice a year/£215.00, single issues £125.00

Comparison for the eleven standard UK regions, and eight national lifestyle groups. Reports comparing specific towns are prepared to order.

34078 COUNCIL OF MORTGAGE LENDERS ANNUAL REPORT

Building Societies Association/Council of Mortgage Lenders

UK - English annual/£5.00

Basic data on the housing finance industry.

34079 COUNTRY BOOKS: UK

United Kingdom Iron and Steel Statistics Bureau

UK - English annual/£75.00 ISSN No. 0307-7608

Annual summary tables showing production, materials consumed, apparent consumption, imports and exports of 130 products by quality and market.

34080 COUNTRY FORECASTS: UK

Business International Ltd

UK - English quarterly/£360.00 P.A.

Contains an executive summary, sections on political and economic outlooks and the business environment, a fact sheet, and key economic indicators. Includes a quarterly Global Outlook.

34081 COUNTRY REPORT - UNITED KINGDOM

Business International Ltd

UK - English quarterly/£150.00 P.A. ISSN No. 0269-5492

Covers main political and economic events over the past quarter, with statistical appendices. An annual country profile is included.

34082 COUNTRY RISK SERVICE - UNITED KINGDOM

Business International Ltd

UK - English quarterly/£1,835.00 P.A. for seven countries

Monitors the financial and political conditions in the UK and gives short term forecasts and risk assessments. There is a minimum subscription to seven countries.

34083 DATAMONITOR REPORTS: CONSUMER DURABLES

Datamonitor

UK - English annual/£300.00 per report

Market research type reports which look at the UK consumer durable and household products market.

34084 DATAMONITOR REPORTS: COSMETIC PRODUCTS

Datamonitor

UK - English annual/£300.00 per report

Market research type reports which look at the UK cosmetic product markets.

34085 DATAMONITOR REPORTS: DRINKS INDUSTRY

Datamonitor

UK - English annual/£300.00 per report

Market research type reports which look at the UK alcoholic and non-alcoholic drinks industry.

34086 DATAMONITOR REPORTS: FINANCIAL SERVICES

Datamonitor

UK - English annual/£600.00 per report

Market research type reports which look at the UK financial services sector.

34087 DATAMONITOR REPORTS: FOOD INDUSTRY

Datamonitor

UK - English annual/£300.00 per report

Market research type reports which look at the UK food industry and food products.

34088 DATAMONITOR REPORTS: LEISURE PRODUCTS
Datamonitor

UK - English annual/£300.00 per report

Market research type reports which look at the UK leisure products markets; sports goods, toys and games.

34089 DATAMONITOR REPORTS: UK RETAILING
Datamonitor

UK - English annual/£300.00 per report

Market research type reports which look at the UK retailing sector.

34090 DATAQUEST MARKET RESEARCH REPORTS
Dataquest

UK - English annual/US$500.00 - US $6,000.00 per report

A series of market research reports on high technology markets. Typical titles include "User wants & needs" and "Case studies in service quality".

34091 DEBENHAM TEWSON AND CHINNOCKS PROPERTY REPORTS
Debenham Tewson and Chinnocks

UK - English/price varies with report

Property research type reports which look at the UK commercial property market. Reports cost about £10-£50 each.

34092 DEFENCE STATISTICS (STATEMENT OF THE DEFENCE ESTIMATES VOLUME II)
HMSO - Her Majesty's Stationery Office

UK - English annual/£8.00

Sections cover defence finance, trade in defence equipment, equipment programme, manpower, welfare, and the defence services provided.

34093 DELIVERIES OF YARN
Textile Statistics Bureau

UK - English monthly/£40.00 P.A.

Deliveries classified by spinners, doublers, and main consuming industries.

34094 DIGEST OF ENVIRONMENTAL PROTECTION AND WATER STATISTICS
HMSO - Her Majesty's Stationery Office

UK - English annual/£13.50

34095 DIGEST OF THE UK ENERGY STATISTICS
HMSO - Her Majesty's Stationery Office

UK - English annual/£16.50

Deals with UK energy production and consumption. Separate sections deal with production and consumption of individual fuels, oil and gas reserves, fuel prices, and foreign trade in fuels. Annexes give longer runs of data and data on renewable sources of energy. Contains tables, commentary and charts.

34096 DIGEST OF WELSH STATISTICS
Welsh Office

UK - English annual/£6.00

Broad coverage of economic and social statistics.

34097 DIRECTORS REPORT
Henley Centre for Forecasting

UK - English monthly/£210.00 P.A. ISSN No. 0952-5467

A digest giving economic and social trends with forecasts.

34098 DIRECTORS REWARDS
The Reward Group

UK - English annual/£180.00

Full analysis of directors renumeration and benefits. Published in association with the Institute of Directors.

34099 DISTRIBUTIVE TRADES SURVEY
Confederation of British Industry (CBI)

UK - English monthly/£295.00 (£170.00 to CBI members) P.A.

Covers 500+ outlets, detailing results for 21 sectors of business. Data on volume of sales, orders placed with suppliers, stocks, employment, business confidence, and investment. Reflects short-term expectations in the retail and wholesale trades.

34100 DOING BUSINESS IN NORTHERN IRELAND
Price Waterhouse

UK - English irregular/1992 330-340pp free to clients

A business guide which includes sections on the investment climate, doing business, accounting and auditing, taxation and general country data. Updated by supplements.

34101 DOING BUSINESS IN THE ISLE OF MAN
Ernst & Young International

UK - English irregular/1991 110-130pp free to clients

A business guide which includes an executive summary, the business environment, foreign investment, company structure, labour, taxation, financial reporting and auditing, and general country data.

34102 DOING BUSINESS IN THE UK
Ernst & Young International

UK - English irregular/1991 110-130pp free to clients

A business guide which includes an executive summary, the business environment, foreign investment, company structure, labour, taxation, financial reporting and auditing, and general country data.

34103 DOING BUSINESS IN THE UK
Price Waterhouse

UK - English irregular/1991 330-340pp free to clients

A business guide which includes sections on the investment climate, doing business, accounting and auditing, taxation and general country data. Updated by supplements.

34104 DOUBLING STATISTICS
Textile Statistics Bureau

UK - English monthly/£40.40 P.A.

Production, employment, spindles in place and active are detailed.

34105 DRINKS POCKET BOOK
NTC Publications Ltd

UK - English annual/176pp £26.00

Data on the drink industry.

34106 ECONOMIC INDICATORS: FORECASTS FOR COMPANY PLANNING
Staniland Hall Associates

UK - English quarterly/£290.00 P.A.

Forecasts for the main economic indicators including growth, inflation etc.

34107 ECONOMIC OUTLOOK
Gower Publishing Company Ltd

UK - English monthly/£190.00 P.A. ISSN No. 0140-489X

Gives a four year business forecast for the UK economy.

34108 ECONOMIC PROGRESS REPORT
The Treasury Office

UK - English bi-monthly/free

Published by the Treasury, it contains articles on economic subjects, analysis, background and economic policy.

34109 ECONOMIC PROSPECTS
Co-operative Union Ltd

Information services - statistics/UK - English annual/62pp £55.00

Economic and statistical information.

34110 ECONOMIC REPORT ON SCOTTISH AGRICULTURE
Scottish Office Library

UK - English annual/£4.50

Looks at products, costs and levels of activity.

34111 ECONOMIC SITUATION REPORT
Confederation of British Industry (CBI)

UK - English monthly/£250.00 (£150.00 to CBI members) P.A.

Contains economic analysis, surveys of business people, regional reports, European and world outlooks, forecasts and official economic and financial indicators.

34112 ECONOMIC TRENDS
HMSO - Her Majesty's Stationery Office

UK - English monthly/annual subscription £140.00 P.A. single issues £ 11.50 ISSN No. 0013-0400

Commentary and selection of tables and charts provide a broad background to trends in the UK economy. The Annual Supplement presents very long runs of quarterly figures for all the statistical series.

34113 ECONOMICS BULLETIN
Chemical Industries Association

UK - English quarterly/£60.00 P.A.

Monitors, analyses and forecasts the economic performance of the UK chemical industry. A main table of chemical industry basic economic indicators is included.

34114 ECONOMIST INTELLIGENCE UNIT: RESEARCH REPORTS: RETAIL AND CONSUMER GOODS
Business International Ltd

UK - English/£100.00 - £500.00 per report

Market research reports which look at European retailing and consumer goods.

34115 EMPLOYMENT GAZETTE
HMSO - Her Majesty's Stationery Office

UK - English monthly/£48.00 P.A. ISSN No. 0309-5045

Articles, charts, and tables on the labour force, employment, unemployment, hours worked, earnings, labour costs, retail prices, stoppages due to disputes, & training.

34116 ENERGY TRENDS
HMSO - Her Majesty's Stationery Office

UK - English monthly/annual subscription only - price on application ISSN No. 0308-1222

Contains monthly and quarterly tables of production, consumption, and trade in fuels, aggregated consumption and fuel prices with tables, commentary and graphs.

34117 ENGLISH HERITAGE MONITOR
English Tourist Board

UK - English annual/£15.00

Numbers of historic buildings and conservation areas and details of properties open to the public, admission charges and visitor trends.

34118 ENGLISH LANGUAGE COURSE VISITORS TO THE UK
British Tourist Authority

UK - English irregular/£15.00

Reports on visitors attending English language courses in the UK. Data derived from questions included in the Employment Department's International Passenger Survey.

34119 ENGLISH TOURIST BOARD ANNUAL REPORT
English Tourist Board

UK - English annual/1991 £10.00

Reports on the Board's activities, gives data on tourism in the UK and prospects for the future.

34120 EUROPEAN ADVERTISING & MEDIA FORECAST
NTC Publications Ltd

UK - English quarterly/£950.00 P.A.

34121 EUROPEAN INVESTMENT REGION SERIES
Business International Ltd

UK - English/£195.00 per report

Covers the economic prospects for a single investment region, with a comprehensive regional profile included. The reports appear as part of the EIU Special Report Series.

34122 EUROPEAN SOCIAL FUND: ANNUAL REPORT OF ACTIVITIES
Office for Official Publications of the European Communities

UK - English/French annual/price varies

Details financial support from the social fund.

34123 EXPORTS
Textile Statistics Bureau

UK - English quarterly/£12.70 per report

Quarterly series showing cumulative totals from the beginning of the year. Divided by type and country detail (weight). Series are; man-made staple fibre, waste and filament yarn; cotton yarn; spun man-made fibre yarn; woven cotton piecegoods; woven man-made fibre piecegoods; knitted man-made fibre piecegoods.

34124 FAMILY SPENDING
Office of Population Censuses and Surveys

UK - English annual/136pp £19.50

Detailed analysis of income and expenditure by type of household for the UK and includes some regional analyses. Latest report available for 1990.

34125 FARM BUSINESS DATA NORTHERN IRELAND
Department of Agriculture Northern Ireland DANI

UK - English annual/£3.00

34126 FARM INCOMES IN SCOTLAND
Scottish Office Library

UK - English annual/£2.00

34127 FARM INCOMES IN THE UNITED KINGDOM
HMSO - Her Majesty's Stationery Office

UK - England annual/122pp £20.00

Published by MAFF it gives detailed analyses of developments in the income, assets and liabilities of agriculture in aggregate and at farm level.

34128 FARM INCOMES IN WALES
Welsh Office

UK - English annual/£5.00

34129 FARM MANAGEMENT POCKET BOOK
Agricultural Economics Unit

UK - English annual 1986/216pp £6.00

Data relating to farm management.

34130 FIFTY YEARS OF THE NATIONAL FOOD SURVEY 1940-1990
HMSO - Her Majesty's Stationery Office

UK - English/1991 £15.00 ISBN No. 0112429092

34131 FINANCIAL AND ACCOUNTING REWARDS
The Reward Group

UK - English annual/£180.00

A report of salaries and benefits paid to accounting staff in companies.

34132 THE FINANCIAL CONSUMER
NTC Publications Ltd

UK - English annual/128pp £16.95

A statistical profile of the financial consumer and the nature and size of UK consumer markets for financial products.

34133 FINANCIAL STATEMENT AND BUDGET REPORT
HMSO - Her Majesty's Stationery Office

UK -English annual/£9.25

Published by the Treasury on Budget day, it sets out the Government's economic and financial strategy and reviews the performance and prospects of the economy. It describes tax measures and estimates public income and expenditure.

34134 FINANCIAL STATISTICS
HMSO - Her Majesty's Stationery Office

UK - English monthly/£8.95 single issues, £110.00 annual subscription ISSN No. 0015-203X

Key financial and monetary statistics for the UK. Annual subscription includes Financial Statistics Handbook.

34135 FINANCIAL SURVEYS
ICC Information Group Ltd

UK - English annual/£350.00 per report

Directory type reports which cover one industrial or commercial sector and give brief financial details of up to 1,000 companies per report.

34136 FINANCING FOREIGN OPERATIONS: UK
Business International Ltd

UK - English annual/£695.00 for complete European series

Business overview, currency outlook, exchange regulations, monetary system, short/medium/long-term and equity financing techniques, capital incentives, cash management, investment trade finance and insurance.

34137 FINISHING STATISTICS
Textile Statistics Bureau

UK - English monthly/£40.00 P.A.

Goods invoiced and employment by process and fibre.

34138 FOOD POCKET BOOK
NTC Publications Ltd

UK - English annual/128pp £16.95

A compendium of statistics including original data from leading food research organisations.

34139 FRAMEWORK FORECASTS FOR THE UK
Henley Centre for Forecasting

UK - English monthly/£1,125.00 P.A. ISSN No. 0305-562

Monthly forecasts for the main economic activity indicators.

34140 FT-SE SHARE INDICES FIVE YEAR HISTORY
London Stock Exchange

UK - English annual/50pp £10.00

Provides the latest facts and figures as well as historic information dating back to the start of the index in 1984.

34141 GENERAL BUSINESS STATISTICS
Association of British Insurers

UK - English annual/20pp £100.00 (price includes a series of titles)

Detailed premium and revenue account figures for UK general business by accounting class and key totals for worldwide business of UK companies. Detailed commentary on trends in UK results over the past five years, although many figures given for the current year only.

34142 GENERAL HOUSEHOLD SURVEY
Office of Population Censuses and Surveys

UK - English annual/£18.50

A continous sample survey of households relating to a wide range of social and socio-economic policy areas. Latest report for 1989.

34143 GERMAN AND BRITISH ECONOMY
German Chamber of Industry & Commerce

UK - English monthly/£40.00

Monthly statistics on Anglo-German trade.

34144 GUIDELINES FOR EXPORTERS OF SELECTED AGRICULTURAL PRODUCTS
Commonwealth Secretariat

UK - English irregular/price varies with report

A series of guides for exporters to the UK covering such products as cut flowers and vegetables. Also contains general market information.

34145 HAMBRO COMPANY GUIDE
The Hambro Company Guide

UK - English quarterly/£99.00 P.A. ISSN No. 0144-2015

Covers UK stock market companies with five year financials, share performance and contact details.

34146 HANDBOOK OF MARKET LEADERS
Extel Financial

UK - English half-yearly/683pp £175.00 P.A. ISSN No. 0308-9673

Includes the constituent companies of the FT Actuaries All-Share Index, which represents 90% of the total market capitalisation of all UK quoted companies. Basic financial details, company descriptions, and company rankings are covered.

34147 HEALTH AND PERSONAL SOCIAL SERVICES STATISTICS FOR ENGLAND
HMSO - Her Majesty's Stationery Office

UK - English annual/£11.50

34148 HEALTH AND SAFETY STATISTICS
HMSO - Her Majesty's Stationery Office

UK - English annual/£9.00 P.A. (single issue£4.15)

Was issued as a £4.15 supplement to the Employment Department's "Employment Gazette". (Ring HSE at +44 (0)787 881165 for update).

34149 HEAVY GOODS VEHICLES IN GREAT BRITAIN
HMSO - Her Majesty's Stationery Office

UK - English annual/£8.25

34150 HENDERSON TOP 1000 CHARITIES
The Hambro Company Guide

UK - English irregular/£75.00

Covers UK charities giving comparitive data on income, funds and expenditure.

34151 HM CUSTOMS AND EXCISE REPORT
HMSO - Her Majesty's Stationery Office

UK - English annual/£10.80

Includes statistics of all taxes adminstered by Customs and Excise, including the duties on tobacco, alcoholic drink, hydrocarbon oil, betting and gaming, customs duties, agricultural levies, matches and mechanical lighters, car tax and VAT.

34152 HORWATH-ETB ENGLISH HOTEL OCCUPANCY SURVEY
English Tourist Board

UK - English monthly with an annual summary/£300.00 P.A.

Provides data about occupancy levels and the proportion of overseas arrivals at English hotels.

34153 HOSPITAL STATISTICS ENGLAND

HMSO - Her Majesty's Stationery Office

UK - English annual/price on application

34154 HOUSE PRICE INDEX BULLETIN

Halifax Building Society

UK - English quarterly/price on application

Looks at national and regional house prices, includes tables, graphs and analysis.

34155 HOUSEHOLD FISH CONSUMPTION IN GREAT BRITAIN

Sea Fish Industry Authority

UK - English quarterly/£55.00 P.A. ISSN No. 0262-3269

Provides an analysis of the sales of fish by species for household consumption in Britain and in each of the major television regions in Britian. The data is divided on the basis of fresh/chilled and frozen sales, data for the preceding year is given for comparison.

34156 HOUSEHOLD FOOD CONSUMPTION AND EXPENDITURE: REPORT OF THE NATIONAL FOOD SURVEY COMMITTEE 1990. WITH A STUDY OF THE TRENDS OVER THE PERIOD 1940-1990

HMSO - Her Majesty's Stationery Office

UK - English annual/£21.00 ISBN No. 0112429106

Looks at the average expenditure on food and gives historical comparisons.

34157 HOUSING AND CONSTRUCTION STATISTICS GREAT BRITAIN

HMSO - Her Majesty's Stationery Office

UK - English annual/£22.00

Construction industry and house-building statistics.

34158 HOUSING AND CONSTRUCTION STATISTICS GREAT BRITAIN - QUARTERLY

HMSO - Her Majesty's Stationery Office

UK - English quarterly/£38.00 P.A.

Construction industry and house-building statistics. A quarterly publication in two parts.

34159 HOUSING FINANCE

Building Societies Association/Council of Mortgage Lenders

UK - English quarterly/£60.00 P.A. ISSN No. 0955-3800

Covers the UK mortgage market and includes 25 tables of statistics relevant to the institutions operating in the market.

34160 HOUSING FINANCE FACT BOOK

Building Societies Association/Council of Mortgage Lenders

UK - English annual/£20.00

An annual report on the housing finance industry.

34161 IMPORTS

Textile Statistics Bureau

UK - English monthly/£40.40 per report

Monthly series showing cumulative totals from the beginning of the year. Divided by type and country detail (weight). Series are; man-made staple fibre, waste and filament yarn; cotton yarn; spun man-made fibre yarn; woven cotton piecegoods; woven man-made fibre piecegoods; knitted man-made piecegoods.

34162 IMPORTS (TIME SERIES)

Textile Statistics Bureau

UK - English monthly/£40.40 P.A.

Monthly totals for imports of cotton yarn, spun man-made fibre yarn, woven cotton piecegoods (total and for re-export), woven cotton made up goods (estimated square metres), woven made up fibre piecegoods (total and seven year major sub-divisions).

34163 IMPORTS OF MAN-MADE FIBRE: EXPORTS OF MAN-MADE CONTINUOUS FILAMENT YARN

Textile Statistics Bureau

UK - English monthly/£42.70 P.A.

Two tables show imported and exported yarns by commodity type, with major country, weight value and average value, with weight also being shown as a cumulative total.

34164 IN DEPTH COUNTRY APPRAISALS ACROSS THE EC - CONSTRUCTION AND BUILDING SERVICES - UK

Building Services Research and Information Association

UK - English 1991/£125.00

Part of an 11 volume survey covering the construction industry in EC countries.

34165 INDUSTRIAL PERFORMANCE ANALYSIS

ICC Information Group Ltd

UK - English annual/£95.00

An analysis of industrial sectors using business ratios such as return on capital employed.

34166 INDUSTRIAL TRENDS SURVEY/MONTHLY TRENDS ENQUIRY

Confederation of British Industry (CBI)

UK - English monthly/£395.00 (£240.00 to CBI members) P.A.

Latest short-term current and expected trends in UK manufacturing, monitoring 50 sectors. The quarterly survey covers business optimism, order books, capacity, employment, prices, stocks, investment, training and output. The monthly survey reports orders, stocks, prices and output.

34167 INFOTECH REPORTS

Infotech Services Ltd

UK - English annual/C£195.00 per report

A series of market research reports in a financial ratio format which looks at a variety of consumer and industrial market sectors.

34168 INLAND REVENUE STATISTICS

HMSO - Her Majesty's Stationery Office

UK - English annual/£17.95

Statistics related to all taxes administered by the Inland Revenue, and analysis of personal incomes. Includes data on non-domestic rateable values, property transactions, agricultural land prices, the Business Expansion Scheme, Personal Equity Plans, employee shares, profit related pay, and pension fund surpluses.

34169 INPUT-OUTPUT TABLES FOR THE UK 1984

HMSO - Her Majesty's Stationery Office

UK - English irregular/1988 84pp £19.95 ISBN No. 0115600396

A snapshot of the economy for one time period, illustrating the connections between primary inputs and final demand. A floppy disc is included.

34170 INSTITUTE OF GROCERY DISTRIBUTION REPORTS

Institute of Grocery Distribution

UK - English annual/about £200.00 per report

A series of market research reports on retailing and distribution in the UK. Prices vary with each report. Leading retailers are also profiled.

34171   INTERNATIONAL ROAD HAULAGE BY UK REGISTERED VEHICLES
HMSO - Her Majesty's Stationery Office
UK - English annual/£16.50

34172   INVESTING, LICENSING AND TRADING CONDITIONS ABROAD: UK
Business International Ltd
UK - English annual/£825.00 for complete European series
Loose-leaf format covering the state role in industry, competition rules, price controls, remittability of funds, taxation, incentives and capital sources, labour and foreign trade.

34173   INVESTMENT INTENTIONS SURVEY
Chemical Industries Association
UK - English annual/£40.00
A survey covering investment intentions of CIA member companies. Covers actual expenditure in the previous year together with investment intentions over the coming three years.

34174   IP STATISTICAL SERVICE
Institute of Petroleum
UK - English/£25.00 P.A.
Included are Oil Data sheets with updates quarterly, UK Petroleum Industry Statistics, World Oil statistics annual booklet, Petroleum Statistics card annual, and Oil Data sheet monthly.

34175   JORDAN SURVEYS
Jordan & Sons Ltd
UK - English annual/£135.00 - £350.00 per report
A series of reports which rank companies according to financial criteria, on an industry by industry basis. Typical titles include Britain's top construction companies and Britain's top Japanese owned companies.

34176   KEY BRITISH ENTERPRISES
Dun & Bradstreet International
UK - English annual/£495.00 single issue, £396.00 on subscription
ISSN No. 0142-5048
Three volumes covering 50,000 companies, giving financial details, business activity, contact details, and SIC codes.

34177   KEY DATA
HMSO - Her Majesty's Stationery Office
UK - English annual/£3.95
Contains over 300 tables, maps, and charts, and covers a wide range of social and economic data. References to sources are given.

34178   KEY INDICATORS
Sea Fish Industry Authority
UK - English quarterly/£25.00 P.A.
ISSN No. 0953-8348
Contains data and statistical analyses of UK fish landings, fishing fleet structure, imports and exports, and sales.

34179   KEY NOTE REPORTS: AGRICULTURE
ICC Information Group Ltd
UK - English annual/£170.00 per report
A series of market research reports on the agricultural industry.

34180   KEY NOTE REPORTS: BUILDING MATERIALS
ICC Information Group Ltd
UK - English annual/£170.00 per report
A series of market research reports on the building materials and DIY markets. Typical titles include "DIY" and "Electrical contractors".

34181   KEY NOTE REPORTS: CHEMICAL INDUSTRIES
ICC Information Group Ltd
UK - English annual/£170.00 per report
A series of market research reports on the chemical and related industries. Typical titles include "Plastics processing" and "Adhesives".

34182   KEY NOTE REPORTS: CONSUMER DURABLES
ICC Information Group Ltd
UK - English annual/£170.00 per report
A series of market research reports on the market for consumer durables. Typical titles include "Cutlery" and "Home furnishings".

34183   KEY NOTE REPORTS: COSMETICS
ICC Information Group Ltd
UK - English annual/£170.00 per report
A series of market research reports on the cosmetics and personal care industries. Typical titles include "Hair care products" and "Soaps and detergents".

34184   KEY NOTE REPORTS: ELECTRONICS INDUSTRY
ICC Information Group Ltd
UK - English annual/£170.00 per report
A series of market research reports on the market for electronic products. Typical titles include "Videotex" and "Printed circuits".

34185   KEY NOTE REPORTS: ENGINEERING INDUSTRY
ICC Information Group Ltd
UK - English annual/£170.00 per report
A series of market research reports on the engineering industry. Typical titles include "Aerospace" and "Process plant".

34186   KEY NOTE REPORTS: FINANCE
ICC Information Group Ltd
UK - English annual/£170.00 per report
A series of market research reports on the finance industry and professional services. Typical titles include "Estate agents" and "Debt collecting & factoring".

34187   KEY NOTE REPORTS: FOOD AND DRINKS INDUSTRY
ICC Information Group Ltd
UK - English annual/£170.00 per report
A series of market research reports on the food and drinks industries. Typical titles include "Fast food outlets" and "Chilled foods".

34188   KEY NOTE REPORTS: LEISURE
ICC Information Group Ltd
UK - English annual/£170.00 per report
A series of market research reports on leisure. Typical titles include "Football clubs" and "Tourism in the UK".

34189   KEY NOTE REPORTS: PRINTING AND PACKAGING INDUSTRIES
ICC Information Group Ltd
UK - English annual/£170.00 per report
A series of market research reports on the printing and packaging industries. Typical titles include "Packaging - glass" and "Disposable paper products".

34190   KEY NOTE REPORTS: PUBLISHING INDUSTRY
ICC Information Group Ltd
UK - English annual/£170.00 per report
A series of market research reports on the publishing industry. Typical titles include "Womens magazines" and "Business press".

34191   KEY NOTE REPORTS: RETAILING
ICC Information Group Ltd
UK - English annual/£170.00 per report
A series of market research reports on retailing. Typical titles include "CTN's" and "Retailing chemists".

34192 KEY POPULATION AND VITAL STATISTICS

HMSO - Her Majesty's Stationery Office

UK - English annual/£10.30

Basic demographic data broken down by age, sex etc.

34193 KEYNOTE MARKET REVIEWS

ICC Information Group Ltd

UK - English annual/£295.00 per report

Market research type reports which cover specific industrial sectors, including commissioned consumer research data.

34194 LABOUR MARKET QUARTERLY REPORT

Employment Department

UK - English quarterly/free

Commentary, tables and charts on current labour market trends and the implications for training policy. Includes special features on specific topics.

34195 LEISURE FORECASTS

Leisure Consultants

UK - English annual/1973 £180.00

Published in two volumes. Forecasts for a broad range of leisure activities. Other occasional studies are also published on leisure topics.

34196 LEISURE FUTURES

Henley Centre for Forecasting

UK - English quarterly/£1,125.00 P.A. ISSN No. 0263-7774

Forecasts for the UK leisure market and leisure products.

34197 LEISURE INTELLIGENCE

Mintel International Group Ltd

UK - English quarterly/£895.00 P.A.

Market research on leisure markets and activities, with chapters on each product or market.

34198 LLOYDS BANK ANNUAL REVIEW

Lloyds Bank Plc

UK - English annual/270pp £25.00 ISSN No. 0953-5004

Economic review of the past year. Available from - Pinter Publications, 25 Floral Street, London WC2E 9DS.

34199 LLOYDS BANK ECONOMIC BULLETIN

Lloyds Bank Plc

Patrick Foley/UK - English monthly/ 4pp £12.00 P.A. ISSN No. 0261-0175

General economic bulletin covering growth, inflation, balance of payments etc.

34200 LLOYDS BANK ECONOMIC PROFILE OF GREAT BRITAIN

Lloyds Bank Plc

Economics Department/UK - English annual/26pp free

Economic review of the past year. Avialable from - Pinter Publications, 25 Floral Street, London WC2E 9DS.

34201 LOCAL GOVERNMENT FINANCIAL STATISTICS, ENGLAND

HMSO - Her Majesty's Stationery Office

UK - English annual/£7.50

34202 LONDON STOCK EXCHANGES NEWS

London Stock Exchange

UK - English quarterly/8pp free

Full coverage of the Exchange's activities and current issues affecting members.

34203 LONDON WEIGHTING

The Reward Group

UK - English annual/£125.00

Analysis of payments for the private sector in London and other large towns.

34204 LONG-TERM ADVERTISING EXPENDITURE FORECAST

NTC Publications Ltd

UK - English quarterly/£690.00 P.A.

Forecasts for advertising expenditure and covers the effects of political and economic factors on expenditure.

34205 LONG-TERM BUSINESS STATISTICS

Association of British Insurers

UK - English annual/40pp £100.00 (includes a series of titles)

34206 MACE MARKET RESEARCH REPORTS

Computer Technology Research Corporation

UK - English annual/C£190.00 per report

A series of market research reports on the market for information technology in various sectors.

34207 MACHINE TOOL STATISTICS

Machine Tool Technologies Association

UK - English annual/£50.00

Statistical review by the leading trade association in machine tools.

34208 MAN-MADE FIBRE PRODUCTION

Textile Statistics Bureau

UK - English monthly/£40.40 P.A.

Production and deliveries of staple fibre and filament yarn to home trade and export.

34209 MARKET ASSESSMENT REPORTS: CONSUMER DURABLES

Market Assessment Publications Ltd

UK - English annual/£495.00 per report

Market research type reports which look at the UK market for consumer durables.

34210 MARKET ASSESSMENT REPORTS: COSMETICS

Market Assessment Publications Ltd

UK - English annual/£495.00 per report

Market research type reports which look at the UK market for cosmetic and healthcare products.

34211 MARKET ASSESSMENT REPORTS: DRINKS INDUSTRY

Market Assessment Publications Ltd

UK - English annual/£495.00 per report

Market research type reports which look at the UK market for drinks.

34212 MARKET ASSESSMENT REPORTS: FOOD INDUSTRY

Market Assessment Publications Ltd

UK - English annual/£495.00 per report

Market research type reports which look at the UK market for food and food products.

34213 MARKET ASSESSMENT REPORTS: HOUSEHOLD GOODS

Market Assessment Publications Ltd

UK - English annual/£495.00 per report

Market research type reports which look at the UK market for household non-durable goods.

34214 MARKET ASSESSMENT REPORTS: LEISURE INDUSTRY AND PRODUCTS

Market Assessment Publications Ltd

UK - English annual/£495.00 per report

Market research type reports which look at the UK market for leisure products such as sports goods.

34215 MARKET ASSESSMENT REPORTS: OFFICE EQUIPMENT
Market Assessment Publications Ltd

UK - English annual/£495.00 per report

Market research type reports which look at the UK market for office equipment.

34216 MARKET ASSESSMENT REPORTS: PERSONAL FINANCE
Market Assessment Publications Ltd

UK - English annual/£495.00 per report

Market research type reports which look at the UK market for personal finance products.

34217 MARKET FORECASTS
Market Assessment Publications Ltd

UK - English annual/£250.00 per report

An annual survey of 200 UK markets with forecasts for growth.

34218 MARKET INTELLIGENCE
Mintel International Group Ltd

UK - English monthly/£895.00 P.A.

Market research on mainly consumer items or markets, published in a magazine format with chapters on each product or market.

34219 MARKET RESEARCH GB
Euromonitor Plc

UK - English monthly/£405.00 P.A.
ISSN No. 0308-3447

Six reports every month, mainly on consumer markets and products, in a market research journal format.

34220 MARKET RESEARCH REPORTER
Headland Press

UK - English 20 times per year/ £139.00 P.A.

Covers consumer, retail and industrial sectors.

34221 MARKET SUMMARIES
British Tourist Authority

UK - English 1992/£200.00 full set or £10.00 per country survey.

Digest of tourism facts and figures from the International Passenger Survey, on overseas traffic to the UK from 24 of Britain's major markets. Countries covered include, USA, Canada, Japan, Hong Kong, India, Australia, New Zealand, S.Africa, France, Germany, Belgium, Netherlands, Switzerland, Austria, Italy, Spain, Portugal, Greece, Denmark, Finland, Norway, Sweden, Israel and Brazil.

34222 MARKETING POCKET BOOK
NTC Publications Ltd

UK - English annual/144pp £14.95

Contains economic and demographic data, data on the consumer distribution channels, advertising expenditure, and the media.

34223 MARKETPOWER MARKET RESEARCH REPORTS: FOOD INDUSTRY
Marketpower

UK - English/£350.00 - £9,000.00 per report

Market research reports on the food service and food industries.

34224 MEDIA POCKET BOOK
NTC Publications Ltd

UK - English annual/128pp £16.95

Data on all media channels, with the emphasis on advertising.

34225 MERCHANT FLEET STATISTICS
HMSO - Her Majesty's Stationery Office

UK - England annual/£16.50

34226 MINTEL SPECIAL REPORTS
Mintel International Group Ltd

UK - English/prices vary for each report

Market research on mainly consumer items or markets with some industrial coverage.

34227 MONTHLY DIGEST OF STATISTICS
HMSO - Her Majesty's Stationery Office

UK - English monthly/£80.00 annual subscription, £6.95 single issue
ISSN No. 0308-6666

Basic information on 20 subjects including population, employment and prices, social services, production and output, energy, engineering, construction, transport, catering, and national and overseas finance.

34228 MONTHLY STATISTICAL REVIEW
Society of Motor Manufacturers and Traders Ltd

UK - English monthly/£90.00 P.A.

Statistics on production by manufacturer and type. UK registrations analysed by marque and taxation class. UK imports and exports by product type. Quarterly international production and export data.

34229 MONTHLY STATISTICS DIGEST
Building Societies Association/Council of Mortgage Lenders

UK - English monthly/£48.00 P.A.
ISSN No. 0261-6416

Covers mortgages and savings and includes a commentary.

34230 MORTGAGE MONTHLY
Building Societies Association/Council of Mortgage Lenders

UK - English monthly/£12.00 P.A.

Reports and press releases covering the housing and mortgage markets, plus basic data on the housing finance industry.

34231 MORTGAGE WEEKLY
Building Societies Association/Council of Mortgage Lenders

UK - English weekly/£250.00 P.A.

News reports and press releases covering the housing and mortgage markets, plus basic data on the housing finance industry.

34232 MOTOR INDUSTRY GREAT BRITAIN - WORLD AUTOMOTIVE STATISTICS
Society of Motor Manufacturers and Traders Ltd

UK - English annual/260pp £82.00 P.A.

Divided into five sections, it includes statistics on production, registrations, vehicles in use, and overseas trade. Covers both cars and commercial vehicles.

34233 MOTOR VEHICLE REGISTRATION INFORMATION SYSTEM (MVRIS)
Society of Motor Manufacturers and Traders Ltd

UK - English/price on application

Gives detailed statistics on new registrations of cars and commercial vehicles. Standard monthly and individually tailored reports can be obtained. Produced by the SMMT MVRIS department.

34234 MOTORPARC STATISTICS SERVICE (MPSS) - VEHICLES IN USE
Society of Motor Manufacturers and Traders Ltd

UK - English/price on application

Standard and tailored reports available on application. Uses the annual census information of the Driver and Vehicle Licensing Centre (DVLC). Gives statistics by vehicle category, marque, model and age at post code level for Great Britain.

34235 MSI REPORTS: BUILDING MATERIALS
Marketing Strategies for Industry
UK - English irregular/£90.00 - £450.00 per report
Market research type reports which look at the UK market for various building materials and products. Reports cost about £90-£450 each depending on depth of report.

34236 MSI REPORTS: CONSUMER BEHAVIOUR
Marketing Strategies for Industry
UK - English irregular/£90.00 - £450.00 per report
Market research type reports which look at UK consumer behaviour. Reports cost about £90-£450 each depending on depth of report.

34237 MSI REPORTS: CONSUMER DURABLES
Marketing Strategies for Industry
UK - English irregular/£90.00 - £450.00 per report
Market research type reports which look at the UK consumer durable markets. Reports cost about £90-£450 each depending on the depth of report.

34238 MSI REPORTS: DRINKS INDUSTRY
Marketing Strategies for Industry
UK - English irregular/£90.00 - £450.00 per report
Market research type reports which look at the UK market for alcoholic drink and non-alcoholic drink products. Reports cost about £90-£450 each depending on depth of report.

34239 MSI REPORTS: ENGINEERING AND ENGINEERING PRODUCTS
Marketing Strategies for Industry
UK - English irregular/£90.00 - £450.00 per report
Market research type reports which look at the UK markets for engineering products and processes. Reports cost about £90-£450 each depending on depth of report.

34240 MSI REPORTS: FOOD INDUSTRY
Marketing Strategies for Industry
UK - English irregular/£90.00 - £450.00 per report
Market research type reports which look at the UK market for food products. Reports cost about £90-£450 each depending on depth of report.

34241 MSI REPORTS: HEALTHCARE
Marketing Strategies for Industry
UK - English irregular/£90.00 - £450.00 per report
A series of market research type reports which look at the UK market for healthcare products. Reports cost about £90-£450 depending on depth of report.

34242 MSI REPORTS: HORTICULTURE INDUSTRY
Marketing Strategies for Industry
UK - English irregular/£90.00 - £450.00 per report
Market research type reports which look at the UK markets for gardening and horticultural products. Reports cost about £90-£450 each depending on depth of report.

34243 MSI REPORTS: LEISURE INDUSTRY
Marketing Strategies for Industry
UK - English irregular/£90.00 - £450.00 per report
Market research type reports which look at the UK leisure industry and leisure products. Reports cost about £90-£450 each depending on depth of report.

34244 MSI REPORTS: RETAILING AND DISTRIBUTION
Marketing Strategies for Industry
UK - English irregular/£90.00 - £450.00 per report
Market research type reports which look at the UK retailing and distribution sectors. Reports cost about £90-£450 each depending on depth of report.

34245 NATIONAL FACTS OF TOURISM
British Tourist Authority
UK - English irregular/1992 £00.50
Reference card summarizing data on British Tourism.

34246 NEW EARNINGS SURVEY
HMSO - Her Majesty's Stationery Office
UK - English annual in six parts/ £63.00 P.A.
Earnings of employees by industry, occupation, region etc at April each year.

34247 NHBC PRIVATE HOUSE-BUILDING STATISTICS
National House Building Council
UK - English quarterly/20pp £50.00 P.A.

Statistics on private house-building in the UK taken from NHBC's 25,000 members and covering 90% of all private house-building.

34248 NORTHERN IRELAND AGRICULTURAL CENSUS STATISTICS
Department of Agriculture Northern Ireland DANI
UK - English annual/price on application

34249 NORTHERN IRELAND ANNUAL ABSTRACT OF STATISTICS
Northern Ireland Office
UK - English annual/price on application
Abstract of statistics covering main statistical series.

34250 OECD ECONOMIC SURVEY - UNITED KINGDOM
Organisation for Economic Co-operation and Development (OECD)
UK - English annual/£102.00 ISSN No. 0376-6438
Comprehensive economic survey with forecasts.

34251 OTR-PEDDER REPORTS: COMPUTERS AND COMPUTER APPLICATIONS
OTR-Pedder
UK - English annual/£2,500 per report
A series of market research reports on computers and related markets. Typical titles include "GP computers" & "Software package trends".

34252 OVERSEAS CONFERENCE VISITORS TO THE UK IN 1990
British Tourist Authority
UK - English irregular/1991 £15.00
Basic facts about business visitors to the UK who attended a conference during the course of their stay. Includes some past data for comparative purposes.

34253 OVERSEAS TRADE ANALYSED IN TERMS OF INDUSTRIES (BUSINESS MONITOR MQ10)
HMSO - Her Majesty's Stationery Office
UK - English quarterly/£21.00 P.A.
Gives an analysis of commodities imported and exported, broken down by the industry producing the products.

34254 OVERSEAS TRADE FAIR/ EXHIBITION VISITORS TO THE UK
British Tourist Authority
UK - English irregular/£10.00
Details of market size, average spending and length of stay, country and area of residence, and seasonal distribution.

34255 OVERSEAS TRADE STATISTICS OF THE UNITED KINGDOM (BUSINESS MONITOR MM20/MA20)
HMSO - Her Majesty's Stationery Office
UK - English monthly/£385.00 P.A.
Gives detailed statistics of exports and imports cumulated throughout the year. Annual volumes are also produced. The monthly volume is known as MM20. December edition of monthly volume is available, price £50.00. The Annual Revised edition, known as MA20, is also available, price £50.00.

34256 PERFORMANCE RANKINGS GUIDE
The Hambro Company Guide
UK - English twice a year/£135.00 P.A.
Looks at UK corporate performance and gives comparitive league tables by industrial sector.

34257 PERSONAL FINANCE INTELLIGENCE
Mintel International Group Ltd
UK - English quarterly/£995.00 P.A.
Market research on personal finance products and markets, with chapters on each product or market.

34258 PERSONNEL REWARDS
The Reward Group
UK - English annual/£200.00
Details of salaries and benefits for all personnel and related jobs. Produced in association with the Institute of Personnel Management.

34259 PIRA REPORTS
Packaging Industry Research Association (PIRA)
UK - English/£350.00 - £650.00 per report
A series of market research reports covering the packaging industry, giving details of markets and companies.

34260 PLANNING CONSUMER MARKETS
Henley Centre for Forecasting
UK - English quarterly/£1,050.00 P.A. ISSN No. 0308-7751

Review of consumer spending, future trends and markets.

34261 PLANNING FOR SOCIAL CHANGE
Henley Centre for Forecasting
UK - English annual/£8,000.00 ISSN No. 0140-2779
Socio-economic reviews and forecasts of consumer attitudes, motivation and behaviour.

34262 POPULATION ESTIMATES SCOTLAND
General Register Office (Scotland)
UK - English annual/£2.30 ISSN No. 0066-3964
Also available on floppy disc.

34263 POPULATION PROJECTIONS
Office of Population Censuses and Surveys
UK - English biennial/£11.30
Latest projections, published on microfiche.

34264 POPULATION TRENDS
Office of Population Censuses and Surveys
UK - English quarterly/£31.60 annual subscription, £8.15 single issue ISSN No. 0307-4463
Regular series and tables on a variety of population and medical topics.

34265 PORT STATISTICS
British Ports Federation
UK - English annual/price on application

34266 PROPERTY INDEX
Jones Lang Wootton
UK - English annual/price on application
Data collected in four aggregate sectors; shops, offices, industrial and warehouses, and agricultural.

34267 QUALITY OF MARKETS COMPANIES BOOK
London Stock Exchange
UK - English annual/240pp £35.00
1,000 largest listed UK companies, 500 most active Stock Exchange companies, turnover and market value, newly admitted companies, and overseas companies trading on the exchange.

34268 QUALITY OF MARKETS MONTHLY FACT SHEET
London Stock Exchange
UK - English monthly/16pp £35.00 P.A.

Monthly round-up of activity. Includes total market turnover, daily market movements, new issues, traded options, gilts and fixed interest.

34269 QUARTERLY COUNTRY TRADE REPORTS: UK
United Kingdom Iron and Steel Statistics Bureau
UK - English quarterly/£200.00 P.A. for four countries ISSN No. 0307-7608
Cumulative trade statistics of imports and exports of steel products.

34270 QUARTERLY GENERAL BUSINESS CLAIMS
Association of British Insurers
UK - English quarterly/£100.00 (price includes a series of titles)
Figures on gross incurred claims received during the latest quarter. Split between fire, theft and weather damage claims and by commercial and domestic business. Figures for business interruption and domestic subsidence claims. Figures given for latest eight quarters.

34271 QUARTERLY NEW LONG-TERM BUSINESS
Association of British Insurers
UK - English quarterly/£100.00 P.A. (price includes a series of titles)
Detailed figures on new business written during the latest quarter. Brief comments on major influences.

34272 QUARTERLY STATISTICAL REVIEW
Textile Statistics Bureau
UK - English quarterly/£103.95 P.A.
Main indicators of activity and foreign trade in textiles and clothing. Includes overseas statistics for major countries only in the spun yarn and woven cloth sectors.

34273 QUARTERLY SURVEY OF ADVERTISING EXPENDITURE
NTC Publications Ltd
UK - English quarterly/145pp £295.00 P.A. ISSN No. 0951-7766
Advertising expenditure divided by national and regional media.

34274 QUARTERLY TRANSPORT STATISTICS
Department of Transport
UK - English quarterly/annual subscription £12.00, single issue £5.00.

34275 REGION OF RESIDENCE OF ITALIAN VISITORS TO THE UK
British Tourist Authority
UK - English irregular/£10.00

Italian visitor market analysed by region of residence in Italy, including details of market size, average spending, length of stay, and propensities to travel.

34276 REGIONAL DIRECTORIES OF KEY BUSINESS PROSPECTS

Jordan & Sons Ltd

UK - English annual/price on application

A series of reports which rank companies according to financial criteria, on a regional basis; North East, Wales etc.

34277 REGIONAL MARKETING POCKET BOOK

NTC Publications Ltd

UK - English annual/£14.95

Contains economic and demographic data, data on the consumer, distribution channels, advertising expenditure, and the media.

34278 REGIONAL SALARY AND WAGES SURVEY

The Reward Group

UK - English twice a year/£180.00 P.A.

Payments and analysis of salaries, wages, and conditions of service for 17 regions and 37 UK towns.

34279 REGIONAL TOURISM FACTS

English Tourist Board

UK - English annual/£90.00 per full set or £10.00 per individual report

A set of twelve regional reports detailing tourism facts in graphs, diagrams, tables and text for each English tourism region.

34280 REGIONAL TRENDS

Office of Population Censuses and Surveys

UK - English annual/200pp £24.75 ISSN No. 0261-1783

The primary source of official statistics about the standard regions of the UK. Includes data on population, law and crime, earnings, industry and education.

34281 THE REGISTRAR GENERAL'S ANNUAL REPORT FOR NORTHERN IRELAND

General Register Office (Northern Ireland)

UK - English annual/price on application

Population statistics.

34282 REGISTRAR GENERAL'S ANNUAL REPORT FOR SCOTLAND

General Register Office (Scotland)

UK - English annual/price on

application ISSN No. 0080-7869

Population figures for Scotland, broken down by age, sex etc.

34283 REPORT OF THE COMMISSIONERS OF HM REVENUE

HMSO - Her Majesty's Stationery Office

UK - English annual/£13.70

Includes statistics of all taxes administered by the Inland Revenue, in particular, tax collected, recovered by investigation, remitted, and costs of collection.

34284 REPORTS

German Chamber of Industry & Commerce

UK/German - English/price varies with report

Individual reports on specific industries are available on request.

34285 RESEARCH AND DEVELOPMENT

The Reward Group

UK - English annual/£190.00

Analyses basic pay and total renumeration payments in R & D organisations and departments.

34286 RETAIL BUSINESS MARKET REPORTS

Business International Ltd

UK - English monthly/£485.00 P.A. ISSN No. 0034-6012

Covers consumer market surveys, economic and technological reviews, and product sector overviews.

34287 RETAIL BUSINESS TRADE REVIEWS

Business International Ltd

UK - English quarterly/£190.00 P.A. ISSN No. 0951-9742

Covers one retail sector on an annual basis and profiles the leading retailers in that sector.

34288 RETAIL INTELLIGENCE

Mintel International Group Ltd

UK - English every 2 months/ £945.00 P.A.

Market research on mainly consumer items or markets, with chapters on each product or market.

34289 RETAIL MONITOR INTERNATIONAL

Euromonitor Plc

UK/International - English monthly/ £350.00

34290 RETAIL POCKET BOOK

NTC Publications Ltd

UK - English annual/160pp £22.00

Compendium of information about retailing in the UK with some international statistics included.

34291 RETAIL RANKINGS

Corporate Intelligence Research Publications

UK - English annual/350pp £175.00

Market research type report which looks at the retailing sector in the UK and ranks the top 650 plus retailers.

34292 RETAIL RATIOS

Corporate Intelligence Research Publications

UK - English annual/1992 150pp £160.00

Financial and other ratios applied to the top 550 UK retailers.

34293 RETAIL RESEARCH REPORT

Corporate Intelligence Research Publications

UK - English 10 reports per year/ £495.00

Market research type report which looks at the retailing sector in the UK.

34294 RETAIL SERVICE RANKINGS

Corporate Intelligence Research Publications

UK - English annual/175pp £145.00

Market research type report which looks at the service sector in the UK and ranks the top 450 retailers.

34295 REWARD - THE MANAGEMENT SALARY SURVEY

The Reward Group

UK - English twice a year/£310.00 P.A. single issues £180.00

Survey of management and technical staff analysing salaries for over 140 main job occupations.

34296 ROAD GOODS VEHICLES ON ROLL-ON-ROLL-OFF FERRIES TO MAINLAND EUROPE

Department of Transport

UK - English quarterly/£5.00 P.A.

34297 SALARY SURVEY FOR SMALLER BUSINESSES

Institute of Management

UK - English annual/£160.00

Condensed version of the 1992 National Management Salary Survey, with a special emphasis on the interests of smaller firms.

34298 SALES AND MARKETING REWARDS

The Reward Group

UK - English annual/£200.00

A survey giving details of salaries and benefits for marketing and sales positions. Published in association with the Institute of Marketing.

34299 SCOTLAND'S TOP 2000 COMPANIES

Jordan & Sons Ltd

UK - English annual/£190.00

Ranks companies by turnover, gives contact details, activities and brief descriptions of the company.

34300 SCOTTISH ABSTRACT OF STATISTICS

Scottish Office Library

UK - English annual/£18.00

A broad selection of statistics that are available covering Scotland. Major socio-economic areas are covered. Latest edition available is No. 19. 1990.

34301 SCOTTISH ECONOMIC BULLETIN

Scottish Office Library

UK - English 6 monthly/£10.00 P.A.

Contains an economic review, articles on the Scottish economy and regular statistics.

34302 SCOTTISH FISHING FLEET

Scottish Office Library

UK - English annual/£6.00

Statistical coverage including numbers and types operating.

34303 SCOTTISH LOCAL GOVERNMENT FINANCIAL STATISTICS

Scottish Office Library

UK - English annual/£5.00

Summary volumes for 1975-76 and 1987-88 are also available price £5.00.

34304 SCOTTISH SEA FISHERIES STATISTICAL TABLE

Scottish Office Library

UK - English annual/£11.00 ISSN No. 0080-8202

Details numbers operating, catches etc.

34305 SCOTTISH TRANSPORT STATISTICS

Scottish Office Library

UK - English annual/£6.00

Major forms of transport are detailed statistically.

34306 SEA FISHERIES STATISTICAL TABLES

HMSO - Her Majesty's Stationery Office

UK - English annual/50pp £10.00

Broad picture of the UK fishing industry. Information is included on landings of fish, consumption level estimates, fishermen and fishing vessels, and international trade.

34307 SEABORNE TRADE STATISTICS OF THE UNITED KINGDOM

HMSO - Her Majesty's Stationery Office

UK - English annual/£18.50

34308 SIGHTSEEING IN THE UK

English Tourist Board

UK - English annual/£19.00

Analysis of the use and capacity of attractions based on a survey of all main tourist sights. Includes numbers of visitors, new attractions, opening periods, admission charges, overseas visitors, demand relative to capacity, employment, capital expenditure and revenue trends.

34309 SMALL BUSINESS TRENDS REPORT

Durham University Business School

UK - English annual/250pp £105.00

Analysis of research and statistics on small businesses and related trends.

34310 SOCIAL SECURITY STATISTICS

HMSO - Her Majesty's Stationery Office

UK - English annual/£15.90

34311 SOCIAL TRENDS

HMSO - Her Majesty's Stationery Office

UK - English annual/256pp £24.75

Statistics on a wide range of areas including leisure, education. health, transport and housing.

34312 SOURCES OF NEW PREMIUM INCOME

Association of British Insurers

UK - English annual/5pp £100.00 (price includes a series of titles)

Shows the percentage of new premium income for long-term business sold through the various distribution channels. Comparable figures from 1989.

34313 SPINNING AND WASTE SPINNING

Textile Statistics Bureau

UK - English monthly/£40.40 P.A.

Includes end-use deliveries on a quarterly basis. Production, fibre consumption, employment, spindles in place and active, hours and shifts worked are detailed.

34314 STANILAND HALL ASSOCIATES MARKET RESEARCH REPORTS

Staniland Hall Associates

UK - English/£125.00 - £200.00

A range of market research reports, examples being those on the leisure and DIY markets.

34315 STATISTICAL DIGEST FOR THE FURNITURE INDUSTRY

Furniture Industry Research Association

UK - English annual/60pp £140.00 (FIRA members £60.00)

Output of furniture production, imports, exports, retail sales and price indices.

34316 STATISTICAL NEWS

HMSO - Her Majesty's Stationery Office

UK - English quarterly/£5.50 single issue, £20.00 annual subscription ISSN No. 0017-3630

Comprehensive account of current developments in British statistics to help all who use or would like to use official statistics. Shorter notes give news of the latest developments in many fields including international statistics.

34317 STATISTICAL REVIEW OF NORTHERN IRELAND AGRICULTURE

Department of Agriculture Northern Ireland DANI

UK - English annual/£10.00

34318 STOCK EXCHANGE FACT SERVICE

London Stock Exchange

UK - English irregular/£95.00

Includes a selection of publications - Stock Exchange Quarterly, FT-SE Share indices five year history. Quality of markets monthly fact sheet, and Quality of markets companies book.

34319 STOCK EXCHANGE OFFICIAL YEARBOOK

London Stock Exchange

UK - English annual/1135pp £185.00

Covers the securities market, trading, settlement and transfers, taxation, trustees, domestic and overseas securities, companies and public corporations, and lists registrars.

34320 STOCK EXCHANGE QUARTERLY

London Stock Exchange

UK - English quarterly/96pp £60.00 P.A.

A comprehensive statistical section and analysis of the Exchange's markets and their performance.

34321 SUPPLY ESTIMATES - SUMMARY AND GUIDE

HMSO - Her Majesty's Stationery Office

UK - English annual/price on application

Details of estimated central government cash requirements for the forthcoming financial year.

34322 TIME RATES OF WAGES AND HOURS OF WORK
Employment Department
UK - English annual/£43.00
Loose-leaf publication with monthly updates.

34323 TIMES 1000
Time Books
UK - English annual/£29.00 ISSN No. 0082-4429
Gives basic financial data on the top 1000 UK companies ranked by performance.

34324 TOP MARKETS
Market Assessment Publications Ltd
UK - English annual/£250.00
An annual survey of the top UK markets in executive summary format.

34325 TOURISM AND THE EUROPEAN COMMUNITY
British Tourist Authority
UK - English 1990/£10.00
Covers the impact of the single market on UK tourism. Includes analysis of abolition of frontier controls, travel liberalization, VAT and excise duties, and working conditions.

34326 TOURISM INTELLIGENCE QUARTERLY
British Tourist Authority
UK - English quarterly/£75.00
Statistical data relating to UK and overseas tourism.

34327 TOY INDUSTRY IN THE UK
British Toy and Hobby Association
UK - English annual/£98.00
Market research type report on British toys and hobbys.

34328 TRADE BULLETIN - UK IMPORTS AND EXPORTS OF FISH AND FISH PRODUCTS
Sea Fish Industry Authority
UK - English monthly/£25.00 P.A. ISSN No. 0963-9446
Gives the quantity and value of imports and exports of fish intended for human consumption based on HM Customs and Excise data. The periods covered are the latest month and the year to date, with figures for the same period of the preceding year provided for comparison.

34329 TRAFFIC IN GREAT BRITAIN
Department of Transport
UK - English quarterly/£8.00 P.A. Annual subscription £27.00, quarterly £8.00.

34330 TRAINING STATISTICS
HMSO - Her Majesty's Stationery Office
UK - English annual/£11.25
Statistical tables and charts compiled from a wide range of sources. Technical summaries and notes on sources included.

34331 TRANSPORT OF GOODS BY ROAD IN GREAT BRITAIN
HMSO - Her Majesty's Stationery Office
UK - English annual/£12.00
Covers levels, volume and type of goods transported.

34332 TRANSPORT STATISTICS FOR LONDON 1980-1990
HMSO - Her Majesty's Stationery Office
UK - English 1991/£10.90 ISBN No. 0115510427

34333 TRANSPORT STATISTICS GREAT BRITAIN
HMSO - Her Majesty's Stationery Office
UK - English annual/£24.00

34334 UK AIRLINES - ANNUAL OPERATING, TRAFFIC AND FINANCIAL STATISTICS (CAP 596)
Civil Aviation Authority
UK - English annual/£12.50

34335 UK AIRLINES - MONTHLY OPERATING AND TRAFFIC STATISTICS
Civil Aviation Authority
UK - English monthly/£43.00 P.A. ISSN No. 0265-0266

34336 UK AIRPORTS - ANNUAL STATEMENTS OF MOVEMENTS, PASSENGERS, AND CARGO - (CAP 592)
Civil Aviation Authority
UK - English annual/£11.00
Statistical survey of all types of air transport.

34337 UK AIRPORTS - MONTHLY STATEMENTS OF MOVEMENTS, PASSENGERS, AND CARGO
Civil Aviation Authority
UK - English monthly/£43.00 P.A. ISSN No. 0265-0258

34338 UK BALANCE OF PAYMENTS (CSO PINK BOOK)
HMSO - Her Majesty's Stationery Office
UK - English annual/80pp £11.75
The principal annual publication for balance of payments statistics, covering visible trade, invisibles, investments, capital transactions, and official financing. Also covers specific aspects of the balance of payments, with data for the last 11 years and full notes and definitions.

34339 UK CHEMICAL INDUSTRY ENVIRONMENTAL PROTECTION SPENDING SURVEY
Chemical Industries Association
UK - English annual/£40.00
Covers CIA member companies capital and current spending on environmental protection.

34340 UK ELECTRICITY
Electricity Association
Mrs D. Dasgupta/UK - English annual/1991 50pp £17.00
Electricity industry in the UK with European comparisons.

34341 UK EXHIBITION INDUSTRY: THE FACTS
British Tourist Authority
UK - English annual/£150.00
Compiled in association with the Exhibition Industry Federation. Provides statistics on the overall volume and value, plus details of the composition of the industry.

34342 UK IRON AND STEEL INDUSTRY: ANNUAL STATISTICS
United Kingdom Iron and Steel Statistics Bureau
UK - English annual/£60.00 ISSN No. 0952-5505
47 detailed statistical tables relating to the UK iron and steel industry with historical comparisons and trade information.

34343 UK MARKET STATISTICS
Association of British Insurers
UK - English annual/20pp £100.00 (price includes a series of titles)
Estimate of total premium income of UK insurance industry. Figures include all companies authorized in the UK, Lloyds, Mutual Clubs and Associations, and Friendly Societies. Data is split between general business accounting classes and long-term business.

34344   UK NATIONAL ACCOUNTS (CSO BLUE BOOK)
HMSO - Her Majesty's Stationery Office

UK - English annual/157pp £13.95

The principal annual publication for the national accounts statistics providing detailed estimates of national product, income and expenditure for the UK. It covers value-added by industry, personal sector, public operations, central and local government, capital formation and financial accounts.

34345   UK PRINTING AND PUBLISHING INDUSTRY STATISTICS
Packaging Industry Research Association (PIRA)

UK - English annual/£450.00

34346   UK RETAIL MARKETING SURVEY
Institute of Petroleum

UK - English annual/price on application

Released with the March issue of "Petroleum Review".

34347   UK RETAIL REPORTS
Corporate Intelligence Research Publications

UK - English/1992 70-140pp £145.00 per report

Six market research type reports which cover discount food retailing, charity shops, garden centres, factory shops, petrol forecourt shops, and convenience stores.

34348   UK TOURIST: STATISTICS 199–
English Tourist Board

UK - English annual/£55.00

Covers the volume, value and characteristics of UK residents tourism. Based on the UK Tourism Survey.

34349   UK'S 10,000 LARGEST COMPANIES
ELC International

UK - English annual/£170.00

Marketing and financial data on companies, giving the industrial sector, region, and a series of company rankings.

34350   UNITED KINGDOM MINERALS YEARBOOK
British Geological Survey

UK - English annual/£27.50 ISSN No. 0957-4697

Production and consumption and trade statistics for UK minerals. 1991 edition gives data from 1986 to 1990 and a commentary for 1991.

34351   VERDICT RETAIL REPORTS
Verdict Research

UK - English/£125.00 per report

A series of market research reports on different types of retailers.

34352   VISITORS TO TOURIST ATTRACTIONS
British Tourist Authority

UK - English annual/£11.00

Details of attendances at tourist attractions in the UK. Includes historic houses, gardens, museums, galleries, wildlife parks etc. Data is published 1-2 years in arrears.

34353   WATERBORNE FREIGHT IN THE UNITED KINGDOM
HMSO - Her Majesty's Stationery Office

UK - English annual/price on application

34354   WEAVING STATISTICS
Textile Statistics Bureau

UK - English monthly/£40.40 P.A.

Production, employment, looms in place and active, hours and shifts worked, and yarn consumed are detailed.

34355   WELSH AGRICULTURAL STATISTICS
Welsh Office

UK - English annual/£5.00

Broad coverage of Welsh agriculture.

34356   WELSH ECONOMIC TRENDS
Welsh Office

UK - English biennial/£9.50

Economic series for Wales. A supplement is also available - price £5.00. Data is published 1-2 years in arrears.

34357   WELSH HOUSING STATISTICS
Welsh Office

UK - English annual/£4.00

34358   WELSH LOCAL GOVERNMENT FINANCIAL STATISTICS
Welsh Office

UK - English annual/£7.00

34359   WELSH TRANSPORT STATISTICS
Welsh Office

UK - English annual/£6.00

Covers all major modes of transport and shows levels and volume for each.

YUGOSLAVIA

Country profile: Yugoslavia

Official name: Not applicable *

Area: 225,804 sq.km.

Bordered in the north by Hungary and Romania, in the south by Albania and Greece, the east by Bulgaria, and the west by the Adriatic Sea .

Population: 23.46m (1991)

Capital: Not applicable *

Language(s):   Slovene, Macedonian, Serbo-Croatian

Currency: Not applicable *

National bank: Not applicable *

Government and Political structure: Yugoslavia was a federation of six republics: Bosnia and Herzegovina, Croatia, Macedonia, Montenegro, Serbia, and Slovenia, with two autonomous provinces within Serbia; Kosovo and Vojvodina.

* At the time of writing the former state of Yugoslavia was in a state of flux. The sources and data detailed refer to the pre-1992 period unless otherwise indicated.

Yugoslavia

35001 COUNTRY FORECASTS: YUGOSLAVIA
Business International Ltd

Yugoslavia - English quarterly/ £360.00 P.A.

Contains an executive summary, sections on political and economic outlooks and the business environment, a fact sheet, and key economic indicators. Includes a quarterly Global Outlook.

35002 COUNTRY PROFILE - YUGOSLAV REPUBLICS
Business International Ltd

Yugoslav republics - English annual/44pp £50.00

Annual survey of political and economical background.

35003 COUNTRY REPORT - YUGOSLAV REPUBLICS
Business International Ltd

Yugoslav republics - English quarterly/29pp £150.00 P.A. (4 issues + a country profile)

Analysis of economic and political trends every quarter. Contains an executive summary, sections on political and economic structure, outlook for the next 12-18 months, statistical appendices, and a review of key political developments. Includes an annual Country Profile.

35004 COUNTRY RISK SERVICE: SLOVENIA
Business International Ltd

Yugoslavia - English quarterly/ minimum subscription £1,855.00 P.A. for seven countries

35005 COUNTRY RISK SERVICE: YUGOSLAV REPUBLICS
Business International Ltd

Yugoslavia - English quarterly/ minimum subscription £1,855.00 for seven countries

Contains credit risk ratings, risk appraisals, cross-country databases, data on diskette, and two year projections.

35006 DOING BUSINESS WITH EASTERN EUROPE: SERBIA-MONTENEGRO
Business International Ltd

Serbia-Montenegro - English quarterly/£540.00 P.A.

Part of a ten volume set, updated every three months which gives a broad picture of the market and business practice.

35007 ECONOMIC REVIEW
Privredni Pregled

Yugoslavia/Serbia - English monthly/1955 US$18.00 P.A. ISSN No. 0013-0303

Business economics and economic conditions are covered.

35008 EKONOMIST
Savez Ekonomista Jugoslavije

Oto Norcic/Yugoslavia/Serbia - Contents pages and summaries in English/Russian quarterly/1948 YD11,000 P.A. ISSN No. 0013-3191

Business and economic review.

35009 EKONOMSKA MISAO
Savez Ekonomista Srbije

Lbujomir Madzar/Yugoslavia/ Serbia - Serbo/Croatian/ Summaries in English and Russian quarterly/1968 US$140.00 P.A. ISSN No. 0013-323X

Reviews the Yugoslav economy and provides background data.

35010 EKONOMSKI ANALI
Ekonomski Fakultet u Beogradu

Zarko Bulajic/Yugoslavia/Serbia - Serbo/Croatian quarterly/1951 YD100.00 P.A. ISSN No. 0013-3264

Economic review with some business coverage also.

35011 FINANSIJE
Privredni Pregled

Bogoljub Lazarevic/Yugoslavia/ Serbia - bi-monthly/1956 US$41.70 P.A. ISSN No. 0015-2145

Review of business, economics, banking and finance.

35012 GLOBAL FORECASTING SERVICE - YUGOSLAVIA
Business International Ltd

Yugoslavia - English quarterly/23pp

Economic forecasts.

35013 JOURNAL OF YUGOSLAV FOREIGN TRADE
Export Press

D. Kostic/Yugoslavia/Serbia - English/Russian quarterly/1954 US$8.00 P.A. ISSN No. 0022-5452

Covers business, economics and international commerce.

35014 JUGOSLOVENSKI PREGLED
Jugoslovenski Pregled

Llija Borovnjak/Yugoslavia/Serbia - Serbo/Croatian monthly/US$30.00 P.A. ISSN No. 0022-6114

General economic background. An English edition is also published under the title "Yugoslav Survey" ISSN No. 0044-1341.

35015 NASE GOSPODARSTVO/ OUR ECONOMY: REVIEW OF CURRENT PROBLEMS IN ECONOMICS
Ekonomkso - Poslovna Fakulteta

Davor Savin/Yugoslavia/Slovenia - Slovenian with summaries in English bi-monthly/1955 US$30.00 P.A. ISSN No. 0547-3101

Contains broad economic background details.

Part Two
PUBLISHING BODIES

168 HOUNS LTD

47 Tsarigadska Chausse
Blvd
1504 Sofia
Bulgaria
Tel: +359 (0)44 34 77
Fax: +359 (0)46 32 54

Entry: 08001

ABC PUBLISHING GROUP

POB 4034
W-6100 Darmstadt
Germany
Tel: +49 (0)6151 38 920
Telex: 4 19 257
Fax: +49 (0)6151 33 164

Entries: 15002 15134 15135
15136 15137 15138
15139 15140 15141
15142 15143

ABECOR REPORTS

Order from:
BARCLAYS BANK
ECONOMICS
DEPARTMENT
1, Wimborne Road
Poole
Dorset BH15 2BB
UK
Tel: +44 (0)1202 671212
Telex: 417221 BARPL
Fax: +44 (0)1202 671212 X
4145

Entry: 16001

**ACADEMIC INTERNATIONAL
PRESS**

PO Box 111
Gulf Breeze
FL32561
USA

Entries: 09039 09040

**ADMINISTRATION FÉDÉRALE DES
CONTRIBUTIONS: DIVISION
STATISTIQUE DE
DOCUMENTATION**

Eigerstrasse 65
CH-3003 Berne
Switzerland
Tel: +41 (0)31 61 73 75/
0041(0)31 61 73 77
Fax: +41 (0)31 61 71 67

Entries: 32030 32039

ADVERTISING ASSOCIATION

Abford House
15 Wilton Road
London SW1V 1NJ
UK
Tel: +44 (0)171 828 2771
Fax: +44 (0)171 931 0376

Entry: 34016

**AFFÄRSVÄRLDEN AND THE
SWEDISH ASSOCIATION FOR
SHARE PROMOTION**

S-10612 Stockholm
Sweden
Tel: +46 (0)8 796 65 00
Fax: +46 (0)8 20 21 57

Entry: 31124

AGPOL POLERPOSTPRESS

Marszalkowska 124
00-950 Warszawa
PO Box 726
Poland
Tel: +48 (0)22 27 37 28

Entry: 27053

AGRA EUROPE (LONDON) LTD

25 Frant Road
Tunbridge Wells
Kent
TN2 5JT
UK
Tel: +44 (0)1892 533813
Fax: +44 (0)1892 544895

Entry: 02001

**AGRICULTURAL ECONOMICS
UNIT**

Wye College
Ashford
Kent TN25 5AH
UK
Tel: +44 (0)1233 812401
Telex: 94017832 WYEC G
Fax: +44 (0)1233 813498

Entries: 34129

AIB BANK

Bankcentre
PO Box 452
Ballsbridge
Dublin 4
Republic of Ireland
Tel: +353 (0)1 600 311
Fax: +353 (0)1 682 508

Entry: 20098

**ALBANIAN CHAMBER OF
COMMERCE**

Rruga Konferenca e Pezes 6
Tirana
Albania
Tel: +355 (0)42 22934
Telex: 2179

Entry: 04006

AMT FÜR VOLKWIRTSCHAFT

Kirchstr 7
FL-9490 Vaduz
Liechtenstein
Tel: +4175 661 11
Fax: +4175 66889

Entries: 22001 22002 22003
22004 22007 22008
22010 22013 22014

22017 22021 22022
22023

**ANGLO-SWEDISH TRADE CO-
OPERATION**

Scandinavian News AB
Box 405
S-20124 Malmo 1
Sweden
Tel: +46 (0)40 10 48 75
Telex: 33646 NEWS S
Fax: +46 (0)40 97 35 89

Entry: 31127

ANSWERS RESEARCH LTD

140 Borden Lane
Sittingbourne
Kent ME9 8HW
UK
Tel: +44 (0)1795 423 778
Fax: +44 (0)1795 788 744

Entries: 01011 01012 01013
01014 01015 02013
34019 34020 34021

**ASSOCIATION OF BRITISH
INSURERS**

51 Gresham Street
London EC3V 7HQ
UK
Tel: +44 (0)171 600 3333
Fax: +44 (0)171 696 8999

Entries: 34068 34141
34205 34270 34271
34312 34343

**ASSOCIATION OF CHAMBERS OF
COMMERCE OF IRELAND**

Jude Publications Ltd
2-6 Tara Street
Dublin 2
Republic of Ireland
Tel: +353 (0)1 713 500
Fax: +353 (0)1 713 074

Entry: 20055

**ASSOCIATION OF CONSULTING
ENGINEERS**

Order from:
Thomas Telford Ltd
Thomas Telford House
1 Heron Quay
London E14 9XF
UK

Entry: 03062

**ASSOCIATION OF FINNISH
TEXTILE INDUSTRIES**

Etelaranta 10
00130 Helsinki
Finland
Tel: +358 (9)0 661561
Telex: 125854
Fax: +358 (9)0 653305

Entry: 13090

**ASSOCIATION OF JOINT
VENTURES INTERNATIONAL
CONSORTIA AND
ORGANIZATIONS**
125190 Moscow
A-190 a/a 157
Russia
Tel: + 7 95 943 9481
Fax: + 7 95 943 0020

Entries: 09027   09030

**ASSOCIAZIONE COSTRUTTARI:
MACCHINE ATTREZZATURE PER
L'UFFICIO E PER IL
TRATTAMENTO DELLE
INFORMAZION**
Via Larga 23
l'Ufficio e per il Trattamento
delle Informazion
20122 Milano
Italy
Tel: + 39 (0)2 583 04141
Fax: + 39 (0)2 583 04457

Entry: 21042

**ASSOCIAZIONE COTONIERA
LINIERA E DELLE FIBRE AFFINI**
Viale Sarca 223
20126 Milano
Italy
Tel: + 39 (0)2 6610 3838
Telex: 312479 ACOTEX I
Fax: + 39 (0)2 6610 3863/
6610 3865

Entries: 21026   21055   21056

**BADEN-WÜTTEMBERGISCHE
WERTPAPIERBÖRSE ZU
STUTTGART**
Königstr. 28 (Im Königsbau)
Postfach 10 04 41
D-7000 Stuttgart 1
Germany
Tel: + 49 (0)711 290183
Fax: + 49 (0)711 2268119

Entry: 15007

BANCA COMMERCIALE ITALIANA
Piazza della Scala 6
20121 Milano
Italy
Tel: + 39 (0)2 885 01
Telex: 310080 BCIMOI
Fax: + 39 (0)2 885 03026

Entry: 21034

BANCA D'ITALIA
Via Nazionale 91
00184 Roma
Italy
Tel: + 39 (0)6 479 22404
Telex: 630045

Entries: 21003   21005   21010
          21044   21062   21064

BANCO DE ESPAÑA
Alcalá 50
28014 Madrid
Spain
Tel: + 34 (9)1 446 90 55
Telex: 49461 ESPABE
Fax: + 34 (9)1 531 00 59

Entries: 30008   30009   30052

BANCO DE PORTUGAL
Rua Francisco Ribeiro 2-2í
1100 Lisbon
Portugal
Tel: + 351 (0)1 522 05 3
Telex: 18565 BAGAL P
Fax: + 351 (0)1 523 93 8

Entries: 28019   28055   28066
          28072

BANCO DI NAPOLI SPA
Via Toledo 177
80132 Napoli
Italy
Tel: + 39 (0)81 791 1111

Entry: 21043

BANCO DI ROMA
Via Tupini 180
00144 Roma
Italy
Tel: + 39 (0)6 5445 1
Telex: 616184
Fax: + 39 (0)6 5445 3154

Entry: 21045

BANK BRUXELLES LAMBERT
Avenue Marnix 24
1050 Bruxelles
Belgium
Tel: + 32 (0)2 547 2111
Fax: + 32 (0)2 547 3844

Entries: 07014   07020

BANK GDANSK
IVO/Warszawa
No 300009 - 19132
Poland

Entry: 27057

BANK MEES & HOPE
548 Herengracht
1017 CG Amsterdam
The Netherlands
Tel: + 31 (0)20 527 9111
Telex: 11424
Fax: + 31 (0)20 527 4592

Entry: 25021

BANK OF AMERICA
World Information Services
PO Box 37000
San Francisco
California 94137
USA
Tel: + 1 415 622 1446
Fax: + 1 415 622 0909

Entries: 01042   01046   01051

BANK OF CYPRUS LTD
86-90 Phaneromenis Street
PO Box 1472
Nicosia
Cyprus
Tel: + 357 (0)2 464064
Telex: 2451 KYPRIAKIA

Entries: 10006   10008   10017
          10021

BANK OF ENGLAND
Economics Division
London EC2R 8AH
UK
Tel: + 44 (0)171 601 4030

Entries: 34022   34023   34024

**BANK OF FINLAND (SUOMEN
PANKKI)**
PO Box 160
00101 Helsinki
Finland
Tel: + 358 (9)0 1831
Telex: 121224 SPFB SF
Fax: + 358 (9)0 1748 72

Entries: 13005   13006   13007
          13031   13037   13082

BANK OF GREECE
21 El Venizelou Avenue
10250 Athens
Greece
Tel: + 30 (0)32 02 446
Telex: 215102
Fax: + 30 (0)32 33 025

Entry: 17028

BANK VERLAG
Melatengürtel 113
5000 Köln 30
Germany
Tel: + 49 (0)221 5490 0
Fax: + 49 (0)221 54 90 120

Entries: 15052   15066

BANQUE DE FRANCE
48 Rue Croix-des-Petitis
Champs
BP 14001
75001 Paris
France
Tel: + 33 1 42 92 39 08
Fax: + 33 1 42 92 39 40

Entries: 14026   14029   14032
          14053   14054   14055
          14056   14078   14083
          14084   14086   14088

14093	14101	14102
14103	14104	14106
14107	14109	14113
14123	14124	14125
14139	14142	

BANQUE INTERNATIONALE À LUXEMBOURG

2 Boulevard Royal
L-2953 Luxembourg
Tel: + 352 45 90 1
Telex: 3626 BIL LU
Fax: + 352 45 90 20 10

Entries: 23003   23008

BANQUE NATIONALE DE BELGIQUE

Boulevard de Berlaimont 14
B-1000 Bruxelles
Belgium
Tel: + 32 (0)2 221 2111
Telex: 21389 BKNLE B
Fax: + 32 (0)2 221 3101

Entries: 07011   07066

BARCLAYS BANK PLC

Economics Department, PO
Box No.12
Barclays House
1, Wimborne Road
Poole, Dorset BH15 2BB
UK
Tel: + 44 (0)1202 671212
Fax: + 44 (0)1202 671212 x
4145

Entries:	01001	01129	01205
	01288	07027	19006

BASEL STOCK EXCHANGE

Aeschenplatz 7
CH-4002 Basel
Switzerland
Tel: + 41 (0)61 272 0555
Telex: 962524
Fax: + 41 (0)61 272 0626

Entry: 32009

BDF-INFOSERVICE GMBH

Breitenbachstrasse 1
6000 Frankfurt am Main 93
Germany
Tel: + 49 (0)69 7919 0
Telex: 411627
Fax: + 49 (0)69 7919 227

Entries: 15088   15223

BELENOS PUBLICATIONS LTD

50 Fitzwilliam Square West
Dublin 2
Republic of Ireland
Tel: + 353 (0)1 760 869
Fax: + 353 (0)1 619 781

Entry: 20015

BELGIAN FOREIGN TRADE OFFICE

World Trade Centre
Boulevard Emile Jacqmain
162 (Boîte 36)
B-1210 Bruxelles
Belgium
Tel: + 32 (0)2 219 44 50
Telex: 21205 BEXPO B
Fax: + 32 (0)2 217 61 23

Entries: 07009   23005

BELGO - LUXEMBOURG CHAMBER OF COMMERCE IN GREAT BRITAIN

6 John St
London WC1N 2ES
UK
Tel: + 44 (0)171 831 3508
Fax: + 44 (0)171 831 9151

Entries:	07043	07089	07090
	23021		

BELGRAVE GROUP LTD

Belgrave House
15 Belgrave Road
Rathmines
Dublin 6
Republic of Ireland
Tel: + 353 (0)1 965 711
Fax: + 353 (0)1 964142

Entry: 20069

BIS GROUP

40-44 Rothesay Road
Luton LU1 1QZ
Bedfordshire
UK
Tel: + 44 (0)1582 405 678
Fax: + 44 (0)1582 454 828

Entry: 02014

BLACKWELL PUBLISHERS

108 Cowley Road
Oxford
OX4 1JF
UK
Tel: + 44 (0)1865 791 100
Fax: + 44 (0)1865 3791 347

Entries:	01161	01319	03128
	27048		

BLOC

Box 826
Farmingdale
New York 11737
USA
Tel: + 1 800 288 7076

Entry: 02015

BOARD OF CUSTOMS

Box 512
00101 Helsinki
Finland
Tel: + 358 (9)0 6141
Telex: 121559 TULHS
Fax: + 358 (9)0 614 2813

Entries: 13046   13047

BOLSA DE VALORES DE LISBOA

R. dos Fanqueiros 10
1110 Lisbon
Portugal
Tel: + 351 (0)1 888 27 38
Telex: 44751 BVISB P
Fax: + 351 (0)1 864 23 1

Entries:	28001	28003	28005
	28018	28045	28053
	28054	28074	

BORSA VALORI DI MILANO

Palazzo della Borsa
Piazza Affari 6
20123 Milano
Italy
Tel: + 39 (0)2 8534 4627
Fax: + 39 (0)2 8780 90

Entries: 21035   21038   21041

BRITISH COAL CORPORATION

Hobart House
Grosvenor Place
London
SW1X 7AE
UK
Tel: + 44 (0)171 235 2020
Telex: 882161
Fax: + 44 (0)171 235 5632

Entry: 34038

BRITISH GEOLOGICAL SURVEY

Keyworth
Nottingham NG12 5GG
UK
Tel: + 44 (0)115 9363205
Telex: 378173 BGSKEY G
Fax: + 44 (0)115 9363200

Entries: 01325   34350

BRITISH PORTS FEDERATION

Victoria House
Vernon Place
London WC1B 4LL
UK

Entry: 34265

BRITISH PORTUGUESE CHAMBER OF COMMERCE

R. de Estrela 8
Lisbon
Portugal
Tel: + 351 (0)1 396 15 86
Telex: 12787 BRICHA P
Fax: + 351 (0)1 601 51 3

Entry: 28004

BRITISH SOVIET CHAMBER OF COMMERCE

60A Pembroke Road
London W8 6NZ
UK
Tel: +44 (0)171 602 7692
Telex: 88 13515 BRISOV G
Fax: +44 (0)171 371 4788

Entry: 09002

BRITISH TOURIST AUTHORITY

Thames Tower
Black's Road
Hammersmith
London W6 9EL
UK
Tel: +44 (0)181 846 9000
Fax: +44 (0)181 563 0302

Entries:	34043	34046	34118
	34221	34245	34252
	34254	34275	34325
	34326	34341	34352

BRITISH TOY AND HOBBY ASSOCIATION

80 Camberwell Road
London SE5 0EG
UK
Tel: +44 (0)171 701 7271
Fax: +44 (0)171 708 2437

Entry: 34327

BRITISH-SWISS CHAMBER OF COMMERCE

Freiestrasse 155
CH 8032 Zürich
Switzerland
Tel: +41 (0)1 422 3131
Fax: +41 (0)1 422 3244

Entries: 32010   32011   32066

BUILDING SERVICES RESEARCH AND INFORMATION ASSOCIATION

Old Bracknell Lane
West Bracknell
Berks RG12 7AH
UK
Tel: +44 (0)1344 426511
Fax: +44 (0)1344 487575

Entries:	02018	03118	07040
	12024	14063	15116
	17024	20045	21023
	25029	28041	30046
	34047	34048	34164

BUILDING SOCIETIES ASSOCIATION/COUNCIL OF MORTGAGE LENDERS

BSA/CML Bookshop
3 Saville Row
London
W1X 1AF
UK
Tel: +44 (0)171 437 0655
Fax: +44 (0)171 287 0109

Entries:	01169	02098	34049
	34050	34051	34052
	34074	34078	34159

34160   34229   34230
34231

BULGARIAN CHAMBER OF COMMERCE AND INDUSTRY

11-a Stamboliiski Blvd
Sofia 1040
Bulgaria
Tel: +359 (0)87 26 31
Fax: +359 (0)87 32 09

Entries: 08005   08016

BUNDESANSTALT FÜR AGRARWIRTSCHAFT

A-113 Wien
Schweizertalstrasse 36
Austria
Tel: +43 (0)222 877 36 51
Fax: +43 (0)222 877 36 51 59

Entries: 05056   05071   05089
05090

BUNDESANSTALT FÜR ARBEIT

Regensburger Str. 104
8500 Nürnberg 30
Germany
Tel: +49 (0)911 179 0
Telex: 622348 bad
Fax: +49 (0)911 179 2123

Entries: 15050   15087

BUNDESKAMMER DER GEWERBLICHEN WIRTSCHAFT

Wiedner Haupstrasse 63
1045 Wien
Austria
Tel: +43 (0)222 501 05 4120
Telex: 111871 BUKA
Fax: +43 (0)222 502 06 246

Entries:	05001	05005	05006
	05011	05030	05031
	05041	05065	05072
	05080	05095	

BUNDESKANZLERAMT

Bundespressedienst
Ballhausplatz 2
A-1014 Wien
Austria
Tel: +43 (0)222 53115/0
Telex: 115585 bpdbk a
Fax: +43 (0)222 53115/ 2880

Entries: 05007   05008   05086

BUNDESMINISTERIUM FÜR ARBEIT UND SOZIALORDNUNG

Postfach 140280
5300 Bonn 1
Germany
Tel: +49 (0)228 527 5113/ 14
Telex: 17/228 3650 Kennung BMA Bonn
Fax: +49 (0)228 527 2254

Entries: 15054   15056

BUNDESMINISTERIUM FÜR ERNÄHRUNG, LANDWIRTSCHAFT UND FORSTEN

Rochusstrasse 1
5300 Bonn 1
Germany
Tel: +49 (0)228 529 1
Telex: 886 844
Fax: +49 (0)228 529 4262

Entries:	15005	15023	15055
	15123	15132	15133
	15173	15200	15207
	15210		

BUNDESMINISTERIUM FÜR WIRTSCHAFT

Postfach 14 02 60
5300 Bonn 1
Germany
Tel: +49 (0)228 615 1
Telex: 886 747
Fax: +49 (0)228 615 4151

Entries: 15176   15229

BUNDESMINISTERIUM FÜR WIRTSCHAFTLICHE ANGELEGENHEITEN

Stubenring 1
A-1011 Wien
Austria
Tel: +43 (0)222 71100
Telex: 111145 regeb a
Fax: +43 (0)222 713 79 95

Entry: 05052

BUNDESSTELLE FÜR AUSSENHANDELSINFORMATION

Agrippastr. 87-93
Postfach 108007
D-5000 Köln 1
Germany
Tel: +49 (0)221 2057 0
Telex: 08 882 735 BFA D
Fax: +49 (0)221 2057 212/ 275

Entries:	15053	15075	15082
	15096	15178	15216

BUSINESS AND RESEARCH ASSOCIATES

9 Market Street
Disley
Stockport
SK12 2AA
UK
Tel: +44 (0)1633 765 202
Fax: +44 (0)1633 765 933

Entries: 03021   34054

BUSINESS FOUNDATION LTD

Warszawa Ul
Wspolna 1/3
Poland
Tel: +48 (0)22 21 99 93/21 84 20
Telex: 817088
Fax: +48 (0)22 28 05 49

Entry: 27029

BUSINESS INTERNATIONAL LTD

40 Duke Street
London
WIA 1DA
UK
Tel: +44 (0)171 493 6711
Telex: 266353
Fax: +44 (0)171 491 2107

Entries:	01022	01043	01047
	01049	01052	01072
	01073	01074	01075
	01172	01180	01255
	01276	01305	01327
	02019	02024	02036
	02038	02039	02040
	02041	02049	02050
	02063	02067	02076
	02079	02101	02104
	03022	03086	03089
	03092	04003	04004
	05020	05022	05057
	06004	06005	07024
	07026	07028	07036
	07038	07042	08007
	08011	09005	09010
	09011	09012	09013
	09014	09015	09018
	09020	09025	10013
	11001	11003	11004
	11016	11017	11021
	12008	12010	12022
	12028	13019	13021
	13048	13060	14034
	14035	14037	14057
	14059	14073	15047
	15049	15078	15126
	17008	17010	17011
	17017	17018	17025
	18005	18007	18008
	18012	18031	19003
	19005	20027	20029
	20058	21006	21012
	21014	21017	21018
	21028	21030	21031
	23012	23020	24026
	25013	25015	25026
	25027	25032	26009
	26011	26020	26024
	26028	27010	27012
	27013	27019	27024
	27038	28014	28016
	28017	28039	28052
	28061	28062	29001
	29004	29006	30023
	30025	30026	30043
	30044	30053	30065
	31024	31026	31047
	31055	31068	32020
	32021	32023	32033
	32037	32042	33013
	33014	33015	33016
	33029	33060	33061
	33066	34080	34081
	34082	34114	34121
	34136	34172	34286
	34287	35001	35002
	35003	35004	35005
	35006	35012	

BUSINESS SPAIN

5 Cochrane Rd
Wimbledon
London SW19 3QP
UK
Tel: +44 (0)181 543 3702
Fax: +44 (0)181 542 9185

Entry: 30014

BUSINESS WEEK INTERNATIONAL

14 Avenue D'ouchy
CH-1006 Lausanne
Switzerland
Tel: +41 (0)20 617 4411
Fax: +41 (0)20 617 2919

Entries: 01023   09004

CÁMARA DE COMERCIO E INDUSTRIA DE MADRID

Pl de Independencia 1
28001 Madrid
Spain
Tel: +34 (9)1 538 35 00
Telex: 27307 COIM-E
Fax: +34 (9)1 538 36 77

Entry: 30066

CBD RESEARCH LTD

15 Wickham Road
Beckenham
Kent BR3 2JS
UK
Tel: +44 (0)181 650 7745
Fax: +44 (0)181 650 0768

Entry: 01265

CE FRITZES

10647 Stockholm
Sweden
Tel: +46 (0)16 086 90 9090
Fax: +46 (0)16 08 205021

Entries: 31129   31130

CENTRAL ASSOCIATION OF FINNISH FOREST INDUSTRIES

Etelaesplanaoi 2
SF - 00130 Helsinki
Finland
Tel: +358 (9)0 13261
Telex: 121823 SMKL sf
Fax: +358 (9)0 174479

Entry: 13040

CENTRAL ASSOCIATION OF THE FINNISH CLOTHING INDUSTRY

Etelaranta 10
00130 Helsinki
Finland
Tel: +358 (9)0 661 689
Fax: +358 (9)0 635 108

Entry: 13013

CENTRAL BANK OF CYPRUS

PO Box 5529
Nicosia
Cyprus
Tel: +357 (0)2 445281
Telex: 2424
Fax: +357 (0)2 472012

Entries: 10003   10007

CENTRAL BANK OF ICELAND

Kalkofnsvegur 1
101 Reykjavik
Iceland
Tel: +354 (9)1 699 600
Telex: 2020
Fax: +354 (9)1 621 802

Entries: 19001   19008   19009

CENTRAL BANK OF IRELAND

PO Box No. 559
Dame Street
Dublin 2
Republic of Ireland
Tel: +353 (0)1 716 666
Telex: 31041
Fax: +353 (0)1 716 561

Entries:	20007	20021	20063
	20093		

CENTRAL BANK OF MALTA

Castille Place
Valetta CMR 01
Malta
Tel: +356 24 74 80
Telex: MW 1262
Fax: +356 24 30 51

Entries: 24002   24022

CENTRAL BANK OF THE REPUBLIC OF TURKEY

40 Boulevard des Banques
Ulus
Ankara
Turkey
Tel: +90 (9)4 124 30 50
Telex: 42321

Entries: 33002   33036   33041

CENTRAL BUREAU OF STATISTICS

Skipperg 15
PO Box 8131 Dept 0033
N-Oslo 1
Norway
Tel: +47 2 86 4500
Fax: +47 2 86 4973

Entries:	26001	26007	26015
	26016	26017	26018
	26021	26023	26026
	26030	26031	26033
	26034	26035	26036
	26037	26042	26046
	26047	26048	26052
	26053	26054	26055
	26056	26057	

THE CENTRAL CHAMBER OF COMMERCE OF FINLAND

Fabianinkatu 14
P.O. Box 1000
SF-00101 Helsinki
Finland
Tel: +358 (9)0 650133
Telex: 123814 chamb sf
Fax: +358 (9)0 650303

Entries: 13012   13062

CENTRAL EUROPEAN

Nestor House
Playhouse Yard
London EC4V 5EX
UK
Tel: +44 (0)171 779 8682
Telex: 290706
Fax: +44 (0)171 779 8689

Entry: 02020

CENTRAL OFFICE OF STATISTICS

Auberge d'Italie
Merchants Street
Valetta CMR 02
Malta
Tel: +356 23 05 70
Fax: +356 24 84 83

Entries:	24001	24009	24010
	24012	24015	24020
	24021	24023	24025

CENTRAL PLANNING BUREAU

Van Stolkweg 14
2585 JR Den Haag
The Netherlands
Tel: +31 (0)70 338 3380
Fax: +31 (0)70 338 3350

Entries:	25008	25020	25036

CENTRAL STATISTICAL BOARD

Str. Stravropoleos no.6
Bucharesti
Romania

Entry: 29017

CENTRAL STATISTICAL OFFICE OF POLAND

Al Niepodleglosci 208
00-925 Warszawa
Poland
Tel: +48 (0)22 25 90 78
Telex: 81 45 81a
Fax: +48 (0)22 25 90 78

Entries:	27001	27002	27003
	27004	27005	27006
	27007	27008	27009
	27014	27020	27021
	27022	27023	27027
	27028	27031	27033
	27035	27036	27039
	27040	27041	27042
	27043	27044	27045
	27047	27049	27054
	27055	27056	27058
	27059	27060	27061
	27063	27064	27065
	27066	27067	27068

CENTRAL STATISTICS OFFICE

St Stephen's Green House
Earlsfort Terrace
Dublin 2
Republic of Ireland
Tel: +353 (0)1 767 531
Fax: +353 (0)1 682 221

Entries:	20003	20004	20005
	20010	20011	20012
	20016	20017	20018
	20019	20020	20025
	20030	20033	20035

	20038	20043	20044
	20046	20048	20049
	20050	20051	20052
	20053	20054	20056
	20070	20071	20072
	20073	20074	20075
	20076	20081	20082
	20083	20087	20089
	20090	20091	20092
	20094	20097	20101
	20103	20105	20107
	20108	20109	20110
	20114	20116	20117
	20118	20119	20120
	20121		

CENTRE D'ÉTUDES ET DE RECHERCHES SUR LES QUALIFICATIONS

Masson
120 Bld Saint Germain
75006 Paris
Cedex 06
France
Tel: +33 1 46 34 21 60
Telex: 202671F
Fax: +33 1 45 87 2999

Entry: 14148

CENTRE DE RECHERCHE ET D'INFORMATION SOCIO-POLITIQUE - CRISP

Rue du Congrès 35
1000 Bruxelles
Belgium
Tel: +32 (0)2 218 32 26

Entries:	07029	07047

CENTRE FRANÇAIS DU COMMERCE EXTÉRIEUR

Librairie du Commerce
International
10 Avenue d'Iena
Paris 16e
France
Tel: +33 45 05 30 00
Telex: LICOMIN 613 413
OFCE
Fax: +33 43 36 47 98

Entries:	14031	14110

CENTRE OF PLANNING AND ECONOMIC RESEARCH

22 Hippokratous Street
106 80 Athens
Greece
Tel: +30 (0)36 27 321
Fax: +30 (0)36 30 122

Entries:	17013	17015	17021
	17030	17031	17032
	17035	17036	17037
	17038	17039	17040
	17041	17043	

CHAMBRE DE COMMERCE DU GRAND-DUCHÉ DE LUXEMBOURG

7 Rue Alcide de Gasperi
L-1615
Luxembourg
Tel: +352 43 58 53
Telex: 60174 CHCOM LU
Fax: +352 43 83 26

Entry: 23014

CHAMBRE DE COMMERCE ET D'INDUSTRIE DE BRUXELLES

500 Avenue Louise
B-1050 Bruxelles
Belgium
Tel: +32 (0)2 648 5002
Telex: 22082
Fax: +32 (0)2 640 9328

Entry: 07033

CHAMBRE DE COMMERCE ET D'INDUSTRIE DE GENÈVE

4 Bd du Théâtre
Case Postale 65
1211 Genève 11
Switzerland
Tel: +41 (0)22 311 53 33
Fax: +41 (0)22 310 03 63

Entries:	32029	32040	32074

LA CHAMBRE DE COMMERCE ET D'INDUSTRIE DE PARIS

27 Avenue de Friedland
75382 Paris Cedex 08
France
Tel: +33 42 89 72 41
Fax: +33 42 89 72 86

Entries:	14108	14112

CHARITIES AID FOUNDATION

114-118 Southampton Row
London WC1B 5AA
UK
Tel: +44 (0)171 831 0134
Fax: +44 (0)171 831 0134

Entries:	01025	01167	34006
	34071		

CHEMICAL INDUSTRIES ASSOCIATION LTD

Kings Buildings
Smith Square
London SW1P 3JJ
UK
Tel: +44 (0)171 834 3399
Telex: 916672
Fax: +44 (0)171 834 4469

Entries:	01018	01026	34113
	34173	34339	

CIVIL AVIATION AUTHORITY

Order from:
Printing and Publication
Services

Grenville House
37 Gratton Road
Cheltenham
GL50 2BN
UK
Tel: +44 (0)1242 584139
Telex: 883092
Fax: +44 (0)1242 263993

Entries: 34334   34335   34336
34337

CO-OPERATIVE UNION LTD
Order from:
Information services -
Statistics
Holyoake House
Hanover Street
Manchester
M60 0AS
UK
Tel: +44 (0)161 832 4300
Fax: +44 (0)161 831 7684

Entries: 34076   34109

COMITÉ BELGE DE LA DISTRIBUTION
Rue Marianne 34
1180 Bruxelles
Belgium
Tel: +32 (0)2 345 99 23
Fax: +32 (0)2 346 02 04

Entry: 07030

COMMERCIAL BANK OF GREECE
Economic Research Division
10 Sophocleous Street
10235 Athens
Greece
Tel: +30 (0)32 32 854

Entry: 17014

COMMONWEALTH SECRETARIAT
Marlborough House
Pall Mall
London SW1Y 5HX
UK
Tel: +44 (0)171 839 3411

Entries: 01034   01035   01160
34144

COMPUTER TECHNOLOGY RESEARCH CORPORATION
Brenfield House
Bolney Road
Ansty
West Sussex RH17 5AW
UK
Tel: +44 (0)1444 459 151
Fax: +44 (0)1444 454 061

Entry: 34206

CONFÉDÉRATION NATIONALE DE LA CONSTRUCTION
Rue du Lombard 34-42
1000 Bruxelles
Belgium
Tel: +32 (0)2 510 46 11
Fax: +32 (0)2 513 30 04

Entry: 07022

CONFEDERATION OF BRITISH INDUSTRY (CBI)
Centre Point
103 New Oxford Street
London WC1
UK
Tel: +44 (0)171 379 7400 X
2209

Entries: 11013   18010   27016
34069   34070   34099
34111   34166

CONFEDERATION OF IRISH INDUSTRY
Confederation House
Kildare Street
Dublin 2
Republic of Ireland
Tel: +353 (0)1 779 801
Telex: 24711

Entries: 20022   20023   20034
20080

CONSEIL CENTRAL DE L'ÉCONOMIE
Avenue de la Joyeuse
Entrée 17-21
1040 Bruxelles
Belgium
Tel: +32 (0)2 33 88 11
Fax: +32 (0)2 33 89 12

Entry: 07054

CONSORCIO PARA EL FOMENTO DE LA RIQUEZA GANADERA
C/Santa Catalina 15
40003 Segovia
Spain
Tel: +34 (9)11 417 290
Fax: +34 (9)11 417 233

Entry: 30050

COPENHAGEN STOCK EXCHANGE
Nikolas Plads 6
PO Box 1040
DK 1007 Copenhagen K
Denmark
Tel: +45 (0)33 93 33 66
Telex: 16496
Fax: +45 (0)33 12 86 13

Entries: 01218   12006   12011
12031   12032   12033
12036   12046

CORONAKIS PRESS LTD
10 Fokidos Street
115 26 Athens
Greece
Tel: +30 (0)77 06 922
Fax: +30 (0)77 120 79

Entry: 17020

CORPORATE INTELLIGENCE RESEARCH PUBLICATIONS
51 Doughty Street
London WC1N 2LS
UK
Tel: +44 (0)171 696 9006
Fax: +44 (0)171 696 9004

Entries: 02026   02065   02086
02115   02116   03090
34291   34292   34293
34294   34347

COUNTERTRADE OUTLOOK
Box 7188
Fairfax Station
Virginia 22039
USA
Tel: +1 703 425 1322
Telex: 263 128 CTO UR
Fax: +1 703 425 7911

Entries: 01041   01062

CREDIT SUISSE
Paradeplatz 8
CH 8021 Zürich
Switzerland
Tel: +41 (0)1 333 1111
Telex: 812412
Fax: +41 (0)1 332 5555

Entries: 32001   32004   32013
32024

CREDITANSTALT - BANKVEREIN
Schottengasse 6
1010 Wien
Austria
Tel: +43 (0)222 531 31 0
Fax: +43 (0)222 533 32 82

Entries: 05016   05058   05081

CYPRUS CHAMBER OF COMMERCE AND INDUSTRY
PO Box 1455
Nicosia
Cyprus
Tel: +357 (0)2 449500
Telex: 2077 CHAMBER
Fax: +357 (0)2 458630

Entries: 10015   10025

CYPRUS EMPLOYERS AND INDUSTRIALISTS FEDERATION
PO Box 1657
Nicosia
Cyprus
Tel: +357 (0)2 445102
Telex: 4834
Fax: +357 (0)2 459959

Entries: 10002   10037

CYPRUS POPULAR BANK LTD

PO Box 2032, Nicosia
Popular Bank Building
39 Archbishop Makarios III
Avenue
Nicosia
Cyprus
Tel: +357 (0)2 450000
Telex: 2494 LAIKIH CY
Fax: +357 (0)2 450631

Entry: 10009

CYPRUS PORTS AUTHORITY

Crete 23
PO Box 2007
Nicosia
Cyprus
Tel: +357 (0)2 450100
Telex: 2833 CYPACY
Fax: +357 (0)2 365420

Entries: 10004   10032   10034

CZECHOSLOVAKIA CHAMBER OF COMMERCE AND INDUSTRY

Argentisk 38
170 05 Praha 7
Czech Republic

Entry: 11025

DAFSA

Le Ponant
25, rue Leblanc
75513 Paris Cedex 15
France
Tel: +33 40 60 36 00
Fax: +33 40 60 36 10

Entry: 01055

DANISH BACON AND MEAT COUNCIL

Axeltorv 3
DK 1609 Copenhagen V
Denmark
Tel: +45 (0)33 11 60 50
Fax: +45 (0)33 11 68 14

Entry: 12048

DANMARKS NATIONALBANK

Havnegade 5
DK-1093
Copenhagen K
Denmark
Tel: +45 (0)33 14 14 11
Telex: 27051
Fax: +45 (0)33 14 59 02

Entries: 12005   12014   12035
12043

DANMARKS STATISTIK

Sejrøgade 11
2100 Copenhagen ø
Denmark
Tel: +45 (0)39 17 39 17
Telex: 16236 DASTAT
Fax: +45 (0)31 18 48 01

Entries: 12002   12003   12004
12021   12025   12026

12030	12037	12039
12040	12044	12047
12049	12050	12051
12052	12053	

DANSK HANDELSBLAD A/S

Fenrisvej 11
8230 Aabyhøj
Denmark
Tel: +45 (0)86 15 80 11
Fax: +45 (0)86 15 82 52

Entries: 12013   12020

DANSK INDUSTRI, CONFEDERATION OF DANISH INDUSTRIES

DK-1787 Copenhagen V
Denmark
Tel: +45 (0)33 77 33 77
Telex: 1122 17 DI DK
Fax: +45 (0)33 77 33 00

Entry: 12054

DANSKE ELVRKERS FORENING

Rosenørns Alle 9
DK-1970 Frederiksberg C
Denmark
Tel: +45 (0)31 39 01 11
Fax: +45 (0)31 39 59 58

Entry: 12017

DATAMONITOR

106 Baker Street
London W1M 1LA
UK
Tel: +44 (0)171 625 8448
Fax: +44 (0)171 625 5080

Entries:	02027	02028	02029
	02030	02031	02032
	02033	34083	34084
	34085	34086	34087
	34088	34089	

DATAQUEST

Roussel House
Broadwater park, Denham
Uxbridge
Middlesex UB8 5HP
UK
Tel: +44 (0)1895 835 050
Fax: +44 (0)1985 835 261

Entry: 34090

DE LA SALLE BROTHERS PUBLICATIONS

St Benild School
Church Street
Sliema
Malta
Tel: +356 31 03 94
Fax: +356 32 03 94

Entry: 24019

DEBENHAM TEWSON AND CHINNOCKS

44 Brook Street
London W1A 4AG
UK
Tel: +44 (0)171 408 1161
Fax: +44 (0)171 491 4593

Entry: 34091

DEPARTMENT OF AGRICULTURE NORTHERN IRELAND DANI

Dundonald House
Upper Newtownards Road
Belfast
BT4 3SF
N. Ireland
UK
Tel: +44 (0)1232 650 111

Entries: 34125   34248   34317

DEPARTMENT OF INFORMATION

Auberge de Castille
Valletta
Malta
Tel: +356 22 54 21
Telex: 1512 ACGEN MW
Fax: +356 22 99 25

Entries: 24014   24016

DEPARTMENT OF STATISTICS AND RESEARCH

Ministry of Finance
Byron Avenue No. 19
Postal Code 162
Nicosia
Cyprus
Tel: +357 (0)2 403286
Telex: 3399 MINEIN CY

Entries:	10001	10010	10011
	10014	10016	10018
	10023	10024	10026
	10028	10029	10030
	10031	10033	10035
	10036	10038	10039
	10040	10041	10042
	10043	10044	10045
	10046	10047	10048
	10049		

DEPARTMENT OF TRADE AND INDUSTRY EXPORT PUBLICATIONS

PO Box 55
Stratford-upon-Avon
Warwickshire
CV37 9GE
UK
Tel: +44 (0)1789 296212
Fax: +44 (0)1789 299096

Entries:		01048	01140
	01257	05017	05018
	05021	05035	05042
	05053	05060	05068
	05084	05088	05091
	05101	07025	07039
	08008	08021	09008
	09023	11002	11018
	11020	12009	12023
	13020	13049	14036
	14062	15014	15042
	15048	15076	15085

15094	15171	15215
15227	17009	17012
17019	17023	18006
18018	18030	19004
19011	20028	20042
21013	21019	23011
23018	25014	25028
26010	26025	26027
27011	27032	28015
28040	29002	29009
30024	30045	31025
31058	32022	32038
33045	33056	33057
33058	33059	33063
33064	33065	33067
33069	33070	33071
33072		

DEPARTMENT OF TRANSPORT

Order from:
Publications Sales Unit
Building
1 Victoria Road
South Ruislip
Middlesex HA4 0NZ
UK
Tel: +44 (0)171 276 8188

Entries: 34274   34296   34329

DEUTSCHE BUNDESBANK

Postfach 10 06 02
6000 Frankfurt 1
Germany
Tel: +49 (0)69 158 1
Telex: 414431
Fax: +49 (0)69 560 1071

Entries: 15009   15177   15203

DEUTSCHES WIRTSCHAFTS WISSENSCHAFTLICHES INSTITUT FÜR FREMDENVERKEHR AN DER UNIVERSITÄT MÜNCHEN E.V.

Herman-Sack-Str. 2
8000 München 2
Germany
Tel: +49 (0)89 26 70 81
Fax: +49 (0)89 26 76 13

Entries: 15129   15199   15202

DIAMOND HIGH COUNCIL

Hoveniersstraat 22
2018 Antwerp
Belgium
Tel: +32 (0)3 222 05 11
Fax: +32 (0)3 222 07 24

Entry: 07006

DIRECÇÍO GENERAL DO TURISMO

Avenue António Augusto de
Aguiar 86
1000 Lisbon
Portugal
Tel: +351 (0)1 575 01 5
Telex: 13408 PORTUR P
Fax: +351 (0)575220

Entries: 28056   28057   28058

DIRECTION GÉNÉRALE DES DOUANES ET DROITS INDIRECTS

161 Chemin de Lestang
31057 Toulouse Cedex
France
Tel: +33 62 11 23 00
Telex: 521256
Fax: +33 62 11 24 80

Entries: 14019   14105   14132
         14136   14137

DIRECTION GÉNÉRALE DES DOUANES: SECTION STATISTIQUES

Monbijoustrasse 40
CH-3003 Berne
Switzerland
Tel: +41 (0)31 61 6525/61 6575
Telex: 911100
Fax: +41 (0)31 61 7872

Entries: 32015   32070   32072
         32078

THE DIVISION FOR SALE BY THE SUPERINTENDANT FOR DOCUMENTS

Library of Congress
Information Office
Washington, D.C. 20540
U.S.A.

Entry: 29013

LA DOCUMENTATION FRANÇAISE

29 Quai Voltaire
75344 Paris
Cedex 07
France
Tel: +33 1 40 15 70 00
Fax: +33 1 40 157230

Entries:		
14021	14024	14030
14046	14049	14051
14052	14076	14077
14079	14087	14090
14097	14098	14099
14114	14117	14120
14122	14128	14130
14140	14141	14146

DRESDNER BANK

Jürgen-Ponto-Platz 1
D-6000 Frankfurt/Main
Germany
Tel: +49 (0)69 2630
Telex: 415 240
Fax: +49 (0)69 263 4831

Entries: 15061   15219

DREWRY SHIPPING CONSULTANTS

11 Heron Quay
London
E14 4JF
UK
Tel: +44 (0)171 538 0191
Fax: +44 (0)171 987 9396

Entry: 01065

DUN & BRADSTREET INTERNATIONAL

Holmers Farm Way
High Wycombe
Buckinhamshire
HP12 4UL
UK
Tel: +44 (0)1494 424 295
Fax: +44 (0)1494 422260

Entries: 01050   01174   11010
         34176

DUNCEER & HUMBLOT GMBH

PO Box 41 03 29
1000 Berlin 41
Germany
Tel: +49 (0)30 79 00 06 0
Fax: +49 (0)30 79 00 06 31

Entries:		
15101	15104	15125
15128	15153	15165
15194	15195	15226
15231	15232	15236
15237		

DURHAM UNIVERSITY BUSINESS SCHOOL

Mill Hill Lane
Durham City
DH1 3LB
UK
Tel: +44 (0)191 374 2000
Telex: 537351 DURLIBG
Fax: +44 (0)191 374 3748

Entry: 34309

EAST EUROPE BUSINESS FOCUS

Third Floor, Brigade House
Parsons Green Lane
London SW6 4TH
UK

Entry: 02037

ECONOMIC AND SOCIAL RESEARCH INSTITUTE

4 Burlington Road
Dublin 4
Irish Republic
Republic of Ireland
Tel: +353 (0)1 760 115
Fax: +353 (0)1 688 231

Entries: 20079   20096   20106

ECONOMIST BOOKS

25 St James Street
London SW1A 1HG
UK
Tel: +44 (0)171 839 7000
Fax: +44 (0)171 839 2968

Entry: 01020

THE ECONOMIST INTELLIGENCE UNIT

40 Duke Street
London
WIA 1DW
UK
Tel: +44 (0)171 493 6711
Fax: +44 (0)171 499 9767

Entries:	06003	08010	09007
	09009	10012	11008
	11009	18015	18027
	18034	27030	27050
	29003		

ECONOMISTA

Calle del Conde de Aranda 8
Apdo. 1024, Madrid
28001 Spain
Tel: +34 (9)1 435 04 62

Entry: 30031

EKONOMISK LITTERATUR AB

Box 56
S-16211 Vallingby
Sweden
Tel: +46 (0)8 381 500
Fax: +46 (0)8 890 990

Entries:	02064	12015	26040
	31128		

EKONOMKSO - POSLOVNA FAKULTETA

Razlagova 14
Maribov
Slovenia

Entry: 35015

EKONOMSKI FAKULTET U BEOGRADU

Kamenicka 6
11000 Belgrade
Serbia

Entry: 35010

ELC INTERNATIONAL

109 Oxbridge Road
Ealing
London W5 5TL
UK
Tel: +44 (0)181 566 2288

Entry: 34349

ELECTRA PRESS PUBLICATIONS

4 Stadiou Street
Athens 105 64
Greece
Tel: +30 (0)32 33 203
Fax: +30 (0)32 35 160

Entries: 17016   17022

ELECTRICITY ASSOCIATION

Business Information Centre
30 Millbank
London SW1P 4RD
UK
Tel: +44 (0)171 834 2333
Fax: +44 (0)171 828 9945

Entry: 34340

ELSEVIER SCIENCE PUBLISHERS LTD

Crown House
Laiton Road
Barking
Essex 1G11 8JU
UK
Tel: +44 (0)181 594 7272

Entries:	01070	01082	01320
	02057		

EMPLOYMENT & TRAINING CORPORATION

Hal Far
ZRQ 06
Malta
Tel: +356 68 49 45
Fax: +356 68 33 96

Entry: 24003

EMPLOYMENT DEPARTMENT

Moorfoot
Sheffield
S1 4PQ
UK
Tel: +44 (0)114 2594952

Entries: 34194   34322

ENGLISH TOURIST BOARD

Thames Tower
Black's Road
Hammersmith
London W6 9EL
UK
Tel: +44 (0)181 846 9000

Entries:	34037	34041	34117
	34119	34152	34279
	34308	34348	

EOLAS

Ballymun Road
Dublin 9
Republic of Ireland
Tel: +353 (0)1 370 101

Entry: 20115

ERC STATISTICS INTERNATIONAL

5-11 Shorts Gardens
London
WC2H 9AT
UK
Tel: +44 (0)171 497 2312
Fax: +44 (0)171 497 2313

Entries:	01098	02059	02060
	02061	02062	

ERLANDSSON LTD

Reprovagen 6
5-183 64 Taby
Sweden
Tel: +46 (0)8 631 10 60
Fax: +46 (0)8 756 57 40

Entry: 31126

ERNST & YOUNG INTERNATIONAL

Becket House
1 Lambeth Palace Road
London SE1 7EU
UK
Tel: +44 (0)171 928 200
Telex: 885234
Fax: +44 (0)171 928 1345

Entries:	01064	02035	03042
	07031	14043	15058
	25016	27017	30027
	32025	34101	34102

ESTRUCTURA GRUPO DE ESTUDIOS ECONÓMICOS SA

c/Gran Via 32
2 Planta, 28013 Madrid
Spain
Tel: +34 (9)1 521 01 64
Telex: 44461

Entry: 30062

EUROMONEY BOOKS

Order from:
Plymbridge Distributors Ltd
Estover
Plymouth
Pl6 7PZ
UK
Tel: +44 (0)1752 695 745
Fax: +44 (0)1752 695 668

Entries: 01044

EUROMONITOR

87-88 Turnmill Street
London EC1M 5QU
UK
Tel: +44 (0)171 251 8024
Fax: +44 (0)171 608 3149

Entries:	01171	01251	01315
	01317	01330	02016
	02017	02022	02023
	02066	02068	02071
	02072	02078	02103
	02114	30064	34219
	34289		

EUROPA PUBLICATIONS

18 Bedford Square
London WC1B 3JN
UK
Tel: +44 (0)171 580 8236
Telex: 21540 EUROPA G
Fax: +44 (0)171 636 1664

Entries:	01099	02044	02120
	03071		

EUROPEAN BANK FOR RECONSTRUCTION AND DEVELOPMENT

1 Exchange Square
London EC2A 2EH
UK

Entry: 02010

EUROPEAN FREE TRADE ASSOCIATION

9-11 Rue de Varembé
CH-1211 Genève 20
Switzerland

Tel: +41 (0)22 749 1111
Telex: 414102 EFTA CH
Fax: +41 (0)22 733 9291

Entries:		
	01077	01078
01079	01080	01119
02051	02052	02053
02054	02055	02056
02087	03050	03109
28002		

EUROPEAN INVESTMENT BANK

Order from:
Information/Public Relations
Division
L-2950
Luxembourg

Tel: +352 43 79 42 19

Entry: 03085

EUROSTAT

HMSO Publications Centre
51 Nine Elms Lane
London SW8 5DR
UK

Tel: +44 (0)171 873 9090
Telex: 29 71 138
Fax: +44 (0)171 873 8463

Entries:	03001	03004	03006
	03007	03009	03011
	03013	03017	03023
	03024	03025	03034
	03035	03037	03038
	03039	03043	03044
	03045	03046	03048
	03052	03055	03056
	03061	03063	03069
	03079	03084	03091
	03094	03095	03096
	03097	03098	03099
	03100	03103	03106
	03111	03114	03120
	03124	03126	03127
	03131	03132	03135
	03136	03137	03138
	03139	03143	03146
	03148	03149	03152
	03153	03158	03160
	03163	03166	03167
	03168	03169	03171
	03173	03174	

EXPORT PRESS

Francuska 27
Belgrade
Serbia

Entry: 35013

EXTEL FINANCIAL

Fitzroy House
13-17 Epworth Street
London
EC2A HDL
UK

Tel: +44 (0)171 251 3333

Entry: 34146

FABRIMÉTAL

Rue des Drapiers 21
B-1050 Bruxelles
Belgium

Tel: +32 (0)2 510 23 11
Telex: 210 78
Fax: +32 (0)2 510 23 01

Entries:	07008	07034	07063
	07071	07074	07093

FACHVERBAND DER ERDÖLINDUSTRIE ÖSTERREICHS

Erdbergstrasse 72
A-1031 Wien
Austria

Tel: +43 (0)222 713 23 48
Telex: 13 21 38
Fax: +43 (0)222 713 05 10

Entries:	05059	05079

FEBELGRA - FÉDÉRATION BELGE DES INDUSTRIES GRAPHIQUES A.S.B.L

Rue Belliard 20, Boîte 16
1040 Bruxelles
Belgium

Tel: +32 (0)2 513 36 38
Fax: +32 (0)2 512 56 76

Entry: 07035

FÉDÉRATION DE L'INDUSTRIE CIMENTIÈRE

Rue César Franck 46
1050 Bruxelles
Belgium

Tel: +32 (0)2 645 52 11
Fax: +32 (0)2 640 06 70

Entry: 07061

FÉDÉRATION DE L'INDUSTRIE DU BÉTON

Boulevard A. Reyers 207-209
1040 Bruxelles
Belgium

Tel: +32 (0)2 735 80 15
Fax: +32 (0)2 734 77 95

Entry: 07018

FÉDÉRATION DE L'INDUSTRIE DU GAZ (FIGAZ)

Avenue Palmerston 4
1040 Bruxelles
Belgium

Tel: +32 (0)2 237 11 11
Fax: +32 (0)2 230 44 80

Entries:	07003	07052	07064

FÉDÉRATION DES ENTREPRISES DE MÉTAUX NON FERREUX

Rue Montoyer 47
B-1040 Bruxelles
Belgium

Tel: +32 (0)2 506 4111
Fax: +32 (0)2 511 7553

Entries:	07062	07070

FÉDÉRATION DES INDUSTRIES TRANSFORMATRICES DE PAPIER ET CARTON (FETRA)

Chaussée de Waterloo 715
Boîte 25
B-1180 Bruxelles
Belgium

Tel: +32 (0)2 344 19 62
Fax: +32 (0)2 344 86 61

Entries:	07065	07072

FEDERATION OF DANISH CO-OPERATIVES

Vester Farimagsgade 3
1601 Copenhagen V
Denmark

Tel: +45 (0)33 12 14 19
Fax: +45 (0)33 12 61 48

Entry: 12001

FEDERATION OF FINNISH INSURANCE COMPANIES

Bulevardi 28
SF 00120
Helsinki
Finland

Tel: +358 (9)0 680 401
Fax: +358 (9)0 680 4 0216

Entries:	13041	13056	13057
	13058		

FEDERATION OF FINNISH METAL ENGINEERING AND ELECTROTECHNICAL INDUSTRIES

P.O. Box 10
00131 Helsinki
Finland

Tel: +358 (9)0 19231
Telex: 124997 FIMET
Fax: +358 (9)0 624462

Entry: 13044

FEDERATION OF SWEDISH COMMERCE AND TRADE

P. O. Box 5512
S-11485 Stockholm
Sweden

Tel: +46 (0)8 666 1100
Telex: 19673 alvdor-s
Fax: +46 (0)8 662 74 57

Entry: 31134

**LA FÉDÉRATION
PROFESSIONELLE DES
PRODUCTEURS ET
DISTRIBUTEURS D'ÉLÉCTRICITÉ
DE BELGIQUE (FPE)**

Avenue de Tervueren 34
d'Éléctricité de Belgique
B-1040 Bruxelles
Belgium
Tel: +32 (0)2 733 96 07

Entry: 07037

**FÉDÉRATION DE L'INDUSTRIE DU
VERRE**

Rue Montoyer 47
1040 Bruxelles
Belgium
Tel: +32 (0)2 509 15 20
Fax: +32 (0)2 514 23 45

Entries: 07016   07041   07086

FIGYELÖ CO

1355 Budapest
PO Box 18 V Alkotmany U
10
Hungary
Tel: +36 (0)1 11 18 50
Telex: 22 66 13
Fax: +36 (0)1 53 01 06

Entry: 18014

**FINANCIAL SUPERVISORY
AUTHORITY**

P. O. Box 7831
10398 Stockholm
Sweden
Tel: +46 (0)8 787 80 00
Fax: +46 (0)8 468 24 1335

Entries: 31015   31065   31138

**FINANCIAL TIMES - SURVEYS
DEPARTMENT**

Surveys Department
1 Southwark Bridge
London SE1 9HL
UK
Tel: +44 (0)171 873 3000

Entries: 01108   01109   29012

**FINANCIAL TIMES MAGAZINES
LTD**

Greystoke Place
Fetter Lane
London
EC4A 1ND
UK
Tel: +44 (0)171 405 6969
Fax: +44 (0)171 242 9249

Entry: 34025

**FINANSRÅDET - DANISH BANKERS
ASSOCIATION**

7 Amaliegade
DK-1256 Copenhagen K
Denmark
Tel: +45 (0)33 12 02 00
Fax: +45 (0)33 93 40 57

Entry: 12019

FINLAND - MINISTRY OF LABOUR

P.O. Box 524
00101 Helsinki
Finland
Tel: +358 (9)0 1856 1
Fax: +358 (9)0 1856 427

Entry: 13042

FINNAIR OY

Dagmarinkatu 4
00100 Helsinki
Finland
Tel: +358 (9)0 818 7211,
Sw.board 818 81
Telex: 124946
Fax: +358 (9)0 818 7457

Entry: 13002   13036

**THE FINNISH BANKERS
ASSOCIATION**

P.O. Box 1009 (Museoleatu
8A)
SF-00101 Helsinki
Finland
Tel: +358 (9)0 440211
Fax: +358 (9)0 498030

Entries: 13001

**FINNISH FOREIGN TRADE
ASSOCIATION**

P.O. Box 908
SF-00101 Helsinki
Finland
Tel: +358 (9)0 69591
Fax: +358 (9)0 690028

Entry: 13045

**FINNISH SHIPOWNERS
ASSOCIATION**

Satamakatu 4A
00160 Helsinki 16
Finland
Tel: +358 (9)0 170 401
Fax: +358 (9)0 669 251

Entry: 13070

FINNISH STATE RAILWAYS

Box 488
00101 Helsinki
Finland
Tel: +358 (9)0 7071
Telex: 12301151
Fax: +358 (9)0 7073 500

Entry: 13078

FINNISH TOURIST BOARD

PO Box 53
SF-00521 Helsinki
Finland
Tel: +358 (9)0 403 011
Telex: 122690 MEK SI
Fax: +358 (9)0 4030 1333

Entry: 13091

FINTEL PUBLICATIONS LTD

Clifton House
Lower Fitzwilliam Street
Dublin 2
Republic of Ireland
Tel: +353 (0)1 613 788
Fax: +353 (0)1 615 200

Entry: 20013

FISKERIDIREKTORATET

Postboks 185/186
5001 Bergen
Norway

Entry: 26022

FLEGON PRESS

37B New Cavendish Street
London W1
UK

Entry: 09036

**FRANCE - BRITISH CHAMBER OF
COMMERCE AND INDUSTRY**

110 Rue de Longchamp
75116 Paris
France
Tel: +33 1 44 05 32 88
Fax: +33 1 44 05 32 99

Entries: 14039   14040   32076

**FRANKFURTER ALLGEMEINE
ZEITUNG GMBH
INFORMATIONSDIENSTE**

PO Box 10 08 08
6000 Frankfurt am Main 1
Germany
Tel: +49 (0)6196 9606 01
Fax: +49 (0)6196 9606 49

Entry: 15084

**FRENCH CHAMBER OF
COMMERCE IN GREAT BRITAIN**

197 Knightsbridge 2nd Floor,
Knightbridge House
London SW7 1RB
UK
Tel: +44 (0)171 225 5250
Telex: 269132
Fax: +44 (0)171 225 5557

Entry: 14065

FRENCH ELECTRICAL AND ELECTRONICS INDUSTRIES ASSOCIATION - FIEE

11-17 Rue Hamelin
75116 Paris
France
Tel: + 33 1 45 05 70 70
Telex: SYCELEC 611045 F
Fax: + 33 1 45 53 03 93

Entries: 14064 14082

FROST & SULLIVAN LTD

4 Grosvenor Gardens
London
SW1W 0DH
UK
Tel: + 44 (0)171 730 3438
Fax: + 44 (0)171 730 3343

Entries: 01027 02088 02089
02090 02091 02092
02093 02094 02095
02096

FT MANAGEMENT REPORTS

Customer Services
PO Box 6
Cambourne TR14 9EQ
UK
Tel: + 44 (0)1209 711 928

Entries: 01120 01121 01122
01123 01124 01125
01126 01127 01128
02097

FURNITURE INDUSTRY RESEARCH ASSOCIATION

Maxwell Road
Stevenage
Herrfordshire SG1 2EW
UK
Tel: + 44 (0)1438 313433
Telex: 827653 FIRA G
Fax: + 44 (0)1438 727607

Entries: 34005 34013 34014
34315

GALE RESEARCH INTERNATIONAL

PO Box 699
North Way
Andover
Hampshire SP10 5YE
UK
Tel: + 44 (0)1264 334 446
Fax: + 44 (0)1264 334 158

Entries: 01139 02077

GARANTI BANK

40 Mete Caddesi
Taksim 80060
Istanbul
Turkey
Tel: + 90 (9)1 251 08 80
Telex: 24538 gafo tr
Fax: + 90 (9)1 251 45 49

Entry: 33025

GENERAL REGISTER OFFICE (NORTHERN IRELAND)

Oxford House
49-55 Chichester Street
Belfast
BT1 4HL
N. Ireland
UK
Tel: + 44 (0)232 235211 x 2341

Entry: 34281

GENERAL REGISTER OFFICE (SCOTLAND)

Ladywell House
Ladywell Road
Edinburgh
UK
Tel: + 44 (0)131 334 0380

Entries: 34262 34282

GENEVA STOCK EXCHANGE

Rue de la Confédération 8
Case Postale 228
CH-1211 Genève 11
Switzerland
Tel: + 41 (0)22 310 06 84
Telex: 423559 BGGE CH
Fax: + 41 (0)22 310 03 81

Entries: 32005 32031 32035
32036

GERMAN CHAMBER OF INDUSTRY & COMMERCE

Economics and Marketing
Department
Mecklenburg House
16 Buckingham Gate
London SW1E 6LB
UK
Tel: + 44 (0)171 233 5656
Telex: 919442 German G
Fax: + 44 (0)171 233 7835

Entries: 15001 15041 15044
15071 15077 15083
15086 15217 15218
15222 34143 34284

GOVERNMENT PUBLICATIONS

Sales Office
Sun Alliance House
Molesworth House
Dublin 2
Republic of Ireland
Tel: + 353 (0)1 710 309
Fax: + 353 (0)1 478 0645

Entries: 20006 20008 20009
20014 20024 20032
20036 20037 20039
20041 20057 20077
20099 20100 20104
20111 20112 20113

GOWER PUBLISHING COMPANY LTD

Gower House
Croft Road
Aldershot
Hants GU11 3HR
UK
Tel: + 44 (0)1252 331551
Fax: + 44 (0)1252 344405

Entry: 34107

HADLEIGH MARKETING SERVICES

47A Palace Road
London
N8 8QL
UK
Tel: + 44 (0)181 340 1186
Fax: + 44 (0)181 340 4969

Entries: 01310 02073 02074
02083 34039 34040
34042 34045

HALIFAX BUILDING SOCIETY

Order from:
Group Planning & Research
Trinity Road
Halifax
HX1 2RG
UK
Tel: + 44 (0)1422 333333
Telex: 517441
Fax: + 44 (0)1422 332043

Entry: 34154

THE HAMBRO COMPANY GUIDE

Order from:
Hemmington Scott
Publishing Ltd
City Innovation Centre
26-31 Whiskin Street
London
EC1R 0BP
UK
Tel: + 44 (0)171 278 7769
Fax: + 44 (0)171 278 9808

Entries: 34145 34150 34256

HEADLAND PRESS

1 Henry Smith's Terrace
Headland
Cleveland TS24 0PD
UK
Tel: + 44 (0)1429 231 902
Fax: + 44 (0)1429 861 403

Entry: 34220

HENLEY CENTRE FOR FORECASTING

2 Tudor Street
Blackfriars
London EC4Y 0AA
UK
Tel: + 44 (0)171 353 9961
Fax: + 44 (0)171 353 2899

Entries: 01130 34097 34139
34196 34260 34261

HMSO - HER MAJESTY'S STATIONERY OFFICE

Order from:

HMSO Publications Office
PO Box 276
London SW8 5DT
UK
Tel: +44 (0)171 873 0011
Fax: +44 (0)171 873 8200

Entries:	34010	34012	34015
	34017	34018	34027
	34030	34053	34055
	34056	34057	34058
	34059	34060	34061
	34062	34063	34064
	34065	34066	34072
	34092	34094	34095
	34112	34115	34116
	34127	34130	34133
	34134	34147	34148
	34149	34151	34153
	34156	34157	34158
	34168	34169	34171
	34177	34192	34201
	34225	34227	34246
	34253	34255	34283
	34306	34307	34310
	34311	34316	34321
	34330	34331	34332
	34333	34338	34344
	34353		

HUNGARIAN CENTRAL STATISTICAL OFFICE

Budapest
Keleti Károly u 5-7
Hungary 1024
Tel: +36 (0)202 4011, 202 4881, 202 4490
Telex: 22 43 08 STATI H
Fax: +36 (0)202 115 9085

Entries:	18001	18019	18025
	18026	18029	18035
	18037	18039	18040
	18041	18042	

HUNGARIAN CHAMBER OF COMMERCE

Budapest
PO Box 106
Hungary H-1389
Tel: +36 (0)1 53 33 33
Telex: KAMARA BUDAPEST 22-4745
Fax: +36 (0)1 53 12 85

Entries:	18002	18003	18013
	18016	18017	18021
	18022	18024	18028
	18032	18033	

HUNGARIAN FOREIGN TRADING CO

1389 Budapest XIII
Jasz u 103-105
Hungary

Entry: 18023

ICC INFORMATION GROUP LTD

Field House
72 Oldfield Road
Hampton
Middlesex TW12 2HQ
UK
Tel: +44 (0)181 783 0922
Fax: +44 (0)181 783 1940

Entries:	02102	34067	34135
	34165	34179	34180
	34181	34182	34183
	34184	34185	34186
	34187	34188	34189
	34190	34191	34193

ICELAND REVIEW

Hoefdabakki 9
PO Box 12122
132 Reykjavik
Iceland
Tel: +354 (9)1 675 700
Fax: +354 (9)1 674 066

Entries: 19012   19017

ICEP: INVESTIMENTAS COMÉRCIO E TURISMO DE PORTUGAL

Av 5 de Outubro 101
1000 Lisbon Codex
Portugal
Tel: +351 (0)1 793 01 03
Telex: 12199 ICEP P
Fax: +351 (0)7935028

Entries:	28006	28008	28010
	28038	28065	28073
	28076		

IFO - INSTITUT FÜR WIRTSCHAFTSFORSCHUNG E.V.

Poschingerstr. 5
8000 München 86
Germany
Tel: +49 (0)89 9224 0
Telex: 522 269
Fax: +49 (0)89 985369

Entries:	15045	15064	15074
	15097	15098	15099
	15100	15102	15103
	15105	15106	15107
	15108	15109	15110
	15111	15112	15113
	15114	15115	15124
	15152		

I.M.A. NEVA MEDIA

App. 7044
1 Morskoy Slavy Pl
199 106 St Petersburg
Russia
Tel: +7 812 355 1644
Telex: 121 585 MRW SU
Fax: +7 812 355 1643

Entry: 09033

IMD

PO Box 915
CH 1001
Lausanne
Switzerland
Tel: +41 (0)21 618 0204
Fax: +41 (0)21 618 0204

Entries: 01083   01306

INDUSTRIAL DEVELOPMENT AUTHORITY

Wilton Park House
Wilton Place
Dublin 2
Republic of Ireland
Tel: +353 (0)1 686 633
Fax: +353 (0)1 603 703

Entry: 20047

INFORM. KATALOG LTD

Sumavska 31
612 64 Brno
Czech Republic
Tel: +42 (0)5 759256
Fax: +42 (0)5 758252

Entry: 11012

INFORMATION MOSCOW

Ul. Naberezhnaya 18
Babovka
Moscow
Russia
C.I.S
Tel: +7 95 443 4088

Entry: 09024

INFOTECH SERVICES LTD

Lenton House
Lenton Lane
Nottingham
NG7 2NR
UK
Tel: +44 (0)1602 524 000
Fax: +44 (0)1602 524 001

Entry: 34167

INSIGHT INTERNATIONAL PUBLISHING LTD

200 Brent Street
London NW4 1BH
UK
Tel: +44 (0)181 203 1883
Fax: +44 (0)181 203 4740

Entry: 02100

INSTITUT FÜR ANGEWANDTE WIRTSCHAFTSFORSCHUNG

0b dem Himmelreich 1
W-7400 Tuebingen
Germany
Tel: +49 (0)7071 9896 0
Fax: +49 (0)7071 9896 99

Entry: 15175

INSTITUT FÜR WELTWIRTSCHAFT

Düsternbrooker Weg 120
D-2300 Kiel 1
Germany
Tel: +49 (0)431 8814 1
Telex: 212 479 weltw d
Fax: +49 (0)431 85853

Entries: 15057 15146 15147
15148 15149 15150
15228

INSTITUT MONÉTAIRE LUXEMBOURGEOIS

63 Avenue de la Liberté
L-2938
Luxembourg
Tel: +352 40 29 29 203
Telex: 2766
Fax: +352 49 21 80

Entries: 23007 23017 23023
23031

INSTITUT NATIONAL D'ÉTUDES DÉMOGRAPHIQUES - INED

27 Rue du Commandeur
75675 Paris
Cedex 14
France
Tel: +33 1 43 20 13 45
Fax: +33 1 43 27 72 40

Entries: 14118 14119

INSTITUT NATIONAL DE LA STATISTIQUE ET DES ÉTUDES ÉCONOMIQUES

18 Blvd Adolphe Pinard
75675 Paris
Cedex 14
France
Tel: +33 41 17 50 50
Telex: 204924 F INSEEGD
Fax: +33 41 17 66 66

Entries: 14018 14022 14023
14025 14027 14038
14041 14042 14047
14048 14050 14066
14068 14069 14070
14071 14072 14074
14080 14081 14100
14111 14126 14127
14131 14145 14147

INSTITUT NATIONAL DE STATISTIQUE: MINISTÈRE DES AFFAIRES ÉCONOMIQUES

Rue de Louvain 44
1000 Bruxelles
Belgium
Tel: +32 (0)2 513 96 50
Fax: +32 (0)2 513 75 20

Entries: 07002 07004 07005
07012 07019 07021
07046 07051 07053
07056 07058 07059
07067 07068 07069
07075 07078 07079
07080 07081 07082
07083 07084 07085
07087 07088 07094
23034

INSTITUTE OF BANKERS IN IRELAND

Nassau House
Nassau Street
Dublin 2
Republic of Ireland
Tel: +353 (0)1 715 311
Fax: +353 (0)1 796 680

Entry: 20061

INSTITUTE OF CHARTERED ACCOUNTANTS IN IRELAND

87-89 Pembroke Road
Ballsbridge
Dublin 4
Republic of Ireland
Tel: +353 (0)1 680 400
Telex: 30567
Fax: +353 (0)1 680 842

Entry: 20001

INSTITUTE OF GROCERY DISTRIBUTION

Letchmore Heath
Watford
Hertfordshire
WD2 8DQ
UK
Tel: +44 (0)1923 857141
Fax: +44 (0)1923 852531

Entries: 03121 34170

INSTITUTE OF MANAGEMENT

Order from:
Burston Distribution
Services
Management House
Cottingham Road
Corby
Northants NN17 1TT
UK
Tel: +44 (0)1536 204223

Entry: 34297

INSTITUTE OF PETROLEUM

61 New Cavendish Street
London W1M 8AR
UK
Tel: +44 (0)171 636 1004
Telex: 264380
Fax: +44 (0)171 255 1472

Entries: 01182 34174
34346

INSTITUTE OF PUBLIC ADMINISTRATION

57-61 Landsdowne Road
Dublin 4
Republic of Ireland
Tel: +353 (0)1 686 233
Fax: +353 (0)1 269 8644

Entries: 20002 20059

INSTITUTO NACIONAL DE ESTADÍSTICA

Paseo de la Castellana, 183
28046 Madrid
Spain
Tel: +34 (9)1 538 94 38
Fax: +34 (9)1 579 27 13

Entries: 30001 30002 30012
30013 30015 30016
30017 30018 30019
30020 30021 30029
30032 30033 30034
30035 30036 30037
30038 30039 30040
30041 30042 30047
30048 30049 30056
30057 30058 30059
30067

INSTITUTO NACIONAL DE ESTATÍSTICA

Av. António José de Almeida
1078 Lisbon Cedex
Portugal
Tel: +351 (0)1 847 00 50
Telex: 63738 PCDINE P
Fax: +351 (0)1 808 09 3

Entries: 28007 28009 28011
28012 28020 28021
28022 28023 28024
28025 28026 28027
28028 28029 28030
28031 28032 28033
28034 28035 28036
28037 28042 28043
28044 28046 28047
28048 28049 28050
28051 28063 28064
28068 28069 28070
28071 28075

INSTITUTO NAZIONALE DI STATISTICA

Via Cesare Balbo 16
00100 Roma
Italy
Tel: +39 (0)6 482 7666
Telex: 610338-626282
ISTAT
Fax: +39 (0)6 467 32273

Entries: 21001 21002 21004
21007 21008 21009
21020 21021 21024
21027 21029 21033
21036 21046 21047
21048 21049 21050
21051 21052 21053
21054 21057 21058
21059 21060 21061

INTERNATIONAL ATOMIC ENERGY AGENCY

Vienna International Centre
Wagramerstrasse 5
PO Box 100, A-1400
Wien
Austria
Tel: +43 (0)222 431 23 60
Telex: 112645
Fax: +43 (0)222 431 23 45
64

Entries: 01089 01219 01231

INTERNATIONAL BUSINESS CLUB

H-1300 Budapest
PO Box 135
Hungary
Fax: + 36 (0)1 12 96 410

Entry: 18020

INTERNATIONAL CIVIL AVIATION AUTHORITY

Statistics Section
1000 Sherbrooke Street
West
Suite 400
Montreal, Quebec
Canada H3A 2R2

Entries: 01028   01059

INTERNATIONAL COCOA ORGANISATION

22 Berners Street
London W1P 3DB
UK
Tel: + 44 (0)171 637 3211
Telex: 28173 ICOCOA G
Fax: + 44 (0)171 631 0114

Entries: 01008   01030   01031
01244

INTERNATIONAL DAIRY SECRETARIAT

41 Square Vergote
1040 Bruxelles
Belgium
Tel: + 32 (0)2 733 98888
Fax: + 32 (0)2 733 0413

Entry: 01040

INTERNATIONAL ECONOMIC REVIEW

3718 Locust Walk
University of Pennsylvania
Philadelphia
PA 19104-6297
USA
Tel: + 1 215 898 5841
Fax: + 1 215 573 2072

Entry: 01162

INTERNATIONAL FEDERATION OF THE PHONOGRAPHIC INDUSTRY

54 Regent Street
London
W1R 5PJ
UK
Tel: + 44 (0)171 434 3521
Fax: + 44 (0)171 439 9166

Entry: 01331

INTERNATIONAL HERALD TRIBUNE

63 Long Acre
London WC2E 9JH
UK
Tel: + 44 (0)171 836 4802
Telex: 262009
Fax: + 44 (0)171 240 2254

Entry: 01168

INTERNATIONAL INDUSTRIAL INFORMATION LTD

PO Box 12
Monmouth
Gwent
NP5 3TL
UK
Tel: + 44 (0)1600 890274
Fax: + 44 (0)1600 890774

Entry: 09032

INTERNATIONAL LEAD AND ZINC STUDY GROUP

Metro House
58 St James's Street
London SW1A 1LD
UK
Tel: + 44 (0)171 499 9373
Telex: 299819 G IL2S 9
Fax: + 44 (0)171 493 3723

Entries: 01024   01071   01187
01202   01241   01249
01312   01313   01314

INTERNATIONAL MONETARY FUND

Publications Services
700 19th Street N.W.
Washington DC 20431
USA
Tel: + 1 202 623 7430
Telex: RCA 248331 IMF UR
Fax: + 1 202 623 7201

Entries: 01016   01060   01061
01104   01135   01142
01143   01144   01145
01157   01164   01165
01166   01209   01260
01316   04001   09022
29015

INTERNATIONAL STATISTICAL INSTITUTE

428 Prinses Beatrix laan
P O Box 950
2270 A2 Voorburg
Sweden
Tel: + 46 (0)31 70 3375737
Telex: 32260 151 nl
Fax: + 46 (0)31 70 3860025

Entry: 31017

INTERNATIONAL STEEL STATISTICS BUREAU

Canterbury House
2 Sydenham Road
Croydon CR9 2LZ
UK
Tel: + 44 (0)181 686 9050
Fax: + 44 (0)181 680 8616

Entries: 01176   01336   01338

INTERNATIONAL TEA COMMITTEE LTD

Sir John Lyon House
5 Timber Street
London EC4V 3NH
UK
Tel: + 44 (0)171 248 4672
Telex: 887911 TELTEA G
Fax: + 44 (0)171 248 3011

Entries: 01005   01206

INTERNATIONAL TEXTILE MANUFACTURERS FEDERATION

Am Schanojengrab 29
CH-8039
Postfach
Zurich
Switzerland
Tel: + 41 (0)1 201 7050
Telex: 56798
Fax: + 41 (0)1 201 71 34

Entries: 01053   01158   01173
01178   01262

INTERNATIONAL WHEAT COUNCIL

1 Canada Square
Canary Wharf
London E15 5AE
UK
Tel: + 44 (0)171 513 1122
Fax: + 44 (0)171 712 0071

Entries: 01136   01248   01250
01321

IRISH BROKER

Holyrood Publications Ltd
139A Lower Drumcondra
Road
Dublin 9
Republic of Ireland
Tel: + 353 (0)1 374 311
Fax: + 353 (0)1 360624

Entry: 20062

IRISH EXPORTER

Jude Publications Ltd
2-6 Tara Street
Dublin 2
Republic of Ireland
Tel: + 353 (0)1 713 500
Fax: + 353 (0)1 713 074

Entry: 20064

IRISH TIMES LTD
13, D'Olier Street
Dublin 2
Republic of Ireland
Tel: +353 (0)1 679 2022
Fax: +353 (0)1 679 3910

Entry: 20067

**IRISH TOURIST BOARD - BORD
FAILTE EIREANN**
Baggot Street Bridge
Dublin 2
Republic of Ireland
Tel: +353 (0)1 765 871
Fax: +353 (0)1 764 764

Entry: 20068

ISLANDBANKI HF
15-155 Reykjavik
Iceland
Tel: +354 (9)1 608 500
Telex: 2047 ISBA IS
Fax: +354 (9)1 687 784

Entry: 19014

**ISTANBUL CHAMBER OF
COMMERCE**
Ragip Gümüçpala Caddesi
84
34378 Eminönü
P.K.377 Istanbul
Turkey
Tel: +90 (9)1 511 41 50
Telex: 22682 Oda tr
Fax: +90 (9)1 526 21 97

Entry: 33039

ISVESTIA
c/o Dawsons
1 Cresswell Park
London SE3 9RD
UK
Tel: +44 (0)181 297 1544

Entry: 09026

JEMMA PUBLICATIONS LTD
Marino House
53 Glashule Road
Sandycove
County Dublin
Republic of Ireland
Tel: +353 (0)1 280 0000
Fax: +353 (0)1 284 4041

Entry: 20078

JONES LANG WOOTTON
22 Hanover Square
London W1A 2NB
UK
Tel: +44 (0)171 493 6040
Fax: +44 (0)171 408 0220

Entries: 34004   34266

JORDAN & SONS LTD
PO Box 260
Bristol BS99 7XZ
UK
Tel: +44 (0)117 9230600
Telex: 449119 SEARCH G
Fax: +44 (0)117 9230063

Entries: 34029   34031   34032
34033   34034   34035
34175   34276   34299

JUGOSLOVENSKI PREGLED
Mose Pijade 8-1
PO Box 677
11001 Belgrade
Serbia

Entry: 35014

**JUPITER
VERLAGSGESELLSCHAFT MBH**
Robertgasse 2
A-1020 Wien
Austria
Tel: +43 (0)222 214 22 94
Fax: +43 (0)222 216 07 20

Entries: 05040   05067

**KAMER VOOR HANDEL EN
NIJVERHEID VAN HET
ARRONDISEMENT LEUVEN**
Tiense Vest 61
300 Leuven
Belgium
Tel: +32 (0)16 22 26 89
Fax: +32 (0)16 23 78 28

Entries: 07015   07050

KANSALLIS-OSAKE-PANKKI
Aleksaunterinkatu 42
00100 Helsinki
Finland
Tel: +358 (9)0 1631
Telex: 124412
Fax: +358 (9)0 163 2968

Entries: 13003   13061   13065

KOGAN PAGE
120 Pentonville Road
London N1 9BR
UK
Tel: +44 (0)171 278 0433
Telex: 263088 KOGAN G
Fax: +44 (0)171 837 6348

Entries: 08014   09016   11015
18009   27015   29005

**LANDBOUW - ECONOMISCH
INSTITUUT**
P.O. Box 29703
2502 LS Den Haag
The Netherlands
Tel: +31 (0)70 330 8330
Fax: +31 (0)70 361 5624

Entries: 25001   25034   25076

LANDSBANKI ISLANDS
PO Box 170
Reykjavik
Iceland
Tel: +354 (9)1 606 600
Telex: 2030
Fax: +354 (9)1 298 82

Entries: 19002   19007

**LANGUAGE COMPREHENSIVE
SERVICES**
Moravska 904
120 00 Praha
Czech Republic
Tel: +42 (0)2 251 8928

Entries: 11006   11007

**LATVIAN CHAMBER OF
COMMERCE AND INDUSTRY**
21 Brivibas Blvd
226189 Riga
Latvia
Tel: +371 (0)132 332269
Telex: 161100 PKP SU
Fax: +371 (0)132 332276

Entry: 06001

LEISURE CONSULTANTS
Lint Grovis, Foxearth
Sudbury
Suffolk
CO10 7JX
UK
Tel: +44 (0)1787 375777

Entry: 34195

LIECHTENSTEIN VERLAG AG
PO Box 133
FL-9490 Vaduz
Liechtenstein
Tel: +4175 224 14
Telex: 889326
Fax: +4175 24340

Entries: 22005   22015

**LIECHTENSTEINISCHE
FREMDENVERKEHRSZENTRALE**
Postfach 139
FL-9490 Vaduz
Liechtenstein
Tel: +4175 232 14 43
Telex: 889488
Fax: +4175 2310806

Entry: 22009

LLOYD'S REGISTER OF SHIPPING
71 Fenchurch Street
London EC3M 4BS
UK
Tel: +44 (0)171 709 9166
Telex: 888370
Fax: +44 (0)171 488 4796

Entries: 01010   01189   01190
01197   01258

LLOYDS BANK PLC
Order from:
Economics Department
Hays Lane House
1 Hays Lane
London SE1 2HA
UK
Tel: +44 (0)171 407 1000
Fax: +44 (0)171 357 4378

Entries: 34198   34199   34200

LO-LANDSORGANISATIONEN I DANMARK (THE DANISH CONFEDERATION OF TRADE UNIONS)
Rosenørns Alle 12
1634 Copenhagen V
Denmark
Tel: +45 (0)31 35 35 41
Fax: +45 (0)35 37 37 41

Entries: 12027   12029

LONDON STOCK EXCHANGE
Old Broad Street
London
EC2N 1HP
UK
Tel: +44 (0)171 797 1000
Telex: 886557

Entries: 34140   34202   34267
         34268   34318   34319
         34320

LORD CHANCELLOR'S DEPARTMENT
Information Management
Unit
6th floor, Trevelyan House
30 Great Peter Street
London WC2R 1LB
UK
Tel: +44 (0)171 438 7370

Entry: 34007

MACHINE TOOL TECHNOLOGIES ASSOCIATION
62 Bayswater Road
London W2 3PS
UK
Tel: +44 (0)171 402 6671
Telex: 27829 MTTAGB G
Fax: +44 (0)171 724 7250

Entries: 34026   34207

MACMILLAN DISTRIBUTION LTD
Houndmills
Basingstoke
Hants RG21 2XS
UK
Tel: +44 (0)1256 29242
Telex: 858493
Fax: +44 (0)1256 84 2084

Entry: 02021

MALTA CHAMBER OF COMMERCE
Exchange Buildings
Republic Street
Valletta VLT 05
Malta
Tel: +356 24 72 33
Fax: +356 24 52 23

Entries: 24011   24024

MALTA EXPORT TRADE CORPORATION
Trade Centre
Industrial Estate
San Gwann
Malta
Tel: +356 44 61 86
Telex: 1966 EXPORT
Fax: +356 49 66 87

Entry: 24018

MALTA FEDERATION OF INDUSTRY
Development House
St Anne Street
Floriana
Malta
Tel: +356 23 44 28
Telex: MW 1380 AGIO

Entries: 24004   24017

MALTA INTERNATIONAL BUSINESS AUTHORITY
Palazzo Spinola
PO Box St Julians 29
STJ 01
Malta
Tel: +356 34 42 30
Telex: 1692 MIBA MW
Fax: +356 34 43 34

Entry: 24005

MARKET ASSESSMENT PUBLICATIONS LTD
2 Duncan Terrace
London N1 8BZ
UK
Tel: +44 (0)171 278 9517
Fax: +44 (0)171 278 6246

Entries: 34209   34210   34211
         34212   34213   34214
         34215   34216   34217
         34324

THE MARKET RESEARCH SOCIETY
Marketing Department
15 Northburgh Street
London
EC1V 0AH
UK
Tel: +44 (0)171 490 4911

Entry: 01045

MARKETING STRATEGIES FOR INDUSTRY
Heathmans House
19 Heathmans Road
Parsons Green
London SW6 4TJ
UK
Tel: +44 (0)171 371 0955
Fax: +44 (0)171 371 5284

Entries: 02107   02108   34235
         34236   34237   34238
         34239   34240   34241
         34242   34243   34244

MARKETPOWER
84 Uxbridge Road
London W13 8RA
UK
Tel: +44 (0)181 840 5252
Fax: +44 (0)181 840 6173

Entries: 02105   02106   34223

MEDIOBANCA
Via Filodrammatici 10
20121 Milano
Italy
Tel: +39 (0)2 882 91
Fax: +39 (0)2 882 9367

Entries: 21016   21022   21025
         21032

MEFA - THE ASSOCIATION OF DANISH PHARMACEUTICAL INDUSTRY
Strødamvej 50A
DK-2100 Copenhagen ø
Denmark
Tel: +45 (0)39 27 00 88
Fax: +45 (0)39 27 00 50

Entry: 12018

METAL BULLETIN BOOKS LTD
Park House
Park Terrace
Worcester Park
Surrey KT4 7HY
UK
Tel: +44 (0)181 330 4311
Telex: 21383 METBUL G
Fax: +44 (0)181 337 8943

Entries: 01103   01147   01148
         01183   01198   01199
         01200   01261   01307
         01324   01326   01332
         01335   02082

MIDDLE EAST ECONOMIC DIGEST
21 John Street
London
WC1N 2BP
UK
Tel: +44 (0)171 404 5513
Telex: 27165 MEEDGAR G
Fax: +44 (0)171 831 9537

Entry: 33068

MINES SERVICE

6 Demetsana Street
Nicosia
Cyprus

Entry: **10005**

**MINISTÈRE DE L'AGRICULTURE:
INSTITUT ÉCONOMIQUE
AGRICOLE (IEA)**

Manhattan Center, Office
Tower
Avenue du Boulevard 21
20c étape, B-1210 Bruxelles
Belgium
Tel: +32 (0)2 211 7600

Entry: **14017**

**MINISTÈRE DE L'AGRICULTURE:
SERVICE CENTRAL DES
ENQUÊTES ET ÉTUDES**

4 Avenue de Saint Mandé
75570 Paris
Cedex 12
France
Tel: +33 43 44 46 33
Telex: 43406579
Fax: +33 49 55 85 11

Entries:		
14001	**14003**	**14004**
14005	**14006**	**14007**
14008	**14009**	**14010**
14011	**14012**	**14013**
14014	**14015**	**14016**
14020	**14045**	**14060**
14061	**14089**	**14096**

**MINISTÈRE DE L'EMPLOI ET DU
TRAVAIL**

Rue Belliard 51
B-1040 Bruxelles
Belgium
Tel: +32 (0)2 233 44 71
Telex: 22937 ARBTRA
Fax: +32 (0)2 233 44 27

Entry: **07073**

**MINISTÈRE DE L'EQUIPEMENT, DU
LOGEMENT ET DES
TRANSPORTS: DIRECTION DES
TRANSPORTS TERRESTRES (DTT)**

l'Arche de la Défense - Paroi
sud
92055 Paris la Défense,
Cedex 04
France
Tel: +33 1 40 81 21 22
Telex: 610835 F
Fax: +33 1 40 81 37 95

Entry: **14085**

**MINISTÈRE DE L'INDUSTRIE DU
COMMERCE EXTÉRIEUR (SESSI)**

83-85 Boulevard de
Montparnasse
75270 Paris
Cedex 06
France
Tel: +33 45 56
Fax: +33 45 56 40 71

Entries:		
14028	**14067**	**14075**
14092	**14094**	**14095**
14133	**14134**	**14135**
14138	**14144**	

**MINISTÈRE DES AFFAIRES
ÉCONOMIQUES**

Direction Générale des
Études de la Documentation
Rue de l'Industrie 6
1040 Bruxelles
Belgium
Tel: +32 (0)2 506 63 07
Fax: +32 (0)2 513 46 57

Entries:		
07007	**07045**	**07049**
07092		

**MINISTÈRE DES AFFAIRES
ÉCONOMIQUES -
ADMINISTRATION DES MINES**

Rue de Mot 28-30
B-1040 Bruxelles
Belgium
Tel: +32 (0)233 61 11
Fax: +32 (0)230 56 62

Entries:		
07001	**07076**	**07077**

**MINISTÈRE DES AFFAIRES
ÉCONOMIQUES: ADMINISTRATION
DE L'ENERGIE**

J-A de Mostraat 30
1040 Bruxelles
Belgium
Tel: +32 (0)2 233 61 11
Telex: 13509 ENERGI

Entries:	
07010	**07017**

**MINISTÈRE DES
COMMUNICATIONS ET DE
L'INFRASTRUCTURE**

Rue d'Arlon 104
B-1040 Bruxelles
Belgium
Tel: +32 (0)2 233 12 11
Telex: 322 61880 VERTRA
B
Fax: +32 (0)2 231 18 33

Entries:	
07048	**07057**

**MINISTERIO DE ECONOMIA Y
HACIENDA: DIRECCÍON GENERAL
DE SEGUROS**

Castellana 44
28046 Madrid
Spain
Tel: +34 (9)1 339 70 00
Fax: +34 (9)1 339 71 13

Entries:		
30006	**30007**	**30055**

**MINISTERIO DE ECONOMIA Y
HACIENDA: SECRETARÍA
GENERAL DE COMERCIO**

Castellana 162
Planta 16
Madrid 28046
Spain

Entry: **30051**

**MINISTERIO DEL INTERIOR:
DIRECCIÓN GENERAL DE
TRAFICO**

Josefa Valcarcel No 28
28071 Madrid
Spain
Tel: +34 (9)1 742 31 12
Fax: +34 (9)1 741 81 34

Entries:	
30010	**30011**

MINISTRY OF FINANCE

St Calcedonius Square
Floriana CMR 02
Malta
Tel: +356 22 54 21
Fax: +356 22 04 37

Entries:	
24007	**24008**

**MINISTRY OF FINANCE
ECONOMICS DEPARTMENT**

Box 295
SF 00171 Helsinki
Finland
Tel: +358 (9)0 1601
Telex: 123241

Entries:	
13022	**13069**

**MINISTRY OF INDUSTRY AND
TRADE, INFORMATION AGENCY**

H-1024 Budapest
Margit Krt 85
Hungary
Tel: +36 (0)1 75 51 28
Fax: +36 (0)1 75 32 95

Entry: **18004**

MINISTRY OF TOURISM

Order from:
General Directorate of
Investments
I.Inönü Bulvari
Bahçelievler
Ankara
Turkey
Tel: +90 (9)4 212 83 00
Fax: +90 (9)4 212 83 97

Entries:		
33005	**33006**	**33010**
33024	**33052**	

MINISTRY OF TRADE AND INDUSTRY ENERGY DEPARTMENT

Aleksanterinkatu 10
00170 Helsinki
Finland
Tel: +358 (9)0 1601
Telex: 125452 +
Fax: +358 (9)0 1603 666

Entry: 13023

MINISTRY OF TRANSPORT AND COMMUNICATIONS

P.O. Box 235
SF-00131 Helsinki
Finland
Tel: +358 (9)0 173 61
Fax: +358 (9)0 173 6292

Entries: 13089   13095   13103

MINTEL INTERNATIONAL GROUP LTD

18-19 Long Lane
London EC1A 9HE
UK
Tel: +44 (0)171 606 4533
Fax: +44 (0)171 606 5932

Entries: 34197   34218   34226
          34257   34288

MOSCOW NARODNY BANK LTD

81 King William Street
London EC4P 4SS
UK
Tel: +44 (0)171 623 2066
Fax: +44 (0)171 283 4840

Entry: 09001

MULTI MEDIA INTERNATIONAL

PO Box 6805
2001 Jh Haarlem
Netherlands

Entry: 09028

NARODOWY BANK POLSKI

00-950 Warszawa
Ul - Swietokrzysta
Poland
Tel: +48 (0)22 20 03 21
Telex: 814 681
Fax: +48 (0)22 20 62 41

Entry: 27037

NATIONAL BANK OF GREECE S.A.

86 Eolou Street
102 32 Athens
Greece
Tel: +30 (0)32 10 411
Telex: 21 4931 38/21 5385
86 NBG GR
Fax: +30 (0)32 28 187

Entry: 17003

NATIONAL BOARD OF AGRICULTURE

Planning Department/
Statistics
P.O. Box 250
SF-00171 Helsinki
Finland
Tel: +358 (9)0 134 211
Fax: +358 (9)0 134 21573

Entries: 13028   13050   13066
          13100

NATIONAL BOARD OF FORESTRY

S-551 83 Jonkoping
Sweden
Tel: +46 (0)36 15 56 00
Fax: +46 (0)36 16 61 70

Entry: 31113

NATIONAL BOARD OF HEALTH AND WELFARE

S-106 30 Stockholm
Sweden
Tel: +46 (0)8 783 30 00
Fax: +46 (0)8 783 30 70

Entry: 31004

NATIONAL BOARD OF NAVIGATION FINLAND - STATISTICS DIVISION

P.O. Box 158
00141 Helsinki
Finland
Tel: +358 (9)0 18081
Telex: 121471
Fax: +358 (9)0 1808640

Entries: 13043   13083   13084
          13085

NATIONAL ECONOMIC AND SOCIAL COUNCIL

Upper Castle Yard
Dublin Centre
Dublin 2
Republic of Ireland
Tel: +353 (0)1 713 155
Fax: +353 (0)1 713 589

Entry: 20084

NATIONAL ECONOMIC INSTITUTE

Kalkofnsvegi 1
150 Reykjavik
Iceland
Tel: +354 (9)1 699 500
Fax: +354 (9)1 626 540

Entry: 19013

NATIONAL HOUSE BUILDING COUNCIL

Publications Department
58 Portland Place
London W1N 4BU
UK
Tel: +44 (0)171 580 9381
Fax: +44 (0)171 636 6378

Entries: 02109   34247

NATIONAL STATISTICAL INSTITUTE

10, 6th September St
1000- Sofia
Bulgaria
Tel: +359 (0)87 96 38
Fax: +359 (0)87 78 25

Entries: 08002   08003
          08006   08012   08013
          08015   08017   08018
          08019   08020   08022
          08023   08024   08025
          08026   08027   08028
          08029   08031   08032

NATIONAL STATISTICAL SERVICE OF GREECE

Publications and Information
Division
14-16 Lycourgou Street
Athens 101 66
Greece
Tel: +30 (0)32 44 746
Telex: 216734 ESYE
Fax: +30 (0)32 42 217

Entries: 17001   17002   17004
          17005   17006   17026
          17027   17029   17033
          17042   17044   17045
          17047   17048   17049
          17050   17051   17052

NATIONALOKONDMISCHE GESELLSCHAFT

Hohenstaufengasse 9
A-1010 Wien
Austria
Tel: +43 (0)222 40103/2676
Fax: +43 (0)222 532 14 98

Entry: 05036

DE NEDERLANDSCHE BANK

Westeinde 1
Postbus 98
1000 AB Amsterdam
The Netherlands
Tel: +31 (0)20 524 9111
Telex: 11355 DNB AM NL
Fax: +31 (0)20 524 34 29

Entry: 25003

NEDERLANDS INSTITUUT VOOR HET BANK EFFECTENBEDRIJT (NIBE)

Herengracht 205
1016 BE Amsterdam
The Netherlands
Tel: +31 (0)20 520 8520
Fax: +31 (0)20 622 9446

Entry: 25005

NETHERLANDS BRITISH CHAMBER OF COMMERCE

307 High Holborn
London WC1V 7LS
UK
Tel: +44 (0)171 405 1358
Fax: +44 (0)171 405 1689

Entries: 25006 25009 25010
25011 25017 25018
25025 25031 25037
25064 25065 25078

NETHERLANDS CENTRAL BUREAU OF STATISTICS

SDU/Publishers
P.O. Box 20014
2500 EA Den Haag
The Netherlands
Tel: +31 (0)70 337 3800
Fax: +31 (0)70 387 1429

Entries: 25002 25004 25007
25019 25022 25023
25024 25030 25033
25035 25038 25039
25040 25041 25042
25043 25044 25045
25046 25047 25048
25049 25050 25051
25053 25054 25055
25056 25057 25058
25059 25060 25061
25063 25066 25067
25068 25069 25070
25071 25072 25073
25074 25075 25077

NEUER WEG

Piata Presei Libere 1
Bucharesti
R-71341
Romania
Tel: +40 (0)1 817 23
Telex: 11618

Entry: 29011

NOMOS VERLAGS-GESELLSCHAFT MBH & CO KG

Waldseestrasse 3-5
Postfach 610
7570 Baden-Baden
Germany
Tel: +49 (0)722 1 2104 0
Telex: 7 81 201
Fax: +49 (0)722 1 210427

Entries: 15011 15051 15060
15063 15065 15073
15090 15091 15166
15179 15193 15196

THE NORDIC STATISTICAL SECRETARIAT

Sejrøgade 11
DK-2100 Copenhagen ø
Denmark
Tel: +45 (0)39 17 39 82
Fax: +45 (0)31 18 48 01

Entries: 12038 12055 19018
19019 19023

NORGES BANK

Postboks 1179 Sentrum
0107 Oslo 1
Norway
Tel: +47 2 31600
Fax: +47 241 31 05

Entry: 26013

NORGES REDERIFORBUND

Raadhusgt 25
Postboks 1452 Vika
N-0116 Oslo
Norway
Tel: +47 241 60 80
Fax: +47 241 50 21

Entry: 26051

DEN NORSKE BANK

P.O. Box 1171 Sentrum
0107 Oslo
Norway
Tel: +47 248 10 50
Fax: +47 248 18 70

Entries: 26003 26014 26045

NORTH ATLANTIC TREATY ORGANISATION

Economics and Information
Directorates
B-1110 Bruxelles
Belgium
Tel: +32 (0)2 728 4111
Telex: 23 867
Fax: +32 (0)2 728 4117

Entries: 01216 01217

NORTH RIVER PRESS

Box 309
Croton-on-Hudson
N.Y. 10520
US
Tel: +1 914 941 7175
Fax: +1 914 941 9579

Entry: 09035

NORTHERN IRELAND OFFICE

Order from:
Policy Planning and
Research Unit
Department of Finance and
Personnel
Parliament Buildings
Stormont
Belfast BT4 3SW
UK
Tel: +44 (0)1232 763 210

Entry: 34249

NORWEGIAN BANKERS ASSOCIATION

Postboks 1489
Dronning Maudsgt 15
Vika
0116 Oslo
Norway
Tel: +47 283 31 60
Fax: +47 283 07 43

Entries: 26019 26043

THE NORWEGIAN CONFEDERATION OF TRADE UNIONS

Youngsgaten 11
0181 Oslo 1
Norway
Tel: +47 240 10 50
Fax: +47 240 17 43

Entry: 26002

NORWEGIAN TRADE COUNCIL

N-0243 Oslo
Norway
Tel: +47 292 63 00
Telex: 78532 n
Fax: +47 292 64 00

Entry: 26039

LE NOUVEL ÉCONOMISTE SA

65 Avenue des Champs
Elysées
75008 Paris
France
Tel: +33 1 40 74 80 00
Telex: 648991
Fax: +33 1 42 25 94 73

Entry: 14091

NTC PUBLICATIONS LTD

PO Box 69
Henley-on-Thames
Oxfordshire
RG9 1GB
UK
Tel: +44 (0)1491 574671
Fax: +44 (0)1491 571188

Entries: 34008 34009 34044
34105 34120 34132
34138 34204 34222
34224 34273 34277
34290

N.V. TRENDS

Research Park Zellik, De
Haak
1731 Zellik
Bruxelles
Belgium
Tel: +32 (0)2 467 57 00
Fax: +32 (0)2 467 57 58

Entry: 07091

OBERÖSTERREICH TOURISTIK

Schillerstrasse 50
A-4010 Linz
Austria
Tel: +43 (0)732 2123
Telex: 222175
Fax: +43 (0)600229

Entry: 05033

OESTERREICHISCHE NATIONALBANK

Abteilung für
Öffentlichkeitsarbeit
Otto-Wagner-Platz 3
1090 Wien
Austria
Tel: +43 (0)222 40420
Telex: 114669
Fax: +43 (0)222 40420/
7240

Entries: 05009   05013   05045
          05098

OESTERREICHISCHES STATISTISCHES ZENTRALAMT

Hintere Zollamtsstrasse 2b
A-1033 Wien
Postfach 9000
Austria
Tel: +43 (0)222 711 28/
7414
Fax: +43 (0)222 711 28/
7728

Entries:	05004	05010	05012
	05014	05015	05023
	05024	05025	05026
	05027	05028	05029
	05034	05037	05038
	05039	05043	05044
	05046	05047	05048
	05049	05050	05051
	05054	05055	05061
	05062	05063	05064
	05066	05069	05070
	05074	05075	05076
	05077	05078	05082
	05083	05087	05092
	05093	05094	05096
	05097	05099	05102
	05103	05104	05105
	05107	05108	05109
	05110		

OFFICE FÉDÉRALE DE LA STATISTIQUE

Hallwylstrasse 15
3003 Berne
Switzerland
Tel: +41 (0)31 61 86 60
Fax: +41 (0)31 61 78 56

Entries:	32003	32012	32014
	32016	32017	32018
	32028	32032	32045
	32046	32047	32048
	32049	32050	32052
	32053	32054	32055
	32057	32064	32065
	32067	32068	32071
	32073	32084	32085
	32086	32087	

OFFICE FÉDÉRALE DE LA SUISSE

Hallwylstrasse 15
3003 Berne
Switzerland
Tel: +41 (0)31 61 86 60
Telex: 32526
Fax: +41 (0)31 61 78 56

Entry: 32002

OFFICE FOR OFFICIAL PUBLICATIONS OF THE EUROPEAN COMMUNITIES

HMSO Publications Centre
51 Nine Elms Lane
London SW8 5DR
UK
Tel: +44 (0)171 873 9090
Telex: 29 71 138
Fax: +44 (0)171 873 8463

Entries:	02025	02045	02069
	02070	02085	03002
	03003	03005	03008
	03010	03012	03014
	03015	03016	03018
	03019	03020	03026
	03027	03028	03029
	03030	03031	03032
	03033	03036	03040
	03047	03049	03051
	03053	03054	03057
	03058	03059	03060
	03064	03065	03066
	03067	03068	03070
	03072	03073	03074
	03075	03076	03077
	03078	03080	03081
	03082	03083	03087
	03088	03093	03101
	03102	03104	03105
	03107	03108	03110
	03112	03113	03115
	03116	03117	03119
	03122	03123	03125
	03129	03130	03133
	03134	03140	03141
	03142	03144	03145
	03147	03150	03151
	03154	03155	03156
	03157	03159	03161
	03162	03164	03165
	03170	03172	03175
	03176	03177	08004
	08009	11011	13027
	18038	29016	34122

OFFICE OF POPULATION CENSUSES AND SURVEYS

10 Kingsway
London
WC2B 6JP
UK
Tel: +44 (0)171 242 0262

Entries:	34001	34002	34003
	34124	34142	34263
	34264	34280	

ORBIS INFORMATION SERVICES / RIS/

45 Vinohradska
120 00 Praha
Czech Republic
Tel: +42 (0)2 252 064
Fax: +42 (0)2 254 385

Entry: 11005

ORGANISATION FOR ECONOMIC CO-OPERATION AND DEVELOPMENT (OECD)

2 Rue André-Pascal
75775 Paris Cedex 16
France
Tel: +33 1 4524 8200
Telex: 620160 OCDE
Fax: +33 1 45 24 85 00

Order from:
(in UK try HMSO)

Entries:	01003	01007	01017
	01019	01029	01037
	01039	01058	01067
	01076	01081	01087
	01088	01090	01091
	01092	01093	01096
	01105	01106	01110
	01111	01114	01146
	01149	01152	01153
	01154	01184	01185
	01192	01193	01194
	01195	01201	01207
	01211	01215	01220
	01221	01222	01223
	01224	01225	01226
	01227	01228	01230
	01243	01245	01246
	01247	01252	01253
	01256	01263	01266
	01267	01269	01274
	01275	02058	02099
	02113	07055	14115
	19020	20088	21037
	23028	28059	30060
	32058	34250	

OSLO BØRS

Post Boks 460 Sentrum
0105 Oslo
Norway
Tel: +47 234 1700
Fax: +47 241 6590

Entries:	26004	26005	26006
	26029	26032	26041
	26044	26049	

OTR-PEDDER

34 Duncan Road
Richmond
Surrey TW9 2JD
UK
Tel: +44 (0)181 940 4300
Fax: +44 (0)181 948 1531

Entry: 34251

OVERSEAS DEVELOPMENT ADMINISTRATION

Eaglesham Road
East Kilbride
Glasgow
G75 8EA
UK
Tel: +44 (0)1355 843246

Entry: 34036

OVUM LTD

7 Rathbone Street
London W1P 1AF
UK
Tel: +44 (0)171 255 2670
Fax: +44 (0)171 255 1995

Entry: 02111

OY NOVOMEDIA LTD

Vapaalantie 2A
01650 Vantaa
Finland
Tel: +358 (9)0 840144
Fax: +358 (9)0 840110

Entry: 13038

OY TALENTUM AB
Ratavartijankatu 2, P.O. Box 920
SF 00101 Helsinki
Finland
Tel: +358 (9)0 148801
Fax: +358 (9)0 148382

Entries: 13087   13088

PACKAGING INDUSTRY
RESEARCH ASSOCIATION (PIRA)

Publishing and information
services
Randalls Road
Leatherhead
Surrey KT22 7RU
UK
Tel: +44 (0)1372 376 161
Fax: +44 (0)1372 377 526

Entries: 02080   02081   02112
            34259   34345

PERGAMON PRESS PLC

Headington Hill Hall
Oxford
OX3 0BW
UK
Tel: +44 (0)1865 64881
Telex: 83177
Fax: +44 (0)1865 60285

Entry: 09019

PLANNING BUREAU,
GOVERNMENT OF CYPRUS

Apellis Str.
Nicosia
Cyprus
Tel: +357 (0)2 302153
Telex: 4521 PLANBUR CY
Fax: +357 (0)2 366810

Entries: 10020   10022   10027

POLISH CHAMBER OF COMMERCE

Ul Trebocka 4
00-074 Warszawa
Poland
Tel: +48 (0)22 26 02 21
Telex: 814361 PIHZ PL
Fax: +48 (0)22 27 46 73

Entries: 27025   27026   27046
            27062

POLISH FOREIGN TRADE

PO Box 726
Marszalkowska 124
00-950 Warszawa
Poland
Tel: +48 (0)22 26 77 30

Entry: 27052

POSTS AND
TELECOMMUNICATIONS OF
FINLAND

Mannerheimintie 11
00100 Helsinki
Finland
Tel: +358 (9)0 195 4715
Fax: +358 (9)0 195 4855

Entry: 13072

POSTVERKET

General Directorate of Posts
S-105 00 Stockholm
Sweden
Tel: +46 (0)8 781 10 00
Telex: 13356 POSTGEN S
Fax: +46 (0)8 21 96 11

Entries: 31012   31099

THE PRAGUE POST

Dlovha tr 2
110 00 Praha 1
Czech Republic
Tel: +42 (0)2 231 7820
Fax: +42 (0)2 231 4756

Entry: 11023

PRESSE- UND
INFORMATIONSAMT

Regierungsgebäude
FL-9490 Vaduz
Liechtenstein
Tel: +4175 667 21
Fax: +4175 66460

Entry: 22018

PRICE WATERHOUSE

Southwark Towers
32 London Bridge St
London SE1 9SY
UK
Tel: +44 (0)171 9393000

Entries: 01063   02034   03041
            05032   07032   09017
            10019   11014   12016
            14044   15059   18011
            20031   21015   22006
            24013   26012   27018
            30028   31030   32026
            33017   34100   34103

PRIVREDNI PREGLED

Marsala Birjuzova 3-5
11000 Belgrade
Serbia

Entries: 35007   35011

PROBUS EUROPE

Sheraton House
Castle Park
Cambridge
CB3 0AX
UK
Tel: +44 (0)1223 462244
Fax: +44 (0)1223 460178

Entries: 11019   27034

PROFESSIONAL AND BUSINESS
INFORMATION PLC

Munro House
14 St Cross Street
London EC1N 8YY
UK
Tel: +44 (0)171 430 2020
Fax: +44 (0)171 430 1773

Entry: 02075

RAPID A.C.

28 Rijna 13
112 79 Praha 1
Czech Republic

Entry: 11022

REED BUSINESS PUBLISHING LTD

Quadrant House
The Quadrant, Sutton
Surrey
SM2 5AS
UK
Tel: +44 (0)181 652 3500

Entry: 01170

THE RESEARCH INSTITUTE OF
THE FINNISH ECONOMY, ETLA

Lönnrotinkatu 4 B
00120 Helsinki
Finland
Tel: +358 (9)0 609 900
Fax: +358 (9)0 601 753

Entry: 13039

RESOURCES CSFR

Lazarska 3
110 00 Praha 1
Czech Republic
Tel: +42 (0)2 201 325
Fax: +42 (0)2 201 812

Entry: 11024

THE REWARD GROUP

Reward House, Diamond
Way
Stone Business Park
Stone
Staffordshire ST15 0SD
UK
Tel: +44 (0)1785 813 566
Fax: +44 (0)1785 817 007

Entries: 34028   34073   34077
            34098   34131   34203
            34258   34278   34285
            34295   34298

RHEINISCH - WESTFÄLISCHE BÖRSE ZU DÜSSELDORF

Postfach 10 42 62
Ernst - Schneider-Platz 1
4000 Düsseldorf
Germany
Tel: +49 (0)211 13 89 0
Fax: +49 (0)211 133 287

Entry: 15130

RIA-NOVOSTI

3 Rosary Gardens
London SW7 4NW
UK
Tel: +44 (0)171 370 1873
Fax: +44 (0)171 244 7875

Entries: 09006   09029   09031

RICERCHE & STUDI

Piazza Paolo Ferrari 6
20121 Milano
Italy
Tel: +39 (0)2 86 46 2394 -
8646 2348
Fax: +39 (0)2 86 22 67

Entry: 21040

RICHARD COTTRELL AND ASSOCIATES

Dean House
Clanage Road
Bower Ashton
Bristol BS3 2JX
UK
Tel: +44 (0)117 9663404
Fax: +44 (0)117 9639502

Entry: 27051

RIKSFORSAKRINGSVERKET (NATIONAL INSURANCE BOARD)

S-103 51 Stockholm
Sweden
Tel: +46 (0)8 786 9000
Fax: +46 (0)8 112 789

Entry: 31114

ROMANIA LIBERIA

Piata Presei Libere 1
Bucharesti
R-71341
Romania
Tel: +40 (0)1 817 23
Telex: 11618

Entry: 29008

ROYAL DANISH MINISTRY OF FOREIGN AFFAIRS

Asiatisk Plads 2
1448 Copenhagen K
Denmark
Tel: +45 (0)33 92 00 00
Telex: 31292 ETR DK
Fax: +45 (0)31 54 05 33

Entry: 01057

THE ROYAL INSTITUTION OF CHARTERED SURVEYORS

12 Great George street
London SW1P 3AD
UK
Tel: +44 (0)171 222 7000
Fax: +44 (0)171 222 9430

Entry: 18036

THE ROYAL LIBRARY

P.O.B. 2149
DK-1016 Copenhagen K
Denmark
Tel: +45 (0)33 93 01 11
Telex: 15009
Fax: +45 (0)33 32 98 46

Entry: 12034

SAASTO PANKKILIITTO

~PL 47
00100 Helsinki
Finland
Tel: +358 (9)0 133 987
Fax: +358 (9)0 133 4935

Entries: 13079   13081

SAVEZ EKONOMISTA JUGOSLAVIJE

Nusiceva 6-111
11000 Belgrade
Serbia
Tel: +381 11 334417

Entry: 35008

SAVEZ EKONOMISTA SRBIJE

Marsala Tita 16
Belgrade
Serbia

Entry: 35009

SCHWEIZERISCHE NATIONALBANK

Postfach
CH-8002
Zurich
Switzerland
Tel: +41 (0)1 221 37 50
Telex: 813530 SNBCH
Fax: +41 (0)1 221 07 10

Entries: 32007   32056   32063
32069

SCHWEIZERISCHER BAUERNVERBAND: ABTEILUNG DOKUMENTATION

Laurstrasse 10
CH-5200 Brugg
Switzerland
Tel: +41 (0)56 32 51 11
Fax: +41 (0)56 41 53 48

Entries: 32044   32075

SCOTTISH OFFICE LIBRARY

Publications Sales
Room 1-44
New St Andrew's House
Edinburgh EH1 3TG
Scotland
Tel: +44 (0)31 244 4806

Entries: 34011   34110   34126
34300   34301   34302
34303   34304   34305

SEA FISH INDUSTRY AUTHORITY

Sea Fisheries House
18/19 Beaverbank Office
Park
Logie Green Road
Edinburgh EH7 4HG
UK
Tel: +44 (0)131 558 3331
Fax: +44 (0)131 558 1442

Entries: 02084   34155   34178
34328

SERVICE CENTRAL DE LA STATISTIQUE ET DES ÉTUDES ÉCONOMIQUES

19-21 Bld Royal
Boîte Postale 304
L-2013
Luxembourg
Tel: +352 47 81
Telex: 3464 ECOLU
Fax: +352 46 42 89

Entries: 23002   23009   23019
23022   23025   23027
23032   23033   23035

SOCIÉTÉ DE LA BOURSE DE LUXEMBOURG

11 Avenue de la Porte-
Neuve
B.P 165
L-2011 Luxembourg
Tel: +352 47 79 36 1
Telex: 2559 STOEX LU
Fax: +352 47 32 98

Entries: 23001   23004   23006
23010   23015   23016
23024   23026   23029
23030

SOCIÉTÉ DES BOURSES FRANÇAISES: BOURSE DE PARIS

39 Rue Cambon
75001 Paris
France
Tel: +33 1 49 27 10 00
Telex: 215561 F
Fax: +33 1 49 27 14 33

Entries: 14002   14116   14129
14143

SOCIETY OF MOTOR MANUFACTURERS AND TRADERS LTD

Forbes House
Halkin Street
London SW1X 7DS
UK
Tel: +44 (0)171 235 7000
Fax: +44 (0)171 235 7112

Entries: 01004 01208 34228
34232 34233 34234

SOSLAND PUBLISHING COMPANY

4800 Main Street
Suite 100
Kansas City, M064112
U.S.A.
Tel: +1 816 756 1000
Telex: 820182 SOSLAND
Fax: +1 816 756 0494

Entry: 01002

SOVINFORM LTD

Publishing Division
PO Box 107
Larkhill House
Abingdon OX14 1FA
UK

Entry: 09037

SPANISH CHAMBER OF COMMERCE IN GREAT BRITAIN

5 Cavendish Square
London W1M 0DP
UK
Tel: +44 (0)171 637 9061
Telex: 8811583 CAMCOE G
Fax: +44 (0)171 436 7188

Entry: 30030

ST GALLER ZENTRUM FÜR ZUKUNFTSFORSCHUNG

Dufourstrasse 30
9000
St Gallen
Switzerland
Tel: +41 (0)71 24 28 16
Fax: +41 (0)71 24 13 10

Entry: 32051

ST JAMES PRESS

c/o Gale Research Inc.
835 Penobscot Building
Detriot, Michigan
U.S.A.
Tel: +1 313 961 2242/800
345 0392

Entry: 04007

ST MARTINS PRESS INC.

175 Fifth Avenue
New York 10010
U.S.A.
Tel: +1 800 221 7945
Fax: +1 212 420 9314

Entry: 04005

STANILAND HALL ASSOCIATES

PO Box 643
Alderbury House
Upton Park
Slough SL1 2JU
UK
Tel: +44 (0)1753 691 874

Entries: 34075 34106 34314

STATE COMMITTEE FOR STATISTICS OF THE REPUBLIC OF LATVIA

1 Lacplesa Street
Riga 226301
Latvia
Fax: +371 (0)132 273545

Entries: 06006 06007 06008
06009 06010 06011
06012

STATE INSTITUTE OF STATISTICS

Necatibey Caddesi No 114
06100 Ankara
Turkey
Tel: +90 (9)4 117 64 40
Telex: 46 347 DIE TR
Fax: +90 (9)4 125 33 87

Entries: 33001 33004 33007
33008 33009 33011
33012 33021 33022
33023 33026 33027
33028 33032 33033
33034 33035 33037
33038 33040 33043
33046 33047 33048
33049 33050 33053
33054 33055 33062
33074 33075 33076

STATISTICAL BUREAU OF ICELAND

Skuggasund 3
15-150 Reykjavik
Iceland
Tel: +354 (9)1 609 800
Fax: +354 (9)1 628 865

Entries: 19010 19015 19016
19021 19022

STATISTICS FINLAND

P.O. Box 504
SF 00100 Helsinki
Finland
Tel: +358 (9)0 17341
Fax: +358 (9)0 1734 2465

Entries: 13008 13009 13010
13011 13014 13015
13016 13017 13024
13025 13026 13029
13030 13032 13034
13035 13051 13052
13053 13054 13055
13059 13063 13064
13067 13068 13071
13073 13074 13075
13076 13080 13086
13092 13093 13094
13097 13098 13099
13101 13102

STATISTICS OFFICE

6 Convent Place
Gibraltar
Tel: +350 75515/75490
Fax: +350 74524

Entries: 16002 16004 16005
16006 16007 16008

STATISTICS SWEDEN

S-70189 Orebro
Sweden
Tel: +46 (0)49 191 76 800
Fax: +46 (0)49 191 76932

Entries: 31001 31002 31003
31005 31010 31013
31014 31016 31018
31019 31020 31021
31022 31027 31028
31029 31031 31032
31033 31034 31035
31036 31037 31038
31039 31042 31043
31044 31045 31046
31048 31049 31050
31051 31052 31053
31054 31056 31057
31059 31060 31061
31062 31063 31064
31069 31070 31071
31072 31073 31075
31076 31077 31078
31079 31080 31081
31082 31083 31084
31085 31086 31087
31088 31089 31090
31092 31093 31094
31095 31096 31097
31098 31100 31102
31104 31105 31110
31111 31112 31116
31117 31118 31119
31120 31121 31125
31131 31132 31133
31135 31136 31137
31139 31140 31142
31143 31144 31145
31146 31148 31149
31150 31151

STATISTISCHES BUNDESAMT

Gustav-Stresemann-Ring 11
6200 Wiesbaden 1
Germany
Tel: +49 (0)611 75 1
Telex: 61186 StBA
Fax: +49 (0)611 724000

Entries: 15151 15205 15206
15208 15209 15230

STOCKMANN - GRUPPEN A/S

Egebaekgaard
Egebaekvei 98
2850 Naerum
Denmark
Tel: +45 (0)42 80 70 15
Fax: +45 (0)42 80 79 48

Entry: 12041

SUCRO SA
Hernán Cortés 5
46004 Valencia
Spain
Tel: +34 (9)6 352 53 01
Telex: 62905 SUCRO E
Fax: +34 (9)6 352 57 52

Entries: 30003   30004   30068

SVENSKA ARBETSGIVAREFORENINGEN
Sodra Blasieholmshamnen
4A
Box 16120
103 23 Stockholm
Sweden
Tel: +46 (0)8 14 05 00

Entries: 31041   31067   31074
         31107   31108   31109
         31141

SVENSKA ARBETSGIVAREFORENINGEN, SAF (SWEDISH CONFEDERATION OF EMPLOYERS)
Hovslagaregaten 5
5-10330 Stockholm
Sweden
Tel: +46 (0)8 762 60 00
Fax: +46 (0)8 762 68 85

Entries: 31106   31122

SVENSKA FORSAKRINGSFORENINGEN
Tegeluddsvagen 100
115 87 Stockholm
Sweden
Tel: +46 (0)8 783 71 76
Fax: +46 (0)8 663 22 35

Entries: 31066   31123

SVENSKA HANDELBANKEN
Kungsträdgårdsgatan 2
S-10670 Stockholm
Sweden
Tel: +46 (0)8 701 1000
Telex: 11090
Fax: +46 (0)8 611 5071

Entry: 31007

SVERIGES RIKSBANK
S-10337 Stockhlom
Sweden
Tel: +46 (0)8 787 00 00
Telex: 19150
Fax: +46 (0)8 210 531

Entries: 31008   31011   31103
         31115

SWEDBANK
S-10534 Stockholm
Sweden
Tel: +46 (0)8 22 2320
Telex: 12826 Swedenbnk S
Fax: +46 (0)8 11 9013

Entries: 31006   31009

SWEDISH CIVIL AVIATION ADMINISTRATION
S-60179 Norrköping
Sweden
Tel: +46 (0)11 19 2000
Telex: 64250 CIVAIR S
Fax: +46 (0)11 19 2575

Entry: 31147

THE SWEDISH INSTITUTE
Box 7434
S-10391 Stockholm
Sweden
Tel: +46 (0)8 789 20 00
Telex: 10025 SWEDINS S
Fax: +46 (0)8 20 7248

Entry: 31040

SWEDISH STATISTICAL ASSOCIATION
Dept of Statistics
Viktoriag 13
S-41125 Göteborg
Sweden
Tel: +46 (0)31 773 1281
Fax: +46 (0)31 773 1274

Entry: 31091

SWISS BANK CORPORATION
Aescherplatz 6
CH-4002 Basel
Switzerland
Tel: +41 (0)61 288 7574
Fax: +41 (0)61 288 9274

Entries: 32006   32034   32061
         32077

SWISS OFFICE FOR TRADE PROMOTION
Avenue de l'Avant-Poste 4
Case Postale 1128
CH-1011 Lausanne
Switzerland
Tel: +41 (0)21 20 32 31
Telex: 455 425 OSEC CH
Fax: +41 (0)21 20 73 37

Entries: 32060   32080   32083

TEAGASC - THE AGRICULTURAL AND FOOD DEVELOPMENT AUTHORITY
19 Sandymount Avenue
Ballsbridge
Dublin 4
Republic of Ireland
Tel: +353 (0)1 688 188
Fax: +353 (0)1 688 023

Entries: 20040   20060   20065
         20085   20086   20102

TELEMALTA CORPORATION
1 Pender Place
St Andrew's Road
St Julians
Malta
Tel: +356 24 30 00
Telex: 1357 MW TMBRD
Fax: +356 24 20 00

Entry: 24006

TENEO SA
Pza Marqués de Salamanca
8
28006 Madrid
Spain
Tel: +34 (9)1 396 16 43/15 19
Fax: +34 (9)1 575 64 1/564 18 77

Entry: 30005

TEXTILE STATISTICS BUREAU
Reedham House
31 King Street West
Manchester M3 2PF
UK

Entries: 34093   34104   34123
         34137   34161   34162
         34163   34208   34272
         34313   34354

THESSALONIKI PORT AUTHORITY
Thessaloniki
Greece
Tel: +30 (0)59 39 11

Entry: 17046

THOMAS TELFORD
Thomas Telford House
4 Heron Key
London E14 4JD
UK
Tel: +44 (0)171 987 6999
Fax: +44 (0)171 538 9849

Entry: 01038

TIME BOOKS
77-85 Fulham Palace Road
Hammersmith
London W6 8JB
UK

Entry: 34323

THE TREASURY OFFICE
Order from:
Economic Progress Report
(Distribution)
Central Office of Information
Hercules Road
London
SE1 7DU
UK

Entry: 34108

TSE PUBLISHING APS

Vimmelskaftet 42A,5
1161 Copenhagen K
Denmark
Tel: +45 (0)33 14 21 27
Fax: +45 (0)33 93 80 32

Entry: 12045

TURKISH CONFEDERATION OF EMPLOYER ASSOCIATIONS

Mecrutiyet Caddesi 1/4
06650 Kizilay
Ankara
Turkey
Tel: +90 (9)4 125 27 85
Telex: 42122
Fax: +90 (9)4 118 44 73

Entries: 33030   33031   33051

TURKISH CYPRIOT CHAMBER OF COMMERCE

Redretin Demirel Avenue
Lefkosa
Mersin
10-Turkey
Tel: +90 (9)741 787 60/63
Telex: 57137 KTTO TK

Entry: 33020

TÜRKIYE IS BANKASI A.S.

Order from:
Economic Research
Department
Atatürk Bulvari No.191
Ankara
Turkey
Tel: +90 (9)4 418 80 96
Telex: 42082 tab tr
Fax: +90 (9)4 425 07 53

Entries: 33018   33019   33044

TURKIYE SMAI KALKMMA BANKASI A.S. (TSKB)

Meclisi Mebusan Caddesi
No.137 P.K. 17 Karakoy
Istanbul
Turkey

Entry: 33003

TÜSIAD - TURKISH INDUSTRIALISTS' AND BUSINESSMEN'S ASSOCIATION

Mesrutiyet Cad. No:74
80050
Tepebasi
Istanbul
Turkey
Tel: +90 (9)1 249 08 95
Fax: +90 (9)1 249 13 50

Entries: 33042   33073

UKRAINE BUSINESS AGENCY

Vigilant House
120 Wilton Road
London
SW1V 1J2
UK
Tel: +44 (0)171 931 0665
Fax: +44 (0)171 873 8633

Entry: 09038

UNIDANMARK A/S

Economics Department
DK-1786 Copenhagen V
Denmark
Tel: +45 (0)33 33 54 03
Fax: +45 (0)33 11 90 96

Entry: 12012

UNION BANK OF FINLAND

Box 868
00101 Helsinki
Finland
Tel: +358 (9)0 1651
Telex: 127707
Fax: +358 (9)0 1775 08

Entries: 13004   13096

UNION BANK OF SWITZERLAND

PO Box 645
CH-8021 Zürich
Switzerland
Tel: +41 (0)1 234 24 34
Fax: +41 (0)1 234 61 90

Entries:	32008	32027	32041
	32043	32059	32079
	32082	32088	32089

UNION DE EMPRESAS SIDERÚRGICAS (UNESID)

c/Castelló 128 - 3e Planta
28006 Madrid
Spain
Tel: +34 (9)1 562 40 10
Telex: 22228 UNESI E
Fax: +34 (9)1 562 65 84

Entry: 30054

UNION LUXEMBOURGEOISE DES CONSAMMATEURS

55 Rue des Bruyères
1274 Howald
Luxembourg
Tel: +352 49 60 22
Telex: 2966
Fax: +352 49 49 57

Entry: 23013

UNION PROFESSIONNELLE DES ENTREPRISES D'ASSURANCES (UPEA)

Square de Meeurs 29
1040 Bruxelles
Belgium
Tel: +32 (0)2 513 68 45
Telex: 5114241
Fax: +32 (0)2 514 24 69

Entries: 07013   07044

UNIONE INDUSTRIALI PASTAI ITALIANI (U.N.I.P.I.)

Via Po, 102
00198 Roma
Italy
Tel: +39 (0)6 854 3291/841 6473
Telex: 622439 UNIPI I
Fax: +39 (0)6 844 2685

Entry: 21063

UNITED KINGDOM IRON AND STEEL STATISTICS BUREAU

Canterbury House
2 Sydenham Road
Croydon
CR9 2LZ
UK
Tel: +44 (0)181 686 9050
Fax: +44 (0)181 680 8616

Entries:	01177	01337	01339
	05019	05085	07023
	07060	12007	12042
	13018	13033	13077
	14033	14058	14121
	15046	15189	17007
	17034	20026	20095
	21011	21039	25012
	25052	25062	26008
	26038	26050	28013
	28060	28067	30022
	30061	30063	31023
	31101	32019	32062
	32081	34079	34269
	34342		

UNITED NATIONS ECONOMIC COMMISSION FOR EUROPE - UN/ECE

Order from:
Her Majesty's Stationery Office
HMSO Publications Centre
51 Nine Elms Lane
London SW8 5DR
UK
Tel: +44 (0)171 873 9090
Telex: 29 71 138
Fax: +44 (0)171 873 8463

Entries:	01006	01009	01021
	01066	01068	01085
	01086	01096	01115
	01116	01117	01118
	01133	01240	01254
	01268	01273	01282
	01283	02002	02003
	02004	02005	02006
	02007	02008	02009
	02011	02012	02047
	02048	02110	02117
	02118	02119	

UNITED NATIONS PUBLICATIONS

Order from:

(In U.K. first try H.M.S.O. or
Microinfo Ltd.)
Sales Office and Bookshop
CH-1211 Genève 10
Switzerland
Tel: +41 (0)22 734 1473
Fax: +41 (0)22 740 0931

Entries:	01032	01033	01036
	01054	01056	01094
	01101	01102	01137
	01138	01150	01175
	01179	01181	01203
	01204	01210	01212
	01213	01214	01232
	01236	01237	01264
	01271	01272	01281
	01284	01285	01286
	01318	01329	01333
	01340	02046	

US GENERAL ACCOUNTING OFFICE

441 G. Street,NW
Washington DC 20548
U.S.A.
Tel: +1 202 275 5067

Entries: 29007   29010

VERBAND DEUTSCHER REEDER

Esplanade 6
Postfach 30 55 80
2000 Hamburg 36
Germany
Tel: +49 (0)40 3 50 97 0
Telex: 211 407 VDR
Fax: +49 (0)40 350 97 211

Entry: 15131

VERDICT RESEARCH

112 High Holborn
London WC1V 6JS
UK
Tel: +44 (0)171 404 5042
Fax: +44 (0)171 430 0059

Entry: 34351

VERWALTUNGS - UND PRIVAT-BANK

Postfach 885
FL-9490 Vaduz
Liechtenstein
Tel: +4175 235 66 55
Telex: 889200
Fax: +4175 235 65 00

Entries:	22011	22012	22016
	22019	22020	

V/O VNESHTORGEKLAMA

31 Ul Kakhovka
Moscow 113461
Russia
Tel: +7 95 331 8311
Telex: 411265 VTR SU
Fax: +7 95 310 7005

Entry: 09003

W. KOHLHAMMER GMBH

Postfach 42 11 20
Stuttgart am Mainz
D 6500 42
Germany
Tel: +49 (0)6131 5 90 94/5
Telex: 04 187 768 DGV

Entries:	01239	15003	15004
	15006	15008	15010
	15012	15013	15015
	15016	15017	15018
	15019	15020	15021
	15022	15024	15025
	15026	15027	15028
	15029	15030	15031
	15032	15033	15034
	15035	15036	15037
	15038	15039	15040
	15043	15062	15067
	15068	15069	15070
	15072	15079	15080
	15081	15089	15092
	15093	15095	15117
	15118	15119	15120
	15121	15122	15127
	15144	15145	15154
	15155	15156	15157
	15158	15159	15160
	15161	15162	15163
	15164	15167	15168
	15169	15170	15172
	15174	15180	15181
	15182	15183	15184
	15185	15186	15187
	15188	15190	15191
	15192	15197	15198
	15201	15204	15211
	15212	15213	15214
	15220	15221	15224
	15225	15235	

WELSH OFFICE

Order from:

Economic and statisitcal
services division
Crown Building
Cathays Park
Cardiff
CF1 3NQ
UK
Tel: +44 (0)1222 82 5065

Entries:	34096	34128	34355
	34356	34357	34358
	34359		

WELT PUBLISHING CO

1413 K Street NW
Suite 1400
Washington DC 20005
USA

Entry: 02042

WIENER BORSEKAMMER

Wipplingerstrasse 34
A-1011 Wien
Austria
Tel: +43 (0)222 534 99
Telex: 13 2447 wbk a
Fax: +43 (0)222 535 68/57

Entries:	05002	05003	05073
	05106		

WILSON HARTNELL ADVERTISING

12 Lesson Park
Dublin 2
Republic of Ireland
Tel: +353 (0)1 978 822
Telex: 93900
Fax: +353 (0)1 975 163

Entry: 20066

WIRTSCHAFTS UND SOZIALWISSEN - SCHAFTLICHES INSTITUT DES DEUTSCHE GEWERKSCHAFTSBUNDES GMBH

Hans - Bockler - Strasse 39
4000 Düsseldorf 30
Germany
Tel: +49 (0)211 43 750
Fax: +49 (0)211 43 75 74

Entries: 15233   15234

WIRTSCHAFTSSTUDIO DES ÖSTERREICH - ISCHEN GESELLSCHAFTS - UND WIRTSCHAFTSMUSEUMS

Vogelsanggasse 36
A-1050 Wien
Austria
Tel: +43 (0)222 5516210
Fax: +43 (0)222 55162179

**Entries: 01107   01113   01151
01186   05100**

WORLD BANK PUBLICATIONS

Order from:
Microinfo Ltd (UK Agents)
PO Box 3
Alton
Hampshire
GU34 2PG
UK
Tel: +44 (0)1420 86848
Fax: +44 (0)1420 89889

Entries: 01188   01238   01270

WORLD BANK PUBLICATIONS

Order from:
(see Microinfo below)
1818 H.Street N.W.
Washington DC 20433
USA
Tel: +1 202 473 1155
Fax: +1 202 676 0581

**Entries: 01069   01084   01097
01100   01112   01131
01141   01156   01163
01233   01234   01235
01242   01259   01277
01278   01279   01280
01287   01289   01290
01291   01292   01293
01294   01295   01296
01297   01298   01299
01300   01301   01302
01303   01304   01308
01309   01311   01328
01334   04002   09021
29014**

WORLD HEALTH ORGANISATION

1211 Genève 27
Switzerland
Tel: +41 (0)22 791 21 11
Telex: 415416 OMS
Fax: +41 (0)22 788 0401

Entries: 01132   01322   01323

WORLD REPORTS LTD

108 Horseferry Road
London SW1P 2EF
UK
Tel: +44 (0)171 222 3836
Fax: +44 (0)171 233 0185

**Entries: 01134   01155   01159
01191   01229   02043
09034**

Index 1
SOURCES LISTED ALPHABETICALLY

168 Houns BBN | 08001
1991 census age, sex and marital status | 34001
1991 census county monitors | 34002
1991 census county reports | 34003
The 300 largest German firms | 15001
50 centres - a guide to office, industrial and retail rental trends in England, Scotland and Wales | 34004

A

A brief review of the UK domestic furniture market | 34005
A community of twelve: key figures | 03002
A guide to foreign exchange and the money markets | 32001
A picture of the British voluntary sector 1979/89 | 34006
A social portrait of Europe | 03003
ABC Industrial information and references | 15002
Abecor country report - Gibraltar | 16001
Abecor country reports | 01001
Abfallerhebung industrie | 05001
Absatz von bier | 15003
Abschlusse der aktiengesellschaften | 15004
Abstract of statistics | 16002
Abstract of statistics for index of retail prices, average earnings, social security benefits and contributions | 34007
Accommodation statistics | 31001
Accountancy Ireland | 20001
Actions: statistiques boursières mensuelles | 14002
Administration yearbook and diary | 20002
Admission to official stock listing and public offer of transferable securities | 23001
Advertising forecast | 34008
Advertising statistics yearbook | 34009
Agra special reports and market studies | 02001
Agrarbericht (agrar - und ernährungspolitischer bericht der bundesregierung über die lage der landwirtschaft...) | 15005
AGRESTE conjoncture: Commerce extérieur agro-alimentaire | 14003
AGRESTE conjoncture: Fruits | 14004
AGRESTE conjoncture: Grandes cultures | 14005
AGRESTE conjoncture: La conjoncture générale (agriculture et agro-alimentaire) | 14006
AGRESTE conjoncture: Lait et produits laitiers | 14007
AGRESTE conjoncture: Légumes | 14008
AGRESTE conjoncture: Productions animales | 14009
AGRESTE conjoncture: Viticulture | 14010
AGRESTE données chiffrées: statistiques agricole annuelle | 14011
AGRESTE Données chiffrées: statistiques forestières | 14012
AGRESTE: Études | 14013
AGRESTE séries: Animaux hebdo | 14014
AGRESTE séries: Aviculture | 14015
AGRESTE séries: Commerce extérieur bois et dérivés | 14016
Agribusiness worldwide | 01002
Agricultural commodity production and means of production in agriculture in 199... | 27001
Agricultural economic report | 25001
Agricultural economics | 31002
Agricultural enterprises and labour | 31003
Agricultural income | 03004
Agricultural input price index | 20003
Agricultural markets: prices | 03005
Agricultural output price index | 20004
Agricultural policies, markets and trade: monitoring and outlook 1992 | 01003
Agricultural prices - price indices &

absolute prices | 03006
Agricultural prices - prices indices and absolute prices | 03007
Agricultural review for Europe | 02002
The agricultural situation in the Community | 03008
Agricultural statistics | 10001
Agricultural statistics | 26001
Agricultural statistics of Greece | 17001
Agricultural statistics United Kingdom | 34010
Agriculture (statistical yearbook) | 27002
Agriculture, forestry and fisheries | 03009
Agriculture in Europe: development, constraints, and perspectives | 03010
Agriculture in Scotland | 34011
Agriculture in the United Kingdom | 34012
Agriculture - statistical yearbook | 03011
Agriculture: économie de l'agriculture Française en Europe: forces et faiblesses | 14017
Air traffic survey | 16004
Aktienmarkte | 15006
Albania: from isolation towards reform | 04001
Alkoholstatistik | 31004
All about the Norwegian Federation of Trade Unions | 26002
All reports on culture and media | 31005
Amtliches kursblatt der Baden-Wüttembergischen wertpapierbörse zu Stuttgart | 15007
An agricultural strategy for Albania | 04002
An economic outline of the Republic of Latvia | 06001
An overview of the UK bed market | 34013
An overview of the UK office furniture market | 34014
Analysis Sweden | 31006
Analytical tables of external trade: nimexe, exports | 03012
Andelsbladet | 12001
Angestelltenverdienste in industrie und handel | 15008
Animal production - quarterly statistics | 03013
Animals exported by sea and air | 20005
Annales d'économie et de statistique | 14018
Annales des mines de Belgique | 07001
Annuaire abrégé | 14019
Annuaire de l'industrie automobile mondiale - OICA | 01004
Annuaire de statistique agricole | 14020
Annuaire de statistiques industrielles | 14021
Annuaire de statistiques regionales | 07002
Annuaire rétrospectif de la France 1948-1988 | 14022
Annuaire Statistique | 07003
Annuaire statistique de la Belgique | 07004
Annuaire statistique de la France | 14023
Annuaire statistique de la Suisse | 32002
Annuaire statistique de poche | 07005
Annuaire statistique du Luxembourg | 23002
Annuaire Suisse de l'économie forestière et de l'industrie du bois | 32003
Annual abstract of statistics | 24001
Annual abstract of statistics | 34015
Annual bulletin of coal statistics for Europe | 02003
Annual bulletin of electric energy statistics for Europe | 02004
Annual bulletin of gas statistics for Europe | 02005
Annual bulletin of general energy statistics for Europe | 02006
Annual bulletin of housing and building statistics for Europe | 02007
Annual bulletin of statistics: International Tea Committee | 01005
Annual bulletin of steel statistics for Europe | 02008
Annual bulletin of trade in chemical products | 01006

Annual bulletin of transport statistics for Europe | 02009
Annual economic review | 02010
Annual facts sheet | 05002
Annual housing statistics bulletin | 20006
Annual industrial survey | 17002
Annual manufacturing industry statistics | 33001
Annual oil market report | 01007
Annual publication of foreign trade statistics | 25002
Annual report | 19001
Annual report | 19002
Annual report | 23003
Annual report | 26003
Annual report | 26004
Annual report | 15009
Annual report | 09001
Annual report | 17003
Annual report | 33002
Annual report | 33003
Annual report | 05003
Annual report - advertising association | 34016
Annual report | 28001
Annual report | 32004
Annual report | 24002
Annual report | 10002
Annual report - Central Bank of Cyprus | 10003
Annual report | 10004
Annual report - De Nederlandsche Bank | 25003
Annual report | 28002
Annual report | 24003
Annual report | 13001
Annual report | 13002
Annual report | 32005
Annual report | 13003
Annual report | 32004
Annual report | 24004
Annual report | 24005
Annual report of the central bank | 20007
Annual report of the Department of Labour | 20008
Annual report of the Environment Institute | 03014
Annual report of the European Foundation for the Improvement of Living and Working conditions | 03015
Annual report of the European Investment Bank | 03016
Annual report of the International Cocoa Organization | 01008
Annual report of the mines service | 10005
Annual report of the Revenue Commissioners | 20009
Annual report | 32006
Annual report | 32007
Annual report - Société de la Bourse de Luxembourg | 23004
Annual report Svenska Handelsbanken | 31007
Annual report Sveriges Riksbank | 31008
Annual report Swedbank | 31009
Annual report | 24006
Annual report | 32008
Annual report | 13004
Annual review of engineering industries and automation | 01009
Annual review of engineering industries and automation | 02011
Annual review of government funded R&D | 34017
Annual review of the chemical industry | 02012
Annual statistical survey on mines, quarries and salterns | 17004
Annual statistics | 26005
Annual summary of merchant ships completed in the world | 01010
Annuario di contabilità nazionale | 21001
Annuario statistico Italiano | 21002
Antwerp facets | 07006
Anuario estadístico de España - avance | 30001
Anuario estadístico de España: edicion extenso | 30002
Anuario hortofruticola Español | 30003

Anuario hortofruticola mercado interior	30004
Apercu économique trimestriel	07007
Appendix to the monthly bulletin of price statistics	25004
Appropriation accounts	34018
Aquaculture in Sweden	31010
Arbeiterverdienste in der landaitschaft-arbeitnehmerverdienste in industrie und handel	15010
Arbeits - und sozialpolitik	15011
Arbeitskosten in produsierenden gewebe	15012
Arbeitskrafte	15013
Arbeitsstaettenzaehlung	05004
Arbejdsmarked: statistiske efterretninger	12002
Arbetsrapport	31011
Area and population in territorial section	27003
Area, yields and crops production	08002
ARL research reports: consumer durables	01011
ARL research reports: consumer goods	01012
ARL research reports: electronics industries	01013
ARL research reports: food industries	34019
ARL research reports: healthcare industries	01014
ARL research reports: pharmaceutical and healthcare	34020
ARL research reports: pharmaceuticals and healthcare	02013
ARL research reports: veterinary products	01015
ARL research reports: veterinary products	34021
Arsredovisning	31012
As acções no mercado de catações oficials	28003
Asbestos industry in Western Germany - sector report	15014
Assessment of real estate	31013
Aufwendungen der industrie für den umweltschutz	05005
Ausfuhrnach verbrauchs - und kauferlandern und warengruppen	15015
Ausgenahlte zahlen für die agrarwirtschaft betriegsarbeits - und einkosmmensverhaltrusse betriebe	15016
Ausgenahlte zahlen für die bauwirtschaft	15017
Ausgenahlte zahlen zur energiewirtschaft	15018
Auslanderstatistik	22001
Aussenhandel (länder - und warengliederung)	05006
Aussenhandel nach landern und gutergruppen der produktionsstatistiken (spezialhandel)	15019
Aussenhandel nach waren und landern (spezialhandel)	15020
Austria, facts and figures	05007
Austria in figures	05008
The Automobile in the world	07008
Avance resultados	30005
Average earnings and hours worked	20010

B

Balance of international payments	20011
Balance of payments statistics yearbook	01016
Balances y cuentas: Seguros privados	30006
The Baltic States: what price freedom	06003
Bank of Cyprus Bulletin	10006
Bank of England quarterly bulletin	34022

Bank of England statistical abstract	34023
Bank of England working papers	34024
Bank of Finland annual statement	13005
Bank of Finland Bulletin	13006
Bank of Finland yearbook	13007
Bank profitability statistical supplement: financial statement of banks 1981-1990	01017
Bankenstatistiche daten	05009
The Banker	34025
Banking, insurance and building societies - employment and earnings	20012
Banking Ireland	20013
Bankruptcies	13008
Bankruptcies	31014
The banks	13009
Banks and brokers in the Netherlands	25005
Banks	31015
Banks	31016
Bankstatistik	22002
Basel Stock Exchange	32009
Basic facts	34026
Basic horticultural statistics for the United Kingdom 1981-90	34027
Basic international chemical industry statistics	01018
Basic science and technology indicators	01019
The basic statistical data on units employing less than five employees	27004
Basic statistics of the Community	03017
Baumobstflachen	15021
Baustatistik	22003
Baustatistik	05010
Bautatigkeit	15022
Befolkning og valg: statistiske efterretninger	12003
Beiträge zur regionalstatistik - bundesländervergleich	05011
Belgium: economic and commercial information	23005
Belgium: economic and commercial information	07009
Benefits	34028
Bereichszaelungen	05012
Berichte und studien	05013
Berichte über landwirtschaft	15023
Beschaftigte und umsatz in elnzelhandel (messzahlen)	15024
Beschaftigte und umsatz in gastgenerbe (messzahlen)	15025
Beschaftigte und umsatz in grosshandel (messzahlen)	15026
Beschaftigung umsatz, investitionen und kostenstrucktur der unternehmen in der energie - und wasserver	15027
Beschaftigung umsatz und energieversorgung der unternehmen und betriebe im bergbau und im verarbeit	15028
Beschaftigung umsatz und investitionen der unternehmen im bergbau und im verarbeitenden generbe	15029
Beschaftigung umsatz und investitionen der unternehmen in der energie - und wasserversorgung	15030
Beschaftigung, umsatz, wareneingang, lagerbestand und investitionen in einzelhandel	15031
Besitzverhaltrisse grundstucksverkeht fachliche vorbildung der betriebsleiter	15032
Best of British - the top 20,000 companies	34029
Bestandsstatistik der kraftfahrzeuge in Oesterreich im jahre	05014
Betriebliche personalvorsorgestatistik	22004
Betriebsgrossenstruktur	15033
Betriebssysteme und	

standardbetriebseinkommen	15034
Bevolkerung gestern; heute and morgen	15035
Bevolkerungsstruktur und wirtschaftskraft der bundeslander	15036
Bilans de l'énergie 1970 à 1990	14024
Bilans Energétiques	07010
Bildung in zahlenspiegel	15037
Bi-monthly bulletin	32010
Bi-monthly bulletin of the British - Portuguese Chamber of Commerce	28004
BIS reports	02014
Bloc	02015
Bloc-notes de l'Observatoire Économique de Paris	14025
Bodennutzung der betriebe	15038
Bodennutzung und pflanzliche	15039
Bodennutzungserhebung	05015
Boletin de Cotações	28005
Boletin de oportunidades de negócio	28006
Boletin mensal de estatística	28007
Boletín de información trimestral seguros privados	30007
Boletín económico	30008
Boletín estadístico	30009
Boletín informativo - accidentes	30010
Boletín informativo - anuario estadístico general	30011
Boletín mensual de estadística	30012
Boletín trimestral de coyuntura	30013
Bollettino economico	21003
Bollettino mensile di statistica	21004
Bollettino statistico	21005
The book of European forecasts	02016
The book of European regions	02017
Book of vital world statistics	01020
Bourse informations	23006
Brauwirtschaft	15040
Britain: an official handbook	34030
Britain's top European owned companies	34031
Britain's top foreign owned companies	34032
Britain's top Japanese owned companies	34033
Britain's top privately owned companies	34034
Britain's top US owned companies	34035
Britain in the Netherlands	25006
British aid statistics	34036
British coach operators survey	34037
British Coal annual report	34038
British drinks index	34039
British food index	34040
British German trade	15041
British holiday intentions	34041
British horeca index	34042
The British on holiday	34043
The British shopper	34044
British Soviet Business	09002
British supermarket index	34045
British Tourist Authority annual report	34046
The British-Swiss bulletin	32011
BSRIA European reports	02018
BSRIA market studies	34047
BSRIA statistics bulletin	34048
Budget 199-	20014
Budget estimates	24007
Budget speech	24008
Budgets de ménage	32012
Budgets municipal and provincial special administrations and villages	33004
Building cost indices	13010
Building materials and construction products - sector report	15042
Building Societies Association annual report	34049
Building societies yearbook	34050
Building societies yearbook - updating service	34051
Building society annual accounts data	34052
Bulgaria 92	08003

Bulgaria's agriculture: situation, trends and prospects 08004
Bulgarian foreign trade 08005
Bulletin 18001
Bulletin - Central Bank of Cyprus 10007
Bulletin: Credit Suisse 32013
Bulletin de la Banque Nationale de Belgique 07011
Bulletin de statistique 07012
Bulletin des assurances 07013
Bulletin financier 07014
Bulletin - Kamer voor Handel en Nijverheid van Het: Arrondisement Leuven 07015
Bulletin mensuel 14026
Bulletin mensuel d'information 07016
Bulletin mensuel de l'énergie électrique 07017
Bulletin mensuel de statistique 14027
Bulletin mensuel de statistique industrielle 14028
Bulletin of energy prices 03018
Bulletin of statistics 13011
Bulletin of statistics on world trade in engineering products 01021
Bulletin of the economic and social committee 03019
Bulletin of the European Communities 03020
Bulletin of the International Statistical Institute 31017
Bulletin of tourism statistics 33005
Bulletin of yacht statistics 33006
Bulletin trimestriel 23007
Bulletin trimestriel de la Banque de France 14029
Bundesarbeitsblatt 15043
Bus and coach statistics Great Britian 34053
Business and finance 20015
Business and research associates reports: furniture industries 03021
Business and research associates reports: furniture industries 34054
Business directory of the Hungarian chamber of commerce 18002
Business Europe 02019
Business Europe 03022
Business Finland 13012
Business guide to Hungary 18003
Business guide to Cyprus 10008
Business international 01022
Business Italy 21006
Business monitors: annual census of production reports: annual and occasional monitors covering more than industry 34055
Business monitors: annual census of production reports: construction industry 34056
Business monitors: annual census of production reports: energy and water supply industries 34057
Business monitors: annual census of production reports: extraction of mineral and ores other than fuels: manufacture of metals, mineral products and chemicals 34058
Business monitors: annual census of production reports: metal goods, engineering and vehicle industries 34059
Business monitors: annual census of production reports: miscellaneous monitors 34060
Business monitors: annual census of production reports: other manufacturing industries 34061
Business monitors: production monitors: energy industries 34062
Business monitors: production monitors: extraction of minerals and ores other than fuels: manufacture of metals, mineral products and chemicals 34063
Business monitors: production monitors: metal goods,

engineering and vehicle industries 34064
Business monitors: production monitors: other manufacturing industries 34065
Business monitors: service and distributive monitors 34066
Business Moscow 09003
Business of advertising agencies 20016
Business profile series - Cyprus 10009
Business ratio reports 34067
Business Spain 30014
Business survey 25007
Business week international 01023
Business week: Russian language edition 09004
Buying in Portugal - general information 28008
Bygge - og anlaegsvirksohmhed: statistiske efterretninger 12004
Béton 07018
Børsfersfektiver 26006

C

CA Quarterly: facts & figures on Austria's economy 05016
Cahiers: analyses et études 14001
Cahiers économiques 23008
Cahiers Français 14030
Cahiers économiques 23009
Capacity changes in lead and zinc in the 1980's 01024
Caracterizaçío das empresas Portuguesas 28009
Carpets - sector report 05017
Carriage of goods - inland waterways 03023
Carriage of goods - railways 03024
Carriage of goods - roads 03025
Catalogue of companies industrial, commercial and building, industrial sector 18004
Catalogue of statistical sources 34068
Causes and effects of air pollution 33007
Causes de decès 07019
Causes of death 31018
CBI/BSL regional trends survey 34069
CBI/Coopers Deloitte financial services survey 34070
CEDEFOP news - vocational training in Europe 03026
Censo agrario 30015
Censo de edificios - avance de resultados 30016
Censo de locales 30017
Censo de poblacíon - (1) avance de resultados (2) projecto 30018
Censo de viviendas 30019
Census of agriculture 24009
Census of agriculture - provisional estimates 20017
Census of building and construction 20018
Census of industrial production: report and summary tables 24010
Census of industry and business establishments 33008
Census of population 1991 20019
Census of population: administrative division 33009
Census of services 20020
Census report 16005
Central bank statistics supplement 20021
Central economic plan 25008
Central European 02020
Central Government borrowing and debt 12005
CFM News 12006
Chambers of industry and commerce in Germany 15044
Changes in prices in 199.. 27005
Changes in prices in national economy in 1991 and in the first half-year of 1992 27006
Charge fiscale en Suisse 19.. personnes physiques par

communes 32014
Charity Trends 01025
Charity trends 34071
Charter bulletin 33010
Chemical industry main markets 1989-2000 01026
Chemsource '93 01027
Chronique de conjoncture Belge 07020
CII Newsletter 20022
CIRET - studien 15045
The C.I.S. market atlas 09005
The C.I.S. market atlas 06004
Civil aviation statistics of the world 01028
Civil service statistics 34072
Clerical and operative rewards 34073
Clothing industry yearbook 13013
CML quarterly lending figures 34074
Coal and the international energy market 03027
Coal information 01029
Cocoa consumption 01030
Cocoa newsletter 01031
Comecon data 02021
Comércio externo Português 28010
Commentaire annuel 32015
Commerce exterieur de la Grece 17005
Commerce extérieur de la France 14031
Commercial banks 31019
Commercial courier 24011
Commercial gardening & farming 25009
Commercio estero 21007
Commoditiy trade statistics 01032
Commodity trade statistics 01033
Commonwealth notes: facts in brief 01034
Commonwealth of Independent States - founding documents 09006
Commonwealth of Independent States - Global forecasting service 09007
Commonwealth Secretariat: expert group reports 01035
Communications 03028
Communiqué hebdomadaire 07021
The community budget: the facts in figures 03029
Community services in 199.. 27007
Community, social and personal services statistics 10010
Companies and taxes in Liechtenstein 22005
Companies, cooperatives and firms statistics 33011
Comparative tables of the social security schemes in the member states of the European Communities. General scheme (employees industry and commerce) 03030
Comparison of enterprises by basic financial economic indicators, characterizing efficiency 08006
Compendio statistico Italiano 21008
Compendium of social statistics and indicators: social statistics and indicators - series K 01036
Competition policy in OECD countries 01037
Completing the internal market: an area without internal frontiers 03031
Completing the internal market: current status 1st January 1992 03032
Comptage Suisse de la circulation routière 19.. 32016
Compte rendu annuel des operations de la Banque de France 14032
Comptes nationaux de la Suisse 32017
The concentration of industrial production 1990-1991 27008
Concise statistical yearbook 27009
Concise statistical yearbook of Greece 17006
Confectionery - sector report 05018
Confederation of Irish Industry annual report 20023
Construction 07022
Construction and housing statistics 10011
Construction and housing yearbook 13014

Construction industry in Ireland
review and outlook   20024
Construction statistics   26007
Construction statistics   33012
Construction today   01038
Constructions exécutées en 199–
et constructions projetées pour
199– en Suisse   32018
Consumer barometer   13015
Consumer buying expectations   31020
Consumer Europe 199-   02022
Consumer goods food   25010
Consumer goods
non-food   25011
Consumer policy in OECD
countries   01039
Consumer policy in the single
market   03033
Consumer price index   13016
Consumer price index   31021
Consumer price index   20025
Consumer price index   03034
Consumer price index numbers
1914 onwards   31022
Consumer prices in the EEC 1990   03035
Consumer southern Europe 199-   02023
Consumer spending forecasts   34075
Consumption statistics for milk and
milk products   01040
Contabilidad nacional de España   30020
Contabilidad regional de España   30021
Contas nacionais   28011
Contas regionais   28012
Conti economici trimestrali   21009
Contributi all'analisi economica   21010
Cooperative statistics   34076
Corporate enterprises and personal
business in Finland   13017
Cost of living reports - regional
comparisons   34077
Cote officièlle - Bourse de
Luxembourg   23010
Council of Mortgage Lenders
annual report   34078
Countertrade outlook   01041
Country books: Austria   05019
Country books: Belgium -
Luxembourg   07023
Country books: Denmark   12007
Country books: Finland   13018
Country books: France   14033
Country books: German Federal
Republic   15046
Country books: Greece   17007
Country books: Irish Republic   20026
Country books: Italy   21011
Country books: Netherlands   25012
Country books: Norway   26008
Country books: Portugal   28013
Country books: Spain   30022
Country books: Sweden   31023
Country books: Switzerland   32019
Country books: UK   34079
Country data forecasts   01042
Country forecasts   01043
Country forecasts: Austria   05020
Country forecasts: Belgium   07024
Country forecasts: Bulgaria   08007
Country forecasts: Czechoslovakia   11001
Country forecasts: Denmark   12008
Country forecasts: Finland   13019
Country forecasts: France   14034
Country forecasts: Germany   15047
Country forecasts: Greece   17008
Country forecasts: Hungary   18005
Country forecasts: Ireland   20027
Country forecasts: Italy   21012
Country forecasts: Netherlands   25013
Country forecasts: Norway   26009
Country forecasts: Poland   27010
Country forecasts: Portugal   28014
Country forecasts: regional
overview Europe   02024
Country forecasts: Romania   29001
Country forecasts: Spain   30023
Country forecasts: Sweden   31024
Country forecasts: Switzerland   32020
Country forecasts: Turkey   33013

Country forecasts: UK   34080
Country forecasts: Yugoslavia   35001
Country guides   01044
Country notes   01045
Country outlooks   01046
Country profile: Austria   05021
Country profile: Belgium   07025
Country profile: Belgium,
Luxembourg   07026
Country profile: Bulgaria   08008
Country profile: Bulgaria 1991   08009
Country profile: Czech and Slovak
republic   11002
Country profile: Denmark   12009
Country profile: Denmark, Iceland   19003
Country profile: Finland   13020
Country profile: France   14035
Country profile: France   14036
Country profile: Germany   15048
Country profile: Greece   17009
Country profile: Hungary   18006
Country profile: Iceland   19004
Country profile: Ireland   20028
Country profile: Italy   21013
Country profile: Luxembourg   23011
Country profile; Malta, Cyprus   10012
Country profile: Netherlands   25014
Country profile: Norway   26010
Country profile: Poland   27011
Country profile: Portugal   28015
Country profile: Romania   29002
Country profile: Russia   09008
Country profile: Spain   30024
Country profile: Sweden   31025
Country profile: Switzerland   32021
Country profile: Switzerland   32022
Country profile: Yugoslav republics   35002
Country profiles   01047
Country profiles   01048
Country report: Austria   05022
Country report: Belgium   07027
Country report: Belgium,
Luxembourg   23012
Country report: Belgium -
Luxembourg   07028
Country report: Central and
Eastern Europe 1991: Bulgaria,
Poland, Romania, Soviet Union,
Czechoslovakia, Hungary   02025
Country report - Commonwealth of
Independent States   09009
Country report: Czechoslovakia   11003
Country report: Denmark, Iceland   19005
Country report: Denmark, Iceland   12010
Country report: Finland   13021
Country report: France   14037
Country report: Georgia, Armenia,
Azerbaijan, Central Asian
Republics   09010
Country report: Germany   15049
Country report: Greece   17010
Country report: Hungary   18007
Country report: Iceland   19006
Country report: Ireland   20029
Country report: Italy   21014
Country report: Netherlands   25015
Country report: Norway   26011
Country report: Poland   27012
Country report: Portugal   28016
Country report - Romania, Bulgaria,
Albania   29003
Country report - Romania, Bulgaria,
Albania   08010
Country report: Romania, Bulgaria,
Albania   04003
Country report: Russia   09011
Country report: Spain   30025
Country report: Sweden   31026
Country report: Switzerland   32023
Country report: Turkey   33014
Country report: Ukraine, Belarus,
Moldova   09012
Country report: United Kingdom   34081
Country report: Yugoslav republics   35003
Country reports   01049
Country reports   01050
Country risk monitor   01051
Country risk service   01052

Country risk service: Baltic
Republics   06005
Country risk service: Bulgaria   08011
Country risk service: C.I.S.   09013
Country risk service: Cyprus   10013
Country risk service:
Czechoslovakia   11004
Country risk service: Greece   17011
Country risk service: Hungary   18008
Country risk service: Poland   27013
Country risk service: Portugal   28017
Country risk service: Romania   29004
Country risk service: Russia   09014
Country risk service: Slovenia   35004
Country risk service: Spain   30026
Country risk service: Turkey   33015
Country risk service: Ukraine   09015
Country risk service: United
Kingdom   34082
Country risk service: Yugoslav
Republics   35005
Country statements ITMF   01053
The courier   03036
Courrier des statistiques   14038
Courrier hebdomadaire   07029
Credit and finance - sector report   17012
Credit market statistics for the
municipalities and county
councils   31027
Criminal statistics   10014
Crop husbandry   31028
Crop production - quarterly
statistics   03037
Cross border retailing in Europe   02026
Cross channel connections   14039
Cross channel trade   14040
Cultural statistics   31029
Culture   08012
Current economic business   08013
Customs areas of the world -
statistical papers, series M   01054
The Cypriot exporter   10015
Cyprus 1960-1990: time series
date   10016
Cyprus economy in figures   10017
Czechoslovak economic digest   11005
Czechoslovak policy digest   11006
Czechoslovakia economy digest   11007
Czechoslovakia - Global
forecasting service   11008
Czechoslovakia in transition   11009
Czechoslovakia: Major businesses   11010
Czechoslovakia's agriculture:
situation, trends and prospects   11011

D

DAFSA International reports   01055
Daily official list   12011
Danish economic outlook   12012
Dansk Handelsblad   12013
Data on Danish public foreign
borrowing   12014
Datamonitor reports: consumer
durables   34083
Datamonitor reports: cosmetic
products   34084
Datamonitor reports: drinks
industry   34085
Datamonitor reports: European
consumer durables   02027
Datamonitor reports: European
cosmetics   02028
Datamonitor reports: European
drinks industry   02029
Datamonitor reports: European
food industry   02030
Datamonitor reports: European oil
and petrochemical industries   02031
Datamonitor reports: European
paper industry   02032
Datamonitor reports: European
transport industry   02033
Datamonitor reports: financial
services   34086
Datamonitor reports: food industry   34087
Datamonitor reports: leisure

products	34088
Datamonitor reports: UK retailing	34089
Dataquest market research reports	34090
Daten zum arbeitsmarkt	15050
Datenschutz in der Europäischen gemeinschaft	15051
De konsument	23013
Debenham Tewson and Chinnocks property reports	34091
December livestock survey	20030
Defence statistics (Statement of the Defence Estimates Volume II)	34092
Deliveries of yarn	34093
Demographic report	10018
Demographic review of the Maltese islands	24012
Demographic statistics	03038
Demographic yearbook	01056
Demographisches jahrbuch Oesterreichs	05023
Demography (statistical yearbook)	27014
Denmark review	01057
Denmarks 15,000 largest companies	12015
Der aussenhandel Oesterreichs, serie 1A	05024
Der aussenhandel Oesterreichs, serie 1B	05025
Der aussenhandel Oesterreichs, serie 2	05026
Der fremdenverkehr in Oesterreich im jahre...	05027
Der intensivobstbau in Oesterreich	05028
Der weinbau Oesterreich	05029
Deux siècles de travail en France	14041
Development cooperation	01058
Development in the international financial system and the unification of the European internal market: effects on the Greek economy	17013
Dictionnaire des sources statistiques	14042
Die arbeitskosten in der industrie Österreichs	05030
Die bank	15052
Die fünf neuen bundesländer als wirtschaftspartner	15053
Die gesetzliche Unfallversicherung in der Bundesrepublik Deutschland	15054
Die molkereistruktur im bundesgebiet	15055
Die Rentenbestände in der gesetzlichen rentenversicherung	15056
Die unselbständig beschaftigten der gewerblichen wirtschaft (nach der kammersystematik)	05031
Die weltwirtschaft	15057
Digest	28018
Digest of environmental protection and water statistics	34094
Digest of statistics on social protection in Europe - volume 1: old age	03039
Digest of the UK energy statistics	34095
Digest of Welsh statistics	34096
Digests of statistics: International Civil Aviation Authority	01059
Direction of trade statistics - quarterly	01060
Direction of trade statistics - yearbook	01061
Directors report	34097
Directors rewards	34098
Directory of Czech and Slovak companies	11012
Directory of organizations providing countertrade services 1992-93	01062
Disabled persons: statistical data	03040
Distribution d' Aujhourd'hui	07030
Distributive trades survey	34099
Documentary credits, documentary collections, bank guarantees	32024
Doing business in Austria	05032

Doing business in Belgium	07031
Doing business in Belgium	07032
Doing business in Cyprus	10019
Doing business in Czechoslovakia	11013
Doing business in Czechoslovakia	11014
Doing business in Denmark	12016
Doing business in Eastern Europe: Bulgaria	08014
Doing business in Eastern Europe: Czechoslovakia	11015
Doing business in Eastern Europe: Hungary	18009
Doing business in Eastern Europe: Poland	27015
Doing business in Eastern Europe: Romania	29005
Doing business in Eastern Europe: Russia	09016
Doing business in France	14043
Doing business in France	14044
Doing business in Germany	15058
Doing business in Germany	15059
Doing business in Hungary	18010
Doing business in Hungary	18011
Doing business in Italy	21015
Doing business in Liechtenstein	22006
Doing business in Luxembourg	23014
Doing business in Malta	24013
Doing business in Northern Ireland	34100
Doing business in Norway	26012
Doing business in Poland	27016
Doing business in Poland	27017
Doing business in Poland	27018
Doing business in Spain	30027
Doing business in Spain	30028
Doing business in Sweden	31030
Doing business in Switzerland	32025
Doing business in Switzerland	32026
Doing business in the Isle of Man	34101
Doing business in the Netherlands	25016
Doing business in the Republic of Ireland	20031
Doing business in the UK	34102
Doing business in the UK	34103
Doing business in the USSR	09017
Doing business in Turkey	33016
Doing business in Turkey	33017
Doing business in...	01063
Doing business in...	01064
Doing business in...	02034
Doing business in...	03041
Doing business in...	02035
Doing business in...	03042
Doing business with Eastern Europe	02036
Doing business with Eastern Europe: Albania/Bulgaria	04004
Doing business with Eastern Europe: C.I.S.	09018
Doing business with Eastern Europe: Czechoslovakia	11016
Doing business with Eastern Europe: Hungary	18012
Doing business with Eastern Europe: Poland	27019
Doing business with Eastern Europe: Romania	29006
Doing business with Eastern Europe: Serbia-Montenegro	35006
Dokumentation zum bundesdatenschitzgesetz	15060
Domestic trade in Bulgaria	08015
Donau, inn und salzach	05033
Données chiffrées: IAA	14045
Données économiques de l'environnement	14046
Données sociales	14047
Doubling statistics	34104
Dresdner Bank statistical survey	15061
Drewry shipping reports	01065
Drinks pocket book	34105
Dungemittelvorsorgung	15062
Dutch companies and their UK agents/distributors	25017
Dutch companies and their UK associates	25018

E

Earnings - industry and services	03043
East Europe Business Focus	02037
The East European chemical industry: restructuring for East Europe	02038
The East European chemical monitoring service	02039
East European energy: Romania's energy needs persist	29007
East European finance update	02040
East European industrial monitoring service	02041
East/West business and trade	02042
Eastern Europe analyst	02043
Eastern Europe and the Commonwealth of Independent Sates 1992	02044
Eastern Europe and the USSR: the challenge of freedom	02045
East-West joint ventures and investment news	02046
ECE energy series	01066
Echo aus Deutschland	15063
EC-NICs trade	03044
Ecoflash	14048
Economic accounts for agriculture	01067
Economic accounts for agriculture and forestry	03045
Economic and social indicators	10020
Economic bulletin	26013
Economic bulletin	26014
Economic bulletin	17014
Economic bulletin for Europe	01068
Economic bulletin for Europe	02047
Economic change in the Balkan States: Albania, Bulgaria, Romania and Yugoslavia	04005
Economic development institute series	01069
Economic forecasts	01070
Economic growth and lead and zinc consumption	01071
Economic indicators	28019
Economic indicators	25019
Economic indicators: forecasts for company planning	34106
Economic indicators of Turkey	33018
Economic inequalities in Greece: developments and probable effects	17015
Economic news	10021
Economic news of Bulgaria	08016
Economic outlook	19007
Economic outlook	10022
Economic outlook	34107
Economic progress report	34108
Economic prospects	25020
Economic prospects	34109
Economic report	10023
Economic report	33019
Economic report on Scottish agriculture	34110
Economic review	25021
Economic review	35007
Economic review and outlook	20032
Economic series	20033
Economic situation report	34111
Economic statistics quarterly	19008
Economic survey	26015
Economic survey	24014
Economic survey: Finland	13022
Economic survey international	15064
Economic survey of Europe	02048
Economic trends	24015
Economic trends	20034
Economic trends	34112
Economic trends in Switzerland	32027
Economics bulletin	34113
Economics of Bulgaria	08017
Économie et prévision	14049
Économie et statistique	14050
Économie prospective internationale	14051
Economist intelligence unit: research reports: automotive	

industry 02049
Economist intelligence unit: research reports: chemicals and pharmaceuticals 01072
Economist intelligence unit: research reports: commodities 01073
Economist intelligence unit: research reports: paper and packaging 01074
Economist intelligence unit: research reports: retail and consumer goods 02050
Economist intelligence unit: research reports: retail and consumer goods 34114
Economist intelligence unit: research reports: travel and tourism 01075
Economy and finance 03046
The economy of Iceland 19009
ECOTASS - The economist and commercial bulletin of the news agency Tass 09019
The ECU and its role in the process towards monetary union 03047
The ECU bond market in 1991 23015
ECU - EMS information 03048
ECU: Le marché Euro-obligitaire en ECU statistiques mensuelles 23016
The ECU report 03049
Education in OECD countries: a compendium of statistical information 01076
Education in Sweden 31031
Education statistics 10024
The EEA agreement 02051
The EEA agreement 03050
EEA - EFTA fact sheets 02052
Efectos de comercio devueltos impagados 30029
Effectif des véhicules à moteur en Suisse au 30 Septembre 19.. 32028
EFTA 01077
EFTA 02053
EFTA annual report 01078
EFTA annual report 02054
EFTA trade 01079
EFTA trade 02055
EFTA - what it is and what it does 01080
EFTA - what it is and what it does 02056
EG magazin 15065
EIB - information 03051
Einbanddecken und jahrgangssammler 15066
Einburgerungsstatistik 22007
Einfuhr nach horstellungs - und einkaufslandern und warengruppen 15067
Einkommensteuer 15068
Einkommensteuerstatistik 05034
Einnahmen und ausgaben ausgenahlter privater haushalte 15069
Eisenbahnverkehr, strassenbahnverkehr 15070
Ekonomi 33020
Ekonomist 35008
Ekonomska misao 35009
Ekonomski anali 35010
El comercio Hispano-Britannico 30030
El economista 30031
Electric energy supply and district heating 31032
Electricity prices 1985-91 03052
Electricity statistics 26016
Electricity supply in OECD countries 01081
Electronic components - sector report 05035
Electronics and computer industries in Sweden 31033
Elforsyningens Tiårsoversigt 12017
Elsevier advanced technology reports 01082
Elsevier advanced technology reports 02057
Emerging market economies report 01083
Emerging stock markets fact book 01084
Empirica 05036
Emploi-croissance société 14052
Employees of British firms in Germany: income tax and social security 15071
Employment and industrial relations glossaries 03053
Employment and wages and salaries 27020
Employment gazette 34115
Employment in Europe 03054
Employment in national economy 199... 27021
Employment in the private sector 31034
Employment survey 16006
Emporoviomichaniki Merchantindustrialist 10025
Encuesta continua de presupuestos familiares 30032
Encuesta de población activa EPA: principales resultados 30033
Encuesta de población EPA: resultados detallados 30034
Encuesta de población EPA: tablas anuales 30035
Encuesta de salarios en la industria y los servicios 30036
Encuesta industrial 30037
Energiestatistik 22008
Energieverbrauch der haushalte im jahre 05037
Energieversorgung Oesterreichs 05038
Energy 1960-1988 03055
Energy and industry - panorama of the EEC industry - statistical summary 1992 03056
Energy balance sheets 03057
Energy balances and electricity profiles 01085
Energy balances for Europe and North America 1970-2000 01086
Energy balances of OECD countries 01087
Energy balances of OECD countries - historical series 01088
Energy - Complete subscription 03058
Energy, electricity and nuclear power for the period up to 2010 01089
Energy in Europe: annual energy review 03059
Energy in the European Community 03060
Energy - monthly statistics 03061
Energy policies of IEA countries: 1990 review 01090
Energy prices and taxes 01091
Energy review 13023
Energy - statistical yearbook 03001
Energy statistics 26017
Energy statistics 13024
Energy statistics and balances in non-OECD countries 02058
Energy statistics for dwellings and premises 31035
Energy statistics of OECD countries 01092
Energy statistics of OECD countries - historical series 01093
Energy statistics yearbook 01094
Energy supply 31036
Energy supply in the Netherlands 25022
Energy trends 34116
Engineering consultancy in the European Community 03062
English, French-Albanian foreign trade 04006
English heritage monitor 34117
English language course visitors to the UK 34118
English Tourist Board annual report 34119
Enquête sur le comportement des entreprises de l'industrie et du batîment - génie civil 14053
Enquête conjoncturelle 32029
Enquête financière 14054
Enquête mensuelle de conjoncture 14055
Enterprise and income statistics of farm economy 13025
Enterprises 13026
Enterprises financial accounts 31037
Entreprendre 07033
Environment 08018
Environment in Europe and North America: annotated statistics - statistical standards and studies 01096
Environment papers 01097
Environment statistics 03063
Environment statistics 03064
Environmental protection costs in Swedish agriculture 31038
Environmental quarterly 25023
Environmental statistics of the Netherlands 25024
Epilogi - monthly economic review 17016
ERC European market reports: consumer durables and household goods 02059
ERC European market reports: food 02060
ERC European market reports: leisure products 02061
ERC European market reports: pharmaceuticals and healthcare products 02062
ERC international market reports: food and drinks products 01098
Ergebnisse der landwirtschaftlichen statistik im jahre... 05039
Erzeugung von geflugel 15072
España en cifras 30038
Estado das culturas e previsío das colheitas 28020
Estadística de hipotecas 30039
Estadística de sociedades mercantiles 30040
Estadística de ventas a plazos 30041
Estadística Española 30042
Estatísticas agrícolas 28021
Estatísticas da construcçío et habitaçío 28022
Estatísticas da energia 28023
Estatísticas da pesca 28024
Estatísticas da protecçío social, associaçľes sindicais e patronais 28025
Estatísticas da saudé 28026
Estatísticas das contribuiçľes e impostas 28027
Estatísticas das finanças públicas 28028
Estatísticas das sociedades 28029
Estatísticas demográficas 28030
Estatísticas do comércio externo 28031
Estatísticas do turismo 28032
Estatísticas dos salários 28033
Estatísticas dos transportes e comunicaçľes 28034
Estatísticas industriais 28035
Estatísticas monetárias e financeiras 28036
Estatísticas regionais 28037
Estimated output, input and income arising in agriculture 20035
Estimates for the public services 20036
Estimates of receipts and expenditure for the year 20037
Estonia, Latvia, Lithuania 06006
Études 23017
Euroatom supply agency: annual report 03065
Eurobarometer: public opinion in the European Community 03066
Europa transport - observation of the transport markets 03067
Europa world yearbook 01099
Europarecht 15073
Europe 2000: outlook for the development of the Community's territory 03068
Europe in figures 3rd edition 03069
Europe on the Move 03070
Europe retail 02063
Europe's 15,000 largest companies 02064
Europe's top retailers - volume 1 02065
European advertising & media forecast 34120
European advertising, marketing

and media data 02066
European automotive components business 02067
European business planning factors: Key issues for corporate strategy in the 1990's 02068
European communities encyclopedia and directory 03071
European Communities: financial report 03072
European Community: 1992 and beyond 03073
The European community and its Eastern neighbours 02069
The European Community and its Eastern neighbours 03074
The European Community and the Mediterranean countries 02070
The European Community and the Mediterranean countries 03075
The European Community and the third world 03076
European Community direct investment 03077
European Community forest health report 03078
European compendium of marketing information 02071
European consumer packaging marketing directory 02072
European databanks: a guide to official statistics 03079
European drinks index 02073
European economies in graphs and figures 15074
European economy 03080
European economy - supplement - series A: economic trends 03081
European economy - supplement - series B: business and consumer survey results 03082
European file 03083
European food index 02074
European fund directory 02075
European insurance and the single market 1992 03084
European Investment Bank Papers 03085
European investment region series 03086
European investment region series 34121
European investment region series: research reports 02076
European market share reporter 02077
European marketing data and statistics 02078
European motor business 02079
European packaging 02080
European Parliament: EP news 03087
European printing 02081
European regional development fund: annual report from the Commission to the Council, the European Parliament and the Economic and Social Committee 03088
European retail 03089
European Retailing in the 1990s 03090
European social fund: annual report of activities 34122
European statistics - a guide to official sources 03091
European steel stockholding databook 02082
European supermarket index 02083
European supplies bulletin 02084
European update 03092
Eurostat 03093
Eurostatistics: data for short-term economic analysis 03094
Evaluation results for 1990 01100
Exhibitions and fairs in Hungary 92 18013
The export/import year 31039
Exportieren in die Bundesrepublik Deutschland 15075
Exporting consumer goods to Germany - sector report 15076
Exports 34123
External trade 26018
External trade - analytical tables - imports and exports 03095

External trade and balance of payments 03096
External trade and balance of payments - monthly statistics 03097
External trade and balance of payments - statistical yearbook 1991. Recapitulation 1958-90 03098
External trade by mode of transport 03099
External trade - generalized system for tariff preferences (GSP) 03100
External trade - provisional figures 20038

F

Fabrimétal magazine 07034
Fact sheets on Sweden 31040
Fact sheets on the European Parliament and the activities of the European Community 03101
Facts 199.. - Medicine and Health Care 12018
Facts about the Swedish economy 31041
Facts through figures: a statistical portrait of EFTA in the European economic area 13027
Facts through figures: a statistical portrait of EFTA in the European Economic Area 02085
Facts through figures: a statistical portrait of the European Community in the European economic area 03102
Fair trade commission annual report 20039
Family budgets - comparative tables 1988 03103
Family food expenditure survey 31042
Family spending 34124
FAO production yearbook 01101
FAO trade yearbook 01102
Farm business data Northern Ireland 34125
Farm incomes in Scotland 34126
Farm incomes in the United Kingdom 34127
Farm incomes in Wales 34128
Farm management pocketbook 34129
Farm management survey 20040
Farm register 13028
Fascicules de résultats de la Centrale des Bilans 14056
Features of the German market 15077
Febelgra - informations 07035
Federal, Cantonal and Communal taxes: an outline of the Swiss system of taxation 32030
Ferienhandbuch für das fürstentum Liechtenstein 22009
Ferro-Alloy directory and databook 3rd edition 1992 01103
Fifty years of the national food survey 1940-1990 34130
Figures in Turkey 33021
Figyelö (economic weekly) 18014
Fil de la Bourse 32031
Final accounts municipal and provincial special administrations 33022
Finance 13029
Finance (statistical yearbook) 27022
Finance accounts 20041
Finance and development 01104
Finance, insurance, real estate and business services statistics 10026
The finances of economic units in 19– and in the first half-year of 19– 27023
The finances of Europe 03104
Finances publiques en Suisse 19.. 32032
Financial accounts 31043
Financial accounts 31044
Financial aggregates for 1,790 Italian companies (1968-1991) 21016
Financial and accounting rewards 34131
The financial consumer 34132
Financial enterprises 31045

Financial investment companies in the Netherlands 1991 25025
The financial market 31046
Financial market statistics 13030
Financial market trends 01105
Financial market trends and OECD financial statistics 01106
Financial markets 13031
Financial report 24016
Financial reports 199-: European coal and steel community 03105
Financial review 26019
Financial statement and budget report 34133
Financial statements statistics 13032
Financial statistics 34134
Financial statistics of OECD 01107
Financial surveys 34135
Financial times surveys: country surveys 01108
Financial times surveys: industry surveys 01109
Financing and external debt of developing countries 01110
Financing foreign operations: Belgium 07036
Financing foreign operations: Czechoslovakia 11017
Financing foreign operations: France 14057
Financing foreign operations: Germany 15078
Financing foreign operations: Greece 17017
Financing foreign operations: Italy 21017
Financing foreign operations: Netherlands 25026
Financing foreign operations: Norway 26020
Financing foreign operations: Poland 27024
Financing foreign operations: Spain 30043
Financing foreign operations: Sweden 31047
Financing foreign operations: Switzerland 32033
Financing foreign operations: UK 34136
Financing foreign operations: USSR 09020
Finans & Samfund 12019
Finansije 35011
Finishing statistics 34137
Finland 13033
Finland in figures 13034
Finland's balance of payments 13035
The Finnish banking system 13036
Finnish bond issues 13037
Finnish business report 13038
The Finnish economy 13039
The Finnish forest industries: facts and figures 13040
Finnish insurance review 13041
Finnish labour review 13042
Finnish merchant marine 13043
Finnish metal, engineering and electrotechnical industry 13044
Finnish trade review 13045
Firmenbuch Österreich 05040
Fisheries - statistical yearbook 03106
Fishery statistics 26021
Fishing statistics 31048
Fiskets gang 26022
Five-year development plan 10027
Food and agricultural policy reforms in the former USSR: an agenda for transition 09021
Food consumption statistics 01111
Food distribution in Europe in the 1990's 02086
Food pocket book 34138
Forecast of emissions from road traffic in the European Communities 03107
Foreign direct investment in the states of the former USSR 09022
Foreign exchange and money market operations 32034
Foreign investment advisory

service papers 01112
Foreign markets 27025
Foreign owned enterprises 31049
Foreign trade 19010
Foreign trade 13046
Foreign trade 13047
Foreign trade 27026
Foreign trade (statistical yearbook) 27027
Foreign trade by branch of trade 31050
Foreign trade by commodities 01113
Foreign trade by commodities - series C 01114
Foreign trade: import/export 31051
Foreign trade of the Republic of Bulgaria 08019
Foreign trade statistics 33023
Foreign trade with selected commodities 15079
Foreign trade/exports 31052
Foreign trade/imports 31053
Foreign visitors and tourism receipts 33024
Forest health report 1991: technical report on the 1990 survey 03108
The forest resources of developed, temperate-zone regions 01115
Forest resources of the ECE region (Europe, the former USSR, North America) 01116
Forest resources of the temperate zones 01117
Forest resources of the temperate zones: main findings of the UN-ECE/FAO 1990 forest resource assessment 01118
Forestry statistics 26023
Forschung u dokumentation 05041
FPE 07037
Framework forecasts for the UK 34139
France 14058
Free Romania 29008
The free trade agreements of the EFTA countries and the European Communities 01119
The free trade agreements of the EFTA countries and the European Communities 02087
The free trade agreements of the EFTA countries and the European Communities 03109
Fremdenverkehrsstatistik 22010
Fresh fruits, vegetables and flowers from Portugal 28038
Frit Købmandskab 12020
Frost & Sullivan reports: automotive industries 02088
Frost & Sullivan reports: biotechnology, energy and the environment 02089
Frost & Sullivan reports: consumer goods 02090
Frost & Sullivan reports: defence and security industries 02091
Frost & Sullivan reports: electronics 02092
Frost & Sullivan reports: food and drinks industries 02093
Frost & Sullivan reports: healthcare, pharmaceuticals and chemicals 02094
Frost & Sullivan reports: leisure goods 02095
Frost & Sullivan reports: paper and packaging industries 02096
FT Management reports 02097
FT Management reports: automotive 01120
FT Management reports: banking & finance 01121
FT Management reports: energy 01122
FT Management reports: insurance 01123
FT Management reports: marketing 01124
FT Management reports: packaging 01125
FT Management reports: pharmaceuticals 01126
FT Management reports: property 01127
FT Management reports: telecommunications 01128

FT-SE share indices five year history 34140
Fuel-power economy 199.. 27028
Fuels 31054

G

Gambling in the single market: a study of the current legal and market situation 03110
Garanti Bank annual report 33025
Gardening equipment - sector report 05042
Gas and water statistics 33026
Gas prices 1985-91 03111
Gebarungsuebersichten 05043
Gebiet und bevolkerung 15080
Geld und kreditwessen, privatversicherung 1988 05044
Gemuseanbauflachen 15081
General business statistics 34141
General erhvervsstatistik og handel: statistiske efterretninger 12021
General government accounts and statistics 03112
General household survey 34142
General outline of import charges and import regulations 15082
General report on the activities of the European Communities 03113
General trade index and business guide 27029
Geneva Stock Exchange 32035
Genève et la Suisse: un mariage d'amour et de raison 32036
German and British economy 15083
German and British economy 34143
German brief 15084
German DIY market 15085
German subsidiary companies in the United Kingdom 15086
Geschäftsbericht 05045
Geschäftsbericht der BA 15087
Gesellschaftsformen und steuerbelastung im furstentum liechtenstein 22011
Gewerbestatistik 05046
Gewerbesteuerstatistik 05047
Global currency outlook 01129
Global economic 01130
Global economic prospects and the developing countries 01131
Global forecasting service: Belgium 07038
Global forecasting service: (Denmark) 12022
Global forecasting service: (Finland) 13048
Global forecasting service: France 14059
Global forecasting service: Hungary 18015
Global forecasting service: Italy 21018
Global forecast service: (Netherlands) 25027
Global forecasting service: (Norway) 26024
Global forecasting service: Poland 27030
Global forecasting service: Portugal 28039
Global forecasting service: Spain 30044
Global forecasting service: (Sweden) 31055
Global forecasting service: Switzerland 32037
Global forecasting service: Yugoslavia 35012
Global health situation and projections: estimates 01132
Global outlook 2000: an economic, social and environmental perspective 01133
Gold and silver survey 01134
Government finance statistics yearbook 01135
Government financing of research and development 1980-1990 03114

Grain market report 01136
Graph agri 9- 14060
Graph agri 9- : le fascicule régional 14061
Greece in the 1990s: taking its place in Europe 17018
Greece: reform of the credit and financial sector 17019
Greece's weekly for business and finance 17020
Greek balance of payments: the effects of full membership and the internal market program 17021
The Greek economy in figures 17022
Green Europe 03115
Grenzüberschreitender güterkraftverkehr 15088
Gross domestic product 31056
Gross domestic product in 199... 27031
Gross national product results quarterly 33027
Gross - und einzelhandelsstatistik 05048
Grosszaehlung: ausgewaehlte masszahlen nach gemeinden 05049
Grunderwerb 05050
Guide to the Council of the European Communities 03116
Guidelines for exporters of selected agricultural products 34144
Guterverkehr der verkehrszweige 15089

H

Haeuser - und wohnungszaehlung 05051
Hambro company guide 34145
Handbuch für volkswirtschaftliche berstung 15090
Handbook of industrial statistics 01137
Handbook of international trade and development statistics 01138
Handbook of market leaders 34146
Handbook of the Hungarian economy 1992 18016
Handbook of the Nations 01139
Handbuch für internationale zusammenarbeit 15091
Handbuch für investoreninformation 05052
Handel mit den ostblocklandern 15092
Haushaltsgeld - woher, wohin? 15093
Health and hospital statistics 10028
Health and personal social services statistics for England 34147
Health and safety statistics 34148
Health in figures 31057
Health services 08020
Heavy goods vehicles in Great Britain 34149
Henderson top 1000 charities 34150
Heti vilaggazdassy (world market weekly) 18017
Hints to exporters 01140
Hints to exporters visiting Austria 05053
Hints to exporters visiting Belgium 07039
Hints to exporters visiting Bulgaria 08021
Hints to exporters visiting Czechoslovakia 11018
Hints to exporters visiting Denmark 12023
Hints to exporters visiting Finland 13049
Hints to exporters visiting France 14062
Hints to exporters visiting Germany 15094
Hints to exporters visiting Greece 17023
Hints to exporters visiting Hungary 18018
Hints to exporters visiting Iceland 19011
Hints to exporters visiting Ireland 20042
Hints to exporters visiting Italy 21019
Hints to exporters visiting Luxembourg 23018
Hints to exporters visiting Norway 26025
Hints to exporters visiting Poland 27032
Hints to exporters visiting Portugal 28040
Hints to exporters visiting Romania 29009
Hints to exporters visiting Spain 30045
Hints to exporters visiting Sweden 31058
Hints to exporters visiting Switzerland 32038

Hints to exporters visiting the Netherlands 25028
Hints to exporters visiting the USSR 09023
Hire-purchase and credit-sales 20043
HM Customs and Excise report 34151
Hochsee - und kustenfischerei bodenseefischerei 15095
Horticulture enterprise register 13050
Horwath-ETB English hotel occupancy survey 34152
Hospital statistics England 34153
Hotel occupancy report 16007
House price index bulletin 34154
Household budget survey 13051
Household budget survey 20044
Household fish consumption in Great Britain 34155
Household food consumption and expenditure: report of the national food survey committee 1990. With a study of the trends over the period 1940-1990 34156
Household income and expenditure survey 10029
Housing 13052
Housing 08022
Housing and construction statistics Great Britain 34157
Housing and construction statistics Great Britain - quarterly 34158
Housing construction 13053
Housing construction 31059
Housing construction 27033
Housing finance 34159
Housing finance fact book 34160
Housing finance in Europe 02098
How much do local public services cost in Sweden 31060
How to approach the German market 15096
Hungarian annual statistical yearbook 18019
Hungarian business book 18020
Hungarian business book 1992 18021
Hungarian economic review 18022
The Hungarian economy 18023
Hungarian financial and stock exchange almanac 18024
Hungarian statistical pocket book 18025
Hungary 19– 18026
Hungary in the 1990s; sowing the seeds of recovery 18027

I

I consumi deele famigile 21020
I conti degli Italiani 21021
Iceland review 19012
The Icelandic economy: developments 1991 and outlook for 1992 19013
Ideas for joint venture 18028
IFC and the environment 01141
IFL studien zur arbeitsmarktforschung 15097
IFO - Branchenservice 15098
IFO - Digest 15099
IFO Konjunkturperspektiven 15100
IFO - Schnelldienst 15101
IFO schnelldienst 15102
IFO - Spiegel der wirtschaft: struktur und konjunktur in bild und zahl 15103
Ifo - Studien 15104
IFO studien zur agrarwirtschaft 15105
IFO studien zur bauwirtschaft 15106
IFO studien zur energiewirtschaft 15107
IFO studien zur europäischen wirtschaft 15108
IFO studien zur finanzpolitik 15109
IFO studien zur industrieurtschaft 15110
IFO studien zur

regional - und stadtökonomie 15111
IFO studien zur strukturforschung 15112
IFO studien zur umweltökonmie 15113
IFO studien zur verkehrswirtschaft 15114
IFO - Wirtschaftskonjuktur 15115
Il calepino dell'azionísta (1956-1990) 21022
IMF papers on policy analysis and assessment 01142
IMF staff papers 01143
IMF survey 01144
IMF working papers 01145
Implementation of the reform of the structural funds annual report 03117
Importing from Czechoslovakia 11019
Importing from Poland 27034
Imports 34161
Imports (time series) 34162
Imports and exports statistics 10030
Imports of man-made fibre: exports of man-made continuous filament yarn 34163
Impôt fédérale direct: Statisique 32039
In depth country appraisals across the EC - construction and building services 03118
In depth country appraisals across the EC - construction and building services - Belgium 07040
In depth country appraisals across the EC - construction and building services - Denmark 12024
In depth country appraisals across the EC - construction and building services - France 14063
In depth country appraisals across the EC - construction and building services - Germany 15116
In depth country appraisals across the EC - construction and building services - Greece 17024
In depth country appraisals across the EC - construction and building services - Ireland 20045
In depth country appraisals across the EC - construction and building services - Italy 21023
In depth country appraisals across the EC - construction and building services - Netherlands 25029
In depth country appraisals across the EC - construction and building services - Portugal 28041
In depth country appraisals across the EC - construction and building services - Spain 30046
In depth country appraisals across the EC - construction and building services - UK 34164
Income and consumption 13054
Income and property statistics 26026
Income and wealth distribution 31061
Income distribution statistics 13055
Index der grosshandels - verkaufspreise 15117
Index der grundstoftpreise 15118
Index der tariflohne und - gehalter 15119
Index of employment 20046
Indicadores de Coyuntura 30047
Indicadores de produçao vegetal 28042
Indicateurs rapides 23019
Indicatori mensili 21024
Indicators of industrial activity 01146
Indice de precios de consumo - boletín trimestral 30048
Indices and data on investment in listed Italian securities (1947-1991) 21025
Indices de precios industriales 30049
Indikatoren zur wirtsschaftsentwicklung 15120

Indizes der produktion und der arbeitsproduktiutat, produktion ausgenahlter erzeuguisse im produzie 15121
Indizes der auftragseingangs des umsatzes und des auftragsbestands fur das verabeitende bewerbe 15122
Indkomst, forbrug og priser: statistiske efterretninger 12025
Industri og energi: statistiske efterretninger 12026
Industria Cotoniera Liniera e delle Fibre Affini 21026
Industrial Development Authority annual report 20047
Industrial disputes 20048
Industrial employment 20049
Industrial employment and hours worked 20050
Industrial employment and hours worked - details for supplementary NACE sub-sectors 20051
Industrial investment prospects and purposes in 199– 25030
Industrial minerals 01147
Industrial minerals directory 1991 01148
Industrial performance analysis 34165
Industrial policy in OECD countries: annual review 01149
Industrial policy in OECD countries: annual review 02099
Industrial production 20052
Industrial production index 20053
Industrial production indexes quarterly 33028
Industrial production - quarterly statistics 03119
Industrial products 25031
Industrial statistics 10031
Industrial statistics - consumption of purchased energy 31062
Industrial statistics - regional statistics 31063
Industrial statistics yearbook 01150
Industrial structure statistics 01151
Industrial structure statistics 01152
Industrial trends - monthly 03120
Industrial trends survey/monthly trends enquiry 34166
Industrial turnover index 20054
Industrie du Verre - evolutie 1980-1991 07041
Industrie - und gewerbestatistik 05054
Industries électriques et électroniques 14064
Industriestatistik 05055
Industry (statistical yearbook) 27035
Industry and commerce 20055
Industry today 24017
Indíce de preços no consumidor 28043
Indíce de produçio industrial 28044
Info 14065
Información agropecuaria 30050
Información comercial Española: boletín económico 30051
Information about economic situation of the country in 199.. 27036
Information bulletin 27037
Information, computer and communications policy - ICCP series 01153
Information leaflet on Limassol and Larnaca ports (title varies from year to year) 10032
Information Moscow 09024
Information technology in Sweden 31064
Informationen internationale preise 05056
Informationen über die fishwirtschaft des auslandes 15123
Informations 18029
Informations rapides 14066
Informations statistiques 32040
Informaçio semanal 28045
Informe anual 30052

Infotech reports   34167
Ingénierie, études et conseils   14067
Inland revenue statistics   34168
Input-output tables   20056
Input-output tables for the UK 1984   34169
Input-output-studien   15124
Inquiérito ao emprego: informaçao
  antecipada   28046
Inquiérito às férias dos
  Portugueses   28047
Inquiérito às receitas e despesas
  familiares   28048
Inquiérito de conjuntura ao
  investimento   28049
Inquiérito mensal de conjurtura à
  indústria transformadora   28050
Inquérito ao transporte rodoviario
  de mercadorias   28051
INSEE première   14068
INSEE résultats: consommation -
  mode de vie   14069
INSEE résultats: démographie et
  emploi   14070
INSEE résultats: emploi-revenus   14071
INSEE résultats: économie
  générale   14072
Insight: East European business
  report   02100
Institute of Grocery Distribution
  Reports   03121
Institute of Grocery Distribution
  Reports   34170
Insurance and other financial
  services: structural trends   01154
Insurance annual report   20057
Insurance companies   31065
Insurance companies in Finland   13056
The insurance firm - a bibliography   31066
Insurance in Finland   13057
Insurance in Finnish society   13058
Insurance schemes on the Swedish
  labour market: Statutory
  schemes: Schemes under
  collective agreements:
  Insurance schemes   31067
Integrated tariff of the European
  Communities   03122
Interest rate service   01155
Interest rates   13059
The internal market: a challenge for
  the wholesale trade   03123
The internal market and the social
  dimension   12027
International business
  opportunities service (IBOS)   01156
International capital markets:
  developments and prospects   01157
International cotton industry
  statistics (ICIS)   01158
International currency review   01159
International customs journal:
  Czechoslovakia   11020
International customs journal:
  Hungary   18030
International customs journal:
  Norway   26027
International development policies   01160
International economic outlook   32041
International economic outlook   01161
International economic review   01162
International Financial Corporation
  technical or discussion papers   01163
International financial statistics   01164
International financial statistics -
  supplement   01165
International financial statistics -
  yearbook   01166
International giving and
  volunteering survey   01167
International Herald Tribune   01168
International housing finance
  sourcebook   01169
International management   01170
International marketing data and
  statistics 199-   01171
International motor business   01172
International production cost
  comparison, spinning/weaving

(IPCC)   01173
International risk and payment
  review   01174
International road haulage by UK
  registered vehicles   34171
International sea-borne trade
  statistics yearbook   01175
International steel statistics   01176
International steel statistics:
  summary tables   01177
International textile machinery
  shipment statistics (ITMSS)   01178
International tin statistics   01179
International tourism reports   01180
International trade in services   03124
International trade: Romania trade
  data   29010
International trade statistics
  yearbook   01181
Internationales
  gewerbearchiv
  zeitschrift für klein - und mittel
  unternehmen   15125
Introducing Italy   21027
Inventory of taxes levied in the
  Member States of the European
  Communities by the State and
  local authorities (lander,
  departments, regions, districts,
  provinces, communes).   03125
Investing, licensing and trading
  conditions abroad: Austria   05057
Investing, licensing and trading
  conditions abroad: Belgium   07042
Investing, licensing and trading
  conditions abroad: C.I.S.   09025
Investing, licensing and trading
  conditions abroad:
  Czechoslovakia   11021
Investing, licensing and trading
  conditions abroad: Denmark   12028
Investing, licensing and trading
  conditions abroad: Europe   02101
Investing, licensing and trading
  conditions abroad: Finland   13060
Investing, licensing and trading
  conditions abroad: France   14073
Investing, licensing and trading
  conditions abroad:
  Germany   15126
Investing, licensing and trading
  conditions abroad: Greece   17025
Investing, licensing and trading
  conditions abroad: Hungary   18031
Investing, licensing and trading
  conditions abroad: Ireland   20058
Investing, licensing and trading
  conditions abroad: Italy   21028
Investing, licensing and trading
  conditions abroad: Netherlands   25032
Investing, licensing and trading
  conditions abroad: Norway   26028
Investing, licensing and trading
  conditions abroad: Poland   27038
Investing, licensing and trading
  conditions abroad: Portugal   28052
Investing, licensing and trading
  conditions abroad: Spain   30053
Investing, licensing and trading
  conditions abroad: Switzerland   32042
Investing, licensing and trading
  conditions abroad: Turkey   33029
Investing, licensing and trading
  conditions abroad: UK   34172
Investionen fur
  umweltschutz im
  produzierenden gewerbe   15127
Investment activity in 199-, Part I
  investment outlays   27039
Investment activity in 199-, Part II
  buildings investments in a
  private economy   27040
Investment brief Hungary 1992   18032
Investment intentions survey   34173
Investing, licensing and trading
  conditions abroad: Luxembourg   23020
Investing, licensing and trading
  conditions abroad: Sweden   31068

Investment newsletter   32043
Investments   31069
Investors guide to Hungary 199–   18033
IP statistical service   01182
IP statistical service   34174
Ireland: a directory   20059
Irish agriculture in figures   20060
Irish banking review   20061
Irish Broker   20062
Irish economic statistics   20063
Irish Exporter   20064
Irish journal agricultural and food
  research   20065
The Irish market: facts and figures
  199..   20066
The Irish Times   20067
Irish tourist board annual report   20068
Irish travel trade news   20069
Iron and manganese ore databook   01183
The iron and steel industry   01184
Iron and steel - monthly statistics   03126
Iron and steel - statistical yearbook   03127
Islandbanki - annual report   19014
Isveren/Employer   33030
Isvestia   09026
Italian statistical abstract   21029
Italy: country profile   21030
Italy: country report   21031

J

Jahrbuch der absatz - und
  verbrauchsforschung   15128
Jahrbuch der Wiener borse   05058
Jahrbuch für
  fremdenverkehr   15129
Jahresbericht   32044
Jahresbericht   22012
Jahresbericht   05059
Jahresbericht -
  Arbeitsgemeinschaft der
  Deutschen
  Wertpapierbörsen   15130
Jahresbericht der VDR   15131
Jahresbericht
  forschung im
  geschäftsbereich des
  bundesministers für ernährung,
  landwirtschaft und
  forsten   15132
Jahresbericht über die deutsche
  fischwirtschaft   15133
Jahresschau der Deutschen
  industrie: die bekleidungs und
  wasche industrie   15134
Jahresschau der Deutschen
  industrie: die chemische
  industrie   15135
Jahresschau der Deutschen
  industrie: die eisen, blech und
  metall verarbeitende industrie   15136
Jahresschau der Deutschen
  industrie: die eisen, staht und
  ne-metall industrie   15137
Jahresschau der Deutschen
  industrie: die elektro-industrie   15138
Jahresschau der Deutschen
  industrie: die giesserei industrie   15139
Jahresschau der Deutschen
  industrie: die kunstoff-industrie   15140
Jahresschau der Deutschen
  industrie: die möbel industrie   15141
Jahresschau der Deutschen
  industrie: die nahrungs und
  genussmittel-industrie   15142
Jahresschau der Deutschen
  industrie: die textil-industrie   15143
Jewellery - sector report   05060
Joint venture companies   27041
Joint ventures in the C.I.S.   09027
Jordan surveys   34175
The journal: Belgo Luxembourg
  Chamber of Commerce   07043
The journal: Belgo - Luxembourg
  Chamber of Commerce in Great
  Britain   23021
Journal of common market studies   03128

Journal of official statistics 31070
Journal of Yugoslav foreign trade 35013
Jugoslovenski Pregled 35014

K

Kansallis-Osake-Pankki economic review 13061
Kaufuerte fur bauland 15144
Kaufuerte fur landwirtschaftlichen grundbesitz 15145
Kauppakamari 13062
Key British enterprises 34176
Key data 34177
Key figures 26029
Key indicators 34178
Key note Euroviews 02102
Key note reports: agriculture 34179
Key note reports: building materials 34180
Key note reports: chemical industries 34181
Key note reports: consumer durables 34182
Key note reports: cosmetics 34183
Key note reports: electronics industry 34184
Key note reports: engineering industry 34185
Key note reports: finance 34186
Key note reports: food and drinks industry 34187
Key note reports: leisure 34188
Key note reports: printing and packaging industries 34189
Key note reports: publishing industry 34190
Key note reports: retailing 34191
Key population and vital statistics 34192
Keynote market reviews 34193
Kiel reports 15146
Kiel working papers 15147
Kieler diskussionsbeiträge 15148
Kieler Kurzberichte 15149
Kieler studien 15150
Konjunktur aktuell 15151
Konjunkturindikatoren 15152
Konjunkturpolitik 15153
Konjunkturtest 22013
Konsumerhebung, hauptergebnisse 05061
Konten und standardtabellen 15154
Korperschaftsteuer 15155
Kostenstruktur bei handelsvertretern und handelsmaklern 15156
Kostenstruktur bei rechtsanwalten und anwaltsnotaren, bei wirtschaftsprufern, steuerberatern 15157
Kostenstruktur der nichtbundeseigenen eisenbahnen, des stadt schnellbahn - strassenbahn - und omnibusse 15158
Kostenstruktur der unternehmen im verbrauchsgoter produzieren den gewerbe und in nahrungs - und genussmittel 15159
Kostenstruktur der unternehmen in bangewerbe 15160
Kostenstruktur der unternehmen in bergbau, grundstoff - und produktionsgutergewerbe 15161
Kostenstruktur der gewerbuehen guterkraftverkehrs, der speditionen und lagereien, der binnensehiften 15162
Kostenstrukturim einzelhandel 15163
Kostenstrukturim gastgewerbe 15164
Kovo export 11022
Kraftfahrzeugzulassungen nach zulassungsbehoerden 05062
Krankenversicherungsstatistik 22014
Kredit und kapital 15165
Kritische justiz 15166

L

L'agriculture Suisse - format de poche 32045
L'agriculture Suisse: images, nombres, commentaires 32046
L'Assurance en Belgique 07044
L'Économie Belge en 19.. 07045
L'Espace économique Français 14074
L'Europe de l'energie: Objectif 1992 et perspectives 2010 03129
L'Implantation étrangère dans l'industrie 14075
L'indice des prix en gros 32047
L'indice Suisse des prix à la consommation 32048
L'industrie pétrolière en 19– 14076
L'économie Française en 19–: les résultats, les réformes 14077
La balance des paiements de la France 14078
La balance touristique de la Suisse 32049
La construction des logements en Suisse 19.. 32050
La dispersion des performances des entreprises en 1989: tableau de bord des secteurs industriels 14079
La France des entreprises 14080
La France et ses régions 14081
La industria siderúrgica Española en el año 19- 30054
La lettre des industries électriques et électroniques 14082
La lettre mensuelle régionale 14083
La monnaie en 19– 14084
La note d'information de la Direction des Transports Terrestres (DTT) 14085
La Note financière annuelle 14086
La situation de l'industrie en 19– : résultats détaillées de l'enquête annuelle d'entreprise 14087
La situation du système productif en 19– 14088
Labour costs survey 1988 20070
Labour costs: survey 199-: initial results 03130
Labour force preliminary estimate 20071
Labour force statistics 13063
Labour force statistics 01185
Labour force statistics 01186
Labour force survey 25033
The labour force survey 31071
Labour force survey 17026
Labour force survey 20072
Labour force survey 1983-89 03131
Labour force survey - results 1990 03132
Labour market 13064
The labour market in figures 31072
Labour market quarterly report 34194
Labour market statistics 26030
The labour market tendency 31073
Labour relations in 18 countries 31074
Labour relations in Denmark 12029
Labour statistics 10033
Labour statistics and labour costs 33031
Lagebeurteilung der Bauwirtschaft 32051
Land - und forstwirtschaftliche arbeitskraefte 05063
Landbouweifers (agricultural figures) 25034
Landbrug: statistiske efterretninger 12030
Landwirtschaftlich genutzte flachen 15167
Landwirtschaftliche maschinenzaehlung 05064
Lange reihen zur wirtschaftsentwicklung 15168
The large market of 1993 and the opening-up of public procurement 03133
Larger manufacturing enterprises 31075
Last ten years on Lisbon Stock Exchange 28053
Latvia - A guidebook for a businessman 06007

Latvia in figures 06008
Latvia today 06009
Le bulletin 14089
Le bulletin du STATEC 23022
Le compte du tourisme en 19– : rapport de la Commission des Comptes du Tourisme 14090
Le Luxembourg et sa monnaie 23023
Le nouvel économiste 14091
Le principali societa Italiane (1966-1990) 21032
Le regioni in cifre 21033
Le tourisme Suisse en chiffres 32052
Lead and zinc statistics: 1960-1988 01187
Leisure forecasts 34195
Leisure futures 34196
Leisure intelligence 34197
Les 4 pages 14092
Les cahiers économiques et monétaires 14093
Les chiffres clés de l'industrie dans les régions 14094
Les chiffres clés de l'industries 14095
Les commandes, la production, les chiffres d'affaires et les stocks dans l'industrie et dans le sector principal de la construction 14096
Les Comptes de l'Agriculture Française en 19– 14096
Les Comptes de la sécurité sociale 14097
Les comptes nationaux de la Belgique 07046
Les consommations d'énergie dans l'industrie en 19– : traits fondamentaux du système industriel Français 14098
Les dépenses des touristes étrangères 14099
Les entreprises à l'épreuve des années 80 14100
Les études de la Centrale des Bilans 14101
LES GROUPES D'ENTERPRISES 07047
Les moyens de paiement et circuits de recouvrement 14102
Les principales branches d'activité en 19– 14103
Les principales procédures de financement des besoins des entreprises et des ménages 14104
Les résultats comptables des entreprises Suisses 32054
Les Statistiques de commerce extérieur: importations et exportations en NGP 14105
Les statistiques monétaires et financières trimestrielles 14106
Les statistiques monétaires mensuelles 14107
Les transports en Belgique 07048
Les transports public 32055
Lettre de conjoncture 07049
Lettre mensuelle de conjoncture 14108
Level of living survey 26031
Liechtenstein company law 22015
Liechtenstein in figures 22016
Liechtenstein in zahlen 22017
Liechtenstein - principality in the heart of Europe 22018
Liechtenstein wirtschaftsfragen 22019
Liechtensteinische gesellschaftsformen 22020
Limeniki 10034
List of 50 most active/largest stocks 12031
List of goods for the statistics of foreign trade 25035
List of Italian stocks 21034
List of Listed Companies 12032
List of stockbroking companies 12033
Listed companies in Finland 13065
Live register age-by duration analysis 20073
Live register area analysis 20074
Live register flow analysis 20075
Live register statement 20076

Livestock	31076
Living conditions of the population in 199...	27042
Living standards measurement study (LSMS) working papers	01188
Lloyd's register of shipping annual report	01189
Lloyd's register of shipping statistical tables	01190
Lloyds Bank annual review	34198
Lloyds Bank economic bulletin	34199
Lloyds Bank economic profile of Great Britain	34200
Loans to build the European Community	03134
Local authority estimates	20077
Local government finance	19015
Local government finance	31077
Local government financial statistics, England	34201
Local taxes	31078
Lohnerhabung in der industrie Österreichs	05065
Lohnsteuer	15169
Lohnsteuerstatistik	05066
London currency report	01191
London stock exchanges news	34202
London weighting	34203
Long-term advertising expenditure forecast	34204
Long-term business statistics	34205
Luftverkehr	15170
The Luxembourg franc bond market in 199..: statistics and reflow	23024
Luxembourg in figures	23025
The Luxembourg Stock Exchange: facts and figures	23026

M

Mace market research reports	34206
Machine tool statistics	34207
Macro-economic outlook	25036
Made in Austria	05067
Made in Malta	24018
MAGNUS, including: Dansk økonomisk Bibliografi (Danish Economical Bibliograhy)	12034
Main economic indicators	01192
Main economic indicators - historical statistics	01193
Main science and technology indicators	01194
Major companies in the Netherlands	25037
Malta yearbook	24019
Management	20078
Manedsinformasjon	26032
Man-made fibre production	34208
Manufacturing	31079
Manufacturing enterprises	31080
Manufacturing industry quarterly employment payments production expectation	33032
Manufacturing - Part 1	31081
Manufacturing - Part 2	31082
Manufacturing statistics	26033
Maritime transport	01195
Market assessment reports: consumer durables	34209
Market assessment reports: cosmetics	34210
Market assessment reports: drinks industry	34211
Market assessment reports: food industry	34212
Market assessment reports: household goods	34213
Market Assessment reports: leisure industry and products	34214
Market assessment reports: office equipment	34215
Market assessment reports: personal finance	34216
Market forecasts	34217
Market intelligence	34218

Market research Europe	02103
Market research GB	34219
Market research reporter	34220
Market summaries	34221
Marketing in Europe	02104
Marketing in Europe, special market survey; retailing and wholesaling in Hungary	18034
Marketing medical products - sector report	15171
Marketing pocket book	34222
Marketpower market research reports: food industry	02105
Marketpower market research reports: food industry	34223
Marketpower market research reports: packaging industry	02106
Meat balances of OECD countries	01196
Media pocket book	34224
Medical equipment - sector report	05068
Medium-term review	20079
Memoria estadística seguros privados	30055
Men and women side by side	25038
Mercados de valores mobiliários: estudios e estatísticas	28054
Merchant fleet statistics	34225
Merchant shipbuilding returns	01197
Messzahlen fur bauleistungspreise und preisindizes fur bauwerke	15172
Metal bulletin	01198
Metal bulletin monthly	01199
Metal bulletin prices and data	01200
Méthode des scores de la Centrale des Bilans	14109
Migraciónes	30056
Migration: the demographic aspects	01201
Mikrozensus - jahresergebnisse	05069
Milch und molkereiwirtschaft Bundesrepublik Deutchland, EG-Mitgliedstaaten	15173
Mineralolsteuer	15174
Mining statistics	33033
Mintel special reports	34226
Mitteilungen des Instituts für angewandte wirtschaftsforschung Tuebingen	15175
Monatliche kraftfahrzeugzulassungsstatistik	05070
Monatsbericht	15176
Monatsberichte über die österreichische landwirtschaft	05071
Monatsbezüge der angestellten in der industrie Österreichs	05072
Monetary review	12035
Money and finance	03135
Moniteur du commerce international	14110
Monthly bulletin	28055
Monthly bulletin lead and zinc statistics	01202
Monthly bulletin of agricultural statistics	25039
Monthly bulletin of construction statistics	25040
Monthly bulletin of distribution and service trade statistics	25041
Monthly bulletin of external trade	26034
Monthly bulletin of financial statistics	25042
Monthly bulletin of health statistics	25043
Monthly bulletin of population statistics	25044
Monthly bulletin of price statistics	25045
Monthly bulletin of socio-economic statistics	25046
Monthly bulletin of statistics	26035
Monthly bulletin of statistics	33034
Monthly bulletin of statistics	01203
Monthly bulletin of the Netherlands Central Bureau of Statistics	25047
Monthly bulletin of transport statistics	25048

Monthly commodity price bulletin	01204
Monthly digest of statistics	34227
Monthly digest of Swedish statistics	31083
Monthly economic indicators	10035
Monthly economic indicators	33035
Monthly exchange and interest rate outlook	01205
Monthly first figures for export of goods	31084
Monthly first figures for import of goods	31085
Monthly first figures on the balance of trade	31086
Monthly industrial survey	20080
Monthly information about changes in prices in national economy	27043
Monthly list	12036
Monthly report	32056
Monthly report	15177
Monthly report - la Borsa Valori di Milano	21035
Monthly review and monthly newsletter - Kamer voor Handel en Nijverheid van Het Gewest Gent	07050
Monthly review of agricultural statistics	13066
Monthly statistical bulletin	17027
Monthly statistical bulletin	17028
Monthly statistical bulletin	33036
Monthly statistical bulletin of foreign trade	25049
Monthly statistical bulletin of manufacturing	25050
Monthly statistical reports	05073
Monthly statistical review	34228
Monthly statistical summary: International Tea Committee	01206
Monthly statistics	19016
Monthly statistics digest	34229
Monthly statistics of foreign trade - series A	01207
Monthly summary of foreign trade	33037
Mortgage monthly	34230
Mortgage weekly	34231
Moscow magazine	09028
Motor industry Great Britain - world automotive statistics	34232
Motor registrations	20081
Motor registrations: final, detailed results	20082
Motor registrations: provisional results	20083
Motor vechicle statistics	31087
Motor vehicle registration information system (MVRIS)	34233
Motor vehicle statistics	33038
Motor vehicles in Finland	13067
Motorfahrzeugbestand	22021
Motorparc statistics service (MPSS) - vehicles in use	34234
Motorstat express	01208
Mouvement de la population des communes Belges	07051
Mouvement naturel de la population de la Grece	17029
Movimiento de viajeros en establecimientos turísticos	30057
Movimiento natural de la población	30058
MSI reports: building materials	34235
MSI reports: consumer behaviour	34236
MSI reports: consumer durables	34237
MSI reports: drinks industry	34238
MSI reports: engineering and engineering products	34239
MSI reports: food industry	34240
MSI reports: food industry in Europe	02107
MSI reports: healthcare	34241
MSI reports: horticulture industry	34242
MSI reports: leisure industry	34243
MSI reports: retailing and distribution	34244
MSI reports: retailing in Europe	02108
Multilateral official debt rescheduling; recent experience	01209
Mémento statistique de la Suisse	32057

N

Nachrichten für
aussenhandel | 15178
Narcotic drugs: estimated world
requirements - statistics | 01210
Nase gospodarstvo/our economy:
review of current problems in
economics | 35015
National accounts | 25051
National accounts | 26036
National accounts | 13068
National accounts | 31088
National accounts ESA -
aggregates 1970-90 | 03136
National accounts ESA - detailed
tables by branch 1984-89 | 03137
National accounts ESA - detailed
tables by sector 1970-88 | 03138
National accounts ESA - input-
output tables | 03139
National accounts half-yearly
estimates | 10036
National accounts, Hungary (1988-
1990) | 18035
National accounts of OECD
countries | 01211
National accounts of the Maltese
islands | 24020
National accounts statistics:
analysis of main aggregates | 01212
National accounts statistics: main
aggregates and detailed tables | 01213
National accounts statistics: study
of input-output tables 1970-80 | 01214
National budget for Finland | 13069
National Economic and Social
Council reports | 20084
National economy of Latvia | 06010
National facts of tourism | 34245
National farm survey | 20085
National farm survey provisional
estimates | 20086
National income (statistical
yearbook) | 27044
National income and expenditure | 20087
National policies and agricultural
trade | 01215
Nationalregnskab, offentlige
finanser og betalingsbalance:
statistike efterretninger | 12037
NATO economics colloquium | 01216
NATO facts and figures | 01217
Natural environment in figures | 31089
Natural resources and the
environment | 26037
Navigator | 13070
Net earnings of manual workers in
manufacturing industry in the
Community | 03140
Netherlands | 25052
The Netherlands in Triplo | 25053
Netherlands official statistics | 25054
Neuer Weg | 29011
New earnings survey | 34246
New firms in Sweden | 31090
News from Iceland | 19017
News from the foundation | 03141
Newsletter - FIGAZ | 07052
Newsletter of the Swedish
statistical association | 31091
NHBC Euro-guide overview | 02109
NHBC private house-building
statistics | 34247
Nombre de détenteurs d'autoradios
et de téléviseurs (par commune) | 07053
Nordeuropa - forum | 15179
The Nordic Countries | 19018
The Nordic Countries | 12038
Nordic labour market statistics | 19019
The Nordic securities market | 01218
Northern Ireland agricultural
census statistics | 34248
Northern Ireland annual abstract of
statistics | 34249
Norway | 26038
Norway's foreign trade | 26039

Norway's largest companies | 26040
The Norwegian securities market | 26041
Note de conjoncture | 23027
Note de conjoncture | 14111
Notes d'information | 14112
Notes d'informations | 14113
Notes et études documentaires | 14114
Notes sur la situation économique | 07054
Notiziari ISTAT | 21036
Nuclear power reactors in the world | 01219
Nutzenergieanalyse | 05074
Nutztierhaltung in Oesterreich | 05075
Nyt fra Danmarks statistik | 12039
Números índices de la producción
industrial | 30059

O

O livro branco do turismo | 28056
O turismo em | 28057
O turismo estrangeiro em Portugal | 28058
Occupational choices of Greek
youth: an empirical analysis of
the contribution of information
and socio-economic variables | 17030
Occupational diseases and
occupational accidents | 31092
OECD economic outlook | 01220
OECD economic outlook - historical
statistics 1960-1990 | 01221
OECD economic studies | 01222
OECD economic survey | 20088
OECD economic survey: France | 14115
OECD economic survey - United
Kingdom | 34250
OECD economic surveys and
CCEET economic surveys | 01223
OECD economic surveys: Belgium,
Luxembourg | 07055
OECD economic surveys: Belgium,
Luxembourg | 23028
OECD economic surveys: Iceland | 19020
OECD economic surveys: Italy | 21037
OECD economic surveys: Portugal | 28059
OECD economic surveys: Spain | 30060
OECD employment outlook | 01224
OECD environmental data:
compendium 199- | 01225
OECD Financial statistics | 01226
OECD Nuclear energy data | 01227
OECD observer | 01228
OECD surveys: Switzerland | 32058
Oesterreich - EG:
kaufkraftparitaeten (ein
vergleich von produktivitaet,
lebensstandard und
preisniveaus 1990) | 05076
Oesterreichischer zahlenspiegel | 05077
Oesterreichs volkseinkommen | 05078
Official journal of the European
Communities: series C | 03142
Offshore centres report | 01229
Oil and gas activity | 26042
Oil and gas in Austria | 05079
Oil and gas information 1988-1990 | 01230
økonmiske Analyser | 26057
Okonomisk revy | 26043
Operating experience with nuclear
power stations in member states | 01231
Operation and effects of the
generalized system of
preferences: ninth and tenth
reviews | 01232
Operation of nuclear power stations | 03143
Operations evaluation papers | 01233
Opinions and reports of the
economic and social committee | 03144
Organismes de placement collectif | 23029
Oslo stock exchange information | 26044
Österreichs industrie in zahlen | 05080
Österreichs wirtschaft in zahlen | 05081
Other health and medical care
statistics | 31093
OTR-Pedder reports: computers
and computer applications | 34251
The outcome of the central
government budget for the fiscal

year | 31094
Outlays and results of industry | 27045
Output supply and input demand in
Greek agriculture | 17031
Outstanding exporters | 33039
Overall economic perspective to
the year 2000 | 02110
Overseas conference visitors to the
UK in 1990 | 34252
Overseas trade analysed in terms
of industries (Business monitor
MQ10) | 34253
Overseas trade fair/exhibition
visitors to the UK | 34254
Overseas trade statistics of the
United Kingdom (Business
Monitor MM20/MA20) | 34255
Ovum reports: electronics
industries | 02111

P

Panorama of EC industry 1991-
1992 | 03145
Parc des véhicules à moteur | 07056
Parc des véhicules utilitaires | 07057
The Paris Bourse: organization and
procedures | 14116
Partner from Poland | 27046
Penge - og kapital marked
statistiske efterretninger | 12040
Per Press Supermarkeder og andre
store dagligvarebutikker 1992 | 12041
The performance of Listed Shares | 21038
Performance rankings guide | 34256
Personal finance intelligence | 34257
Personen - und
haushaltseinkommen von
unselbstaendig beschaeftigten
und pensionisten; ergebnisse
des mikrozensus | 05082
Personenverkehr der
strassewerkehrs - unternehmen | 15180
Personnel rewards | 34258
Perspectives de financement de
l'économie Française | 14117
Perspectives de population - le
royaume | 07058
Pesticides in Swedish agriculture | 31095
Petroleum activities in Norway | 26045
Pferde - und rinderrassenerhebung
im jahre... | 05083
Pharmaceutical and health
industries of Europe | 29012
PIRA reports | 02112
PIRA reports | 34259
Planning consumer markets | 34260
Planning for social change | 34261
Planning permissions | 20089
Plirofories & Apopsis | 10037
Pocket yearbook | 25055
Pocket yearbook traffic and
transport statistics | 25056
Pocket book of educational
statistics | 25057
Poland in numbers 199– | 27047
Poland: Industrial development
review | 27048
Poland - selected data | 27049
Poland's next five years; the dash
for capitalism special report no.
2110 | 27050
Policy and advisory service papers | 01234
Polish business voice | 27051
Polish foreign trade | 27052
Polish trade magazine | 27053
Population | 14118
Population | 13071
Population | 08023
Population | 08024
Population and housing census
199– | 31096
Population and human resources
development in Cyprus: issues
and policies | 10038
Population and labour force
projections 1991-2021 | 20090

Population and social conditions 03146
Population and the World Bank:
  implications from eight case
  studies 01235
Population and vital statistics 19021
Population and vital statistics by
  municipality and parish 31097
Population and vital statistics report 01236
Population bulletin of the United
  Nations 01237
Population by sex, age and
  citizenship 31098
Population des communes
  fussionées 07059
Population estimates Scotland 34262
Population et sociétés 14119
Population projections 34263
Population projections for the
  Netherlands 1970-2000 25058
Population statistics Vol 1 26046
Population statistics Vol 2 26047
Population statistics Vol 3 26048
Population trends 34264
Port statistics 34265
Portrait of the regions 03147
Portugal 28060
Portugal: country profile 28061
Portugal: country report 28062
Portugal in figures 28063
Portugal social 1992 28064
Portugal-in 28065
Postens statistika arsbok, statistical
  yearbook 31099
Posts and telecommunications of
  Finland: Statistics 13072
The Prague Post 11023
Preise 15181
Preise und
  preisindizes fur
  die ein - und ausfuhr 15182
Preise und
  preisindizes fur
  die land - und
  forstwirtschaft 15183
Preise und
  preisindizes fur
  die lebenshaltung 15184
Preise und
  preisindizes fur
  gewerbuche produkte
  (erzeugpreise) 15185
Preise und
  prelsindlzes fur
  verkehrsleistungen 15186
Preliminary data on industry 13073
Price analysis 25059
Price list 26049
Price prospects for major primary
  commodities 01238
Prices 13074
Prices and earnings around the
  globe 32059
Prices and price indices in foreign
  countries 01239
Prices of agricultural products and
  selected inputs in Europe and
  North America 01240
Prices of real estate 31100
Prices received by farmers 33040
Principal uses of lead and zinc:
  1960-1990 01241
Problems and prospects of Greek
  exports: prerequisites for their
  development within the unified
  market of the EEC 17032
Problèmes économiques 14120
Proceedings of the World Bank
  annual conference on
  development economics 01242
Procurement by the international
  organizations in Vienna - sector
  report 05084
Producer price indices 13075
Production of milk and milk
  products 20091
Production, prices and incomes in
  EC agriculture 03148
Products and services of

Switzerland 32060
Produktion in
  produzerienden
  gewerbe nach
  wirtschaftzweigen und
  erzengrisgruppen 15187
Produktion in
  produzierenden
  gewerbe des in -
  und auslandes 15188
Profile 9- 32061
Programme for privatization of
  state and municipal enterprises
  in RF in 1992 09029
Property and personal insurance 27054
Property index 34266
Propos 23030
Prospects for East-West European
  transport (ECMT series) 02113
Public economy 13076
Public finance statistics 17033
Public finances of the Republic of
  Latvia 06011
Publishing and press 08025
The pulp and paper industry 01243

Q

Quality of markets companies book 34267
Quality of markets monthly fact
  sheet 34268
Quantity surveyors inquiry 20092
Quarterly accounts 25060
Quarterly bulletin 28066
Quarterly bulletin 33041
Quarterly bulletin 20093
Quarterly bulletin of cocoa statistics 01244
Quarterly bulletin of housing
  statistics 20094
Quarterly bulletin on justice and
  security 25061
Quarterly country reports: Irish
  Republic 20095
Quarterly country trade reports:
  Austria 05085
Quarterly country trade reports:
  Belgium - Luxembourg 07060
Quarterly country trade reports:
  Denmark 12042
Quarterly country trade reports:
  Finland 13077
Quarterly country trade reports:
  France 14121
Quarterly country trade reports:
  German Federal Republic 15189
Quarterly country trade reports:
  Greece 17034
Quarterly country trade reports:
  Italy 21039
Quarterly country trade reports:
  Netherlands 25062
Quarterly country trade reports:
  Norway 26050
Quarterly country trade reports:
  Portugal 28067
Quarterly country trade reports:
  Spain 30061
Quarterly country trade reports:
  Sweden 31101
Quarterly country trade reports:
  Switzerland 32062
Quarterly country trade reports: UK 34269
Quarterly digest of statistics 24021
Quarterly economic commentary 20096
Quarterly foreign trade statistics
  SITC/country 31102
Quarterly general business claims 34270
Quarterly labour force statistics 01245
Quarterly national accounts 01246
Quarterly national accounts ESA 03149
Quarterly new long-term business 34271
Quarterly oil statistics and energy
  balances 01247
Quarterly report - monnaie et
  conjoncture 32063
Quarterly report on vital statistics 20097
Quarterly review 31103

Quarterly review 24022
Quarterly review 20098
Quarterly statistical review 34272
Quarterly survey of advertising
  expenditure 34273
Quarterly transport statistics 34274

R

"R & S" annual directory 21040
Railway statistics 13078
Rapid reports 03150
Rapport annuel 07061
Rapport annuel - Conseil
  Économique et Social 19– 14122
Rapport annuel de Conseil National
  de Crédit 14123
Rapport annuel de L'IML 23031
Rapport annuel de la Commission
  Bancaire 14124
Rapport annuel de la zone Franc 14125
Rapport annuel 07062
Rapport annuel et rapport d'activité 07063
Rapport Annuel - FIGAZ 07064
Rapport annuel 07065
Rapport - Banque Nationale de
  Belgique 07066
Rapport económico 30062
Rapport sur les comptes de la
  nation 14126
Rapporto annuale 21041
Rapporto ASSIN FORM sulla
  situazione dell'informatica in
  Italia 21042
Rassegna economica 21043
Rational approach to labour and
  industry - economic and social
  partnership in Austria 05086
Real estate and housing economy 31104
Real estate guide Hungary 18036
Realsteuervergleich
  gewerbesteuer 15190
Recenseamento agrícola 28068
Recenseamento da populaço e da
  habitaço 28069
Recenseamento das empresas do
  sector de transportes 28070
Recenseamento industrial 28071
Recensement agricole et horticole 07067
Recensement de la circulation
  routière 07068
Recensement de la population 14127
Recensement de la population 23032
Recensement de la population et
  des logements au 1 mars 1981 07069
Recensement fédérale des
  entreprises 32064
Recensement fédérale des
  entreprises agricoles 32065
Rechmungsergebnisse des
  offentlichen gesamthaushalts 15191
Rechrungsergebnisse der
  kommeralen haushalte 15192
Recht der
  biotechnologie 15193
Record of shipments of wheat and
  wheat flour 01248
Recueil de statistiques par
  commune 23033
Recueil statistique 07070
Recueil Statistique 07071
Recueil statistique - FETRA 07072
Recycling lead and zinc: the
  challenge of the 1990's 01249
Reference book - Joint Ventures in
  the C.I.S. 09030
Regards sur l'actualité 14128
Region of residence of Italian
  visitors to the UK 34275
Regional development indicators of
  Greece 17035
Regional directories of key
  business prospects 34276
The regional impact of Community
  policies 03151
Regional marketing pocket book 34277
Regional pocket yearbook 25063

Regional salary and wages survey 34278
Regional tourism facts 34279
Regional trends 34280
Regions 03152
Regions and municipalities in the region of Bulgaria 08026
Regions - statistical yearbook 03153
The Registrar General's annual report for Northern Ireland 34281
Registrar general's annual report for Scotland 34282
Registration of establishments - Vol I: Establishment and Employment by Branch of Economic Activity and District 10039
Registration of establishments - Vol II: Occupational Structure of Employment 10040
Reisegewohnheiten der oesterreicher im jahre... (haupturlaube, kurzurlaube, dienst - und geschaeftsreisen) 05087
Relazione annuale 21044
Report 33042
Report and accounts 12043
Report and accounts 32066
Report of the board of directors 28072
Report of the Commissioners of HM Revenue 34283
Report of the Comptroller and Auditor General and appropriaton accounts 20099
Report of the crop of the year (wheat) 01250
Report of the registrar of friendly societies 20100
Report on competition policy 03154
Report on vital statistics 1988 20101
Reports 34284
Reports for the five-year plan 1988-1992: employment-unemployment 17036
Reports for the five-year plan 1988-1992: energy 17037
Reports for the five-year plan 1988-1992: local government 17038
Reports for the five-year plan 1988-1992: prospects of the Greek monetary system 17039
Reports for the five-year plan 1988-1992: public administration 17040
Reports for the five-year plan 1988-1992 regional policy 17041
Research and development 34285
Research on the "Cost of non-Europe" - basic findings 03155
Research report 20102
Research statistics 31105
Resources CSFR 11024
Restaurants and hotel statistics 10041
Results of agricultural census in 199– 27055
Results of plant production in 199– 27056
Results of sea fishery survey by motor vessels 17042
Results of the business survey carried out among managements in the Community 03156
Retail business market reports 34286
Retail business trade reviews 34287
Retail intelligence 34288
Retail monitor international 01251
Retail monitor international 34289
Retail pocket book 34290
Retail price statistics 33043
Retail rankings 34291
Retail ratios 34292
Retail research report 34293
Retail service rankings 34294
Retail trade Europe 02114
Retailing in Eastern Europe 02115
Retailing in Europe 02116
Retailing in the Netherlands 25064
Retail sales index 20103
Revenue statistics of OECD member countries 01252
Review of economic conditions 33044
Review of economic conditions in

Italy 21045
Review of fisheries in OECD member countries 01253
Review of industrial performance 20104
Review of maritime transport 01254
Revista exportar 28073
Revue annuelle de statistique des bourses Françaises de Valeurs 14129
Revue de la concurrence et de la consommation 14130
Revue du travail 07073
Revue Française des affaires sociales 14131
Reward - the management salary survey 34295
Ria-Novosti reference book 09031
Rilevazione delle forze di lavoro 21046
Road freight transport survey 20105
Road goods vehicles on roll-on-roll-off ferries to mainland Europe 34296
Romania: A country study 29013
Romania: Human resources and the transition to a market economy 29014
The Romanian economic reform program 29015
Rubber trends 01255
Russia express; executive 09032
Russian trade express 09033
RWI - Konjunkturberichte 15194
RWI - Mitteilungen 15195
Résultats annuels 14132
Résultats annuels des enquêtes de branche 14133
Résultats de l'enquête annuelle d'entreprise 14134
Résultats mensuels des enquêtes de branche 14135
Résultats régionalisés 14136
Résultats trimestriels 14137
Résultats trimestriels des enquêtes de branche 14138

S

S & F - Vierteljähresschrift für sicherheit und frieden 15196
Saasto Pankki 13079
SAF tidningen 31106
Saf's kalender 31107
SAF's, LO's, TCO's, OCH SAVO/SR's UPPBYGGNAD 31108
SAF's verksamhetsberattelse 31109
Salary survey for smaller businesses 34297
Sales and marketing rewards 34298
Samfaerdsel og turisme: statistiske efterretninger 12044
Sample survey of income tax returns for individuals at the assessment 31110
Satellite television receiving equipment - sector report 05088
The Scandinavian economies 12045
SCB Economic indicators Part 1: Analysis and Comments 31111
SCB Economic indicators Part 2: Tables 31112
Schaumweinsteuer 15197
Schlachtungen und fleischgeninnung 15198
Schriftenreihe 15199
Schriftenreihe der bundesanstalt für agrarwirtschaft 05089
Schriftenreihe ... reihe C: agrarpolitische berichte der OECD 15200
Schrifttum der agrarwirtschaft 05090
Science, technology and industry review 01256
Scotland's top 2000 companies 34299
Scottish abstract of statistics 34300
Scottish economic bulletin 34301
Scottish fishing fleet 34302
Scottish local government financial

statistics 34303
Scottish sea fisheries statistical table 34304
Scottish transport statistics 34305
Sea fisheries statistical tables 34306
Seaborne trade statistics of the United Kingdom 34307
Second survey on State aids in the European Community in the manufacturing sector and certain other sectors 03157
Security equipment - sector report 05091
Seeschiffahrt 15201
Semi-annual report 12046
Série "Études économiques" 07074
Services 25065
Services 13080
Services and distributive trade statistics 10042
Services and transport - monthly statistics 03158
Setting up business in Turkey 33045
Setting up business in... 01257
Shipbuilding and shiprepairing industry in Greece 17043
Shipping and aviation statistics 24023
Shipping industry and offshore activities 26051
Shipping statistics 17044
Shipyard orders weekly report 01258
Short-term energy outlook for the European Community 03159
Sightseeing in the UK 34308
Sigma 03160
The single financial market - 2nd edition 03161
Sintese anual do mecado de valores mobiliários 28074
Situation financière des régions en 19– 14139
Situation report on the society and economy of Hungary 18037
Situation, tendances et perspectives de l'agriculture en Hongrie 18038
Situation, tendances et perspectives de l'agriculture en Roumanie 29016
Skogsstatistik arsbok 31113
Small business 27057
Small business trends report 34309
Social and economic reports 20106
Social Europe - general review 03162
Social indicators 10043
Social indicators of development 01259
Social insurance statistics - facts 31114
Social protection expenditure and receipts 1980-89 03163
Social security statistics 34310
Social sikring og retsvaesen: statistiske efterretninger 12047
Social statistics 26052
Social trends 34311
Social welfare and health statistics 17045
Sociétés anonymes en Suisse 32067
Some statistics on services 1988 03164
Sonderreihe 15202
Sources of new premium income 34312
Soviet analyst 09034
Soviet independent business directory 09035
Soviet trade directory 09036
Sozialstatistische auswertungen aus der konsumerhebung 05092
Sozialstatistische daten 05093
Spain 30063
Spain: a directory and sourcebook 30064
Spain: country profile 30065
The Spanish economy in figures 30066
Spat banken 13081
Special series 15203
Spinning and waste spinning 34313
Sport statistics 33046
Staff studies for the world economic outlook 01260
Stainless steel databook 2nd edition 01261
Stand der reben und

weinmostemte
weinerzeugung und - bestand 15204
Standard of life of population in the
Republic of Bulgaria 08027
Staniland Hall Associates market
research reports 34314
State of trade reports 01262
Statement of the Bank of Finland 13082
Statistica annuale del commercio
con l'estero 21047
Statistica degli incidenti stradali 21048
Statistica del commercio con
l'estero 21049
Statistical abstract 10044
Statistical abstract 20107
Statistical abstract of Iceland 19022
Statistical bulletin 25066
Statistical bulletin 17046
Statistical bulletin 27058
Statistical bulletin 20108
Statistical bulletin of public finance 17047
Statistical compass 15205
The statistical concept of the town
in Europe 03165
Statistical data on Switzerland 32068
Statistical digest for the furniture
industry 34315
Statistical indicators 33047
Statistical magazine 25067
Statistical monthly review 18039
Statistical news 08028
Statistical news 27059
Statistical news 34316
Statistical pocket book of Budapest 18040
Statistical reference book of the
Republic of Bulgaria 08029
Statistical report of the revenue
commissioners 20109
Statistical report on road accidents
(ECMT series) 01263
Statistical reports of the Board of
Navigation: Domestic
waterborne traffic 13083
Statistical reports of the Board of
Navigation: Merchant Fleet 13084
Statistical reports of the Board of
Navigation: shipping between
Finland and foreign countries 13085
Statistical review of Northern
Ireland Agriculture 34317
Statistical yearbook 28075
Statistical yearbook 31115
Statistical yearbook 27060
Statistical yearbook 01264
Statistical yearbook income and
consumption 25068
Statistical yearbook of
administrative districts of
Sweden 31116
Statistical yearbook of Finland 13086
Statistical yearbook of Greece 17048
Statistical yearbook of Latvia 06012
Statistical yearbook of Norway 26053
Statistical yearbook of Sweden 31117
Statistical yearbook of the
Netherlands 25069
The statistical yearbook of the
Republic of Bulgaria 08031
Statistical yearbook of the Socialist
Republic of Romania 29017
Statistical yearbook of the Swiss
banking industry 32069
Statistical yearbook of Turkey 33048
Statistical yearbook on culture 25070
Statistical yearbook on tourism and
recreation 25071
Statistiche del commercio interno 21050
Statistiche del lavoro 21051
Statistiche del turismo 21052
Statistiche dell'agricoltura,
zootecnia e mezzi di produzione 21053
Statistiche dell'attività edilizià 21054
Statistiche dell'Industria Cotoniera
Liniera e delle Fibre Affini
Internazional: mondo 21055
Statistiche dell'Industria Cotoniera
Liniera e delle Fibre Affini - Italia 21056
Statistiche della caccia, pesca e

cooperazione 21057
Statistiche delle apere pubbliche 21058
Statistiche demografiche 21059
Statistiche forestali 21060
Statistiche industriali 21061
Statisticher
wochendienst 15206
Statistics 12048
Statistics Europe: sources for
social, economic and market
research 01265
Statistics from individual countries,
national statistics from Sweden
and other countries 31118
Statistics from international
organizations 31119
Statistics of coastal and
international sea transportation 33049
Statistics of consumer credit 25072
Statistics of enterprises 25073
Statistics of foreign trade series A:
monthly statistics of foreign
trade 01266
Statistics of foreign trade series C:
foreign trade by commodities 01267
Statistics of motor vehicles 25074
Statistics of owners of quoted
shares 31120
Statistics of port traffic 20110
Statistics of road traffic accidents in
Europe 02117
Statistics of the declared income of
legal entities and its taxation 17049
Statistics of the declared income of
physical persons and its taxation 17050
Statistics of the week 26054
Statistics of vessels 18 gross
tonnages and over 33050
Statistics of world trade in steel 01268
Statistics on regional employment 31121
Statistics on working hours and
absenteeism 31122
Statistik der aktiengesellschaften in
Oesterreich im jahre... 05094
Statistik der
industriellen betriebe 22022
Statistik der kammer und
fachgruppenmitglieder 05095
Statistique Annuelle 32070
Statistique annuelle de la
production 07075
Statistique annuelle définitive 07076
Statistique de l'emploi et de la
population active occupée 32071
Statistique mensuelle 32072
Statistique Suisse des transports 32073
Statistiques 07077
Statistiques agricoles 07078
Statistiques annuelle de traffic
international des ports 07079
Statistiques de l'environement 03166
Statistiques de la construction 14140
Statistiques de la construction et du
logement 07080
Statistiques de la formation
professionelle continue financée
par les entreprises: traitements
des déclarations d'employeurs
no.2483 14141
Statistiques du commerce extérieur
de l'Union Économique Belgo -
Luxembourgeoise 07081
Statistiques du commerce extérieur
de L'Union Économique Belgo -
Luxembourgeoise 23034
Statistiques du commerce extérieur
de l'Union Économique Belgo -
Luxembourgeoise (Annual) 07082
Statistiques du commerce intérieur
et des transports 07083
Statistiques démographiques 07084
Statistiques financières 07085
Statistiques - Fédération de
l'Industrie du Verre 07086
Statistiques historiques 1839-1989 23035
Statistiques industrielles 07087
Statistiques monétaires et
financières 14142

Statistiques quotidiennes et
mensuelles du MONET 14143
Statistiques sociales 07088
Statistiques économiques 32074
Statistische Erhebungen und
Schätzungen über
landwirtschaft und Ernährung 32075
Statistische nachrichten 05096
Statistischer
monatsbericht des
bundesministeriums für
ernährung, landwirtschaft und
forsten 15207
Statistisches handbuch fuer die
republik Oesterreich 05097
Statistisches
jahrbuch 22023
Statistisches
jahrbuch fur die Bundesrepublik
Deutschland 15208
Statistisches
jahrbuch für das Ausland 15209
Statistisches
jahrbuch über ernährung,
landwirtschaft und forsten 15210
Statistisches monatsheft 05098
Statistisk årbog 12049
Statistisk efterretninger 12050
Statistisk månedsoversigt 12051
Statistisk tiårsoversigt 12052
The steel market 02118
The steel market in 199- and the
outlook for 199- 01269
Steuerhaushalt 15211
Stock exchange fact service 34318
Stock exchange official yearbook 34319
Stock exchange quarterly 34320
Strassen, brucken,
parkenrichtungen 15212
Strassenverkehrssicherheit im
jahre... 05099
Structure and activity of industry -
annual survey - main results 03167
Structure and activity of industry -
data by regions 03168
Structure and activity of industry -
data by size of enterprise 03169
Studies of economies in
transformation 01270
Summarized data (foreign trade) 10045
Summary of public expenditure
programme 20111
Summary public capital programme 20112
Summary surveys of foreign trade 15213
Sundry costs 28076
Supplement to the statistical
indicators of short-term
economic changes in ECE
countries: sources and
definitions 01271
Supplementary estimates 20113
Supplementi al Bollettino statistico 21062
Supply estimates - summary and
guide 34321
Survey of the Austrian economy 05100
Survey of the earnings of
permanent male agricultural
workers 20114
Svenska Forsakringsarbok 31123
Sweden business incorporating
Sweden business report, the
Swedish stock market 31124
Sweden in figures 31125
Sweden international 31126
Sweden today 31127
Sweden's largest companies 31128
The Swedish budget 31129
The Swedish economy 31130
Swedish insurance companies 31131
The Swedish public sector 31132
Swedish sea fisheries 31133
Swedish trade 31134
Swiss business 32076
The Swiss capital market: a guide 32077
Swiss foreign trade statistics -
annual report 32078
Swiss stock guide 32079
Swiss textiles 32080

Switzerland 32081
Switzerland in figures 32082
Switzerland your partner 32083
System of national accounts -
   statistical papers, series F 01272

T

Tabakgewerbe 15214
Tableaux de bord des secteurs
   industriels 14144
Tableaux de l'économie Française 14145
Tableaux des consommations
   d'énergie en France 14146
Talouselama/("Business life") 13087
Target 1992 03170
Tavole statistiche 21063
Technology Ireland 20115
Teknikka & Talous/(Technology
   and Business) 13088
Telecommunication statistics 13089
Telecommunications - sector report 15215
Temi di discussione 21064
Tendances de la conjoncture 14147
Textile industry statistics 13090
Timber bulletin 01273
Timber bulletin for Europe 02119
Time rates of wages and hours of
   work 34322
Times 1000 34323
Tips für importeure 15216
Tisk information 33051
Top markets 34324
Tourism 08032
Tourism 33052
Tourism and recreation 27061
Tourism and the European
   Community 34325
Tourism and travel 20116
Tourism and travel quarterly 20117
Tourism - annual statistics 03171
Tourism in Europe: summarizing
   the work of the European
   Parliament in the context of the
   European Year of Tourism 03172
Tourism intelligence quarterly 34326
Tourism, migration and travel
   statistics summary 10046
Tourism policy and international
   tourism in OECD countries 01274
Tourism statistics 13091
Tourism statistics 17051
Tourism statistics 33053
Tourism statistics figures by month
   and quarter 25075
Tourisme en Suisse 32084
Tourisme en Suisse: semestre
   d'hiver 32085
Tourisme en Suisse: Semestre
   d'été 32086
Tourist, migration and travel
   statistics 10047
Tourist survey 16008
Touristes Suisses à l'etranger 19.. 32087
Toy industry in the UK 34327
Toys - sector report 05101
Trade 13092
Trade bulletin - UK imports and
   exports of fish and fish products 34328
The trade directory 07089
Trade directory 24024
Trade enquiries 07090
Trade enquiries from Germany 15217
Trade magazine - Poland 27062
Trade, service, hotel and restaurant 33054
Trade statistics 24025
Trade statistics 20118
Trading Partners USSR 09037
Traffic in Great Britain 34329
Traffic injuries 31135
Training statistics 34330
Transport 13093
Transport and communication
   statistics 26055
Transport and communications
   statistics 17052
Transport and communications

yearbook 13094
Transport and the environment in
   Finland 13095
Transport - annual statistics 1970-
   1989 03173
Transport economy and policy -
   ECMT series 01275
Transport of goods by road in Great
   Britain 34331
Transport statistics 10048
Transport statistics for London
   1980-1990 34332
Transport statistics Great Britain 34333
Transportable capital goods price
   index for agriculture 20119
Transportation and road traffic
   accidents statistics summary 33055
Transporte de viajeros interior
   regular 30067
Travail et emploi 14148
Travel agencies 31136
Travel and tourism analyst 01276
The trend of employment and
   unemployment 1986-1988 20120
Trends and forecasts population,
   education and labour market in
   Sweden 31137
Trends and views on direct
   investment by German-owned
   companies in the UK 15218
Trends - Dresdner Bank analyses
   and forecasts 15219
Trends: financeel ekonomisch
   magazine 07091
Trends in developing economies 01277
Trends in private investment in
   developing countries 01278
Trends in private investment in
   developing countries 1993:
   statistics for 1970-91 01279
Trends in production of lead and
   zinc 01280
Tuinbouwcijfers (horticultural
   figures) 25076
Tungsten statistics: annual report
   of the UNCTAD committee on
   tungsten 01281
Tunisia, Malta - Country report 24026
Turkey - a brief guide: general
   report 33056
Turkey: British connected
   companies in Ankara 33057
Turkey: computer hardware and
   software market 33058
Turkey - consultancy sector: sector
   report 33059
Turkey - country profile 33060
Turkey - country report 33061
Turkey demography survey 33062
Turkey: health report 33063
Turkey - machine tools: sector
   report 33064
Turkey - mining equipment: sector
   report 33065
Turkey monitor 33066
Turkey: natural gas 33067
Turkey: on the road to progress 33068
Turkey - railways: sector report 33069
Turkey - telecommunications:
   sector report 33070
Turkey: the construction sector 33071
Turkey - water and sewerage:
   sector report 33072
Tüsiad members company profiles 33073

U

UBS international finance 32088
Udenrigshandel: statistiske
   efterretninger 12053
Ugebrevet Dansk Industri 12054
UK airlines - annual operating,
   traffic and financial statistics
   (CAP 596) 34334
UK airlines - monthly operating and
   traffic statistics 34335
UK airports - annual statements of

movements, passengers, and
   cargo - (CAP 592) 34336
UK Airports - monthly statements of
   movements, passengers, and
   cargo 34337
UK balance of payments (CSO
   Pink Book) 34338
UK chemical industry
   environmental protection
   spending survey 34339
UK Electricity 34340
UK exhibition industry: the facts 34341
UK iron and steel industry: annual
   statistics 34342
UK market statistics 34343
UK national accounts (CSO Blue
   Book) 34344
UK printing and publishing industry
   statistics 34345
UK retail marketing survey 34346
UK retail reports 34347
UK tourist: statistics 199– 34348
UK's 10,000 largest companies 34349
Ukraine business review 09038
Umsatzsteuer 15220
Umsatzsteuerstatistik 05102
Umuelt in zahlen 15221
Umwelt in Oesterreich, daten und
   trends 05103
Umweltbedingungen des wohnens;
   ergebnisse des mikrozensus 05104
UN/ECE discussion papers 01282
UN/ECE economic studies 01283
UNCTAD commodity yearbook 01284
Understodsforeningar 31138
Unemployment 03174
Unemployment in Poland 27063
UNESCO statistical yearbook 1992 01285
Unitas: economic quarterly review 13096
United Kingdom minerals yearbook 34350
UNCTAD statistical pocket book 01286
Upland area: basic data and
   statistics 03175
Urban management program
   papers 01287
The US - Eastern European trade
   source book 04007
USSR calendar of events annual 09039
USSR facts and figures annual 09040

V

Vademecum des entreprises
   (feuilles mobiles) 07092
Vademecum: Le financement du
   commerce extérieur 07093
Vademecum of health statistics in
   the Netherlands 25077
Valencia fruits 30068
Value added tax in Germany 15222
Véhicules à moteur neufs mis en
   circulation 07094
Venturing into the Netherlands 25078
Verdict retail reports 34351
Verkehrswirtschaftliche zahlen 15223
Vermoegensteuerstatistik 05105
Vienna stock exchange yearbook 05106
Vierteljährliche kassenergebnisse
   der offentlichen
   haushalte 15224
Vierteljahrshefte zur
   auslandsstatitik 15225
Vierteljahrshefte zur
   wirtschaftsforschung 15226
Visitors to tourist attractions 34352
Vocational training - information
   bulletin 03176
Volkszaehlung 05107
Volume index of industrial
   production 13097

W

Wage and salary indices 13098
Wage statistics 26056

Wages	13099
Wages and employment in the private sector	31139
Wages and salaries in Sweden	31140
Wages and total labour costs for workers	31141
Wages, salaries and employment in public sector. Part 2 employees in municipalities	31142
Wages, salaries and employment in the public sector. Part 1 government employees	31143
Water industry - sector report	15227
Waterborne freight in the United Kingdom	34353
Weaving statistics	34354
Weekly exchange & interest rate outlook	01288
Welsh agricultural statistics	34355
Welsh economic trends	34356
Welsh housing statistics	34357
Welsh local government financial statistics	34358
Welsh transport statistics	34359
Weltwirtschaftliches archiv	15228
Western Europe 1993	02120
Wholesale and consumer price indexes: monthly bulletin	33074
Wholesale price index	20121
Wholesale price statistics	33075
Wirtschaft in zahlen	15229
Wirtschaft und statistik	15230
Wirtschaftskonjunktur	15231
Wirtschaftsstatistik der elektrizitaetsversorgungs - unternehmen	05108
Wochenbericht	15232
Wohnungsdaten	05109
Women and men in Sweden equality of the sexes	31144
Women, employment and health	10049
Women in statistics	33076
Work pattern survey	31145
World Bank and the environment	01289
World Bank annual report	01290
World Bank atlas	01291
World Bank CGIAR study papers	01292
World Bank commodity working papers	01293
World Bank country studies	01294
World Bank development committee reports	01295
World Bank discussion papers	01296
World Bank economic review	01297
The World Bank Group and Switzerland	32089

World Bank policy papers	01298
World Bank regional and sectoral series	01299
World Bank research observer	01300
World Bank research program: abstracts of current studies	01301
World Bank staff occasional papers	01302
World Bank symposium series	01303
World Bank technical papers	01304
World commodity forecasts	01305
World competiveness report	01306
World copper databook 2nd edition	01307
World debt tables: country tables 1970-79	01308
World debt tables: external finance for developing countries	01309
World department store index	01310
World development report	01311
World directory: lead and zinc mines and primary metallurgical works	01312
World directory: secondary lead plants	01313
World directory: secondary zinc plants	01314
World economic factbook	01315
World economic outlook; a survey by the staff of the International Monetary Fund	01316
World economic prospects	01317
World economic survey: current trends and policies in the world economy	01318
The world economy	01319
World electronics companies file	01320
World grain statistics	01321
World health statistics annual	01322
World health statistics quarterly	01323
World metal statistics	01324
World mineral statistics 1985-89	01325
World nickel statistics	01326
World outlook	01327
World population projections, 199-: estimates and projection with related demographic statistics	01328
World population prospects - population studies	01329
World retail directory and sourcebook	01330
World sales report - anti-piracy report	01331
World stainless steel statistics 1992	01332
World statistics in brief	01333
World tables	01334
World tin statistics	01335
World trade - stainless, high speed and other alloy steel	01336
World trade: stainless, high speed	

and other alloy steel	01337
World trade steel	01338
World trade steel	01339
The worlds women 1970-1990	31146
WSI informationsdienst arbeit	15233
WSI Mitteilungen	15234

X

XIII Magazine	03177

Y

Yearbook	31147
Yearbook of agricultural statistics	31148
Yearbook of construction statistics	27064
Yearbook of economic statistics	18041
Yearbook of education statistics	31149
Yearbook of farm statistics	13100
Yearbook of housing and building statistics	31150
Yearbook of industrial statistics: volume 1	13101
Yearbook of industrial statistics: volume 2	13102
Yearbook of investment and fixed assets	27065
Yearbook of labour statistics	27066
Yearbook of Nordic statistics	19023
Yearbook of Nordic Statistics	12055
Yearbook of public health statistics	27067
Yearbook of tourism	18042
Yearbook of tourism statistics	01340
Yearbook of transport statistics	27068
Yearbook of transport statistics for Finland	13103
Yearly statistics on regional employment	31151
Your trading partners in Czechoslovakia	11025

Z

Zahlenkomass	15235
Zeitschrift für die gesamte versicherungswissenschaft	15236
Zeitschrift für wirtschafts - und sozialwissenschaften	15237
Zivilluftfart in Oesterreich	05110

Index 2
SUBJECT INDEX

Index 2a
SOURCES LISTED BY SUBJECT AND COUNTRY

ABSENTEEISM
Sweden   31122

ACCIDENTS
Austria   05104
Germany   15054
Sweden   31092

ACCOUNTING
Austria   05055
Ireland   20001

ACCOUNTS
Germany   15154

ADVERTISING
Europe   02066
Ireland   20016
United Kingdom   34008 34009 34016
                    34105 34120 34132
                    34138 34204 34222
                    34224 34273 34277
                    34290

AEROSPACE INDUSTRY
Malta   24023

AGRICULTURAL INDUSTRY
Albania   04002
Austria   05039 05056 05063
                    05071 05075 05083
                    05089 05090
Belgium   07067 07078
Bulgaria   08002 08004
C.I.S.   09021
Cyprus   10001
Czech/Slovak Rep   11011
Denmark   12001 12030 12050
Europe   01240 02001 02002
European Union   03004 03005 03006
                    03007 03008 03009
                    03010 03011 03013
                    03037 03045 03115
                    03148
Finland   13025 13028 13066
                    13100
France   14001 14003 14005
                    14006 14009 14011
                    14013 14014 14017
                    14020 14045 14060
                    14061 14072 14089
                    14096
Germany   15005 15010 15016
                    15023 15105 15132
                    15167 15183 15200
                    15207 15210
Greece   17001 17031
Hungary   18038
International   01002 01003 01067
                    01101 01102 01136
                    01215 01240 01248
                    01250 01292 01321
                    05056
Ireland   20003 20004 20005
                    20017 20030 20035
                    20040 20060 20065
                    20085 20086 20091
                    20102 20114 20119
Italy   21029 21053 21057
Malta   24009
Netherlands   25001 25009 25034
                    25039
Norway   26001
Poland   27001 27002 27055
                    27056
Portugal   28020 28021 28068
Romania   29016
Spain   30003 30004 30015
                    30050
Sweden   31002 31003 31028
                    31038 31076 31095
                    31148
Switzerland   32044 32045 32046
                    32065 32075
Turkey   33040
United Kingdom   34010 34011 34012
                    34027 34110 34125
                    34127 34128 34129
                    34130 34144 34156
                    34179 34248 34317
                    34355

AGRICULTURAL MACHINERY INDUSTRY
Austria   05064

AID
Bulgaria   02025 08009
C.I.S.   02025 02045
Czechoslovakia   02025
Europe   02025 02045 02070
                    03075
European Union   02025 02045 02070
                    03036 03075 03076
                    03157
Hungary   02025
Poland   02025
Romania   02025
United Kingdom   34036

AIR POLLUTION
Turkey   33007

AIR TRANSPORT
Austria   05110
Belgium   07048
Finland   13002
Germany   15170
Gibraltar   16004
International   01028 01059
Sweden   31147
Turkey   33010
United Kingdom   34334 34335 34336
                    34337

AIRLINES
Austria   05110

ALCOHOLIC DRINKS INDUSTRY
Sweden   31004

ALCOHOLISM
Sweden   31004

ARMED FORCES
International   01217

ASSETS
Poland   27065

AUTOMATION
Switzerland   32083

BALANCE OF PAYMENTS
Denmark   12037 12038
European Union   17021
Finland   12038 13035
France   14078
Greece   17021 17022
Iceland   12038
International   01016
Ireland   20011
Italy   21003 21005 21062
Norway   12038
Sweden   12038
United Kingdom   34338

BANKING
Austria   05009 05013 05045
                    05067 05098
Belgium   07011 07066
C.I.S.   09001
Cyprus   10006
Denmark   12019
Finland   13001 13003 13004
                    13005 13006 13007
                    13009 13036 13061
                    13079 13081 13082
                    13096
France   14032 14113 14123
                    14124 14139
Germany   15009 15052 15066
                    15165 15177 15203
Greece   17003 17012 17019
                    17028
Iceland   19002 19014
International   01017 01121 01290
                    01301 01308 01309
Ireland   20007 20012 20013
                    20021 20061 20063
                    20093
Italy   21010 21044 21064
Liechtenstein   22002 22012
Luxembourg   23003 23008
Malta   24002 24022
Netherlands   25003 25005

Norway   26003 26013 26019
Portugal   28072
Spain   30008 30009 30052
Sweden   31007 31008 31009
                    31011 31015 31016
                    31019 31103 31115
                    31124
Switzerland   32004 32006 32007
                    32008 32013 32024
                    32056 32061 32069
                    32089
Turkey   33002 33003 33025
                    33036 33041
United Kingdom   34025

BANKRUPTCY
Finland   13008
Sweden   31014

BARTERING
International   01041 01062

BIOTECHNOLOGY
Europe   02089
European Union   15193
Germany   15193

BOND MARKETS
Finland   13037
Luxembourg   23015 23016 23024

BONDS
Luxembourg   23004

BOOKKEEPING
Austria   05055

BREWING INDUSTRY
Germany   15003 15040 15142
                    15159

BUDGET
Ireland   20014
Sweden   31129
Turkey   33004

BUDGETARY CONTROL
European Union   03029 03072 03117

BUILDING MATERIALS
United Kingdom   34180 34235 34314

BUILDING SOCIETIES
Europe   02098
International   01169
Ireland   20012
United Kingdom   34049 34050 34051
                    34052 34074 34078
                    34159 34160 34229
                    34230 34231

BUSINESS ECONOMICS
Belgium   07033

BUSINESS ENVIRONMENT
Albania   04004 04007
Austria   05032 05057
Baltic States   06007
Belgium   07031 07032 07042
Bulgaria   04004 04007 08001
                    08014
C.I.S.   09003 09004 09016
                    09017 09018 09019
                    09024 09025 09026
                    09028 09032 09033
                    09038 09039
Cyprus   10019
Czech/Slovak Rep   04007 11013 11014
                    11015 11016 11021
                    11023
Denmark   12016 12028
Europe   02015 02019 02020
                    02035 02036 02037
                    02042 02100 02101
European Union   03022 03042 03092
Finland   13060
France   14043 14044 14073
Germany   15058 15059 15098
                    15104 15115 15126
Greece   17025
Hungary   04007 18003 18009
                    18010 18011 18012
                    18021 18023 18031

International	01045	01064	01170
	01257		
Ireland	20031	20058	
Italy	21015	21028	
Liechtenstein	22006		
Luxembourg	23020		
Malta	24013		
Netherlands	25016	25032	
Norway	26012	26028	
Poland	04007	27015	27016
	27017	27018	27019
	27038	27051	
Portugal	28006	28052	28065
Romania	04007	29005	29006
	29008	29011	
Spain	30027	30028	30053
Sweden	31030	31068	
Switzerland	32025	32026	32042
Turkey	33016	33017	33029
	33042	33066	
United Kingdom	34100	34101	34102
	34103	34172	
Yugoslavia	35006		

BUSINESS SERVICES INDUSTRY
| Finland | 13032 |

BUYING
| Austria | 05084 |

CAPITAL MARKETS
| France | 14142 |
| Switzerland | 32077 |

CAR INDUSTRY
| Austria | 05014 |
| Belgium | 07008 |

CARS
Belgium	07056	07094
Spain	30011	
Switzerland	32028	

CEREALS INDUSTRY
France	14005		
International	01248	01250	01321
Italy	21063		
Spain	30050		

CHAMBERS OF COMMERCE
Austria	05095	
Finland	13062	
Germany	15044	
Ireland	20055	
Switzerland	32010	32066

CHARITIES
International	01025	01167	34071
Ireland	20100		
United Kingdom	34006	34071	34150

CHEMICAL INDUSTRY
Europe	02012	02038	02039
	02094		
Germany	15014	15135	
International	01006	01018	01026
	01027	01072	
United Kingdom	34058	34063	34113
	34173	34181	34339

CIVIL SERVICE
| United Kingdom | 34072 |

CLERICAL WORKERS
| United Kingdom | 34073 |

CLOTHING INDUSTRY
Finland	13013
Germany	15134
Switzerland	32080

CO-OPERATIVE SOCIETIES
| United Kingdom | 34076 |

CO-OPERATIVES
| Denmark | 12001 |

COAL MINING INDUSTRY
Europe	02003	
European Union	03027	03105
International	01029	
United Kingdom	34038	

COCOA INDUSTRY
Czech/Slovak Rep	01030		
International	01008	01030	01031
	01244		

COMMODITIES
Germany	15079		
International	01032	01033	01073
	01238	01293	01305

COMMODITY MARKETS
| International | 01204 | 01284 |

COMMUNICATIONS
| Poland | 27068 |

COMMUNICATIONS INDUSTRY
| European Union | 03028 |
| Greece | 17052 |

COMMUNITY LAW
| European Union | 15051 | 15073 |
| Germany | 15051 | 15073 |

COMPANIES
Austria	05040	05067	05094
Belgium	07047	07089	07092
Bulgaria	08006		
C.I.S.	09030	09035	09036
	09037		
Cyprus	10039		
Czech/Slovak Rep	11010	11012	11024
	11025		
Denmark	12015	12032	12033
Europe	02064	02077	
Finland	13017	13026	13065
	13087		
France	14039	14053	14054
	14056	14079	14080
	14088	14100	14109
	14134	14144	
Germany	15001	15004	15086
	15218		
Hungary	18002	18004	18020
International	01055		
Italy	21016	21022	21032
	21034	21035	21040
Liechtenstein	22005		
Luxembourg	07089		
Netherlands	25006	25017	25018
	25025	25037	25073
	25078		
Norway	26040		
Poland	27004	27023	27029
	27041		
Portugal	28009	28029	28071
Spain	30064		
Sweden	31037	31045	31049
	31075	31080	31090
	31124	31128	
Switzerland	32029	32054	32064
	32067	32076	
Turkey	33008	33011	33057
	33073		
United Kingdom	15086	15218	34175
	34323		

COMPANY FORMATION
| Liechtenstein | 22020 |
| Malta | 24005 |

COMPANY LAW
| Liechtenstein | 22005 | 22011 | 22015 |
| Malta | 24005 | | |

COMPANY PERFORMANCE
Italy	21016	21022	21032
Switzerland	32054		
United Kingdom	34067	34135	34145
	34165	34167	34176
	34256	34349	

COMPANY RANKING
United Kingdom	34029	34031	34032
	34033	34034	34035
	34067	34146	34165
	34176	34299	34349

COMPANY TAXATION
| Austria | 05047 |

COMPETITIVENESS
| Europe | 03154 |

COMPETITIVENESS (cont.)
| European Union | 03154 | | |
| International | 01037 | 01083 | 01306 |

COMPUTER INDUSTRY
International	01153	
Italy	21042	
Sweden	31033	
Turkey	33058	
United Kingdom	34090	34251

CONCRETE INDUSTRY
| Belgium | 07018 | 07061 |

CONFECTIONERY TRADE
| Austria | 05018 |

CONFERENCES
| Hungary | 18013 |

CONSTRUCTION
| Finland | 13010 | 13053 |

CONSTRUCTION INDUSTRY
Austria	05010	05109	
Belgium	07022	07040	07080
Cyprus	10011		
Denmark	12004	12024	
Europe	02007	02109	
European Union	03118		
Finland	13014		
France	14063	14140	
Germany	15017	15022	15042
	15106	15116	15122
	15159	15160	15172
Greece	17024		
International	01038		
Ireland	20010	20018	20024
	20045	20046	20089
	20092	20094	
Italy	21023	21054	
Liechtenstein	22003		
Netherlands	25029	25040	
Norway	26007		
Poland	27033	27040	27064
Portugal	28022	28041	
Spain	30016	30046	
Sweden	31059	31150	
Switzerland	32018	32050	32051
	32053		
Turkey	33012	33071	
United Kingdom	34056	34157	34158
	34164	34247	

CONSULTANCY
| Turkey | 33059 |

CONSUMER BEHAVIOUR
European Union	03033	
International	01039	
United Kingdom	34044	34236

CONSUMER CREDIT
Ireland	20043	
Netherlands	25072	
Spain	30029	30041

CONSUMER DURABLES
Europe	02027	02059	
International	01011		
Netherlands	25011		
United Kingdom	34083	34182	34209
	34237		

CONSUMER ECONOMICS
France	14130
Luxembourg	23013
Switzerland	32048

CONSUMER EXPENDITURE
| Finland | 13015 | |
| Sweden | 31020 | 31042 |

CONSUMER GOODS
Europe	02050	02090
Germany	15076	15159
International	01012	
Netherlands	25010	
United Kingdom	15076	34114

CONSUMER MARKETS
| Spain | 30014 | 30064 |

CONSUMER PRICE INDEX
| European Union | 03035 |

Finland	13016		
Germany	15184		
Ireland	20025		
Netherlands	25004		
Portugal	28043		
Spain	30048		
Sweden	31021	31022	
Switzerland	32048		
Turkey	33074		

CONSUMER PROTECTION
European Union	03033

CONSUMER SPENDING
European Union	03082
United Kingdom	34075

CONSUMPTION
Denmark	12050
Finland	13054
Netherlands	25068

COPORATION TAX
Germany	15155

COPPER INDUSTRY
International	01307 01332

CORPORATION TAX
Greece	17049

COSMETICS INDUSTRY
Europe	02028	02062	
United Kingdom	34084	34183	34210

COST OF LIVING
United Kingdom	34077	34203	34278
	34298		

COTTON INDUSTRY
Italy	21026	21055	21056

CREDIT
Austria	05044	
Denmark	12040	
France	14054	14104
Italy	21005	21062
Sweden	31027	31138

CRIME
Cyprus	10014

CULTURE
Bulgaria	08012	08029
Denmark	12050	
Netherlands	25070	
Poland	27062	
Sweden	31005	31029

CURRENCY
Italy	21006
Luxembourg	23023

CURRENCY MARKETS
International	01159 01191

CURRENCY OPTIONS
Switzerland	32034

DAIRY INDUSTRY
France	14007	
Germany	15055	15173
International	01040	
Ireland	20091	

DATA BANKS
European Union	03079

DEBT
Europe	01277		
International	01110	01209	01277
	01308	01309	

DEFENCE
Europe	02091
United Kingdom	34092

DEMOCRACY
C.I.S.	09006

DEMOGRAPHY
Austria	05023	05049	05069
	05093		
Baltic States	06004		
Bulgaria	08023	08024	
C.I.S.	09005		

Cyprus	10018		
Denmark	12003		
European Union	03038	03146	
Germany	15035	15036	15080
Greece	17029		
International	01285		
Ireland	20019	20072	20090
Italy	21059		
Liechtenstein	22007		
Malta	24001	24012	
Norway	26046	26047	26048
Portugal	28030	28069	
Spain	30018		
Sweden	31018	31098	
Turkey	33009	33062	

DEVELOPING COUNTRIES
European Union	03044

DEVELOPMENT
Bulgaria	02025	08009	
C.I.S.	02025	02045	
Czechoslovakia	02025		
Europe	02025	02045	02069
	02070	03074	03075
European Union	02025	02045	02069
	02070	03036	03074
	03075	03076	03157
Hungary	02025		
International	01034	01035	01058
	01104	01138	01160
	01303		
Ireland	20047		
Poland	02025		
Romania	02025		

DEVELOPMENT ECONOMICS
C.I.S.	01270		
Europe	01242		
International	01069	01097	01110
	01112	01163	01188
	01233	01234	01242
	01270	01278	01279
	01287	01290	01295
	01296	01298	01299
	01302	01304	

DISTRIBUTION
Austria	05067
Netherlands	25041
United Kingdom	34066

DISTRIBUTION OF WEALTH
Sweden	31061

DIY
Germany	15085

DOMESTIC TRADE
Bulgaria	08015
Italy	21050

DRINKS INDUSTRY
Europe	02029	02073	02093
International	01098		
United Kingdom	34039	34085	34105
	34187	34189	34211
	34238		

ECONOMIC CONDITIONS
Albania	04001	04002	04003
	04005	04007	08010
	29003		
Austria	05007	05008	05011
	05013	05016	05020
	05021	05022	05032
	05036	05053	05057
	05076	05077	05081
	05096	05097	05100
Baltic States	06001	06003	06004
	06005	06006	06008
	06009	06010	06012
	15179		
Belgium	07002	07004	07005
	07007	07009	07011
	07012	07014	07015
	07020	07021	07024
	07025	07026	07027
	07028	07029	07031
	07032	07036	07038
	07039	07042	07045
	07050	07054	07055

	07066	07074	07091
	23005	23012	23021
	23028		
Bulgaria	04003	04005	04007
	08001	08003	08005
	08007	08008	08010
	08011	08013	08014
	08015	08016	08017
	08021	08028	08029
	08031	29003	29012
C.I.S.	01270	01328	09001
	09002	09004	09005
	09007	09008	09009
	09010	09011	09012
	09013	09014	09015
	09016	09017	09020
	09022	09023	09025
	09028	09031	09032
	09038	09040	29012
Cyprus	10002	10003	10006
	10007	10008	10009
	10012	10013	10016
	10017	10019	10020
	10021	10022	10023
	10025	10027	10035
	10036	10037	10044
	33020		
Czech/Slovak Rep	04007	11001	11002
	11003	11004	11005
	11007	11008	11009
	11014	11015	11017
	11018	11020	11021
	18001		
Czechoslovakia	01223		
Denmark	01057	12008	12009
	12010	12012	12016
	12021	12022	12023
	12028	12035	12043
	12045	12049	12050
	12052	12054	
Europe	01068	01216	01277
	01291	01294	02010
	02016	02017	02019
	02021	02024	02034
	02035	02040	02041
	02044	02047	02048
	02052	02054	02055
	02056	02068	02076
	02078	02085	02087
	02100	02101	02110
	02120	15166	15179
	15219	15225	
European Union	03002	03017	03020
	03022	03031	03034
	03041	03042	03046
	03069	03070	03071
	03077	03080	03081
	03082	03086	03088
	03092	03093	03094
	03102	03109	03112
	03113	03119	03120
	03128	03130	03135
	03136	03137	03138
	03139	03142	03149
	03150	03156	03175
	05076	15065	15074
	15108	15111	15196
	17013		
Finland	12045	13011	13012
	13019	13020	13021
	13022	13027	13032
	13034	13038	13039
	13045	13048	13049
	13060	13061	13073
	13086	13087	13088
	13096	13101	13102
France	14018	14022	14023
	14025	14029	14030
	14032	14034	14035
	14036	14037	14040
	14043	14044	14046
	14048	14049	14050
	14051	14057	14062
	14065	14066	14068
	14073	14074	14077
	14091	14093	14108
	14111	14114	14115
France *(continued)*	14117	14120	14122

	14125	14128	14145
	14147		
Germany	15036	15047	15048
	15049	15057	15058
	15059	15061	15063
	15064	15065	15074
	15077	15078	15083
	15084	15090	15094
	15098	15099	15100
	15101	15102	15103
	15104	15108	15111
	15113	15115	15120
	15125	15126	15128
	15146	15147	15148
	15149	15150	15151
	15152	15153	15165
	15166	15168	15175
	15176	15177	15178
	15179	15194	15196
	15203	15205	15206
	15208	15209	15219
	15225	15226	15228
	15229	15230	15231
	15232	15233	15234
	15235	15237	34143
Gibraltar	16001	16002	16005
Greece	17003	17006	17008
	17009	17010	17011
	17013	17014	17015
	17016	17017	17018
	17020	17022	17023
	17025	17027	17048
Hungary	01223	04007	18001
	18005	18006	18007
	18008	18009	18011
	18014	18015	18016
	18017	18018	18019
	18021	18022	18023
	18025	18026	18027
	18029	18030	18031
	18035	18037	18039
	18040	18041	
Iceland	12010	19001	19002
	19003	19004	19005
	19006	19007	19008
	19009	19011	19012
	19013	19014	19016
	19017	19018	19020
	19022	19023	
International	01001	01020	01022
	01023	01034	01035
	01042	01043	01044
	01046	01047	01048
	01049	01050	01051
	01052	01057	01058
	01063	01064	01068
	01069	01070	01077
	01078	01079	01080
	01083	01099	01100
	01106	01108	01119
	01131	01139	01140
	01141	01146	01157
	01159	01161	01162
	01168	01170	01174
	01192	01193	01203
	01209	01211	01212
	01213	01214	01216
	01217	01220	01221
	01222	01223	01228
	01246	01257	01259
	01260	01264	01265
	01270	01272	01277
	01282	01283	01286
	01287	01290	01291
	01294	01296	01297
	01299	01300	01301
	01302	01303	01304
	01306	01308	01309
	01311	01315	01316
	01317	01318	01319
	01327	01328	01333
	01334	15057	15064
	15074	15146	15147
	15148	15149	15150
	15152	15228	
Ireland	20002	20011	20015
	20021	20022	20023

	20027	20028	20029
	20031	20032	20033
	20034	20042	20044
	20047	20052	20055
	20056	20058	20059
	20063	20066	20067
	20070	20078	20079
	20080	20084	20087
	20088	20093	20096
	20098	20103	20104
	20106	20107	20108
Italy	21002	21003	21004
	21006	21008	21009
	21010	21012	21013
	21014	21015	21017
	21018	21019	21027
	21028	21030	21031
	21033	21037	21041
	21043	21044	21045
	21064		
Liechtenstein	22006	22013	22016
	22017	22019	22023
Luxembourg	07026	07028	07055
	23002	23003	23005
	23008	23009	23011
	23012	23014	23017
	23018	23019	23020
	23021	23022	23025
	23027	23028	23030
	23035		
Malta	10012	24001	24013
	24014	24015	24019
	24021	24022	24026
Netherlands	25003	25007	25013
	25014	25015	25016
	25019	25020	25021
	25026	25027	25028
	25032	25042	25046
	25047	25053	25054
	25055	25063	25066
	25067	25069	25078
Norway	12045	26009	26010
	26011	26012	26013
	26014	26015	26020
	26024	26025	26027
	26028	26035	26036
	26043	26053	26054
	26057		
Poland	01223	04007	18001
	27009	27010	27011
	27012	27013	27015
	27017	27018	27022
	27023	27024	27025
	27030	27031	27032
	27036	27037	27038
	27044	27047	27049
	27050	27051	27052
	27058	27059	27060
Portugal	28004	28007	28008
	28014	28015	28016
	28017	28019	28037
	28040	28052	28055
	28059	28061	28062
	28063	28065	28075
Romania	04003	04005	04007
	08010	29001	29002
	29003	29004	29005
	29007	29008	29009
	29011	29012	29013
	29014	29015	29017
Spain	30001	30002	30008
	30009	30012	30013
	30014	30021	30023
	30024	30025	30026
	30027	30028	30030
	30031	30038	30040
	30042	30043	30045
	30047	30051	30052
	30053	30060	30062
	30064	30065	30066
Sweden	12045	31006	31011
	31024	31025	31026
	31030	31040	31041
	31047	31055	31058
	31063	31068	31070
	31083	31111	31112
	31117	31118	31119
	31125	31126	31127

	31130	31134	
Switzerland	32002	32004	32006
	32007	32010	32011
	32013	32020	32021
	32022	32023	32025
	32026	32033	32036
	32038	32040	32041
	32042	32043	32056
	32057	32058	32061
	32063	32066	32068
	32074	32082	32088
Turkey	33013	33014	33015
	33016	33017	33018
	33019	33020	33021
	33029	33034	33035
	33042	33044	33045
	33048	33056	33060
	33061	33066	33068
United Kingdom	15083	34002	34015
	34022	34023	34024
	34030	34069	34070
	34076	34080	34081
	34082	34096	34097
	34099	34100	34101
	34102	34103	34106
	34107	34108	34109
	34111	34112	34121
	34133	34134	34136
	34143	34166	34169
	34172	34177	34198
	34199	34200	34227
	34249	34250	34284
	34300	34301	34309
	34311	34338	34344
	34356		
Yugoslavia	04005	29012	35001
	35002	35003	35004
	35005	35007	35008
	35009	35010	35011
	35014	35015	

ECONOMIC FORECASTING

Belgium	07038		
C.I.S.	09007		
Czech/Slovak Rep	11008		
France	14059		
Hungary	18015		
Poland	27030		
Portugal	28039		
Spain	30044		
Switzerland	32037		
Yugoslavia	35012		

ECONOMIC INDICATORS

International	01130		
Italy	21009		
Portugal	28011		
Switzerland	32027	32088	
Turkey	33047		
United Kingdom	34139		

ECONOMIC INTEGRATION

European Union	03019	03032	03047
	03068	03073	03101
	03116	03128	03144
	03155	03161	03170

ECONOMIC PERFORMANCE

Denmark	12039	12051	12052
Finland	13039		
Poland	27044		
Switzerland	32040		
United Kingdom	34133		

ECONOMIC PLANNING

Netherlands	25008	
United Kingdom	34133	

ECONOMIC POLICY

Cyprus	10027	
Czech/Slovak Rep	11006	

ECONOMICS

International	01188	

ECU

Luxembourg	23015	23016

EDUCATION

Cyprus	10024		
Denmark	12038	12050	12055
Finland	12038	12055	

Germany 15037
Iceland 12038
International 01076 01285
Netherlands 25057
Norway 12038 12055
Sweden 12038 12055 31031
31137 31149

EEC
Greece 17018

ELECTRICITY INDUSTRY
Austria 05108
Belgium 07017 07037
Denmark 12017
France 14064 14082
Germany 15138
Norway 26016
Sweden 31032
United Kingdom 34340

ELECTRICITY SUPPLY INDUSTRY
European Union 03052
International 01081 01227

ELECTRONICS INDUSTRY
Austria 05035
Europe 02014 02057 02092
02111
France 14064 14082
Germany 15138
International 01013 01082 01320
Sweden 31033
Switzerland 32083
United Kingdom 34184

EMPLOYEES
Austria 05031 05063

EMPLOYERS' ORGANIZATIONS
Sweden 31107 31109

EMPLOYMENT
Austria 05004 05031 05069
Belgium 07049
Cyprus 10033 10040 10049
European Union 03053 03054
France 14027 14041 14052
14055 14071 14080
14148
Germany 15168 15188
Greece 17036
International 01150 01185
Ireland 20012 20046 20048
20049 20050 20051
20071 20073 20074
20075 20076 20120
Italy 21036
Luxembourg 23002 23035
Malta 24003
Poland 27020 27021 27066
Portugal 28033 28046
Spain 30033
Sweden 31034 31071 31072
31107 31108 31109
31121 31139 31142
31143 31145 31151
Switzerland 32071
United Kingdom 34104 34313 34354

ENERGY
Austria 05012 05037 05038
05074 05103
Belgium 07010
Denmark 12017 12026
Europe 01066 01086 02003
02004 02006 02058
02089
European Union 03001 03018 03027
03052 03055 03056
03057 03058 03059
03060 03061 03065
03129 03143 03159
Finland 13023 13024
France 14024 14098 14146
Germany 15018 15027 15030
15107
Greece 17037
International 01066 01085 01086
01087 01088 01089
01090 01091 01092

01093 01094 01122
01247
Liechtenstein 22008
Norway 26017
Poland 27028
Romania 29007
Sweden 31032 31035 31036
31054 31062
United Kingdom 34057 34062 34095
34116

ENERGY CONSUMPTION
Portugal 28023

ENERGY INDUSTRY
Netherlands 25022

ENGINEERING INDUSTRY
Europe 02011
European Union 03062
France 14067
International 01009 01021
United Kingdom 34059 34064 34185
34239

ENGLAND
United Kingdom 34004

ENVIRONMENT
Austria 05103
Bulgaria 08018
Europe 01096 02089
European Union 03014 03063 03064
03107 03108 03166
Finland 13095
Germany 15221
International 01096 01097
01133 01289
Netherlands 25023 25024
Norway 26037
Sweden 31038 31089 31095
United Kingdom 34094 34339

ENVIRONMENTAL ECONOMICS
International 01141 01225

ENVIRONMENTAL PROTECTION
Austria 05005
Germany 15127

EQUAL OPPORTUNITIES
Sweden 31144

EQUALITY
Netherlands 25038

EXCHANGE RATES
European Union 03049
Iceland 19001
International 01129 01205 01288

EXHIBITIONS
United Kingdom 34341

EXPORTING
Austria 05021 05053
Belgium 07025 07039
Bulgaria 08021
C.I.S. 09008 09023
Czech/Slovak Rep 11002 11018 11019
11020 11022
Denmark 12009 12023
Finland 13020 13049
France 14036 14062
Germany 15042 15048 15075
15076 15085 15094
15096
Greece 17009 17023
Hungary 18006 18018 18030
Iceland 19004 19011
International 01048 01140
Ireland 20028 20042
Italy 21013 21019
Luxembourg 23011 23018
Netherlands 25014 25028
Norway 26010 26025 26027
Poland 27011 27032 27034
Portugal 28015 28040 28073
Romania 29002 29009
Spain 30024 30045
Sweden 31025 31058
Switzerland 32022 32038 32060
Turkey 33039 33056 33059

33065 33069 33070
33072
United Kingdom 15076

EXPORTS
Austria 05024 05067
Denmark 12038 12055
European Union 03012 17032
Finland 12038 12055
France 14016 14039 14105
14136
Germany 15075 15182
Greece 17032
Iceland 12038
Italy 21007 21047 21049
Norway 12038 12055 26018
26034
Portugal 28031
Spain 30003
Sweden 12038 12055 31039
31052 31084
Switzerland 32060 32070 32072
32078
Turkey 33039
United Kingdom 34123

FACTORIES
Spain 30017

FAMILY EXPENDITURE
Austria 05061 05092
European Union 03103
Finland 13051
Germany 15069 15093
Italy 21020 21051
Portugal 28048
Spain 30032
Sweden 31042
Switzerland 32012
United Kingdom 34124 34142

FARMS
Finland 13025 13028 13100
Switzerland 32044

FERTILIZERS
Germany 15062

FINANCE
Belgium 07085
Denmark 12055
European Union 03072
Finland 12055 13029
France 14026 14042 14101
14102 14112 14113
14117
Germany 15109 15165
Greece 17012 17019 17028
International 01104 01107 01142
01143 01144 01145
01164 01165 01166
Italy 21004
Luxembourg 23007 23008 23017
23025 23031
Malta 24016
Netherlands 25042
Norway 12055 26013
Poland 27022
Portugal 28055 28066
Spain 30051
Sweden 12055
Switzerland 32013 32024 32063
United Kingdom 34070 34132 34186
Yugoslavia 35011

FINANCE COMPANIES
Cyprus 10026

FINANCIAL INSTITUTIONS
Europe 02040
France 14104
Portugal 28036
United Kingdom 34070 34086

FINANCIAL MARKETS
Hungary 18024
United Kingdom 34134

FINANCIAL PERFORMANCE
France 14084 14086 14123
Italy 21002 21006 21045
Switzerland 32088

FINANCIAL POLICY
European Union　03135
Germany　15109

FINANCIAL STATEMENTS
Sweden　31037

FINANCIAL SYSTEMS
European Union　03048　03161

FISHING INDUSTRY
Europe　02084
European Union　03009　03011　03106
Germany　15095　15123　15133
Greece　17042
Iceland　19017
International　01253　15123
Italy　21057
Norway　26021　26022
Portugal　28024
Sweden　31010　31048　31133
Switzerland　32065
United Kingdom　34155　34178　34302
　　　　34304　34306　34328

FLOOR COVERINGS
Austria　05017

FOOD
International　01111
United Kingdom　34130　34156

FOOD AND DRINKS INDUSTRY
Belgium　07030
Switzerland　32083

FOOD INDUSTRY
Europe　02001　02030　02060
　　　　02074　02093　02105
　　　　02107
European Union　03121
France　14003　14006　14045
Germany　15142　15159　15207
　　　　15210
International　01098
Netherlands　25010
Switzerland　32075
United Kingdom　34019　34040　34087
　　　　34138　34170　34187
　　　　34212　34223　34240

FORECASTING
Albania　04003
Austria　05020　05022
Belgium　07024　07028　23012
Bulgaria　04003　08007　08011
C.I.S.　09013
Cyprus　10013　10022
Czech/Slovak Rep　11001　11003　11004
Denmark　12008　12010　12022
Europe　02016　02024　02110
European Union　03129
Finland　13019　13021　13039
　　　　13048
France　14034　14037
Germany　15047　15049
Greece　17008　17010　17011
Hungary　18005　18007　18008
Iceland　12010
International　01042　01043　01046
　　　　01051　01052　01070
　　　　01131　01133　01161
　　　　01205　01220　01228
　　　　01238　01305　01318
　　　　01327
Ireland　20027　20029　20032
Italy　21012　21014　21018
Luxembourg　07028　23012
Netherlands　25008　25013　25015
　　　　25020　25027　25036
　　　　25053
Norway　26009　26011　26024
Poland　27010　27012　27013
Portugal　28014　28016　28017
Romania　04003　29001　29004
Spain　30023　30025　30026
Sweden　31024　31026　31055
　　　　31137
Switzerland　32020　32023
Turkey　33013　33014　33015
United Kingdom　34008　34075　34080

　　　　34082　34097　34099
　　　　34106　34107　34111
　　　　34120　34139　34196
　　　　34217　34250　34260
　　　　34261
Yugoslavia　35001

FOREIGN EXCHANGE
Switzerland　32001　32034

FOREIGN INVESTMENT
France　14075
Poland　27025　27046
Sweden　31049

FOREIGN TRADE
Belgium　07093
Iceland　19016

FOREIGN WORKERS
Liechtenstein　22001

FOREST PRODUCTS
Italy　21060
Switzerland　32003

FORESTRY INDUSTRY
Austria　05063
C.I.S.　01116　01118
Europe　01273
European Union　03009　03011　03078
　　　　03108
Finland　13040
France　14012　14016　14060
Germany　15132　15183　15207
　　　　15210
International　01115　01116　01117
　　　　01118　01273
Italy　21060
Norway　26023
Sweden　31113
Switzerland　32003　32065

FOUNDRY INDUSTRY
Germany　15139

FREIGHT TRANSPORT
Germany　15162

FRUIT
Austria　05028
France　14004
Germany　15021
Portugal　28038
Spain　30068

FUNDS
Europe　02075
Luxembourg　23029

FURNITURE INDUSTRY
European Union　03021
Germany　15141
United Kingdom　34005　34013　34014
　　　　34054　34315

GAMBLING INDUSTRY
European Union　03110

GAS INDUSTRY
Austria　05079
Belgium　07003　07052　07064
Europe　02005
European Union　03111
International　01230
Norway　26042
Turkey　33026　33067

GEOGRAPHY
European Union　03068

GLASS INDUSTRY
Belgium　07016　07041　07086

GOLD
International　01134

GOVERNMENT
C.I.S.　09006　09040

GOVERNMENT CONTRACTING
European Union　03133

GOVERNMENT DEPARTMENTS
Czech/Slovak Rep　11024

GOVERNMENT EXPENDITURE
Germany　15192
Malta　24007　24008

GROSS NATIONAL PRODUCT
Poland　27031
Sweden　31056
Turkey　33027

HANDICAPPED WORKERS
European Union　03040

HEALTH
Austria　05068
Bulgaria　29012
C.I.S.　29012
Cyprus　10028　10049
Denmark　12018
Europe　02013　02062　02094
Greece　17045
International　01014　01132　01322
　　　　01323
Luxembourg　23002　23035
Romania　29012
Sweden　31057　31093
Turkey　33063
United Kingdom　34020　34147　34153
　　　　34210　34241
Yugoslavia　29012

HEALTH AND SAFETY
United Kingdom　34148

HEALTH ECONOMICS
Italy　21029
Portugal　28026

HEALTH SERVICE
Bulgaria　08020
Netherlands　25043　25077
Poland　27067
Portugal　28026

HEATING AND VENTILATING INDUSTRY
Europe　02018
United Kingdom　34047　34048

HIRE PURCHASE
Ireland　20043
Spain　30041

HORTICULTURAL INDUSTRY
Austria　05028　05042
Belgium　07067
Finland　13050
France　14004　14008
Netherlands　25076
Portugal　28038　28042
Spain　30068
United Kingdom　34027　34242

HOTEL AND CATERING INDUSTRY
Cyprus　10041
Germany　15025　15164
Spain　30057
Switzerland　32085　32086
Turkey　33054
United Kingdom　34042

HOURS OF WORK
Italy　21051

HOUSEHOLD ECONOMICS
Austria　05061　05092
Cyprus　10029
European Union　03103
Finland　13051
France　14069
Ireland　20044
Italy　21020　21051
Norway　26031
Portugal　28048
Spain　30032
Switzerland　32012
United Kingdom　34002　34124　34142

HOUSEHOLD GOODS
Europe　02059
United Kingdom　34213

HOUSING
Austria　05051　05104　05109
Belgium　07069　07080

Bulgaria   08022
Cyprus   10011
Europe   02007 02109
Finland   13052 13053
Germany   15085
Ireland   20006 20094
Poland   27033 27064
Portugal   28022
Spain   30016 30019
Sweden   31059 31096 31150
Switzerland   32050
United Kingdom   34157 34158 34247
                    34357

IMMIGRATION
Liechtenstein   22001

IMPORTING
Czech/Slovak Rep   11019 11022
Germany   15082 15216
Poland   27034

IMPORTS
Austria   05024
Denmark   12038 12055
European Union   03012
Finland   12038 12055
France   14016 14039 14105
                    14136
Germany   15067 15082 15182
                    15216
Iceland   12038
Italy   21007 21047 21049
Norway   12038 12055 26018
                    26034
Portugal   28031
Sweden   12038 12055 31039
                    31053 31085
Switzerland   32070 32072 32078
United Kingdom   34161 34162 34163

INCOME TAX
Austria   05034 05066
Germany   15068 15071 15169
Greece   17049 17050
United Kingdom   15071

INCOMES
Austria   05065 05072 05082
Denmark   12025 12050
Finland   13054 13055
France   14071
Netherlands   25068
Norway   26026
Poland   27054
Spain   30036
United Kingdom   34126 34127 34128

INDUSTRIAL DESIGN
Belgium   07035

INDUSTRIAL DEVELOPMENT
Poland   27048

INDUSTRIAL PERFORMANCE
Cyprus   10031
European Union   03120
International   01137 01150

INDUSTRIAL POLICY
Europe   01149 02099
International   01149 01152

INDUSTRIAL RELATIONS
Austria   05086
Belgium   07073
Denmark   12029
Italy   21051

INDUSTRIAL STRUCTURE
European Union   03145 03167 03168
                    03169
International   01152

INDUSTRIAL TRAINING
France   14141

INDUSTRIES
Austria   05004 05005 05054
                    05067 05080
Cyprus   10002 10025 10031
Europe   02041
European Union   03167

France   14021 14027 14028
                    14040 14055 14072
                    14075 14079 14087
                    14092 14094 14095
                    14101 14103 14133
                    14134 14135 14138
                    14144
Germany   15002 15110 15112
                    15127 15231
Greece   17002
International   01109 01137 01150
                    01152
Ireland   20002 20022 20034
                    20055 20059 20078
                    20080 20104 20115
Italy   21002 21061
Liechtenstein   22022
Malta   24004
Poland   27008 27035 27045
                    27062
Portugal   28035
Spain   30005 30040
Switzerland   32027 32076
Turkey   33066
United Kingdom   34055 34060

INFORMATION INDUSTRY
European Union   03177

INFORMATION RETRIEVAL
Denmark   12034

INFORMATION TECHNOLOGY
International   01153
Sweden   31064
Switzerland   32083
United Kingdom   34090 34206

INLAND WATERWAYS
European Union   03023
Finland   13083

INPUT-OUTPUT ANALYSIS
Germany   15124

INSURANCE
Austria   05044
Belgium   07013 07044
Cyprus   10026
European Union   03084
Finland   13041 13056 13057
                    13058
Germany   15054 15056 15236
International   01123 01154
Ireland   20057 20062
Liechtenstein   22004 22014
Spain   30006 30007 30055
Sweden   31065 31066 31067
                    31123 31131
United Kingdom   34068 34141 34205
                    34270 34271 34312
                    34343

INSURANCE COMPANIES
United Kingdom   34205 34343

INTEREST RATES
Finland   13059
France   14106 14107
International   01129 01155 01288
Switzerland   32034 32088

INTERNATIONAL COOPERATION
Germany   15091

INTERNATIONAL TRADE
Albania   04006
Austria   05006 05021 05024
                    05025 05026 05053
Belgium   07025 07039 07050
                    07081 07082 23021
                    23034
Bulgaria   08005 08008 08019
                    08021
C.I.S.   09002 09008 09023
                    09040
Cyprus   10015 10030 10045
Czech/Slovak Rep   11002 11018 11020
Denmark   12009 12023
Europe   02046 02051 02052
                    02053 02054 02055
                    02056 15213

European Union   03012 03044 03050
                    03095 03096 03097
                    03098 03099 03100
                    03124
Finland   13020 13046 13047
                    13049
France   14003 14016 14019
                    14031 14036 14039
                    14040 14062 14065
                    14096 14105 14110
                    14126 14132 14136
                    14137
Germany   15015 15019 15020
                    15048 15061 15067
                    15079 15092 15094
                    15213 34143
Greece   17005 17009 17023
Hungary   18006 18018 18030
Iceland   19004 19008 19009
                    19010 19011
International   01032 01033 01041
                    01054 01060 01061
                    01062 01079 01113
                    01114 01138 01151
                    01175 01181 01207
                    01232 01266 01267
                    01318 01321 34272
Ireland   20028 20038 20042
                    20064 20110 20118
Italy   21007 21013 21019
                    21024 21047 21049
                    21061
Luxembourg   23011 23014 23018
                    23021 23025 23034
Malta   24025
Netherlands   25002 25014 25028
                    25035 25049
Norway   26010 26018 26025
                    26027 26034 26039
Poland   27011 27026 27027
                    27032 27052 27053
Portugal   28002 28004 28010
                    28015 28040
Romania   29002 29009 29010
Spain   30033 30024 30045
Sweden   31025 31050 31051
                    31052 31053 31058
                    31086 31102 31127
Switzerland   32015 32022 32038
                    32070 32072 32078
Turkey   33023 33037 33045
                    33056 33059 33064
                    33065 33069 33070
                    33072
United Kingdom   34123 34143 34161
                    34162 34163 34253
                    34255 34272 34284
Yugoslavia   35013

INVESTMENT
Austria   05052
C.I.S.   09022
Europe   02076
European Union   03016 03051 03077
                    03085 03134
France   14080
Germany   15045 15218
International   01112 01163 01278
                    01279
Italy   21025
Luxembourg   23029
Netherlands   25025 25030
Poland   27039 27040
Portugal   28049 28076
Sweden   31069
Switzerland   32043
United Kingdom   15218 34173

INVESTMENT INCENTIVES
Hungary   18032 18033

INVESTMENT LAW
Austria   05052

ITALY
United Kingdom   34275

JEWELLERY INDUSTRY
Austria   05060
Belgium   07006

JOINT BUSINESS VENTURES
C.I.S. 09027 09030 09037
Europe 02046
Hungary 18028
Poland 27041

LABOUR
Austria 05086
Denmark 12027 12050
European Union 03131
Finland 13042
France 14131
Germany 15011 15043 15050
15087 15121 15176
International 01186 01245 31074
Ireland 20008 20072 20073
20074 20075 20076
20090
Netherlands 25033
Poland 27066
Sweden 31034 31067 31071
31072 31074 31106
31121 31122 31137
31145
Turkey 33030 33031 33051
United Kingdom 34115 34246 34322

LABOUR COSTS
Austria 05030
European Union 03130
Germany 15012
International 31141
Ireland 20070
Sweden 31141

LABOUR ECONOMICS
Italy 21051

LABOUR MARKET
Belgium 07073
Denmark 12002 12055
Finland 12055 13063 13064
Germany 15097
Iceland 19019
International 01224
Norway 12055 26030
Sweden 12055 31073
United Kingdom 34115

LABOUR RELATIONS
Denmark 12029
Ireland 20048
United Kingdom 34115

LABOUR STATISTICS
European Union 03132
Gibraltar 16006
United Kingdom 34194 34330

LABOUR SUPPLY
Germany 15013
Greece 17026
International 01245
Italy 21029 21046
United Kingdom 34115

LAND
Austria 05050
Germany 15032 15038

LAND OWNERSHIP
Germany 15032

LAND USE
Austria 05015
Germany 15039 15167

LAND VALUES
Germany 15144 15145

LAW
Austria 05052
Germany 15157
Spain 30030

LEAD INDUSTRY
International 01024 01071 01187
01202 01241 01249
01280 01312 01313

LEGISLATION
European Union 15060 15193
Germany 15060 15193

LEISURE
Netherlands 25071
Turkey 33046
United Kingdom 34188 34195 34196
34197 34314

LEISURE INDUSTRY
Europe 02061 02095
United Kingdom 34088 34214 34243
34327

LIBRARIES
Sweden 31005

LOCAL GOVERNMENT
Greece 17038
Sweden 31116
United Kingdom 34201 34303 34358

LOCAL GOVERNMENT FINANCE
Iceland 19015
Ireland 20077
Sweden 31060 31077
Turkey 33004 33022

LOCATION OF INDUSTRY
Cyprus 10039
Spain 30017

MACHINE TOOLS INDUSTRY
Turkey 33064
United Kingdom 34026 34207

MACRO-ECONOMICS
Netherlands 25036

MANAGEMENT
United Kingdom 34098 34295 34297

MANUFACTURING INDUSTRY
Austria 05046 05054 05067
Denmark 12026 12050
Finland 13073 13101 13102
Germany 15028 15029 15122
15188
Malta 24004 24017 24018
Netherlands 25007 25050
Norway 26033
Portugal 28050
Sweden 31075 31079 31080
31081 31082
Turkey 33001 33008 33032
United Kingdom 34061 34065

MARITIME TRANSPORT
Finland 13043
International 01175 01195 01254
Turkey 33049 33050
United Kingdom 34225 34265 34307
34353

MARKET RESEARCH
Europe 02022 02023 02026
02059 02060 02061
02062 02066 02071
02077 02078 02086
02097 02102 02103
02104 02105 02106
02115
European Union 03021 03062 03121
International 01045 01055 01082
01098 01171 01251
34289
Ireland 20066
United Kingdom 34019 34020 34021
34054 34055 34056
34057 34058 34059
34060 34061 34062
34063 34064 34065
34066 34083 34084
34085 34086 34087
34088 34089 34090
34144 34167 34170
34175 34179 34180
34181 34182 34183
34184 34185 34186
34187 34188 34189
34190 34191 34193
34195 34196 34197
34206 34209 34210
34211 34212 34213
34214 34215 34216
34217 34218 34219
34220 34223 34226
34235 34236 34237
34238 34239 34240
34241 34242 34243
34244 34251 34257
34259 34260 34261
34276 34288 34289
34291 34292 34293
34314 34323 34324
34347 34351

MARKETING
Germany 15128
International 01124

MEAT INDUSTRY
Denmark 12048
Germany 15198
International 01196 12048

MEDICAL CARE
Germany 15171

MEDICINE
Austria 05068
Germany 15171
Turkey 33063

MEN
Netherlands 25038

METALS INDUSTRY
Belgium 07034 07062 07063
07070 07071
European Union 03126 03127
Finland 13044
Germany 15136 15137
International 01024 01071 01103
01179 01184 01187
01198 01199 01200
01202 01241 01249
01261 01280 01281
01307 01312 01313
01314 01324 01326
01332 01335
United Kingdom 34058 34059 34063
34064

MIGRATION
Austria 05023
Cyprus 10046 10047
International 01201
Spain 30056 30058

MINERALS RESOURCES
Germany 15174
International 01134 01147 01148
01183 01325
United Kingdom 34350

MINING INDUSTRY
Belgium 07001 07076 07077
Cyprus 10005
Germany 15028 15029 15161
15188
Greece 17004
Turkey 33033 33065
United Kingdom 34058 34063

MONETARY ECONOMICS
Luxembourg 23017
Switzerland 32063

MONETARY SYSTEM
European Union 03047 03049 03104
Greece 17039
Luxembourg 23023

MONEY
Austria 05044
Denmark 12040
France 14026

MONEY MARKETS
C.I.S. 09034
Denmark 01218 12050
Europe 02043 15219
European Union 03161
Finland 13006 13030 13031
France 14106
Germany 15219
International 01105 01106 01218
01226 01229

Italy 21003 21005 21062
Netherlands 25042
Sweden 31006 31045 31046
31115
Switzerland 32001 32034
United Kingdom 34022 34023 34024

MONEY SUPPLY
France 14106 14107 14142

MORTGAGES
Europe 02098
International 01169
Spain 30039
United Kingdom 34049 34050 34051
34052 34074 34078
34159 34160 34229
34230 34231

MOTOR INDUSTRY
Austria 05014 05062 05070
Europe 02049 02067 02079
02088
Finland 13067
International 01004 01120 01172
01208
Ireland 20081 20082 20083
Liechtenstein 22021
Netherlands 25074
Sweden 31087
Turkey 33038
United Kingdom 34059 34064 34228
34232 34233 34234

MUSIC INDUSTRY
International 01331

NATIONAL ACCOUNTING
Belgium 07046
Denmark 12005 12014 12037
12040 12050
European Union 03112 03136 03137
03138 03139 03149
Finland 13068 13069
France 14066 14072 14126
Germany 15191 15192
Hungary 18035
International 01211 01212 01213
01214 01226 01246
01272
Ireland 20014 20111 20112
Italy 21001 21021
Malta 24016 24020
Netherlands 25051 25060
Poland 27044
Spain 30020
Sweden 31043 31044 31088
31094 31129
Switzerland 32017
United Kingdom 34344

NATIONAL DEBT
Denmark 12005 12014

NATIONAL INCOME
Austria 05078
Ireland 20087
Turkey 33027

NATIONAL INSURANCE
Austria 05066

NATURAL GAS
Turkey 33067

NEWSPAPER INDUSTRY
Ireland 20067

NORTHERN IRELAND
United Kingdom 34125 34248 34249
34281 34317

NUCLEAR POWER INDUSTRY
European Union 03065 03143
International 01089 01219 01227
01231

NUTRITION
Germany 15210

OCCUPATIONAL CHOICE
Greece 17030

OCCUPATIONAL DISEASES

Sweden 31092

OECD
Germany 15200

OFFICE AUTOMATION
Italy 21042

OFFICE EQUIPMENT INDUSTRY
United Kingdom 34215

OIL INDUSTRY
Austria 05059 05079
Europe 02031
International 01007 01230 01247
Norway 26042

OPINION POLLS
European Union 03066

OPTIONS
France 14143

ORGANIZATIONS
Baltic States 06007

OUTPUT
Germany 15187 15188

PACKAGING INDUSTRY
Europe 02072 02080 02096
02106 02112
International 01074 01125
Switzerland 32083
United Kingdom 34189 34259

PAPER AND PULP INDUSTRY
Belgium 07065 07072
Europe 02032 02096
International 01074 01243

PASSENGER TRANSPORT
Germany 15180

PASTA INDUSTRY
Italy 21063

PAY
Austria 05065 05066 05072
European Union 03043 03140
Finland 13098 13099
France 14071
Germany 15008 15010 15119
Gibraltar 16006
Ireland 20010 20050 20051
20114
Italy 21024 21036 21051
Norway 26026 26056
Poland 27020
Portugal 28033
Spain 30036
Sweden 31139 31140 31142
31143
Switzerland 32059
United Kingdom 34028 34073 34077
34098 34131 34203
34246 34258 34278
34285 34295 34297
34298 34322

PERSONAL FINANCE
United Kingdom 34216 34257

PERSONNEL
United Kingdom 34258

PETROCHEMICALS
Europe 02031

PETROLEUM INDUSTRY
Austria 05059
France 14076
International 01091 01182
Norway 26045
United Kingdom 34174 34346

PHARMACEUTICAL INDUSTRY
Europe 02013 02062 02094
International 01072 01126 01210
United Kingdom 34020

PLASTICS INDUSTRY
Germany 15140

POLITICS

Albania 04003
Austria 05020 05022
Belgium 07024 07028 23012
Bulgaria 04003 08007 08011
C.I.S. 09013
Cyprus 10013
Czech/Slovak Rep 11001 11003 11004
Denmark 12008 12010
Europe 02024 15219
European Union 03083 03087 15065
Finland 13019 13021
France 14034 14037
Germany 15047 15049 15065
15219
Greece 17008 17010 17011
17020
Hungary 18005 18007 18008
Iceland 12010
Ireland 20027 20029
Italy 21008 21012 21014
Luxembourg 07028 23012
Netherlands 25013 25015
Norway 26009 26011
Poland 27010 27012 27013
Portugal 28014 28016 28017
Romania 04003 29001 29004
Spain 30023 30025 30026
Sweden 31024 31026
Switzerland 32020 32023
Turkey 33013 33014 33015
United Kingdom 34080
Yugoslavia 35001

POPULATION
Austria 05023 05049 05069
05107
Baltic States 06004 06009
Belgium 07019 07051 07058
07059 07069 07084
Bulgaria 08023 08024
C.I.S. 01328 09005 09040
Cyprus 10018 10038
Denmark 12003 12038 12050
12055
Europe 01291
European Union 03038 03146
Finland 12038 12055 13071
France 14023 14042 14070
14118 14119 14127
Germany 15035 15080 15168
Gibraltar 16005
Greece 17022 17029
Iceland 12038 19021
International 01056 01186 01201
01235 01236 01237
01271 01291 01328
01329
Ireland 20019 20072 20090
20097 20101 20107
20108
Italy 21024 21027 21029
21036 21059
Liechtenstein 22007
Luxembourg 23002 23032 23035
Malta 24001 24012
Netherlands 25044 25058
Norway 12038 12055 26046
26047 26048
Poland 27003 27014 27054
Portugal 28030 28069
Spain 30002 30018 30034
30035 30038 30056
30058
Sweden 12038 12055 31061
31096 31097 31098
31137
Turkey 33009 33062
United Kingdom 34001 34002 34003
34124 34142 34192
34262 34263 34264
34280 34281 34282

PORTS
Belgium 07079
Cyprus 10004 10032 10034
Greece 17046
Ireland 20110
United Kingdom 34265

POSTAL SERVICE

Finland 13072
Sweden 31012 31099

POULTRY INDUSTRY
France 14015
Germany 15072

PRICE INDICES
European Union 03034
Spain 30049
Switzerland 32047 32048

PRICES
Austria 05056
Denmark 12025 12050
Europe 01239
Finland 13010 13074 13075
Germany 15117 15118 15176
 15181 15182 15183
 15184 15185 15186
International 01239 05056
Ireland 20121
Italy 21024 21036
Netherlands 25004 25045 25059
Poland 27005 27006 27043
Sweden 31042
Switzerland 32059
Turkey 33040 33074 33075
United Kingdom 34007

PRINTING INDUSTRY
Europe 02081
United Kingdom 34345

PRIVATIZATION
C.I.S. 09029

PRODUCTION
Austria 05001 05012 05046
 05054 05071 05080
Belgium 07049 07075 07087
European Union 03119
Finland 13097
France 14028 14042 14055
 14056 14088
Germany 15121 15153 15161
 15187 15188
Ireland 20049 20050 20051
 20052 20053 20054
 20070 20080
Italy 21024 21036 21061
Malta 24010
Netherlands 25031
Poland 27008 27031
Portugal 28044 28055
Spain 30037 30059
Turkey 33028

PRODUCTS
Germany 15185

PROPERTY
Austria 05050
Germany 15032 15033 15034
Hungary 18036
International 01127
Ireland 20092
Norway 26026
Poland 27065
Sweden 31013 31100 31104
United Kingdom 34004 34091 34154
 34266

PROPERTY TAX
Austria 05105

PUBLIC ACCOUNTING
Switzerland 32032

PUBLIC ADMINISTRATION
Austria 05043
Greece 17040

PUBLIC EXPENDITURE
Austria 05043
Germany 15191
Ireland 20036 20037 20041
 20099 20111 20112
 20113

PUBLIC FINANCE
Baltic States 06011
Belgium 07046 07049

Denmark 12037
Finland 13076 13096
Germany 15061 15195 15224
Greece 17022 17033 17047
International 01135
Italy 21001 21005 21021
 21062
Portugal 28028
Spain 30020
Sweden 31116
Switzerland 32017 32032
United Kingdom 34018 34133 34201
 34303 34321 34358

PUBLIC SECTOR
Germany 15195
Sweden 31132

PUBLIC SERVICES
Ireland 20036
Italy 21058
Poland 27007

PUBLIC TRANSPORT
Switzerland 32055

PUBLISHING INDUSTRY
Bulgaria 08025
United Kingdom 34190

PURCHASING
Austria 05084
Germany 15128

PURCHASING POWER
Austria 05076
European Union 05076
Switzerland 32059

R & D
Austria 05041
European Union 03114
Sweden 31105
United Kingdom 34017 34285

RADIOS
Belgium 07053

RAIL TRANSPORT
Belgium 07048
European Union 03024
Finland 13078
France 14085
Germany 15070
Turkey 33069

REAL ESTATE
Sweden 31100

RECREATION
Poland 27061
Turkey 33046

REGIONAL DEVELOPMENT
Austria 05011
Belgium 07015
European Union 03051 03086 03088
 03117 03147 03151
France 14081 14083 14086
 14094 14139
Germany 15036
Greece 17035
Luxembourg 23033
Portugal 28012
Spain 30021
United Kingdom 34121 34122

REGIONAL DIFFERENCES
Bulgaria 08026
Europe 02017
European Union 03147 03152 03153
 03168
Netherlands 25063
Sweden 31151
United Kingdom 34069 34249 34276
 34278 34280 34300
 34301

REGIONAL PLANNING
Austria 05090

REGIONAL POLICY
European Union 03175 15111
Germany 15111

Greece 17041

RESEARCH
International 01100 01301
Sweden 31105

RESOURCES
Norway 26037

RESTAURANTS
Turkey 33054

RETAIL PRICE INDEX
Turkey 33043

RETAIL PRICING
Germany 15163

RETAILING
Austria 05048
Belgium 07083
Cyprus 10042
Denmark 12013 12020 12041
Europe 02026 02050 02063
 02065 02086 02104
 02108 02114 02115
 02116
European Union 03089 03090 03121
Finland 13092
Germany 15024 15031
Hungary 18034
International 01251 01310 01330
 34289
Ireland 20103
Netherlands 25064
United Kingdom 34044 34089 34099
 34114 34170 34191
 34218 34219 34244
 34286 34287 34288
 34289 34290 34291
 34292 34293 34294
 34346 34347 34351

RISK ANALYSIS
International 01050 01051 01052
 01174
United Kingdom 34082

ROAD ACCIDENTS
Austria 05099
Europe 02117
International 01263
Italy 21048
Spain 30010
Sweden 31135
Turkey 33055

ROAD CONSTRUCTION
Germany 15212

ROAD HAULAGE INDUSTRY
European Union 03025

ROAD TRANSPORT
Austria 05087 05099
Belgium 07048 07068
European Union 03107
France 14085
Germany 15070
International 01263
Ireland 20105
Portugal 28051
Spain 30011
Switzerland 32016
Turkey 33055
United Kingdom 34053 34149 34171
 34296 34329 34331

ROADS
Germany 15212

RUBBER INDUSTRY
International 01255

RURAL INDUSTRIES
Austria 05071 05089 05090

SALES
Ireland 20103

SALESMEN
Germany 15156

SCIENCE
International   01194   01256   01285

SCIENCE POLICY
International   01019

SCOTLAND
United Kingdom   34004   34011   34110
                     34126   34262   34282
                     34299   34300   34301
                     34302   34303   34304
                     34305

SECURITIES
Italy   21025
Luxembourg   23001

SECURITIES MARKET
Norway   26041

SECURITIES MARKETS
Denmark   01218
International   01218
Norway   26006

SECURITY
Austria   05091
Netherlands   25061

SECURITY INDUSTRY
Europe   02091

SERVICE INDUSTRIES
Cyprus   10042
European Union   03124   03158   03164
Finland   13080
Ireland   20020
Netherlands   25041   25065
United Kingdom   34066   34294

SHARE OWNERSHIP
Sweden   31120   31124

SHARE PRICES
Denmark   12011   12036
Germany   15007
Italy   21038
Luxembourg   23010
Norway   26049

SHARES
Germany   15006
Italy   21038
Luxembourg   23004

SHIPBUILDING INDUSTRY
Greece   17043
International   01010   01189   01190
                     01197   01258

SHIPPING
Belgium   07079

SHIPPING INDUSTRY
Finland   13043   13070   13083
                     13084   13085
Germany   15131   15201
Greece   17044   17046
International   01065   01189   01190
Ireland   20110
Malta   24023
Norway   26051

SINGLE MARKET
Europe   03154
European Union   03031   03032   03033
                     03067   03068   03070
                     03073   03080   03083
                     03101   03102   03110
                     03122   03123   03133
                     03145   03150   03151
                     03154   03155   03161

SMALL BUSINESS
Germany   15125
Poland   27057
United Kingdom   34309

SOCIAL ECONOMICS
Austria   05093   05100
Baltic States   06006   06008   06010
                     06012
Belgium   07088
Bulgaria   08003   08029   08031
Cyprus   10044

Europe   02085
European Union   03017   03019   03020
                     03066   03069   03102
                     03113   03144   03162
France   14023   14042   14047
                     14048   14050   14068
                     14119   14122   14131
Germany   15233   15234
Iceland   19012   19022
International   01259   01265   01334
Ireland   20101
Italy   21004   21008   21027
                     21029
Liechtenstein   22018   22023
Luxembourg   23014   23025
Norway   26052
Portugal   28037
Romania   29014   29017
Spain   30002   30038
Sweden   31125
United Kingdom   34096   34177   34311

SOCIAL ENVIRONMENT
Belgium   07004   07005
Denmark   12049
European Union   03002   03003   03040
                     03141   03142   03162
International   01188

SOCIAL INDICATORS
Cyprus   10043
Gibraltar   16002
International   01036
Ireland   20084   20097   20106
                     20107   20108
Netherlands   25046   25069
Portugal   28064

SOCIAL POLICY
Germany   15011

SOCIAL SECURITY
Denmark   12038   12047   12050
                     12055
European Union   03030   03039   03163
Finland   12038   12055
France   14097   14131
Germany   15071
Iceland   12038
Norway   12038   12055
Sweden   12038   12055   31114
United Kingdom   15071   34007   34310

SOCIAL SERVICES
Cyprus   10010
United Kingdom   34147

SOCIAL STRUCTURE
Bulgaria   08029

SOCIAL SURVEYS
European Union   03146
United Kingdom   34311

SOCIAL WELFARE
Greece   17045
Portugal   28025
United Kingdom   34122

SPORTS INDUSTRY
Turkey   33046

STANDARD OF LIVING
Austria   05069   05093
Bulgaria   08027
Norway   26031
Poland   27042

STATISTICAL METHODS
European Union   03091   03093   03160
Netherlands   25067
United Kingdom   34316

STATISTICS
Denmark   12039
France   14038
Poland   27059
Sweden   31017   31091

STEEL INDUSTRY
Austria   05019   05085
Belgium   07023   07060

Denmark   12007   12042
Europe   02008   02082   02118
European Union   03105   03126   03127
Finland   13018   13033   13077
France   14033   14058   14121
Germany   15046   15136   15137
                     15189
Greece   17007   17034
International   01176   01177   01184
                     01261   01268   01269
                     01336   01337   01338
                     01339
Ireland   20026   20095
Italy   21011   21039
Luxembourg   07023   07060
Netherlands   25012   25052   25062
Norway   26008   26038   26050
Portugal   28013   28060   28067
Spain   30022   30054   30061
                     30063
Sweden   31023   31101
Switzerland   32019   32062   32081
United Kingdom   34079   34269   34342

STOCK EXCHANGE
Switzerland   32005

STOCK EXCHANGES
Austria   05002   05003   05058
                     05073   05106
Denmark   12011   12036   12046
Germany   15007   15130
Italy   21035   21041
Luxembourg   23004   23006   23010
                     23026
Norway   26004   26005   26029
                     26032   26044
Switzerland   32009   32031   32035
United Kingdom   34140   34202   34267
                     34268   34318   34319
                     34320

STOCK MARKETS
Denmark   12006   12031   12033
                     12046
Europe   01084
France   14002   14116   14129
                     14143
Germany   15007
Hungary   18024
International   01084
Italy   21034
Portugal   28001   28003   28005
                     28018   28045   28053
                     28054   28074
Sweden   31124
Switzerland   32005   32009   32031
                     32035   32079
United Kingdom   34140   34202   34267
                     34268   34318   34319
                     34320

SUPERMARKETS
Denmark   12041
Europe   02083
United Kingdom   34045

SURVEYS
European Union   03156

TARIFFS
Czech/Slovak Rep   11020
European Union   03122
Hungary   18030
International   01054   01232
Norway   26027

TAXATION
Austria   05034   05047
Denmark   12037
European Union   03125
Germany   15157   15174   15190
                     15197   15211
International   01252
Ireland   20009   20037   20099
                     20109
Portugal   28027
Sweden   31078   31110
Switzerland   32014   32030   32039
United Kingdom   34133   34151   34168
                     34283

TEA INDUSTRY
International   01005 01206

TECHNOLOGY
Finland   13088
Ireland   20115

TELECOMMUNICATIONS
Finland   13072 13089

TELECOMMUNICATIONS INDUSTRY
Austria   05088
Germany   15215
International   01128
Malta   24006
Turkey   33070

TELEVISION
Belgium   07053

TELEVISION INDUSTRY
Austria   05088

TENDER OFFERS
International   01156

TEXTILE INDUSTRY
Finland   13090
Germany   15143
International   01053 01158 01173
                 01178 01262 34272
Italy   21026 21055 21056
Switzerland   32080
United Kingdom   34093 34104 34123
                 34137 34161 34162
                 34163 34208 34272
                 34313 34354

TIMBER TRADE
Europe   02119

TOBACCO INDUSTRY
Germany   15214

TOURISM
Austria   05012 05027 05033
Belgium   07083
Bulgaria   08032
Cyprus   10046 10047
Denmark   12044
European Union   03171 03172
Finland   13091
France   14090 14099
Germany   15129 15199 15202
Hungary   18042
International   01075 01180 01276
                 01340
Italy   21052
Liechtenstein   22009 22010
Malta   24023
Netherlands   25071 25075
Portugal   28032 28047 28056
                 28057 28058
Spain   30057
Sweden   31001 31136
Switzerland   32049 32052 32084
                 32085 32086 32087
Turkey   33005 33006 33010
                 33024 33052 33053
United Kingdom   34037 34041 34043
                 34046 34117 34118
                 34119 34152 34221
                 34245 34252 34254
                 34275 34279 34308
                 34325 34326 34341
                 34348 34352

TOURIST INDUSTRY
Austria   05027 05033
European Union   03171 03172
France   14090 14099
Germany   15129 15199 15202
Gibraltar   16007 16008
Greece   17051

International   01274 01340
Ireland   20068 20069 20116
                 20117
Italy   21052
Poland   27061
Portugal   28032 28047 28056
                 28057 28058
Spain   30057
Sweden   31001 31136
Switzerland   32049 32052 32084
                 32085 32086 32087
Turkey   33005 33006 33010
                 33024 33052 33053
United Kingdom   34037 34041 34043
                 34046 34117 34118
                 34119 34152 34188
                 34221 34245 34252
                 34254 34275 34279
                 34308 34325 34326
                 34348 34352

TOWNS
European Union   03165

TOY INDUSTRY
Austria   05101

TRADE
Austria   05012 05095
Belgium   07043 07090
Denmark   12021 12050 12053
Germany   15041 15053 15096
                 15176 15178 15217
Luxembourg   07043 07090
Malta   24011 24017 24018
                 24024 24025
Poland   27062
Spain   30030 30040 30051
Switzerland   32010 32011 32066
United Kingdom   15041

TRADE ASSOCIATIONS
Austria   05095

TRADE BARRIERS
Ireland   20039

TRADE UNIONS
Denmark   12027 12029
Norway   26002
Turkey   33030 33051

TRAFFIC
Austria   05087 05099
Belgium   07094
Germany   15114 15223

TRAINING
European Union   03026 03176
Malta   24003
United Kingdom   34194 34330

TRANSPORT
Austria   05012 05067
Belgium   07083
Cyprus   10048
Denmark   12038 12044 12050
Europe   02009 02033 02113
                 02117 15088
European Union   03067 03099 03158
                 03164 03173 15088
Finland   12038 13032 13093
                 13094 13095 13103
France   14072
Germany   15088 15089 15158
                 15186 15223
Greece   17052
Iceland   12038
International   01263 01275
Netherlands   25048 25056
Norway   12038 26055
Poland   27007
Portugal   28034 28070

Spain   30067
Sweden   12038
Switzerland   32073
Turkey   33038
United Kingdom   34053 34149 34171
                 34225 34274 34296
                 34305 34307 34329
                 34331 34332 34333
                 34334 34335 34336
                 34337 34353 34359

TRAVEL
Europe   15088
European Union   15088
Germany   15088

TURNOVER
Germany   15220

UNEMPLOYMENT
Denmark   12002
European Union   03174
Greece   17036
Ireland   20073 20074 20075
                 20076 20120
Poland   27063

URBAN DEVELOPMENT
International   01287

VALUE ADDED TAX
Austria   05102
Germany   15222

VEGETABLES
France   14008
Germany   15039 15081
Portugal   28038 28042

VEHICLE FLEETS
Belgium   07057

VETINERARY PRODUCTS INDUSTRY
International   01015
United Kingdom   34021

WALES
United Kingdom   34004 34096 34128
                 34355 34356 34357
                 34358 34359

WASTE
Austria   05001

WATER RESOURCES
Germany   15227

WATER SUPPLY INDUSTRY
Germany   15027 15030 15227
Turkey   33026 33072
United Kingdom   34057

WHOLESALING
Austria   05048
Belgium   07083
Cyprus   10042
European Union   03123
Finland   13092
Germany   15026 15117
Hungary   18034
Ireland   20121

WINE INDUSTRY
Austria   05029
France   14010
Germany   15197 15204

WOMEN
Cyprus   10049
Netherlands   25038
Sweden   31146
Turkey   33076

WORKING CONDITIONS
European Union   03015 03141

Index 2b
SOURCES LISTED BY COUNTRY AND SUBJECT

ALBANIA

agricultural industry	04002		
business environment	04004	04007	
economic conditions	04001	04002	04003
	04005	04007	08010
	29003		
forecasting	04003		
international trade	04006		
politics	04003		

AUSTRIA

accidents	05104		
accounting	05055		
agricultural industry	05039	05056	05063
	05071	05075	05083
	05089	05090	
agricultural machinery industry	05064		
air transport	05110		
airlines	05110		
banking	05009	05013	05045
	05067	05098	
bookkeeping	05055		
business environment	05032	05057	
buying	05084		
car industry	05014		
chambers of commerce	05095		
companies	05040	05067	05094
company taxation	05047		
confectionery trade	05018		
construction industry	05010	05109	
credit	05044		
demography	05023	05049	05069
	05093		
distribution	05067		
economic conditions	05007	05008	05011
	05013	05016	05020
	05021	05022	05032
	05036	05053	05057
	05076	05077	05081
	05096	05097	05100
electricity industry	05108		
electronics industry	05035		
employees	05031	05063	
employment	05004	05031	05069
energy	05012	05037	05038
	05074	05103	
environment	05103		
environmental protection	05005		
exporting	05021	05053	
exports	05024	05067	
family expenditure	05061	05092	
floor coverings	05017		
forecasting	05020	05022	
forestry industry	05063		
fruit	05028		
gas industry	05079		
health	05068		
horticultural industry	05028	05042	
household economics	05061	05092	
housing	05051	05104	05109
imports	05024		
income tax	05034	05066	
incomes	05065	05072	05082
industrial relations	05086		
industries	05004	05005	05054
	05067	05080	
insurance	05044		
international trade	05006	05021	05024
	05025	05026	05053
investment law	05052		
investment	05052		
jewellery industry	05060		
labour costs	05030		
labour	05086		
land use	05015		
land	05050		
law	05052		
manufacturing industry	05046	05054	05067
medicine	05068		
migration	05023		
money	05044		
motor industry	05014	05062	05070
national income	05078		
national insurance	05066		
oil industry	05059	05079	
pay	05065	05066	05072
petroleum industry	05059		
politics	05020	05022	
population	05023	05049	05069
	05107		
prices	05056		
production	05001	05012	05046
	05054	05071	05080
property tax	05105		
property	05050		
public administration	05043		
public expenditure	05043		
purchasing power	05076		
purchasing	05084		
R & D	05041		
regional development	05011		
regional planning	05090		
retailing	05048		
road accidents	05099		
road transport	05087	05099	
rural industries	05071	05089	05090
security	05091		
social economics	05093	05100	
standard of living	05069	05093	
steel industry	05019	05085	
stock exchanges	05002	05003	05058
	05073	05106	
taxation	05034	05047	
telecommunications industry	05088		
television industry	05088		
tourism	05012	05027	05033
tourist industry	05027	05033	
toy industry	05101		
trade associations	05095		
trade	05012	05095	
traffic	05087	05099	
transport	05012	05067	
value added tax	05102		
waste	05001		
wholesaling	05048		
wine industry	05029		

BALTIC STATES

business environment	06007		
demography	06004		
economic conditions	06001	06003	06004
	06005	06006	06008
	06009	06010	06012
	15179		
organizations	06007		
population	06004	06009	
public finance	06011		
social economics	06006	06008	06010
	06012		

BELGIUM

agricultural industry	07067	07078	
air transport	07048		
banking	07011	07066	
business economics	07033		
business environment	07031	07032	07042
car industry	07008		
cars	07056	07094	
companies	07047	07089	07092
concrete industry	07018	07061	
construction industry	07022	07040	07080
economic conditions	07002	07004	07005
	07007	07009	07011
	07012	07014	07015
	07020	07021	07024
	07025	07026	07027
	07028	07029	07031
	07032	07036	07038
	07039	07042	07045
	07050	07054	07055
	07066	07074	07091
	23005	23012	23021
	23028		
economic forecasting	07038		
electricity industry	07017	07037	
employment	07049		
energy	07010		
exporting	07025	07039	
finance	07085		
food and drinks industry	07030		
forecasting	07024	07028	23012
foreign trade	07093		
gas industry	07003	07052	07064
glass industry	07016	07041	07086
horticultural industry	07067		
housing	07069	07080	
industrial design	07035		
industrial relations	07073		
insurance	07013	07044	
international trade	07025	07039	07050
	07081	07082	23021
	23034		
jewellery industry	07006		
labour market	07073		
metals industry	07034	07062	07063
	07070	07071	
mining industry	07001	07076	07077
national accounting	07046		
paper and pulp industry	07065	07072	
politics	07024	07028	23012
population	07019	07051	07058
	07059	07069	07084
ports	07079		
production	07049	07075	07087
public finance	07046	07049	
radios	07053		
rail transport	07048		
regional development	07015		
retailing	07083		
road transport	07048	07068	
shipping	07079		
social economics	07088		
social environment	07004	07005	
steel industry	07023	07060	
television	07053		
tourism	07083		
trade	07043	07090	
traffic	07094		
transport	07083		
vehicle fleets	07057		
wholesaling	07083		

BULGARIA

agricultural industry	08002	08004	
aid	02025	08009	
business environment	04004	04007	08001
	08014		
companies	08006		
culture	08012	08029	
demography	08023	08024	
development	02025	08009	
domestic trade	08015		
economic conditions	04003	04005	04007
	08001	08003	08005
	08007	08008	08010
	08011	08013	08014
	08015	08016	08017
	08021	08028	08029
	08031	29003	29012
environment	08018		
exporting	08021		
forecasting	04003	08007	08011
health service	08020		
health	29012		
housing	08022		
international trade	08005	08008	08019
	08021		
politics	04003	08007	08011
population	08023	08024	
publishing industry	08025		
regional differences	08026		
social economics	08003	08029	08031
social structure	08029		
standard of living	08027		
tourism	08032		

C.I.S.

agricultural industry	09021		
aid	02025	02045	
banking	09001		
business environment	09003	09004	09016
	09017	09018	09019
	09024	09025	09026
	09028	09032	09033
	09038	09039	
companies	09030	09035	09036
	09037		
democracy	09006		
demography	09005		
development economics	01270		
development	02025	02045	
economic conditions	01270	01328	09001
	09002	09004	09005
	09007	09008	09009

	09010	09011	09012
	09013	09014	09015
	09016	09017	09020
	09022	09023	09025
	09028	09031	09032
	09038	09040	29012
economic forecasting	09007		
exporting	09008	09023	
forecasting	09013		
forestry industry	01116	01118	
government	09006	09040	
health	29012		
international trade	09002	09008	09023
	09040		
investment	09022		
joint business ventures	09027	09030	09037
money markets	09034		
politics	09013		
population	01328	09005	09040
privatization	09029		

CYPRUS

agricultural industry	10001		
banking	10006		
business environment	10019		
companies	10039		
construction industry	10011		
crime	10014		
demography	10018		
economic conditions	10002	10003	10006
	10007	10008	10009
	10012	10013	10016
	10017	10019	10020
	10021	10022	10023
	10025	10027	10035
	10036	10037	10044
	33020		
economic policy	10027		
education	10024		
employment	10033	10040	10049
finance companies	10026		
forecasting	10013	10022	
health	10028	10049	
hotel and catering industry		10041	
household economics	10029		
housing	10011		
industrial performance	10031		
industries	10002	10025	10031
insurance	10026		
international trade	10015	10030	10045
location of industry	10039		
migration	10046	10047	
mining industry	10005		
politics	10013		
population	10018	10038	
ports	10004	10032	10034
retailing	10042		
service industries	10042		
social economics	10044		
social indicators	10043		
social services	10010		
tourism	10046	10047	
transport	10048		
wholesaling	10042		
women	10049		

CZECH AND SLOVAK REPUBLICS

agricultural industry	11011		
business environment	04007	11013	11014
	11015	11016	11021
	11023		
cocoa industry	01030		
companies	11010	11012	11024
	11025		
economic conditions	04007	11001	11002
	11003	11004	11005
	11007	11008	11009
	11014	11015	11017
	11018	11020	11021
	18001		
economic forecasting	11008		
economic policy	11006		
exporting	11002	11018	11019
	11020	11022	
forecasting	11001	11003	11004
government departments	11024		
importing	11019	11022	
international trade	11002	11018	11020
politics	11001	11003	11004
tariffs	11020		

CZECHOSLOVAKIA

aid	02025
development	02025
economic conditions	01223

DENMARK

agricultural industry	12001	12030	12050
balance of payments	12037	12038	
banking	12019		
business environment	12016	12028	
co-operatives	12001		
companies	12015	12032	12033
construction industry	12004	12024	
consumption	12050		
credit	12040		
culture	12050		
demography	12003		
economic conditions	01057	12008	12009
	12010	12012	12016
	12021	12022	12023
	12028	12035	12043
	12045	12049	12050
	12052	12054	
economic performance	12039	12051	12052
education	12038	12050	12055
electricity industry	12017		
energy	12017	12026	
exporting	12009	12023	
exports	12038	12055	
finance	12055		
forecasting	12008	12010	12022
health	12018		
imports	12038	12055	
incomes	12025	12050	
industrial relations	12029		
information retrieval	12034		
international trade	12009	12023	
labour market	12002	12055	
labour relations	12029		
labour	12027	12050	
manufacturing industry	12026	12050	
meat industry	12048		
money markets	01218	12050	
money	12040		
national accounting	12005	12014	12037
	12040	12050	
national debt	12005	12014	
politics	12008	12010	
population	12003	12038	12050
	12055		
prices	12025	12050	
public finance	12037		
retailing	12013	12020	12041
securities markets	01218		
share prices	12011	12036	
social environment	12049		
social security	12038	12047	12050
	12055		
statistics	12039		
steel industry	12007	12042	
stock exchanges	12011	12036	12046
stock markets	12006	12031	12033
	12046		
supermarkets	12041		
taxation	12037		
tourism	12044		
trade unions	12027	12029	
trade	12021	12050	12053
transport	12038	12044	12050
unemployment	12002		

EUROPE

advertising	02066		
agricultural industry	01240	02001	02002
aid	02025	02045	02070
	03075		
biotechnology	02089		
building societies	02098		
business environment	02015	02019	02020
	02035	02036	02037
	02042	02100	02101
chemical industry	02012	02038	02039
	02094		
coal mining industry	02003		
companies	02064	02077	
competitiveness	03154		
construction industry	02007	02109	
consumer durables	02027	02059	
consumer goods	02050	02090	
cosmetics industry	02028	02062	
debt	01277		
defence	02091		
development economics	01242		
development	02025	02045	02069
	02070	03074	03075
drinks industry	02029	02073	02093
economic conditions	01068	01216	01277
	01291	01294	02010
	02016	02017	02019
	02021	02024	02034
	02035	02040	02041
	02044	02047	02048
	02052	02054	02055
	02056	02068	02076
	02078	02085	02087
	02100	02101	02110
	02120	15166	15179
	15219	15225	
electronics industry	02014	02057	02092
	02111		
energy	01066	01086	02003
	02004	02006	02058
	02089		
engineering industry	02011		
environment	01096	02089	
financial institutions	02040		
fishing industry	02084		
food industry	02001	02030	02060
	02074	02093	02105
	02107		
forecasting	02016	02024	02110
forestry industry	01273		
funds	02075		
gas industry	02005		
health	02013	02062	02094
heating and ventilating industry	02018		
household goods	02059		
housing	02007	02109	
industrial policy	01149	02099	
industries	02041		
international trade	02046	02051	02052
	02053	02054	02055
	02056	15213	
investment	02076		
joint business ventures	02046		
leisure industry	02061	02095	
market research	02022	02023	02026
	02059	02060	02061
	02062	02066	02071
	02077	02078	02086
	02097	02102	02103
	02104	02105	02106
	02115		
money markets	02043	15219	
mortgages	02098		
motor industry	02049	02067	02079
	02088		
oil industry	02031		
packaging industry	02072	02080	02096
	02106	02112	
paper and pulp industry	02032	02096	
petrochemicals	02031		
pharmaceutical industry	02013	02062	02094
politics	02024	15219	
population	01291		
prices	01239		
printing industry	02081		
regional differences	02017		
retailing	02026	02050	02063
	02065	02086	02104
	02108	02114	02115
	02116		
road accidents	02117		
security industry	02091		
single market	03154		
social economics	02085		
steel industry	02008	02082	02118
stock markets	01084		
supermarkets	02083		
timber trade	02119		
transport	02009	02033	02113
	02117	15088	
travel	15088		

EUROPEAN UNION

agricultural industry	03004	03005	03006

	03007 03008 03009
	03010 03011 03013
	03037 03045 03115
	03148
aid	02025 02045 02070
	03036 03075 03076
	03157
balance of payments	17021
biotechnology	15193
budgetary control	03029 03072 03117
business environment	03022 03042 03092
coal mining industry	03027 03105
communications industry	03028
community law	15051 15073
competitiveness	03154
construction industry	03118
consumer behaviour	03033
consumer price index	03035
consumer protection	03033
consumer spending	03082
data banks	03079
demography	03038 03146
developing countries	03044
development	02025 02045 02069
	02070 03036 03074
	03075 03076 03157
economic conditions	03002 03017 03020
	03022 03031 03034
	03041 03042 03046
	03069 03070 03071
	03077 03080 03081
	03082 03086 03088
	03092 03093 03094
	03102 03109 03112
	03113 03119 03120
	03128 03130 03135
	03136 03137 03138
	03139 03142 03149
	03150 03156 03175
	05076 15065 15074
	15108 15111 15196
	17013
economic integration	03019 03032 03047
	03068 03073 03101
	03116 03128 03144
	03155 03161 03170
electricity supply industry	03052
employment	03053 03054
energy	03001 03018 03027
	03052 03055 03056
	03057 03058 03059
	03060 03061 03065
	03129 03143 03159
engineering industry	03062
environment	03014 03063 03064
	03107 03108 03166
exchange rates	03049
exports	03012 17032
family expenditure	03103
finance	03072
financial policy	03135
financial systems	03048 03161
fishing industry	03009 03011 03106
food industry	03121
forecasting	03129
forestry industry	03009 03011 03078
	03108
furniture industry	03021
gambling industry	03110
gas industry	03111
geography	03068
government contracting	03133
handicapped workers	03040
household economics	03103
imports	03012
industrial performance	03120
industrial structure	03145 03167 03168
	03169
industries	03167
information industry	03177
inland waterways	03023
insurance	03084
international trade	03012 03044 03050
	03095 03096 03097
	03098 03099 03100
	03124
investment	03016 03051 03077
	03085 03134

labour costs	03130
labour statistics	03132
labour	03131
legislation	15060 15193
market research	03021 03062 03121
metals industry	03126 03127
monetary system	03047 03049 03104
money markets	03161
national accounting	03112 03136 03137
	03138 03139 03149
nuclear power industry	03065 03143
opinion polls	03066
pay	03043 03140
politics	03083 03087 15065
population	03038 03146
price indices	03034
production	03119
purchasing power	05076
R & D	03114
rail transport	03024
regional development	03051 03086 03088
	03117 03147 03151
regional differences	03147 03152 03153
	03168
regional policy	03175 15111
retailing	03089 03090 03121
road haulage industry	03025
road transport	03107
service industries	03124 03158 03164
single market	03031 03032 03033
	03067 03068 03070
	03073 03080 03083
	03101 03102 03110
	03122 03123 03133
	03145 03150 03151
	03154 03155 03161
social economics	03017 03019 03020
	03066 03069 03102
	03113 03144 03162
social environment	03002 03003 03040
	03141 03142 03162
social security	03030 03039 03163
social surveys	03146
statistical methods	03091 03093 03160
steel industry	03105 03126 03127
surveys	03156
tariffs	03122
taxation	03125
tourism	03171 03172
tourist industry	03171 03172
towns	03165
training	03026 03176
transport	03067 03099 03158
	03164 03173 15088
travel	15088
unemployment	03174
wholesaling	03123
working conditions	03015 03141

FINLAND

agricultural industry	13025 13028 13066
	13100
air transport	13002
balance of payments	12038 13035
banking	13001 13003 13004
	13005 13006 13007
	13009 13036 13061
	13079 13081 13082
	13096
bankruptcy	13008
bond markets	13037
business environment	13060
business services industry	13032
chambers of commerce	13062
clothing industry	13013
companies	13017 13026 13065
	13087
construction industry	13014
construction	13010 13053
consumer expenditure	13015
consumer price index	13016
consumption	13054
economic conditions	12045 13011 13012
	13019 13020 13021
	13022 13027 13032
	13034 13038 13039
	13045 13048 13049
	13060 13061 13073
	13086 13087 13088

	13096 13101 13102
economic performance	13039
education	12038 12055
energy	13023 13024
environment	13095
exporting	13020 13049
exports	12038 12055
family expenditure	13051
farms	13025 13028 13100
finance	12055 13029
forecasting	13019 13021 13039
	13048
forestry industry	13040
horticultural industry	13050
household economics	13051
housing	13052 13053
imports	12038 12055
incomes	13054 13055
inland waterways	13083
insurance	13041 13056 13057
	13058
interest rates	13059
international trade	13020 13046 13047
	13049
labour market	12055 13063 13064
labour	13042
manufacturing industry	13073 13101 13102
maritime transport	13043
metals industry	13044
money markets	13006 13030 13031
motor industry	13067
national accounting	13068 13069
pay	13098 13099
politics	13019 13021
population	12038 12055 13071
postal service	13072
prices	13010 13074 13075
production	13097
public finance	13076 13096
rail transport	13078
retailing	13092
service industries	13080
shipping industry	13043 13070 13083
	13084 13085
social security	12038 12055
steel industry	13018 13033 13077
technology	13088
telecommunications	13072 13089
textile industry	13090
tourism	13091
transport	12038 13032 13093
	13094 13095 13103
wholesaling	13092

FRANCE

agricultural industry	14001 14003 14005
	14006 14009 14011
	14013 14014 14017
	14020 14045 14060
	14061 14072 14089
	14096
balance of payments	14078
banking	14032 14113 14123
	14124 14139
business environment	14043 14044 14073
capital markets	14142
cereals industry	14005
companies	14039 14053 14054
	14056 14079 14080
	14088 14100 14109
	14134 14144
construction industry	14063 14140
consumer economics	14130
credit	14054 14104
dairy industry	14007
economic conditions	14018 14022 14023
	14025 14029 14030
	14032 14034 14035
	14036 14037 14040
	14043 14044 14046
	14048 14049 14050
	14051 14057 14062
	14065 14066 14068
	14073 14074 14077
	14091 14093 14108
	14111 14114 14115
	14117 14120 14122
	14125 14128 14145
	14147

economic forecasting	14059		
electricity industry	14064	14082	
electronics industry	14064	14082	
employment	14027	14041	14052
	14055	14071	14080
	14148		
energy	14024	14098	14146
engineering industry	14067		
exporting	14036	14062	
exports	14016	14039	14105
	14136		
finance	14026	14042	14101
	14102	14112	14113
	14117		
financial institutions	14104		
financial performance	14084	14086	14123
food industry	14003	14006	14045
forecasting	14034	14037	
foreign investment	14075		
forestry industry	14012	14016	14060
fruit	14004		
horticultural industry	14004	14008	
household economics	14069		
imports	14016	14039	14105
	14136		
incomes	14071		
industrial training	14141		
industries	14021	14027	14028
	14040	14055	14072
	14075	14079	14087
	14092	14094	14095
	14101	14103	14133
	14134	14135	14138
	14144		
interest rates	14106	14107	
international trade	14003	14016	14019
	14031	14036	14039
	14040	14062	14065
	14096	14105	14110
	14126	14132	14136
	14137		
investment	14080		
labour	14131		
money markets	14106		
money supply	14106	14107	14142
money	14026		
national accounting	14066	14072	14126
options	14143		
pay	14071		
petroleum industry	14076		
politics	14034	14037	
population	14023	14042	14070
	14118	14119	14127
poultry industry	14015		
production	14028	14042	14055
	14056	14088	
rail transport	14085		
regional development	14081	14083	14086
	14094	14139	
road transport	14085		
social economics	14023	14042	14047
	14048	14050	14068
	14119	14122	14131
social security	14097	14131	
statistics	14038		
steel industry	14033	14058	14121
stock markets	14002	14116	14129
	14143		
tourism	14090	14099	
tourist industry	14090	14099	
transport	14072		
vegetables	14008		
wine industry	14010		

GERMANY

accidents	15054		
accounts	15154		
agricultural industry	15005	15010	15016
	15023	15105	15132
	15167	15183	15200
	15207	15210	
air transport	15170		
banking	15009	15052	15066
	15165	15177	15203
biotechnology	15193		
brewing industry	15003	15040	15142
	15159		
business environment	15058	15059	15098
	15104	15115	15126

chambers of commerce	15044		
chemical industry	15014	15135	
clothing industry	15134		
commodities	15079		
community law	15051	15073	
companies	15001	15004	15086
	15218		
construction industry	15017	15022	15042
	15106	15116	15122
	15159	15160	15172
consumer goods	15076	15159	
consumer price index	15184		
coporation tax	15155		
dairy industry	15055	15173	
demography	15035	15036	15080
DIY	15085		
economic conditions	15036	15047	15048
	15049	15057	15058
	15059	15061	15063
	15064	15065	15074
	15077	15078	15083
	15084	15090	15094
	15098	15099	15100
	15101	15102	15103
	15104	15108	15111
	15113	15115	15120
	15125	15126	15128
	15146	15147	15148
	15149	15150	15151
	15152	15153	15165
	15166	15168	15175
	15176	15177	15178
	15179	15194	15196
	15203	15205	15206
	15208	15209	15219
	15225	15226	15228
	15229	15230	15231
	15232	15233	15234
	15235	15237	34143
education	15037		
electricity industry	15138		
electronics industry	15138		
employment	15168	15188	
energy	15018	15027	15030
	15107		
environment	15221		
environmental protection	15127		
exporting	15042	15048	15075
	15076	15085	15094
	15096		
exports	15075	15182	
family expenditure	15069	15093	
fertilizers	15062		
finance	15109	15165	
financial policy	15109		
fishing industry	15095	15123	15133
food industry	15142	15159	15207
	15210		
forecasting	15047	15049	
forestry industry	15132	15183	15207
	15210		
foundry industry	15139		
freight transport	15162		
fruit	15021		
furniture industry	15141		
government expenditure	15192		
hotel and catering industry		15025	15164
housing	15085		
importing	15082	15216	
imports	15067	15082	15182
	15216		
income tax	15068	15071	15169
industries	15002	15110	15112
	15127	15231	
input-output analysis	15124		
insurance	15054	15056	15236
international cooperation	15091		
international trade	15015	15019	15020
	15048	15061	15067
	15079	15092	15094
	15213	34143	
investment	15045	15218	
labour costs	15012		
labour market	15097		
labour supply	15013		
labour	15011	15043	15050
	15087	15121	15176
land ownership	15032		

land use	15039	15167	
land values	15144	15145	
land	15032	15038	
law	15157		
legislation	15060	15193	
manufacturing industry	15028	15029	15122
	15188		
marketing	15128		
meat industry	15198		
medical care	15171		
medicine	15171		
metals industry	15136	15137	
minerals resources	15174		
mining industry	15028	15029	15161
	15188		
money markets	15219		
national accounting	15191	15192	
nutrition	15210		
OECD	15200		
output	15187	15188	
passenger transport	15180		
pay	15008	15010	15119
plastics industry	15140		
politics	15047	15049	15065
	15219		
population	15035	15080	15168
poultry industry	15072		
prices	15117	15118	15176
	15181	15182	15183
	15184	15185	15186
production	15121	15153	15161
	15187	15188	
products	15185		
property	15032	15033	15034
public expenditure	15191		
public finance	15061	15195	15224
public sector	15195		
purchasing	15128		
rail transport	15070		
regional development	15036		
regional policy	15111		
retail pricing	15163		
retailing	15024	15031	
road construction	15212		
road transport	15070		
roads	15212		
salesmen	15156		
share prices	15007		
shares	15006		
shipping industry	15131	15201	
small business	15125		
social economics	15233	15234	
social policy	15011		
social security	15071		
steel industry	15046	15136	15137
	15189		
stock exchanges	15007	15130	
stock markets	15007		
taxation	15157	15174	15190
	15197	15211	
telecommunications industry		15215	
textile industry	15143		
tobacco industry	15214		
tourism	15129	15199	15202
tourist industry	15129	15199	15202
trade	15041	15053	15096
	15176	15178	15217
traffic	15114	15223	
transport	15088	15089	15158
	15186	15223	
travel	15088		
turnover	15220		
value added tax	15222		
vegetables	15039	15081	
water resources	15227		
water supply industry	15027	15030	15227
wholesaling	15026	15117	
wine industry	15197	15204	

GIBRALTAR

air transport	16004		
economic conditions	16001	16002	16003
	16005		
labour statistics	16006		
pay	16006		
population	16005		
social indicators	16002		
tourist industry	16007	16008	

GREECE

agricultural industry	17001	17031	
balance of payments	17021	17022	
banking	17003	17012	17019
	17028		
business environment	17025		
communications industry	17052		
construction industry	17024		
corporation tax	17049		
demography	17029		
economic conditions	17003	17006	17008
	17009	17010	17011
	17013	17014	17015
	17016	17017	17018
	17020	17022	17023
	17025	17027	17048
EEC	17018		
employment	17036		
energy	17037		
exporting	17009	17023	
exports	17032		
finance	17012	17019	17028
fishing industry	17042		
forecasting	17008	17010	17011
health	17045		
income tax	17049	17050	
industries	17002		
international trade	17005	17009	17023
labour supply	17026		
local government	17038		
mining industry	17004		
monetary system	17039		
occupational choice	17030		
politics	17008	17010	17011
	17020		
population	17022	17029	
ports	17046		
public administration	17040		
public finance	17022	17033	17047
regional development	17035		
regional policy	17041		
shipbuilding industry	17043		
shipping industry	17044	17046	
social welfare	17045		
steel industry	17007	17034	
tourist industry	17051		
transport	17052		
unemployment	17036		

HUNGARY

agricultural industry	18038		
aid	02025		
business environment	04007	18003	18009
	18010	18011	18012
	18021	18023	18031
companies	18002	18004	18020
conferences	18013		
development	02025		
economic conditions	01223	04007	18001
	18005	18006	18007
	18008	18009	18011
	18014	18015	18016
	18017	18018	18019
	18021	18022	18023
	18025	18026	18027
	18029	18030	18031
	18035	18037	18039
	18040	18041	
economic forecasting	18015		
exporting	18006	18018	18030
financial markets	18024		
forecasting	18005	18007	18008
international trade	18006	18018	18030
investment incentives	18032	18033	
joint business ventures	18028		
national accounting	18035		
politics	18005	18007	18008
property	18036		
retailing	18034		
stock markets	18024		
tariffs	18030		
tourism	18042		
wholesaling	18034		

ICELAND

balance of payments	12038		
banking	19002	19014	
economic conditions	12010	19001	19002
	19003	19004	19005

	19006	19007	19008
	19009	19011	19012
	19013	19014	19016
	19017	19018	19020
	19022	19023	
education	12038		
exchange rates	19001		
exporting	19004	19011	
exports	12038		
fishing industry	19017		
forecasting	12010		
foreign trade	19016		
imports	12038		
international trade	19004	19008	19009
	19010	19011	
labour market	19019		
local government finance	19015		
politics	12010		
population	12038	19021	
social economics	19012	19022	
social security	12038		
transport	12038		

INTERNATIONAL

agricultural industry	01002	01003	01067
	01101	01102	01136
	01215	01240	01248
	01250	01292	01321
	05056		
air transport	01028	01059	
armed forces	01217		
balance of payments	01016		
banking	01017	01121	01290
	01301	01308	01309
bartering	01041	01062	
building societies	01169		
business environment	01045	01064	01170
	01257		
cereals industry	01248	01250	01321
charities	01025	01167	34071
chemical industry	01006	01018	01026
	01027	01072	
coal mining industry	01029		
cocoa industry	01008	01030	01031
	01244		
commodities	01032	01033	01073
	01238	01293	01305
commodity markets	01204	01284	
companies	01055		
competitiveness	01037	01083	01306
computer industry	01153		
construction industry	01038		
consumer behaviour	01039		
consumer durables	01011		
consumer goods	01012		
copper industry	01307	01332	
currency markets	01159	01191	
dairy industry	01040		
debt	01110	01209	01277
	01308	01309	
demography	01285		
development economics	01069	01097	01110
	01112	01163	01188
	01233	01234	01242
	01270	01278	01279
	01287	01290	01295
	01296	01298	01299
	01302	01304	
development	01034	01035	01058
	01104	01138	01160
	01303		
drinks industry	01098		
economic conditions	01001	01020	01022
	01023	01034	01035
	01042	01043	01044
	01046	01047	01048
	01049	01050	01051
	01052	01057	01058
	01063	01064	01068
	01069	01070	01077
	01078	01079	01080
	01083	01099	01100
	01106	01108	01119
	01131	01139	01140
	01141	01146	01157
	01159	01161	01162
	01168	01170	01174
	01192	01193	01203
	01209	01211	01212

	01213	01214	01216
	01217	01220	01221
	01222	01223	01228
	01246	01257	01259
	01260	01264	01265
	01270	01272	01277
	01282	01283	01286
	01287	01290	01291
	01294	01296	01297
	01299	01300	01301
	01302	01303	01304
	01306	01308	01309
	01311	01315	01316
	01317	01318	01319
	01327	01328	01333
	01334	15057	15064
	15074	15146	15147
	15148	15149	15150
	15152	15228	
economic indicators	01130		
economics	01188		
education	01076	01285	
electricity supply industry	01081	01227	
electronics industry	01013	01082	01320
employment	01150	01185	
energy	01066	01085	01086
	01087	01088	01089
	01090	01091	01092
	01093	01094	01122
	01247		
engineering industry	01009	01021	
environment	01096	01097	01133
	01289		
environmental economics	01141	01225	
exchange rates	01129	01205	01288
exporting	01048	01140	
finance	01104	01107	01142
	01143	01144	01145
	01164	01165	01166
fishing industry	01253	15123	
food industry	01098		
food	01111		
forecasting	01042	01043	01046
	01051	01052	01070
	01131	01133	01161
	01205	01220	01228
	01238	01305	01318
	01327		
forestry industry	01115	01116	01117
	01118	01273	
gas industry	01230		
gold	01134		
health	01014	01132	01322
	01323		
industrial performance	01137	01150	
industrial policy	01149	01152	
industrial structure	01152		
industries	01109	01137	01150
	01152		
information technology	01153		
insurance	01123	01154	
interest rates	01129	01155	01288
international trade	01032	01033	01041
	01054	01060	01061
	01062	01079	01113
	01114	01138	01151
	01175	01181	01207
	01232	01266	01267
	01318	01321	34272
investment	01112	01163	01278
	01279		
labour costs	31141		
labour market	01224		
labour supply	01245		
labour	01186	01245	31074
lead industry	01024	01071	01187
	01202	01241	01249
	01280	01312	01313
maritime transport	01175	01195	01254
market research	01045	01055	01082
	01098	01171	01251
	34289		
marketing	01124		
meat industry	01196	12048	
metals industry	01024	01071	01103
	01179	01184	01187
	01198	01199	01200

	01202	01241	01249
	01261	01280	01281
	01307	01312	01313
	01314	01324	01326
	01332	01335	
migration	01201		
minerals resources	01134	01147	01148
	01183	01325	
money markets	01105	01106	01218
	01226	01229	
mortgages	01169		
motor industry	01004	01120	01172
	01208		
music industry	01331		
national accounting	01211	01212	01213
	01214	01226	01246
	01272		
nuclear power industry	01089	01219	01227
	01231		
oil industry	01007	01230	01247
packaging industry	01074	01125	
paper and pulp industry	01074	01243	
petroleum industry	01091	01182	
pharmaceutical industry	01072	01126	01210
population	01056	01186	01201
	01235	01236	01237
	01271	01291	01328
	01329		
prices	01239	05056	
property	01127		
public finance	01135		
research	01100	01301	
retailing	01251	01310	01330
	34289		
risk analysis	01050	01051	01052
	01174		
road accidents	01263		
road transport	01263		
rubber industry	01255		
science policy	01019		
science	01194	01256	01285
securities markets	01218		
shipbuilding industry	01010	01189	01190
	01197	01258	
shipping industry	01065	01189	01190
social economics	01259	01265	01334
social environment	01188		
social indicators	01036		
steel industry	01176	01177	01184
	01261	01268	01269
	01336	01337	01338
	01339		
stock markets	01084		
tariffs	01054	01232	
taxation	01252		
tea industry	01005	01206	
telecommunications industry	01128		
tender offers	01156		
textile industry	01053	01158	01173
	01178	01262	34272
tourism	01075	01180	01276
	01340		
tourist industry	01274	01340	
transport	01263	01275	
urban development	01287		
vetinerary products industry		01015	

IRELAND

accounting	20001		
advertising	20016		
agricultural industry	20003	20004	20005
	20017	20030	20035
	20040	20060	20065
	20085	20086	20091
	20102	20114	20119
balance of payments	20011		
banking	20007	20012	20013
	20021	20061	20063
	20093		
budget	20014		
building societies	20012		
business environment	20031	20058	
chambers of commerce	20055		
charities	20100		
construction industry	20010	20018	20024
	20045	20046	20089
	20092	20094	
consumer credit	20043		
consumer price index	20025		

dairy industry	20091		
demography	20019	20072	20090
development	20047		
economic conditions	20002	20011	20015
	20021	20022	20023
	20027	20028	20029
	20031	20032	20033
	20034	20042	20044
	20047	20052	20055
	20056	20058	20059
	20063	20066	20067
	20070	20078	20079
	20080	20084	20087
	20088	20093	20096
	20098	20103	20104
	20106	20107	20108
employment	20012	20046	20048
	20049	20050	20051
	20071	20073	20074
	20075	20076	20120
exporting	20028	20042	
forecasting	20027	20029	20032
hire purchase	20043		
household economics	20044		
housing	20006	20094	
industries	20002	20022	20034
	20055	20059	20078
	20080	20104	20115
insurance	20057	20062	
international trade	20028	20038	20042
	20064	20110	20118
labour costs	20070		
labour relations	20048		
labour	20008	20072	20073
	20074	20075	20076
	20090		
local government finance	20077		
market research	20066		
motor industry	20081	20082	20083
national accounting	20014	20111	20112
national income	20087		
newspaper industry	20067		
pay	20010	20050	20051
	20114		
politics	20027	20029	
population	20019	20072	20090
	20097	20101	20107
	20108		
ports	20110		
prices	20121		
production	20049	20050	20051
	20052	20053	20054
	20070	20080	
property	20092		
public expenditure	20036	20037	20041
	20099	20111	20112
	20113		
public services	20036		
retailing	20103		
road transport	20105		
sales	20103		
service industries	20020		
shipping industry	20110		
social economics	20101		
social indicators	20084	20097	20106
	20107	20108	
steel industry	20026	20095	
taxation	20009	20037	20099
	20109		
technology	20115		
tourist industry	20068	20069	20116
	20117		
trade barriers	20039		
unemployment	20073	20074	20075
	20076	20120	
wholesaling	20121		

ITALY

agricultural industry	21029	21053	21057
balance of payments	21003	21005	21062
banking	21010	21044	21064
business environment	21015	21028	
cereals industry	21063		
companies	21016	21022	21032
	21034	21035	21040
company performance	21016	21022	21032
computer industry	21042		
construction industry	21023	21054	
cotton industry	21026	21055	21056

credit	21005	21062	
currency	21006		
demography	21059		
domestic trade	21050		
economic conditions	21002	21003	21004
	21006	21008	21009
	21010	21012	21013
	21014	21015	21017
	21018	21019	21027
	21028	21030	21031
	21033	21037	21041
	21043	21044	21045
	21064		
economic indicators	21009		
employment	21036		
exporting	21013	21019	
exports	21007	21047	21049
family expenditure	21020	21051	
finance	21004		
financial performance	21002	21006	21045
fishing industry	21057		
forecasting	21012	21014	21018
forest products	21060		
forestry industry	21060		
health economics	21029		
hours of work	21051		
household economics	21020	21051	
imports	21007	21047	21049
industrial relations	21051		
industries	21002	21061	
international trade	21007	21013	21019
	21024	21047	21049
	21061		
investment	21025		
labour economics	21051		
labour supply	21029	21046	
money markets	21003	21005	21062
national accounting	21001	21021	
office automation	21042		
pasta industry	21063		
pay	21024	21036	21051
politics	21008	21012	21014
population	21024	21027	21029
	21036	21059	
prices	21024	21036	
production	21024	21036	21061
public finance	21001	21005	21021
	21062		
public services	21058		
road accidents	21048		
securities	21025		
share prices	21038		
shares	21038		
social economics	21004	21008	21027
	21029		
steel industry	21011	21039	
stock exchanges	21035	21041	
stock markets	21034		
textile industry	21026	21055	21056
tourism	21052		
tourist industry	21052		

LIECHTENSTEIN

banking	22002	22012	
business environment	22006		
companies	22005		
company formation	22020		
company law	22005	22011	22015
construction industry	22003		
demography	22007		
economic conditions	22006	22013	22016
	22017	22019	22023
energy	22008		
foreign workers	22001		
immigration	22001		
industries	22022		
insurance	22004	22014	
motor industry	22021		
population	22007		
social economics	22018	22023	
tourism	22009	22010	

LUXEMBOURG

banking	23003	23008	
bond markets	23015	23016	23024
bonds	23004		
business environment	23020		
companies	07089		
consumer economics	23013		

currency	23023		
economic conditions	07026	07028	07055
	23002	23003	23005
	23008	23009	23011
	23012	23014	23017
	23018	23019	23020
	23021	23022	23025
	23027	23028	23030
	23035		
ECU	23015	23016	
employment	23002	23035	
exporting	23011	23018	
finance	23007	23008	23017
	23025	23031	
forecasting	07028	23012	
funds	23029		
health	23002	23035	
international trade	23011	23014	23018
	23021	23025	23034
investment	23029		
monetary economics	23017		
monetary system	23023		
politics	07028	23012	
population	23002	23032	23035
regional development	23033		
securities	23001		
share prices	23010		
shares	23004		
social economics	23014	23025	
steel industry	07023	07060	
stock exchanges	23004	23006	23010
	23026		
trade	07043	07090	

MALTA

aerospace industry	24023		
agricultural industry	24009		
banking	24002	24022	
business environment	24013		
company formation	24005		
company law	24005		
demography	24001	24012	
economic conditions	10012	24001	24013
	24014	24015	24019
	24021	24022	24026
employment	24003		
finance	24016		
government expenditure	24007	24008	
industries	24004		
international trade	24025		
manufacturing industry	24004	24017	24018
national accounting	24016	24020	
population	24001	24012	
production	24010		
shipping industry	24023		
telecommunications industry	24006		
tourism	24023		
trade	24011	24017	24018
	24024	24025	
training	24003		

NETHERLANDS

agricultural industry	25001	25009	25034
	25039		
banking	25003	25005	
business environment	25016	25032	
companies	25006	25017	25018
	25025	25037	25073
	25078		
construction industry	25029	25040	
consumer credit	25072		
consumer durables	25011		
consumer goods	25010		
consumer price index	25004		
consumption	25068		
culture	25070		
distribution	25041		
economic conditions	25003	25007	25013
	25014	25015	25016
	25019	25020	25021
	25026	25027	25028
	25032	25042	25046
	25047	25053	25054
	25055	25063	25066
	25067	25069	25078
economic planning	25008		
education	25057		
energy industry	25022		
environment	25023	25024	

equality	25038		
exporting	25014	25028	
finance	25042		
food industry	25010		
forecasting	25008	25013	25015
	25020	25027	25036
	25053		
health service	25043	25077	
horticultural industry	25076		
incomes	25068		
international trade	25002	25014	25028
	25035	25049	
investment	25025	25030	
labour	25033		
leisure	25071		
macro economics	25036		
manufacturing industry	25007	25050	
men	25038		
money markets	25042		
motor industry	25074		
national accounting	25051	25060	
politics	25013	25015	
population	25044	25058	
prices	25004	25045	25059
production	25031		
regional differences	25063		
retailing	25064		
security	25061		
service industries	25041	25065	
social indicators	25046	25069	
statistical methods	25067		
steel industry	25012	25052	25062
tourism	25071	25075	
transport	25048	25056	
women	25038		

NORWAY

agricultural industry	26001		
balance of payments	12038		
banking	26003	26013	26019
business environment	26012	26028	
companies	26040		
construction industry	26007		
demography	26046	26047	26048
economic conditions	12045	26009	26010
	26011	26012	26013
	26014	26015	26020
	26024	26025	26027
	26028	26035	26036
	26043	26053	26054
	26057		
education	12038	12055	
electricity industry	26016		
energy	26017		
environment	26037		
exporting	26010	26025	26027
exports	12038	12055	26018
	26034		
finance	12055	26013	
fishing industry	26021	26022	
forecasting	26009	26011	26024
forestry industry	26023		
gas industry	26042		
household economics	26031		
imports	12038	12055	26018
	26034		
incomes	26026		
international trade	26010	26018	26025
	26027	26034	26039
labour market	12055	26030	
manufacturing industry	26033		
oil industry	26042		
pay	26026	26056	
petroleum industry	26045		
politics	26009	26011	
population	12038	12055	26046
	26047	26048	
property	26026		
resources	26037		
securities market	26041		
securities markets	26006		
share prices	26049		
shipping industry	26051		
social economics	26052		
social security	12038	12055	
standard of living	26031		
steel industry	26008	26038	26050
stock exchanges	26004	26005	26029
	26032	26044	

tariffs	26027		
trade unions	26002		
transport	12038	26055	

POLAND

agricultural industry	27001	27002	27055
	27056		
aid	02025		
assets	27065		
business environment	04007	27015	27016
	27017	27018	27019
	27038	27051	
communications	27068		
companies	27004	27023	27029
	27041		
construction industry	27033	27040	27064
culture	27062		
development	02025		
economic conditions	01223	04007	18001
	27009	27010	27011
	27012	27013	27015
	27017	27018	27022
	27023	27024	27025
	27030	27031	27032
	27036	27037	27038
	27044	27047	27049
	27050	27051	27052
	27058	27059	27060
economic forecasting	27030		
economic performance	27044		
employment	27020	27021	27066
energy	27028		
exporting	27011	27032	27034
finance	27022		
forecasting	27010	27012	27013
foreign investment	27025	27046	
gross national product	27031		
health service	27067		
housing	27033	27064	
importing	27034		
incomes	27054		
industrial development	27048		
industries	27008	27035	27045
	27062		
international trade	27011	27026	27027
	27032	27052	27053
investment	27039	27040	
joint business ventures	27041		
labour	27066		
national accounting	27044		
pay	27020		
politics	27010	27012	27013
population	27003	27014	27054
prices	27005	27006	27043
production	27008	27031	
property	27065		
public services	27007		
recreation	27061		
small business	27057		
standard of living	27042		
statistics	27059		
tourist industry	27061		
trade	27062		
transport	27007		
unemployment	27063		

PORTUGAL

agricultural industry	28020	28021	28068
banking	28072		
business environment	28006	28052	28065
companies	28009	28029	28071
construction industry	28022	28041	
consumer price index	28043		
demography	28030	28069	
economic conditions	28004	28007	28008
	28014	28015	28016
	28017	28019	28037
	28040	28052	28055
	28059	28061	28062
	28063	28065	28075
economic forecasting	28039		
economic indicators	28011		
employment	28033	28046	
energy consumption	28023		
exporting	28015	28040	28073
exports	28031		
family expenditure	28048		
finance	28055	28066	
financial institutions	28036		

fishing industry 28024
forecasting 28014 28016 28017
fruit 28038
health economics 28026
health service 28026
horticultural industry 28038 28042
household economics 28048
housing 28022
imports 28031
industries 28035
international trade 28002 28004 28010
28015 28040
investment 28049 28076
manufacturing industry 28050
pay 28033
politics 28014 28016 28017
population 28030 28069
production 28044 28055
public finance 28028
regional development 28012
road transport 28051
social economics 28037
social indicators 28064
social welfare 28025
steel industry 28013 28060 28067
stock markets 28001 28003 28005
28018 28045 28053
28054 28074
taxation 28027
tourism 28032 28047 28056
28057 28058
tourist industry 28032 28047 28056
28057 28058
transport 28034 28070
vegetables 28038 28042

ROMANIA
agricultural industry 29016
aid 02025
business environment 04007 29005 29006
29008 29011
development 02025
economic conditions 04003 04005 04007
08010 29001 29002
29003 29004 29005
29007 29008 29009
29011 29012 29013
29014 29015 29017
energy 29007
exporting 29002 29009
forecasting 04003 29001 29004
health 29012
international trade 29002 29009 29010
politics 04003 29001 29004
social economics 29014 29017

SPAIN
agricultural industry 30003 30004 30015
30050
banking 30008 30009 30052
business environment 30027 30028 30053
cars 30011
cereals industry 30050
companies 30064
construction industry 30016 30046
consumer credit 30029 30041
consumer markets 30014 30064
consumer price index 30048
demography 30018
economic conditions 30001 30002 30008
30009 30012 30013
30014 30021 30023
30024 30025 30026
30027 30028 30030
30031 30038 30040
30042 30043 30045
30047 30051 30052
30053 30060 30062
30064 30065 30066
economic forecasting 30044
employment 30033
exporting 30024 30045
exports 30003
factories 30017
family expenditure 30032
finance 30051
forecasting 30023 30025 30026
fruit 30068
hire purchase 30041

horticultural industry 30068
hotel and catering industry 30057
household economics 30032
housing 30016 30019
incomes 30036
industries 30005 30040
insurance 30006 30007 30055
international trade 30003 30024 30045
law 30030
location of industry 30017
migration 30056 30058
mortgages 30039
national accounting 30020
pay 30036
politics 30023 30025 30026
population 30002 30018 30034
30035 30038 30056
30058
price indices 30049
production 30037 30059
public finance 30020
regional development 30021
road accidents 30010
road transport 30011
social economics 30002 30038
steel industry 30022 30054 30061
30063
tourism 30057
tourist industry 30057
trade 30030 30040 30051
transport 30067

SWEDEN
absenteeism 31122
accidents 31092
agricultural industry 31002 31003 31028
31038 31076 31095
31148
air transport 31147
alcoholic drinks industry 31004
alcoholism 31004
balance of payments 12038
banking 31007 31008 31009
31011 31015 31016
31019 31103 31115
31124
bankruptcy 31014
budget 31129
business environment 31030 31068
companies 31037 31045 31049
31075 31080 31090
31124 31128
computer industry 31033
construction industry 31059 31150
consumer expenditure 31020 31042
consumer price index 31021 31022
credit 31027 31138
culture 31005 31029
demography 31018 31098
distribution of wealth 31061
economic conditions 12045 31006 31011
31024 31025 31026
31030 31040 31041
31047 31055 31058
31063 31068 31070
31083 31111 31112
31117 31118 31119
31125 31126 31127
31130 31134
education 12038 12055 31031
31137 31149
electricity industry 31032
electronics industry 31033
employers' organizations 31107 31109
employment 31034 31071 31072
31107 31108 31109
31121 31139 31142
31143 31145 31151
energy 31032 31035 31036
31054 31062
environment 31038 31089 31095
equal opportunities 31144
exporting 31025 31058
exports 12038 12055 31039
31052 31084
family expenditure 31042
finance 12055
financial statements 31037
fishing industry 31010 31048 31133

forecasting 31024 31026 31055
31137
foreign investment 31049
forestry industry 31113
gross national product 31056
health 31057 31093
housing 31059 31096 31150
imports 12038 12055 31039
31053 31085
information technology 31064
insurance 31065 31066 31067
31123 31131
international trade 31025 31050 31051
31052 31053 31058
31086 31102 31127
investment 31069
labour costs 31141
labour market 12055 31073
labour 31034 31067 31071
31072 31074 31106
31121 31122 31137
31145
libraries 31005
local government finance 31060 31077
local government 31116
manufacturing industry 31075 31079 31080
31081 31082
money markets 31006 31045 31046
31115
motor industry 31087
national accounting 31043 31044 31088
31094 31129
occupational diseases 31092
pay 31139 31140 31142
31143
politics 31024 31026
population 12038 12055 31061
31096 31097 31098
31137
postal service 31012 31099
prices 31042
property 31013 31100 31104
public finance 31116
public sector 31132
R & D 31105
real estate 31100
regional differences 31151
research 31105
road accidents 31135
share ownership 31120 31124
social economics 31125
social security 12038 12055 31114
statistics 31017 31091
steel industry 31023 31101
stock markets 31124
taxation 31078 31110
tourism 31001 31136
tourist industry 31001 31136
transport 12038
women 31146

SWITZERLAND
agricultural industry 32044 32045 32046
32065 32075
automation 32083
banking 32004 32006 32007
32008 32013 32024
32056 32061 32069
32089
business environment 32025 32026 32042
capital markets 32077
cars 32028
chambers of commerce 32010 32066
clothing industry 32080
companies 32029 32054 32064
32067 32076
company performance 32054
construction industry 32018 32050 32051
32053
consumer economics 32048
consumer price index 32048
currency options 32034
economic conditions 32002 32004 32006
32007 32010 32011
32013 32020 32021
32022 32023 32025
32026 32033 32036
32038 32040 32041
32042 32043 32056

32057 32058 32061
32063 32066 32068
32074 32082 32088
economic forecasting 32037
economic indicators 32027 32088
economic performance 32040
electronics industry 32083
employment 32071
exporting 32022 32038 32060
exports 32060 32070 32072
32078
family expenditure 32012
farms 32044
finance 32013 32024 32063
financial performance 32088
fishing industry 32065
food and drinks industry 32083
food industry 32075
forecasting 32020 32023
foreign exchange 32001 32034
forest products 32003
forestry industry 32003 32065
hotel and catering industry 32085 32086
household economics 32012
housing 32050
imports 32070 32072 32078
industries 32027 32076
information technology 32083
interest rates 32034 32088
international trade 32015 32022 32038
32070 32072 32078
investment 32043
monetary economics 32063
money markets 32001 32034
national accounting 32017
packaging industry 32083
pay 32059
politics 32020 32023
price indices 32047 32048
prices 32059
public accounting 32032
public finance 32017 32032
public transport 32055
purchasing power 32059
road transport 32016
steel industry 32019 32062 32081
stock exchange 32005
stock exchanges 32009 32031 32035
stock markets 32005 32009 32031
32035 32079
taxation 32014 32030 32039
textile industry 32080
tourism 32049 32052 32084
32085 32086 32087
tourist industry 32049 32052 32084
32085 32086 32087
trade 32010 32011 32066
transport 32073

TURKEY
agricultural industry 33040
air pollution 33007
air transport 33010
banking 33002 33003 33025
33036 33041
budget 33004
business environment 33016 33017 33029
33042 33066
companies 33008 33011 33057
33073
computer industry 33058
construction industry 33012 33071
consultancy 33059
consumer price index 33074
demography 33009 33062
economic conditions 33013 33014 33015
33016 33017 33018
33019 33020 33021
33029 33034 33035
33042 33044 33045
33048 33056 33060
33061 33066 33068
economic indicators 33047
exporting 33039 33056 33059
33065 33069 33070
33072
exports 33039
forecasting 33013 33014 33015
gas industry 33026 33067

gross national product 33027
health 33063
hotel and catering industry 33054
industries 33066
international trade 33023 33037 33045
33056 33059 33064
33065 33069 33070
33072
labour 33030 33031 33051
leisure 33046
local government finance 33004 33022
machine tools industry 33064
manufacturing industry 33001 33008 33032
maritime transport 33049 33050
medicine 33063
mining industry 33033 33065
motor industry 33038
national income 33027
natural gas 33067
politics 33013 33014 33015
population 33009 33062
prices 33040 33074 33075
production 33028
rail transport 33069
recreation 33046
restaurants 33054
retail price index 33043
road accidents 33055
road transport 33055
sports industry 33046
telecommunications industry 33070
tourism 33005 33006 33010
33024 33052 33053
tourist industry 33005 33006 33010
33024 33052 33053
trade unions 33030 33051
transport 33038
water supply industry 33026 33072
women 33076

UNITED KINGDOM
advertising 34008 34009 34016
34105 34120 34132
34138 34204 34222
34224 34273 34277
34290
agricultural industry 34010 34011 34012
34027 34110 34125
34127 34128 34129
34130 34144 34156
34179 34248 34317
34355
aid 34036
air transport 34334 34335 34336
34337
balance of payments 34338
banking 34025
building materials 34180 34235 34314
building societies 34049 34050 34051
34052 34074 34078
34159 34160 34229
34230 34231
business environment 34100 34101 34102
34103 34172
charities 34006 34071 34150
chemical industry 34058 34063 34113
34173 34181 34339
civil service 34072
clerical workers 34073
co-operative societies 34076
coal mining industry 34038
companies 15086 15218 34175
34323
company performance 34067 34135 34145
34165 34167 34176
34256 34349
company ranking 34029 34031 34032
34033 34034 34035
34067 34146 34165
34176 34299 34349
computer industry 34090 34251
construction industry 34056 34157 34158
34164 34247
consumer behaviour 34044 34236
consumer durables 34083 34182 34209
34237
consumer goods 15076 34114
consumer spending 34075
cosmetics industry 34084 34183 34210

cost of living 34077 34203 34278
34298
defence 34092
distribution 34066
drinks industry 34039 34085 34105
34187 34189 34211
34238
economic conditions 15083 34002 34015
34022 34023 34024
34030 34069 34070
34076 34080 34081
34082 34096 34097
34099 34100 34101
34102 34103 34106
34107 34108 34109
34111 34112 34121
34133 34134 34136
34143 34166 34169
34172 34177 34198
34199 34200 34227
34249 34250 34284
34300 34301 34309
34311 34338 34344
34356
economic indicators 34139
economic performance 34133
economic planning 34133
electricity industry 34340
electronics industry 34184
employment 34104 34313 34354
energy 34057 34062 34095
34116
engineering industry 34059 34064 34185
34239
England 34004
environment 34094 34339
exhibitions 34341
exporting 15076
exports 34123
family expenditure 34124 34142
finance 34070 34132 34186
financial institutions 34070 34086
financial markets 34134
fishing industry 34155 34178 34302
34304 34306 34328
food industry 34019 34040 34087
34138 34170 34187
34212 34223 34240
food 34130 34156
forecasting 34008 34075 34080
34082 34097 34099
34106 34107 34111
34120 34139 34196
34217 34250 34260
34261
furniture industry 34005 34013 34014
34054 34315
health and safety 34148
health 34020 34147 34153
34210 34241
heating and ventilating industry 34047 34048
horticultural industry 34027 34242
hotel and catering industry 34042
household economics 34002 34124 34142
household goods 34213
housing 34157 34158 34247
34357
imports 34161 34162 34163
income tax 15071
incomes 34126 34127 34128
industries 34055 34060
information technology 34090 34206
insurance companies 34205 34343
insurance 34068 34141
Insurance 34205
insurance 34270 34271 34312
Insurance 34343
international trade 34123 34143 34161
34162 34163 34253
34255 34272 34284
investment 15218 34173
Italy 34275
labour market 34115
labour relations 34115
labour statistics 34194 34330
labour supply 34115
labour 34115 34246 34322
leisure industry 34088 34214 34243

	34327		
leisure	34188	34195	34196
	34197	34314	
local government	34201	34303	34358
machine tools industry	34026	34207	
management	34098	34295	34297
manufacturing industry	34061	34065	
maritime transport	34225	34265	34307
	34353		
market research	34019	34020	34021
	34054	34055	34056
	34057	34058	34059
	34060	34061	34062
	34063	34064	34065
	34066	34083	34084
	34085	34086	34087
	34088	34089	34090
	34144	34167	34170
	34175	34179	34180
	34181	34182	34183
	34184	34185	34186
	34187	34188	34189
	34190	34191	34193
	34195	34196	34197
	34206	34209	34210
	34211	34212	34213
	34214	34215	34216
	34217	34218	34219
	34220	34223	34226
	34235	34236	34237
	34238	34239	34240
	34241	34242	34243
	34244	34251	34257
	34259	34260	34261
	34276	34288	34289
	34291	34292	34293
	34314	34323	34324
	34347	34351	
metals industry	34058	34059	34063
	34064		
minerals resources	34350		
mining industry	34058	34063	
money markets	34022	34023	34024
mortgages	34049	34050	34051
	34052	34074	34078
	34159	34160	34229
	34230	34231	
motor industry	34059	34064	34228
	34232	34233	34234
national accounting	34344		
Northern Ireland	34125	34248	34249
	34281	34317	
office equipment industry	34215		

packaging industry	34189	34259	
pay	34028	34073	34077
	34098	34131	34203
	34246	34258	34278
	34285	34295	34297
	34298	34322	
personal finance	34216	34257	
personnel	34258		
petroleum industry	34174	34346	
pharmaceutical industry	34020		
politics	34080		
population	34001	34002	34003
	34124	34142	34192
	34262	34263	34264
	34280	34281	34282
ports	34265		
prices	34007		
printing industry	34345		
property	34004	34091	34154
	34266		
public finance	34018	34133	34201
	34303	34321	34358
publishing industry	34190		
R & D	34017	34285	
regional development	34121	34122	
regional differences	34069	34249	34276
	34278	34280	34300
	34301		
retailing	34044	34089	34099
	34114	34170	34191
	34218	34219	34244
	34286	34287	34288
	34289	34290	34291
	34292	34293	34294
	34346	34347	34351
risk analysis	34082		
road transport	34053	34149	34171
	34296	34329	34331
Scotland	34004	34011	34110
	34126	34262	34282
	34299	34300	34301
	34302	34303	34304
	34305		
service industries	34066	34294	
small business	34309		
social economics	34096	34177	34311
social security	15071	34007	34310
social services	34147		
social surveys	34311		
social welfare	34122		
statistical methods	34316		
steel industry	34079	34269	34342
stock exchanges	34140	34202	34267

	34268	34318	34319
	34320		
stock markets	34140	34202	34267
	34268	34318	34319
	34320		
supermarkets	34045		
taxation	34133	34151	34168
	34283		
textile industry	34093	34104	34123
	34137	34161	34162
	34163	34208	34272
	34313	34354	
tourism	34037	34041	34043
	34046	34117	34118
	34119	34152	34221
	34245	34252	34254
	34275	34279	34308
	34325	34326	34341
	34348	34352	
tourist industry	34037	34041	34043
	34046	34117	34118
	34119	34152	34188
	34221	34245	34252
	34254	34275	34279
	34308	34325	34326
	34348	34352	
trade	15041		
training	34194	34330	
transport	34053	34149	34171
	34225	34274	34296
	34305	34307	34329
	34331	34332	34333
	34334	34335	34336
	34337	34353	34359
vetinerary products industry		34021	
Wales	34004	34096	34128
	34355	34356	34357
	34358	34359	
water supply industry	34057		

YUGOSLAVIA

business environment	35006		
economic conditions	04005	29012	35001
	35002	35003	35004
	35005	35007	35008
	35009	35010	35011
	35014	35015	
economic forecasting	35012		
finance	35011		
forecasting	35001		
health	29012		
international trade	35013		
politics	35001		